ENERGY AND PROBLEMS OF A TECHNICAL SOCIETY

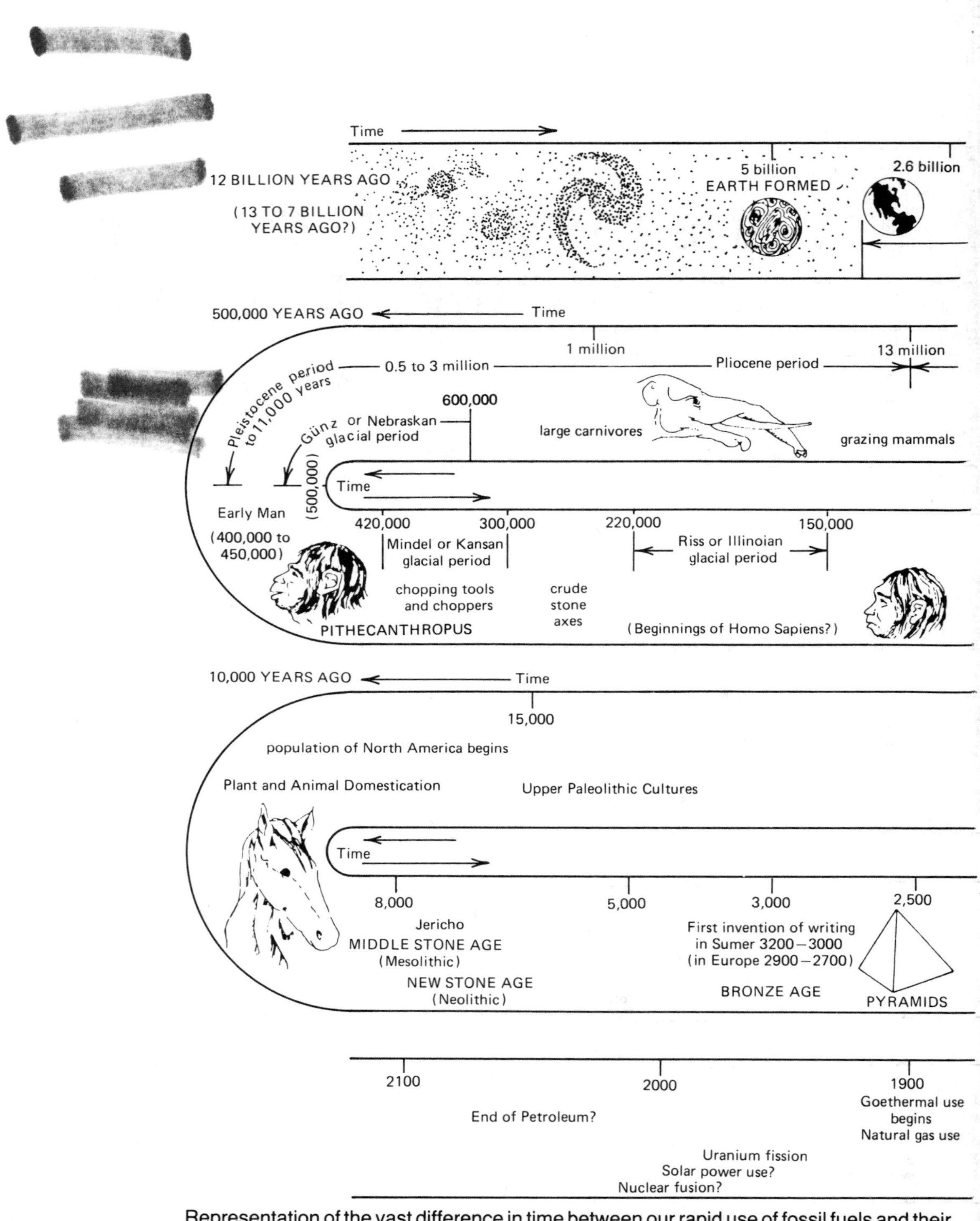

Representation of the vast difference in time between our rapid use of fossil fuels and their early formation. Adapted from Wilson and Jones, *Energy Ecology and the Environment,* copyright © 1974, Academic Press, New York. Reprinted by permission.

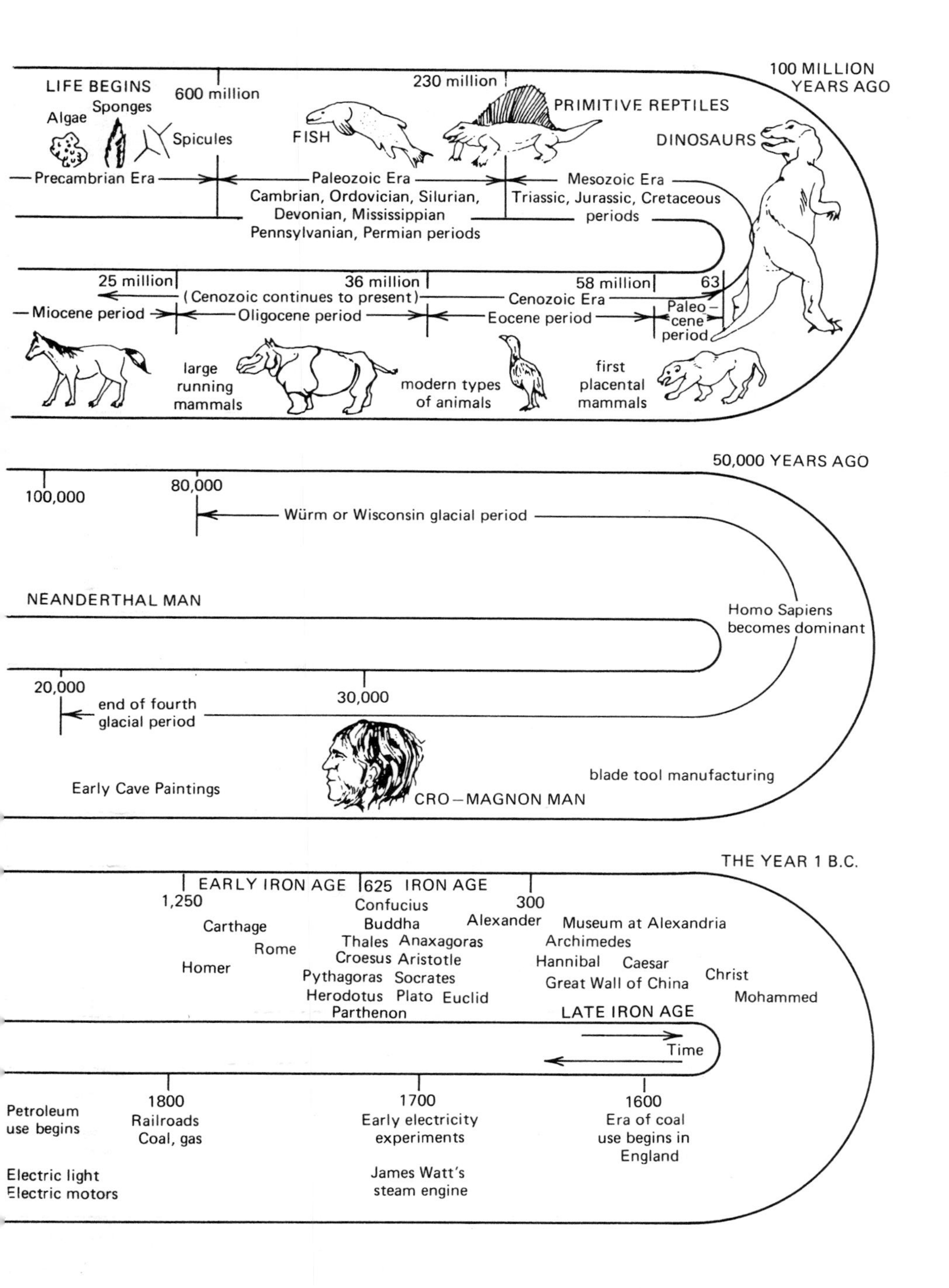

LIFE BEGINS
600 million
230 million
100 MILLION YEARS AGO
Algae
Sponges
Spicules
FISH
PRIMITIVE REPTILES
DINOSAURS
Precambrian Era
Paleozoic Era
Cambrian, Ordovician, Silurian, Devonian, Mississippian Pennsylvanian, Permian periods
Mesozoic Era
Triassic, Jurassic, Cretaceous periods
25 million
36 million
58 million
63
(Cenozoic continues to present)
Cenozoic Era
Miocene period
Oligocene period
Eocene period
Paleo-cene period
large running mammals
modern types of animals
first placental mammals
50,000 YEARS AGO
100,000
80,000
Würm or Wisconsin glacial period
NEANDERTHAL MAN
Homo Sapiens becomes dominant
20,000
30,000
end of fourth glacial period
Early Cave Paintings
blade tool manufacturing
CRO-MAGNON MAN
THE YEAR 1 B.C.
EARLY IRON AGE
1,250
625 IRON AGE
300
Confucius
Carthage
Buddha
Alexander
Museum at Alexandria
Rome
Thales
Anaxagoras
Archimedes
Homer
Croesus
Aristotle
Hannibal
Caesar
Pythagoras
Socrates
Great Wall of China
Christ
Herodotus
Plato
Euclid
Mohammed
Parthenon
LATE IRON AGE
Time
Petroleum use begins
1800
Railroads
Coal, gas
1700
Early electricity experiments
1600
Era of coal use begins in England
Electric light
Electric motors
James Watt's steam engine

ENERGY AND PROBLEMS OF A TECHNICAL SOCIETY

SECOND EDITION

Jack J. Kraushaar
Robert A. Ristinen
Department of Physics
University of Colorado, Boulder

John Wiley & Sons, Inc.
New York • Chichester • Brisbane • Toronto • Singapore

Cover Photo by Otto Rogge/The Stock Market

Acquisitions Editor	Joan Kalkut/Cliff Mills
Marketing Manager	Catherine Faduska
Production Supervisor	Elizabeth Swain
Designer	Laura Nicholls
Manufacturing Manager	Andrea Price
Photo Researcher	Hilary Newman
Illustration	Jaime Perea

This book was set in 10/12 ITC Century Book by Ruttle, Shaw & Wetherill, Inc. and printed and bound by Hamilton Printing. The cover was printed by Phoenix Color Corporation.

Recognizing the importance of preserving what has been written, it is a policy of John Wiley & Sons, Inc. to have books of enduring value published in the United States printed on acid-free paper, and we exert our best efforts to that end.

Library of Congress Cataloging in Publication Data

Kraushaar, Jack J.
Energy and problems of a technical society / Jack J. Kraushaar, Robert A. Ristinen.—2nd ed.
p. cm.
Includes indexes.
ISBN 0-471-57310-8 (alk. paper)
1. Power resources. 2. Power (Mechanics) I. Ristinen, Robert A. II. Title.
TJ163.2.K73 1993
333.79—dc20 92-36414
CIP

Printed in the United States of America

10 9 8 7 6 5 4 3 2 1

PREFACE

This book has grown out of a course offered since 1971 in the Department of Physics at the University of Colorado, Boulder. The first semester of the course is entitled "Energy in a Technical Society," and the second semester, "The Physics of Contemporary Social Problems." The course is intended primarily for nonscience or nonengineering students, although students majoring in science and engineering have occasionally enrolled and have apparently benefited from it.

The principal aim of both the course and the text has been to bring an understanding of the technological problems of our present society to students who normally would not have sufficient background to take otherwise appropriate classes offered in physics or engineering.

The first edition was published in 1984. A revision of the text in 1988 included information more current than that in the 1984 version. This second edition brings the text into the 1990s and covers several now-current topics.

There are a number of important changes in the second edition. New problems have been added at the end of each chapter so there is now a wider variety both in subject matter and difficulty. There has been a general updating of statistical information on essentially all of the topics covered. On the basis of feedback from students and several reviewers, improvements have been made in the way various material is presented. All of the general subjects covered in the first edition have been retained, but major additions have been made in petroleum production and resources, safety and costs of nuclear power, new developments in solar cells, the effects of ionizing radiation, radon in the home, 1990 clean air amendments, greenhouse warming, ozone in the stratosphere, acid rain, and the effects of the end of the cold war on nuclear weapons problems.

The topics presented in the 16 chapters are generally more than can be reasonably covered in a two-semester, three-credit-hour-per-semester course. In teaching the course, particularly in the second semester, we have let the students choose the topics of greatest interest to them and have ignored the material in two or three of the chapters. With some review of fundamentals, the last half of the book can be read and understood by students who have not gone through the first half.

Students attracted to this course have come from every corner of the campus: architecture, journalism, economics, geography, and languages have been well-represented. The only group consistently registering to satisfy a major requirement is from a special degree program in environmental conservation. We do not assume that students taking this course have had any high school or college-level physics, chemistry, or math, nor is any use made of calculus or trigonometry.

It has deliberately not been our goal to use the topics of energy and the environment to teach physics through the back door. No doubt some physics has been gleaned by unsuspecting students, but the emphasis has been entirely on the problem or topic. Only enough physics and some small amounts of other sciences have been introduced as needed to provide an understanding of the problem under discussion.

Although our emphasis has not been on a traditional development of physics, we have endeavored throughout the book to provide more than just a descriptive discussion of the subject matter. The text follows a treatment sufficiently quantitative that the students become involved in doing calculations. It is hoped that this text will help fill the gap between the more lengthy and sophisticated treatments of the problems discussed, where a background in science and mathematics is often required, and the qualitative and descriptive treatments offered by many texts and in the many accounts that steadily appear in magazines and newspapers.

The examples throughout the text and the exercises at the end of each chapter involve no deeper mathematics than elementary algebra. We have found, however, that many students are at first greatly distressed at the prospect of calculating how many British thermal units an hour go through a wall or how much solar energy in joules falls on a collector panel. Most students overcome this apprehension concerning numbers, powers of ten, and units of various kinds to appreciate that a deeper level of understanding can only be achieved by mastery of some fundamental skills in computation. They are then able to enjoy a bit of freedom from blind reliance on the technological experts in our society.

Throughout the text we have chosen to use thoroughly mixed units whenever it seems to serve a natural purpose. Rather than sticking strictly with the British system or one of the metric systems, we have decided to use enough of various common systems of units for the students to gain an ability to operate comfortably in both the laboratory and the lumberyard. We have found that the material on energy, for instance, loses much of its value if the students cannot calculate in joules per square meter as well as understand in a familiar way the R values of common building materials in ($ft^2 \cdot ^\circ F \cdot hr / Btu$).

The topics covered in *Energy and Problems of a Technical Society* have increased in importance with time. The patterns and problems of energy consumption and resource depletion put forth in the first edition have generally continued as predicted on the basis of information available at that time. Petroleum production in the United States is dropping, our energy imports are rising, the threat of nuclear weapons proliferation is greater than ever, the world population is overtaking resources, the global climate may be undergoing change brought on by humans, many of our wildlife species are declining, and we still have no nuclear waste disposal scheme acceptable to the public. These matters have never been in more urgent need of understanding and action by an informed electorate.

Jack J. Kraushaar
Robert A. Ristinen
Boulder, Colorado
1992

ACKNOWLEDGEMENTS

A number of individuals have made substantial contributions to the preparation of this text, and it is a pleasure to acknowledge their help. The students who have taken the course over a 22-year period have served as a constant source of inspiration and guidance as to what is accurate, relevant, and interesting. We especially thank Steve Kelton for his detailed comments on the lecture notes that later became part of this book.

We thank Professor Albert A. Bartlett for his incisive and constructive suggestions on many parts of the book and Professor W. R. Smythe for the comments and corrections that resulted from his using the text at the University of Colorado, Boulder, for six semesters.

The following faculty served as reviewers for the second edition: Professor Steven Gottlieb, Indiana University; Professor Paul N. Houle, East Stroudsburg University; Professor V. Paul Kenney, University of Notre Dame; Professor Don Reeder, University of Wisconsin, Madison; Professor Peter Schroeder, Michigan State University; and Professor George Williams, University of Utah. Their comments and suggestions were most helpful.

The first edition's manuscript was reviewed by: Professor Thomas Griffy, University of Texas; Professor Clyde W. Hibbs, Ball State University; Professor L. W. Seagondollar, North Carolina State University; and Professor Thomas A. Weber, Iowa State University and we appreciate their suggestions for corrections and improvements.

The staff at John Wiley has provided generous and thoughtful assistance in the preparation of the book. In particular for the first edition we thank Robert McConnin, physics editor, for his encouragement from the initial contacts to the completed text. Elyse Rieder of photo research, Deborah Herbert and Lydia Schulman of editing, Linda Indig of production as well as Blanca Ferreris, have been most helpful and their contributions are appreciated.

For the second edition the staff at Wiley have been equally helpful in providing assistance. We thank Joan B. Kalkut, Associate Editor, Elizabeth Swain, production supervisor; Laura Nicholls, designer; Clifford W. Mills, physics editor; Catherine Donovan, administrative assistant; and Hilary Newman, photo researcher.

Finally we are indebted to our wives, Nancy Kraushaar and Elvah Ristinen, for their support throughout the endless hours that have gone into the preparation of the first and second editions. Their thoughtful comments are greatly appreciated. We also thank the daughter-in-law of the Kraushaars, Nancy, for her kindness in reading and commenting on the text for the first edition.

CONTENTS

CHAPTER ONE

ENERGY FUNDAMENTALS

1.1 FOSSIL FUELS IN HUMAN HISTORY

Each of us should be aware that the era in which we live is extraordinarily specialized and is set off from all human history and future on this planet by our use of fossil fuels. These energy resources were laid down over hundreds of millions of years during the earth's evolution, and they are now being consumed in what is essentially an instant in our occupation of our planet. This situation is strikingly illustrated in Figure 1.1, where human consumption of fossil fuels is represented by what appears to be a spike spanning a brief period of time, with the present date somewhere just to the left of the center of the spike. An expanded view of this time period for coal, the largest component of our fossil fuel resources, is shown in Figure 1.2, where the present date is indicated more clearly. Obviously, everything to the right of the present date is prediction, whereas that to the left is established history. The graph may be in error with regard to precise detail, but there can be no doubt that the general form of the prediction is correct. At the approximate time when the curve began to rise above the baseline on the graph, we began to utilize fossil fuels in substantial amounts. Coal, and later petroleum, constituted energy sources vastly more convenient, less expensive, and easier to control than the muscular efforts of humans and animals. New enterprises became feasible, transportation became swift and reliable, and industries of all sorts were founded. It is probably no accident that this point in time corresponds in general to the end of human slavery. For the first time in history, the present era affords the opportunity for large numbers of individuals to spend 20 to 30 years of their lives acquiring formal education, leading to advances in science, invention, and the creative arts and making accessible to ever-greater numbers many of the graces of civilization.

The period from 1850 to the present has presented generally increasing opportunities to citizens of all nations with access to abundant fossil fuels, and worldwide there may be several decades of improving conditions supported by fossil fuel resources still ahead. But on a time scale within the range of prediction, the end of copious fossil fuel resources is clearly in sight. As we move down the right side of the peak, and we surely will, there will be a time of extreme adjustment. Fossil fuels will become critically scarce and expensive.

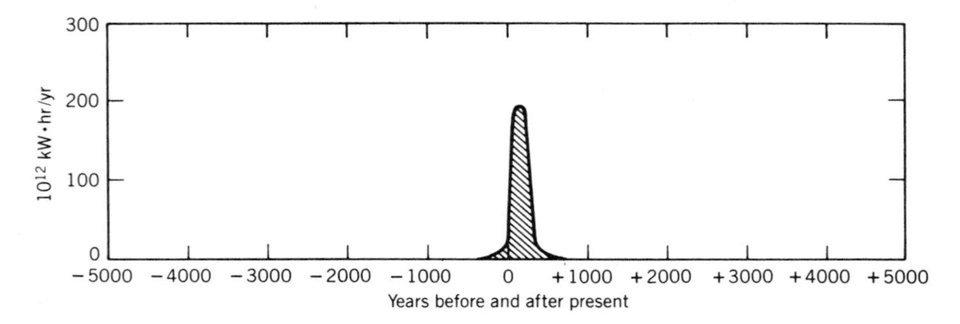

Figure 1.1 The exploitation of the world's fossil fuels will span only a relatively brief time in the 10,000-year period centered around the present time. (Source: Reprinted with permission from M. K. Hubbert, *Resources and Man,* National Academy Press, Washington, D.C., 1969.)

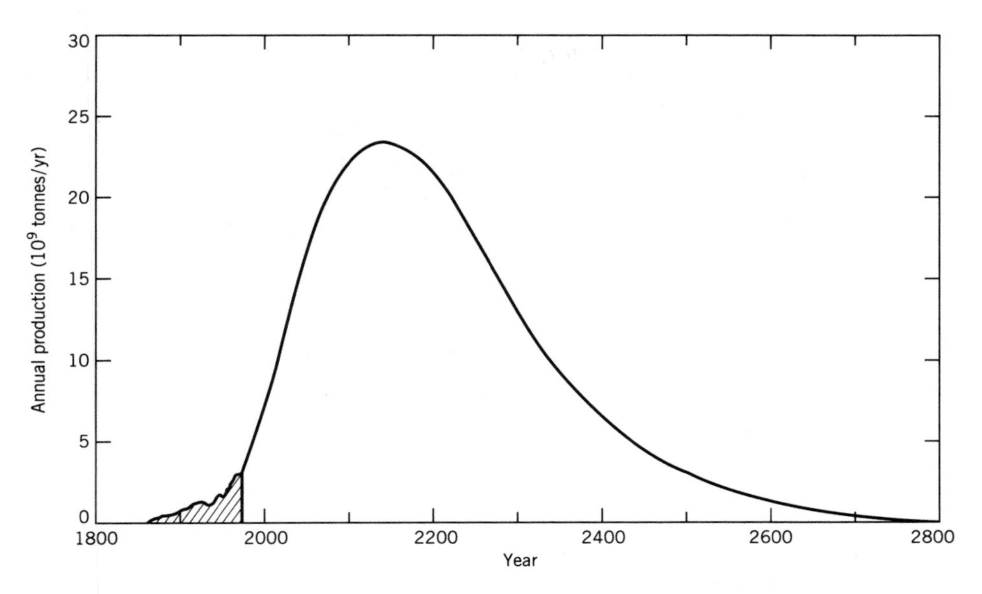

Figure 1.2 The projected world coal production cycle based on a resource estimate of 7.6×10^{12} tonnes. The entire cycle is expected to occupy only a few centuries. For the United States alone the curve would be similar but with reduced annual production (Adapted with permission from M. K. Hubbert, *Resources and Man*, National Academy Press, Washington, D.C., 1969.)

Emerging nations will demand their appropriate share, and international stresses over access to fossil fuel resources can be expected to develop beyond anything yet known. This will be especially true in a time of increasing human population. Humankind must either find a way to escape its dependence on fossil fuels or anticipate a rapid decline in its standard of living accompanied by a reduction in world population.

There is no escaping the reality shown in Figures 1.1 and 1.2. Fossil fuels are formed over very long time periods, and although some new deposits will certainly be discovered, there will be no significant increases in the world inventory over human history. There can be little doubt that the majority of the earth's fossil fuel energy resources will be consumed within a period of a few centuries centered about the present.

Although energy conservation measures may ease the transition away from fossil fuels, the long-term alternative to the impending social stresses lies in our exploiting other energy resources. But the possibilities are limited. There are only four: nuclear energy (fusion and fission), solar energy in all its forms, geothermal energy, and tidal energy. Much of this text will be concerned with the basic science behind these alternatives, but before going on to these matters, there is another aspect of the fossil fuel situation to be considered.

Up to this point, the discussion has focused on the benefits that have come to us from learning to expoit fossil fuels and on the severe problems we face with the exhaustion of our fossil fuel resources. However, as occupants of the earth, we and our successors are perhaps fortunate that the rate at which we consume fossil fuels will soon be in a decline.

For every unit of fossil fuel consumed, there are negative consequences on local, regional, and global scales. One example is the added burden of carbon dioxide in the earth's atmosphere, with its corresponding potential for modifying the world's climate. Other examples could include the acidification of the atmosphere and surface waters, the early deaths of thousands annually from the sulfur dioxide in the air we breathe, the local problems of ozone formation in the air of our cities, and the widespread release of radioactive elements from coal-burning power plants. One could expand the list by pointing to the problems of mining coal from agricultural lands, those of land reclamation and acid drainage, carbon monoxide and other pollutants from auto traffic, thermal pollution of rivers and lakes, the international strife of petroleum politics, and even the tens of thousands of deaths suffered each year by our young people because of extravagant use of the automobile. And when we consume fossil fuels, we are literally burning up valuable chemical feedstocks that should be better preserved for the manufacture of products.

Indeed, even if the shortage of fossil fuels were not forcing it upon us at this time, we might soon find it imperative to seek other energy sources. It is evident that although the exploitation of fossil fuels has brought many of us individual comfort and opportunity, the price has been great.

1.2 BASIC CONCEPTS OF ENERGY

To discuss energy in a useful way, it is necessary to arrive at a definition of energy and ascribe to it physical units, so that the concept of energy can be discussed quantitatively and quantitative calculations can be performed. In physics, energy is defined as "the capacity for doing work." The word *work* is used by each of us every day with many different connotations, but in physics it has a very definite meaning:

$$\text{work} = \text{force} \times \text{displacement}$$

or, "work is the product of force times the displacement through which the force acts." An example of work could be a force, F (perhaps exerted by some person), pushing a box horizontally on a rough floor for some distance, d, as shown in Figure 1.3.

It should be noted that the final work done, W, does not depend on how long a time it took to push the box. The same amount of work would be involved in moving the box 2 meters whether it took 3 seconds or 3 years. To perform this work, energy must have been available. In the case of a person pushing a box, perhaps the energy was in the form of food calories from a bowl of breakfast cereal. In another case, in which the force was exerted by an electrical device, the electrical energy used could have originated in a nuclear reactor in which uranium nuclei were fissioning.

There are two systems of units in common use in the United States, and we shall have occasion to use both of them. In the British system, the force unit is the pound (lb), and displacement is measured in feet (ft), so the work or energy unit is the foot-pound (ft • lb). In the metric system (Système International, or SI), the force unit is the newton (N), and displacement is measured in meters (m), with the work or energy unit being the joule (J). Note that work and the capacity for doing work, energy, have the same units. A system may

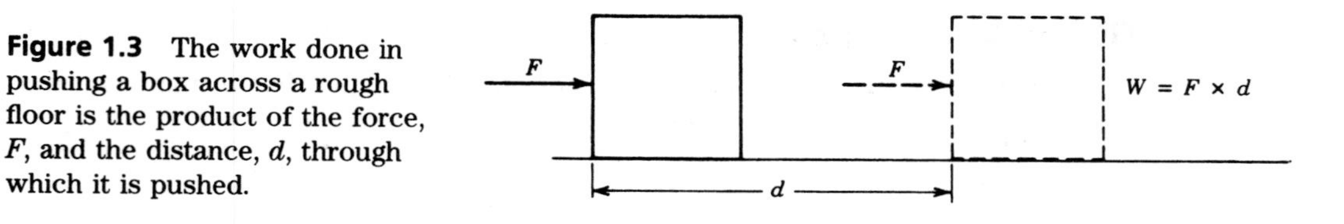

Figure 1.3 The work done in pushing a box across a rough floor is the product of the force, F, and the distance, d, through which it is pushed.

possess energy even when no work is being done. Units may be freely converted from one system of units to another by means of conversion factors. The relationship of the foot-pound to the joule is

$$1 \text{ J} = 0.7376 \text{ ft} \cdot \text{lb}$$

and

$$1 \text{ N} = 0.2248 \text{ lb}$$

System	Force	Displacement	Work (Energy)
British	pound (lb)	foot (ft)	(ft • lb)
Metric	newton (N)	meter (m)	joule (J)

1.3 FORMS OF ENERGY

In a sense, energy is energy—simply the capacity for doing work. Some will argue over whether electricity is a form of energy or simply a means of transporting energy. There are similar arguments concerning hydrogen as a fuel: Is it an energy resource or merely a means of conveniently transporting energy? If a phenomenon is capable of producing work, we shall classify it as a form of energy. The definitions of different forms of energy need not be mutually exclusive; some may rightfully consider what we classify here as chemical energy to be contained in what is called mass energy. The classifications in this book are intended only to be useful and consistent, with no claims made that they are definitive. There are in this context a number of different types of energy. They all correspond to the following definition: Energy is the capacity to do work. Energy may be transformed from one type to another, and much of the discussion of energy is related to the transformation of energy from one form to another so that it can do useful work. There is an important principle stating that the total amount of energy in a closed system remains constant. Energy may change from one form to another, but the total amount in any closed system remains constant. This principle, known as *conservation of energy*, is extremely important for understanding a variety of phenomena, and we shall have many occasions to refer to it. It is one of the laws of physics, and is simply a statement of our observations over a period of more than 100 years that has no known exceptions. We shall see that this principle does not immediately offer a quick solution to the energy crisis. Energy can still be changed from some useful form to some other form that for all practical purposes is useless, even though, formally, energy is conserved in the process.

KINETIC ENERGY

The energy of motion is called kinetic energy. If an object has a mass, m, and a velocity, v, then its kinetic energy is just

$$E_k = \tfrac{1}{2} mv^2$$

The mass of an object, an inherent property of any material object, is a measure of its inertia, or resistance to being set in motion. The mass is the same whether the object is on the earth, on the moon, or in free space. The units for mass are slugs in the British system and kilograms (kg) in the metric system, with

$$1 \text{ slug} = 14.6 \text{ kg}$$

The second quantity in the expression for kinetic energy is velocity (v). Technically, there are two quantities of interest, velocity (which has a stated direction) and speed (which does not), but we do not need to distinguish between them at this point. Velocity is defined as

$$\text{velocity} = \frac{\text{distance traveled}}{\text{time}}$$

In most common terms, velocity is the distance traveled divided by the time it takes to travel the distance. The units of velocity are usually feet per second (ft/sec) or meters per second (m/sec), although when dealing with motor vehicles, miles per hour (mile/hr) and kilometers per hour (km/hr) are often used. A kilometer is 10^3 meters.

If a mass is moving, its energy of motion can be converted into useful work. Consider the hammer head shown in Figure 1.4 to have a mass of 1 kg and a velocity of 10 m/sec just before it strikes a nail:

$$E_k = \tfrac{1}{2} mv^2 = \tfrac{1}{2}(1 \text{ kg})\,(10 \text{ m/sec})^2 = 50 \text{ J}$$

This 50 J of kinetic energy can be used to do work in driving in the nail with some force, F, through a distance, d, such that the work done will be 50 J.

It is necessary to consider the relationship of mass and weight, since both quantities are used in various energy-related problems. Weight (on earth) is the

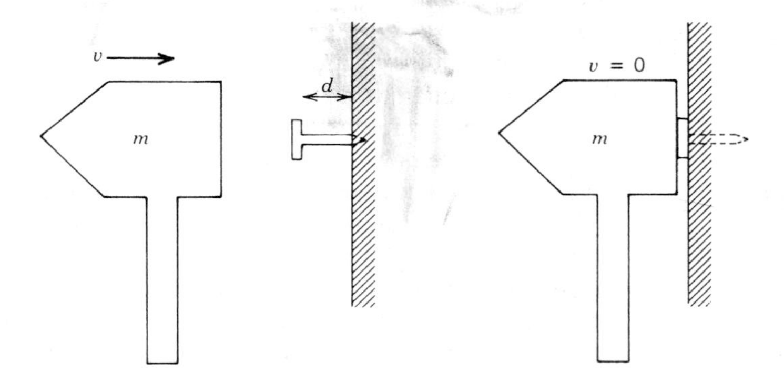

Figure 1.4 A hammer with kinetic energy $\tfrac{1}{2}mv^2$ will have done an amount of work in driving the nail into the wall equal to $\tfrac{1}{2}mv^2$ when it comes to rest.

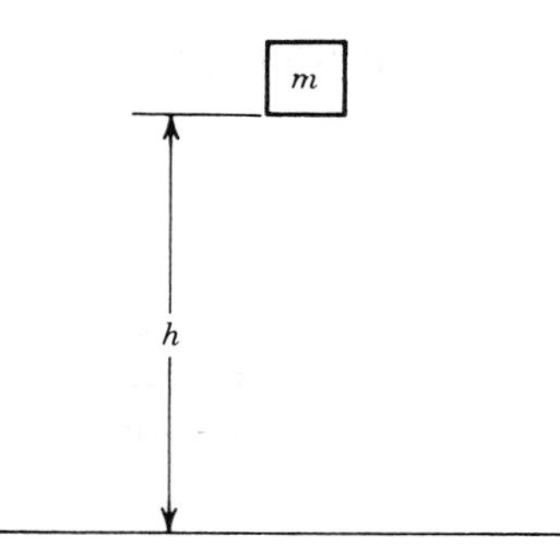

Figure 1.5 A mass held up a distance h over the ground will have a potential energy mgh in the earth's gravitational field relative to the ground.

force with which the earth pulls on a mass, and it has the units of force, namely pounds, or newtons. The relationship for weight is simply

$$\text{weight} = (\text{mass}) \times g$$

where g is the acceleration of a freely falling body due to the gravitational field of the earth. It has the magnitude of 32.2 ft/sec^2 or 9.8 m/sec^2. Acceleration will be discussed in more detail later. It is sufficient now to note that it is the rate of change of velocity. The distinction between mass and weight is important.

POTENTIAL ENERGY

The energy of position in a force field is called potential energy. The most common example of potential energy is a mass at some height, h, above the ground. By virtue of its position, the mass is capable of doing work. Hydroelectric power, for example, comes from the elevated position of a body of water above the turbines. The amount of energy E_p, associated with the mass, m, shown in Figure 1.5 at a height, h, above the ground is mgh, relative to zero potential energy at ground level. Potential energy of this sort must be measured relative to some reference height. It is clear that the mass will gain kinetic energy if it is released and falls toward the earth. In fact, if we ignore air friction, just as it hits the ground the mass will have a kinetic energy equal to its original potential energy, and the potential energy will have gone to zero. Hence, energy has been transformed from one kind to another, but the total amount of energy remains constant. Another example of a simple mechanical system, the pendulum, is shown in Figure 1.6, where a mass, m, is attached to

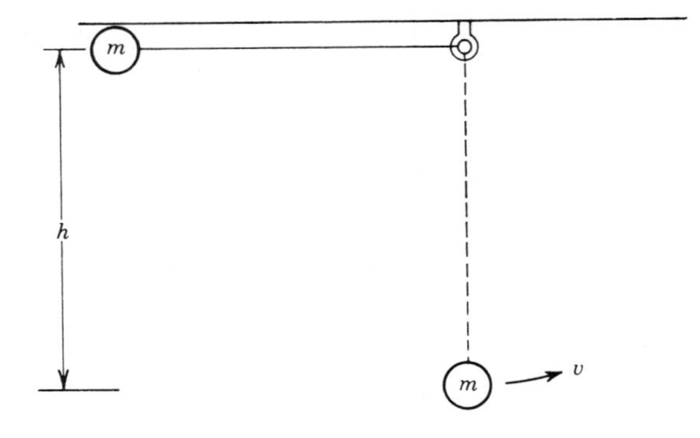

Figure 1.6 The mass of a simple pendulum when released will have its original potential energy, mgh, all converted to kinetic energy, $\frac{1}{2}mv^2$, at the bottom of the swing.

the end of a cord of length h. If it is positioned so that the cord is extended horizontally, the mass will have a potential energy, mgh, that can be converted to kinetic energy, $\frac{1}{2}mv^2$, as it swings down. At the bottom, $\frac{1}{2}mv^2$ is equal to the original mgh.

There are many examples of potential energy other than those involving the gravitational field of the earth. If a spring is compressed, work must have been done on the spring, and the potential energy of the spring is increased. When the spring is released, it is capable of doing work, such as giving kinetic energy to a steel ball in a pinball game.

Example 1.1

A simple pendulum having a mass of 0.5 kg attached to a string 1.5-m long is released from a horizontal starting position. What is its velocity when it reaches its lowest point?

Solution

The starting potential energy, mgh, is converted to kinetic energy as the pendulum swings downward. Thus at the lowest point the velocity can be obtained from

$$mgh = \frac{1}{2}mv^2$$

or

$$v = \sqrt{2gh} = \sqrt{2 \times 9.8 \text{ m/sec}^2 \times 1.5 \text{ m}}$$

$$v = 5.4 \text{ m/sec}$$

MASS ENERGY (NUCLEAR ENERGY)

As a consequence of the *special theory of relativity,* it can be shown that when the mass of some system is reduced by an amount Δm, as in a nuclear reaction, then an amount of energy is released:

$$E_m = \Delta mc^2$$

where c equals the velocity of light (3×10^8 m/sec). This simple equation is the basis of the energy derived when a ^{235}U nucleus fissions, as in a nuclear reactor, or when a deuteron and triton (^{2}H and ^{3}H) fuse in a thermonuclear reaction. At a detailed level, all energy conversion processes, including even chemical reactions, can be analyzed in this way.

CHEMICAL ENERGY

When certain chemicals combine, energy is released, usually in the form of heat. An important chemical reaction is the following:

$$C + O_2 \rightarrow CO_2 + 95 \text{ kcal/mole}$$

A mole is 1 gram molecular weight, which is a mass equivalent to the sum of the atomic weights, expressed in grams. For example, 1 mole of hydrogen gas has a mass of 2 grams, because the hydrogen molecule H_2 consists of two hydrogen atoms, each of which has an atomic weight of 1. In every case, the atomic weight is the same as the number of protons and neutrons in the atomic nucleus. For carbon, C, the atomic weight is 12, and 1 mole has a mass of 12 grams. In the particular case shown above, 12 grams of carbon combine with 32 grams (2×16) of oxygen to form a mole of carbon dioxide (44 grams), and 95 kilocalories (kcal) of heat energy is released. It is interesting to note that 1 mole of any monatomic element always contains 6.02×10^{23} atoms of the element, and 1 mole of any molecular compound always contains 6.02×10^{23} molecules. This number, 6.02×10^{23}, is known as Avogadro's number. The calorie is an energy unit often related to heat transfer, which is discussed in the next section. The energy released from the burning of fossil fuels can be represented by the above equation or similar, somewhat more complicated, equations. Exothermic chemical reactions produce heat. Endothermic reactions require an input of energy in order to proceed.

Another very simple reaction that may become extremely important if we use hydrogen as a fuel is the following:

$$2H_2 + O_2 \rightarrow 2H_2O + 68 \text{ kcal per mole of } H_2$$

Thus, if one has hydrogen and oxygen separately, it is possible to combine them (burn them) and have a source of heat that is pollution-free, producing only pure water.

An important chemical reaction is the combustion (burning) of sugar (glucose) in our bodies:

$$C_6H_{12}O_6 + 6O_2 \rightarrow 6CO_2 + 6H_2O + 690 \text{ kcal/mole}$$

In summary, chemical energy is the most important source of energy for us at the present time. It is the chemical energy in coal, natural gas, oil, wood, and gasoline that heats our homes, powers our cars, and is used to generate electricity, and it is the chemical energy in the food we eat that provides the energy we need for daily life. However, let us remember that the sun's energy is provided through exothermic nuclear processes, releasing what we have designated as mass energy in the form of light.

HEAT ENERGY

Although we have given names to a variety of different types of energy, these differences become less distinct if we start examining the phenomena on a microscopic level. For example, if we have a sealed container (as shown in Figure 1.7) filled with a gas, and heat is added by burning some fuel beneath the container, chemical energy in the fuel will raise the temperature of the gas. We then say that heat has been transferred to the gas. The gas molecules, which originally were bouncing about in the container with some average velocity, will now have their velocities increased by the added heat. The kinetic energy of the gas molecules will be increased, and this increase in kinetic energy is

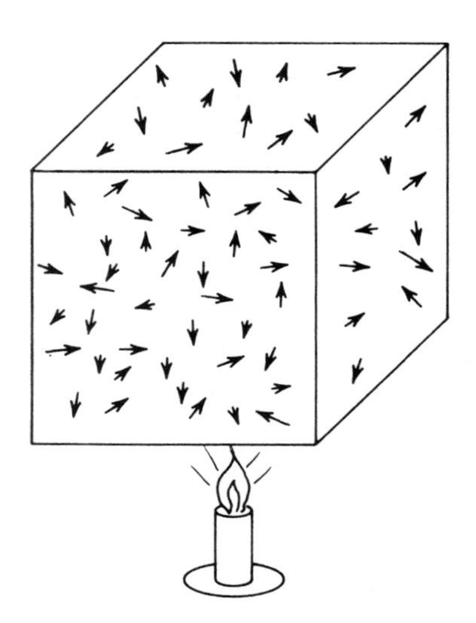

Figure 1.7 The molecules of a gas in a heated container will experience an increase in their velocities and, hence, their kinetic energies.

directly related to the increase in temperature. Heat energy is not thought of in absolute terms but in terms of an amount of heat, ΔQ, transferred into or out of some substance. When heat is added to some substance, two things can happen. There can be a change in the internal energy of the material, ΔU. This corresponds to an increase in the kinetic energy of the molecules or the temperature of the material. The second thing that can happen is that the subtance can expand and perform work on some external system. An example would be a heated gas expanding and pushing a piston of an engine that makes a car move. The symbol ΔW is given to this external work performed. From the *law of conservation of energy* we have

$$\Delta Q = \Delta U + \Delta W$$

The deltas (Δ) in this equation refer to an increment or change. The units of Q, U, and W must all be those of work or energy; for example, foot-pounds or joules. For historical reasons, however, heat changes are usually specified in different terms; that is, either calories or British thermal units (Btu). If heat is added to an amount of water to raise its temperature 1 degree, the effect will be to produce a negligible amount of work; hence, there will just be an increase in the temperature or in the internal energy, ΔU. The calorie (cal) is defined as the amount of heat energy needed to raise the temperature of 1 gram of water 1 degree Celsius (C). More exactly, it is the amount of heat necessary to raise the temperature of water from 14.5°C to 15.5°C. The Celsius scale is specified by setting the freezing point of water equal to 0°C and the boiling point equal to 100°C.

In the metric (SI) system, the kilocalorie (kcal) is the amount of heat energy needed to raise 1 kilogram of water 1°C. There is also a unit called the food calorie used by nutritionists. This is equal to 1 kcal:

$$1 \text{ food cal} = 10^3 \text{ cal} = 1 \text{ kcal}$$

In the British system, the Btu is defined as the amount of heat energy necessary to raise 1 pound of water by 1 degree Fahrenheit (F). The Fahrenheit temperature scale is based on 32°F being the freezing point and 212°F the boiling point of water. Conversion from one temperature scale to another can be accomplished by the following relationships:

$$T_F = 9/5\, T_C + 32$$

$$T_C = 5/9\,(T_F - 32)$$

It has been shown in some classic experiments that there is a mechanical equivalent of heat; that is, there is a direct relationship between joules and calories. The relationship is

$$1 \text{ cal} = 4.184 \text{ J}$$

$$1 \text{ kcal} = 4.184 \times 10^3 \text{ J}$$

Similarly for the British system:

$$1 \text{ Btu} = 778.2 \text{ ft} \cdot \text{lb} = 0.252 \text{ kcal} = 1055 \text{ J}$$

ELECTRICAL ENERGY AND POWER

Before discussing the last energy forms of interest, electricity and electromagnetic radiation, we need to understand the important difference between power and energy. Power is the time rate of delivering or using energy, or

$$\text{power} = \frac{\text{energy (or work)}}{\text{time}}$$

Although the concept seems simple enough, there seems to be no end of confusion between energy and power in newspaper reports and even books and technical articles on energy. By definition, 1 watt (W) is equal to 1 joule per second:

$$1 \text{ W} = 1 \text{ J/sec}$$

And the definition of horsepower (hp) is

$$1 \text{ hp} = 550 \text{ ft} \cdot \text{lb/sec}$$

and

$$1 \text{ hp} = 746 \text{ W} = 0.746 \text{ kW}$$

Car engines are commonly rated in horsepower and appliances in kilowatts, expressing the rate at which they can perform work or the rate at which they use energy. A power by itself says nothing about the energy used. To discuss energy, the power must be multiplied by the time the device is used. For example, the kilowatt hour (kW • hr) is an energy unit frequently used in electrical utility billings. This unit can be converted to joules as follows:

$$1 \text{ kW} \bullet \text{hr} = 10^3 \text{ W} \times 3600 \text{ sec} = 10^3 \frac{\text{J}}{\text{sec}} \times 3.6 \times 10^3 \text{ sec}$$
$$= 3.6 \times 10^6 \text{ J}$$

If wires are connected to the terminal of a battery and then the wires are connected to a light bulb or resistor, as shown in Figure 1.8, it is clear that the energy contained in the battery will be transferred through the wires and dissipated as heat in the resistor.

There are some simple relationships concerning the circuit of Figure 1.8 that we need to know to discuss electrical energy in a useful way. A meter that measures the current, I, has been inserted in the connection to the resistor, and a meter measuring the voltage, V, is connected across the terminals of the resistor. The unit of current is the ampere, the unit of voltage is the volt, the unit of resistance, R, is the ohm. A relationship between current, voltage, and resistance (Ohm's law) states that

$$I \text{ (amperes)} = \frac{V \text{ (volts)}}{R \text{ (ohms)}}$$

The units have been defined relative to the mechanical units of power discussed previously so that the power dissipated in the resistor is

$$\text{power (watts)} = I \times V \text{ (amperes} \times \text{volts)}$$

Using Ohm's law restated as $V = IR$, this becomes

$$\text{power (watts)} = I \times IR = I^2R$$

There are a number of sources of electrical energy besides batteries, which are really electrical energy (or chemical energy) storage devices. By forcing coils of wire to rotate in magnetic fields, electrical energy is generated; this is the basic source of electrical energy for our homes. Perhaps someday practical amounts of electrical energy will come from sunlight falling on a photovoltaic cell or from a high-temperature plasma powered by a fusion reaction. But now our electrical power is mainly generated by rotating coil devices (dynamos) powered by falling water or heat engines (which we discuss later).

Another important device for energy conversion is the electric motor, which

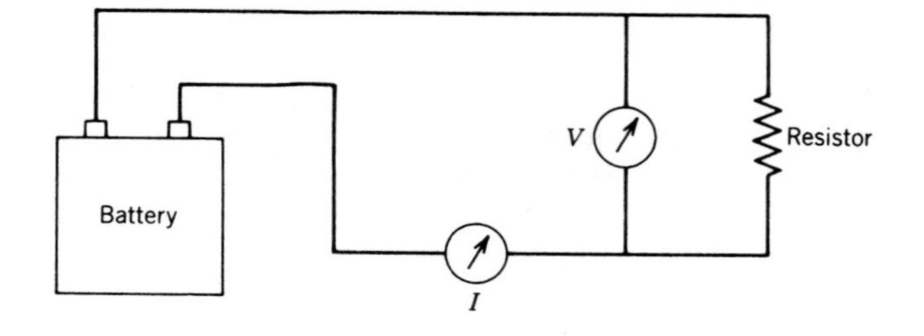

Figure 1.8 A battery is shown connected by wires to a resistor. An ammeter connected in series measures the current, I, and the voltmeter connected across the resistor measures the voltage, V.

makes use of the basic force on a current-carrying conductor in a magnetic field. In a common type of electric motor, the conductors are normally arranged in windings that can rotate, and this permits electric energy to be transformed into rotational kinetic energy. Because electric motors are in general very efficient and adaptable to many needs, they find widespread use in any industrial society.

Example 1.2

On a winter day a home needs 1×10^6 Btu of fuel energy every 24 hours to maintain the interior at 65°F. At what rate is energy being consumed in watts?

Solution

$$P = \frac{E}{t} = \frac{1 \times 10^6 \text{ Btu}}{24 \text{ hr}} \times \frac{1055 \text{ J}}{1 \text{ Btu}} \times \frac{1 \text{ hr}}{3600 \text{ sec}}$$
$$= 12.2 \times 10^3 \text{ J/sec} = 12.2 \text{ kW}$$

ELECTROMAGNETIC RADIATION

Electromagnetic radiation covers a wide variety of phenomena from radio waves to very short wavelength microwaves, infrared radiation, visible light, ultraviolet radiation, and x-rays and gamma rays. All of these various classifications of radiation can transport energy from one point to another by virtue of rapidly varying electric and magnetic fields. For example, the earth's most important source of energy by far is the electromagnetic radiation from the sun.

Figure 1.9 shows the electromagnetic spectrum in which each type of radiation is shown according to its wavelength. The wavelength of radiation is the characteristic distance in space the wave must move before the fluctuating electric or magnetic fields repeat themselves.

There is a convenient relationship between the wavelength and frequency of an electromagnetic wave, namely

$$\lambda f = c$$

where λ is the wavelength in meters, f the frequency in hertz (cycles per seconds), and c the velocity of light (3.0×10^8 m/sec). All electromagnetic radiation travels (in vacuum) at the speed of light. Each section of the electromagnetic spectrum has an important part to play in our use of energy. Of course, radio waves and microwaves can be produced by devices that can be made directly with various electronic components. Wavelengths shorter than those of infrared radiation, however, are so short that this radiation cannot be produced by ordinary circuitry, but rather must be generated by natural systems such as atoms or nuclei. For visible radiation the outermost (valence) electrons of the atom contribute, whereas for x-rays the innermost, tightly bound electrons are the source of the radiation. Gamma rays are produced by the de-excitation of excited atomic nuclei. In dealing with energy in the form of

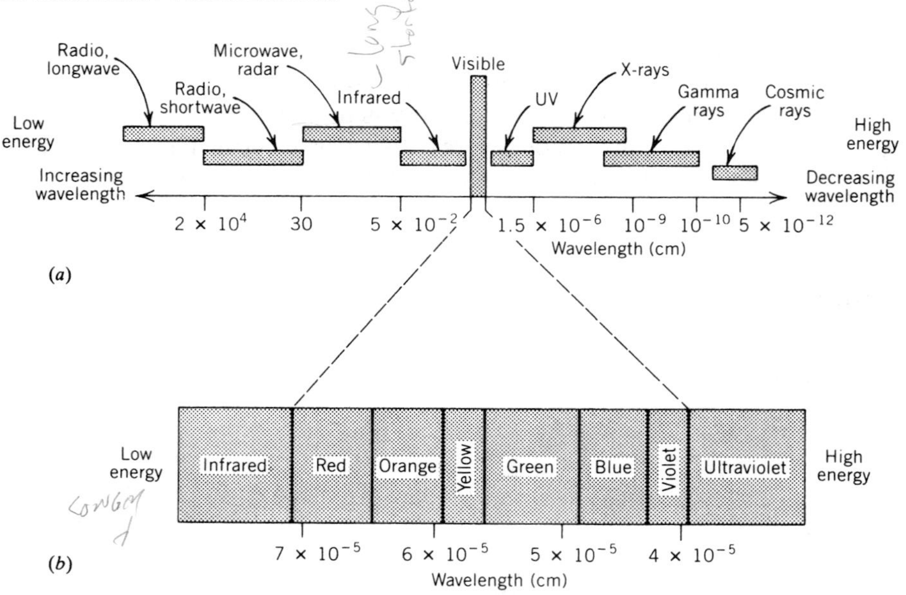

Figure 1.9 (*a*) The electromagnetic spectrum. It varies continuously from the high-energy short-wavelength cosmic ray photons down to long-wavelength radio waves. (*b*) The visible region, just a tiny portion of the entire electromagnetic spectrum, is shown in this expanded view.

electromagnetic radiation, the energy and power units of joules and watts are generally used, although another energy unit, the electron volt (eV), is frequently used in discussing the shorter wavelength radiation such as x-rays and gamma rays.

TRANSFORMATION OF ENERGY FROM ONE FORM TO ANOTHER

In a complex and technical society such as we have in the United States, energy enters our lives in a variety of ways, and there are frequent transformations of energy from one form into another. It is useful to examine some of these processes with the understanding that the total energy must be conserved.

Figure 1.10 shows schematically what takes place in an electric power plant in transforming the chemical energy contained in coal to the electric energy we receive in our homes. The resistor, R, represents the various residential and commercial users of electricity. In this type of power plant there is inevitably some heat loss to the environment due to friction in the bearings of the electric generators as well as through the insulation of the boilers, the electrical resistance of the wires, and the major amount of heat that must be exhausted from the turbines. This latter point is covered in detail in our discussion of heat engines. We also discuss later what happens to this energy that is dumped into the environment.

In addition to the food and manufactured goods that we bring into the house, there are typically two major energy inputs, electricity and natural gas (or heating oil in some sections of the country). A third input is firewood, but it is still a relatively small part of the total energy for most households. Where

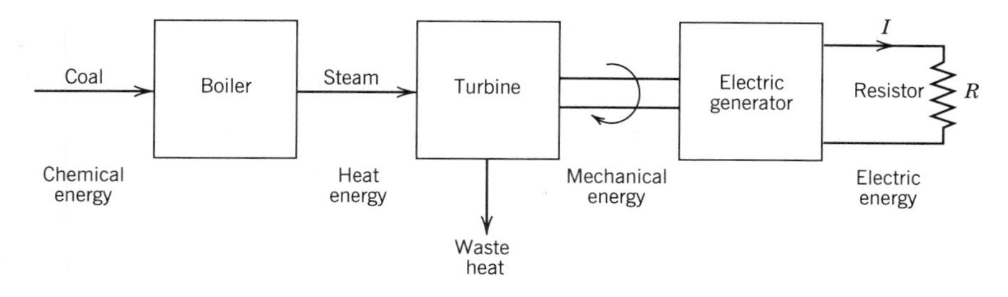

Figure 1.10 A schematic diagram of the various steps in the transformation of the chemical energy in coal to the electrical energy that is finally used in a residence or industry.

does the energy go? The answer, of course, is that virtually all of it goes off in the form of heat, either as the heat that warms the surrounding ground and air or as the heat contained in the wastewater in the sewer line. The electric energy that comes in can perform a number of useful tasks, such as power a vacuum cleaner and provide light, and heat with which to cook the meals. This energy and all the energy that goes into the toaster, TV, electric toothbrush, and stereo does nothing in the end except heat the environment, as does the energy from the natural gas that is used to heat the house, water, and frequently to provide heat for cooking. Sometimes different forms of energy may be developed in the process, such as in the microwave oven, but basically all the energy is ultimately lost as waste heat.

Since energy is conserved, where does the energy go that is put into the atmosphere? The effect of all the houses, factories, automobiles, power plants, and so forth putting heat into the environment will be to raise the temperature of the atmosphere. As we see later when we discuss the heat balance of the earth, the elevated temperatures will mean the atmosphere will radiate an increased amount of energy into outer space. There can also be local effects of heating the environment, such as the heat islands that are formed in cities and the thermal pollution of rivers and lakes. One main overall effect of taking fossil fuels (or ^{235}U) out of the ground and burning them will be to fill the universe with a little more long-wavelength radiation.

Aside from nuclear energy, the major part of the energy we use on earth comes either from the sun directly or is energy that came from the sun tens to hundreds of millions of years ago and is stored as chemical energy in fossil fuels. The only exceptions to this are the energy in the ocean tides, which in a few places is being used to generate electricity, and geothermal energy. The energy from the sun comes to us as radiant energy, mostly in the visible, but partly in the ultraviolet and infrared portions of the electromagnetic spectrum. The source of the energy that drives the sun is the nuclear fusion reaction, for example, the proton–proton cycle through which hydrogen is used up (burned) to form helium with a release of energy. Of course, most of the radiant energy of the sun goes off into space with only a minute fraction being intercepted by our little planet, earth.

Eventually, when we have used up all of the fossil fuels and all of the uranium and thorium that can be obtained economically, we shall be almost entirely dependent on fusion reactions. These will be either fusion reactions

Figure 1.11 The space shuttle *Columbia* rises off its pad. The chemical energy in the fuel is transformed into the kinetic energy of the spacecraft, which then partially becomes potential energy in the earth's gravitational field.

taking place in the sun or fusion reactions taking place on earth between two hydrogen nuclei. Possible additional sources of energy may be tidal and geothermal, a manifestation of radioactive nuclear decay deep within the earth.

1.4 THE HISTORY OF ENERGY CONSUMPTION IN THE UNITED STATES

The energy used by primitive humans in their earliest years was provided by food, which was primarily vegetable matter. As time went on, life became more complex, with hunting, fires for cooking and warmth, and the feeding of work animals. As shown in Table 1.1, the energy consumed per person each day continued to increase as wood was used extensively for heating, cooking, and

Table 1.1 HUMAN ENERGY CONSUMPTION THROUGHOUT HISTORY

	Daily per Capita Consumption (1000 kcal)				
Period	**Food**	**Residential and Commercial**	**Industry and Agriculture**	**Transportation**	**Total**
Primitive	2	—	—	—	2
Hunting	3	2	—	—	5
Primitive agricultural	4	4	4	—	12
Advanced agricultural	6	12	7	1	26
Industrial	7	32	24	14	77
Technological	10	66	91	63	230

Energy consumption represents six stages of human development. Primitives, about 1 million years ago, had only the energy contained in the food they ate. Hunters, who lived about 100,000 years ago, had more food and also used wood for heat and for cooking. About 5000 B.C., primitive agricultural people grew crops and used animals for cultivation. By A.D. 1400, advanced agricultural people in Europe used coal for heating as well as water and wind for power. In the late 19th century, industrial people had added the steam engine as a source of mechanical energy. Modern technological peoples utilize the internal combustion engine, steam turbines, gas turbines, and electricity as their sources of energy. The entries under food include the energy input to the food production and distribution system exclusive of direct solar energy.

Source: Data from Earl Cook, "The Flow of Energy in an Industrial Society," *Scientific American*, September 1971.

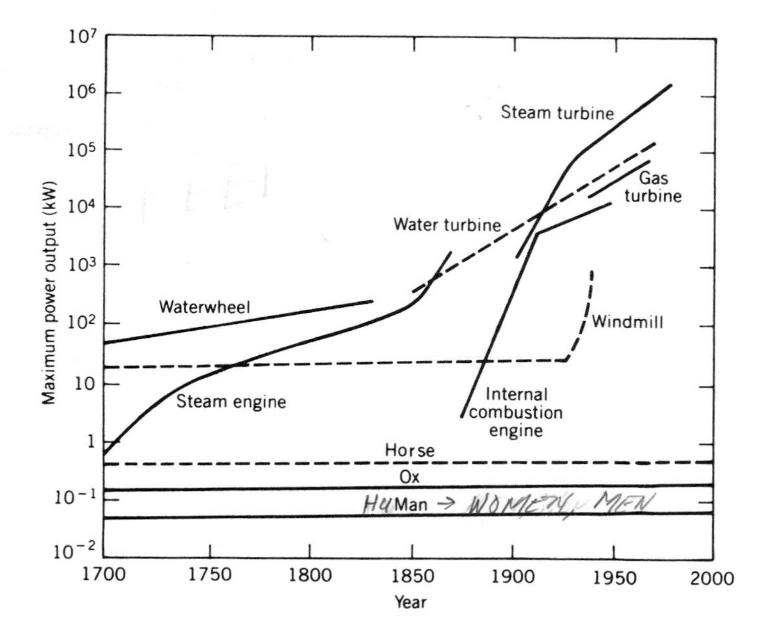

Figure 1.12 The maximum power output of several energy conversion devices. It has generally increased over recent centuries, whereas that of animals has remained fairly constant. (Source: Chauncey Starr, *Energy and Power.* Copyright © September 1971 by Scientific American, Inc. All rights reserved.)

some industries. Later, waterwheels, windmills, and coal added to our sources of energy. The steam engine has a long history, but it was James Watt's improvements in about 1769 that led to the widespread use of the steam engine in various industrial applications and in transportation. At first such engines were fired by wood, but by 1850, as can be seen in Figure 1.12, larger steam engines, and later the development of entirely new engines such as the internal combustion engine, not only hastened consumption of energy but forced reliance upon fossil fuels rather than renewable sources of energy. Goods that were previously made in the home could now be made more efficiently in factories using machinery powered by various engines.

Figure 1.13 shows the changes in the horsepower available to each person in the United States since 1850 for several categories. The curve labeled "non-automotive inanimate" includes many uses that are essential for an industrial country, such as factories, mines, railroads, ships, farms, electric power plants, and aircraft. Up until about 1900 the total horsepower available to each person was less than one, and work animals were a major component. As the automobile became more prevalent the total horsepower available increased dramatically, rising by a factor of about 10 in 20 years, and it continued to rise until the present time. It is worth noting that on average every person in the United States now has the equivalent of about 150 horsepower available to him or her.

Two major milestones in the human use of energy in a technical society occurred in the late 19th century. The development of electric generating equipment and distribution systems made it possible to use electric motors for innumerable tasks previously performed by animals or humans, as well as to use convenient electric lights. The other major development was the improvement of the internal and external combustion engines, which made possible their widespread use in transportation. These developments, more than any others, contributed to the increase in per capita energy consumption in the

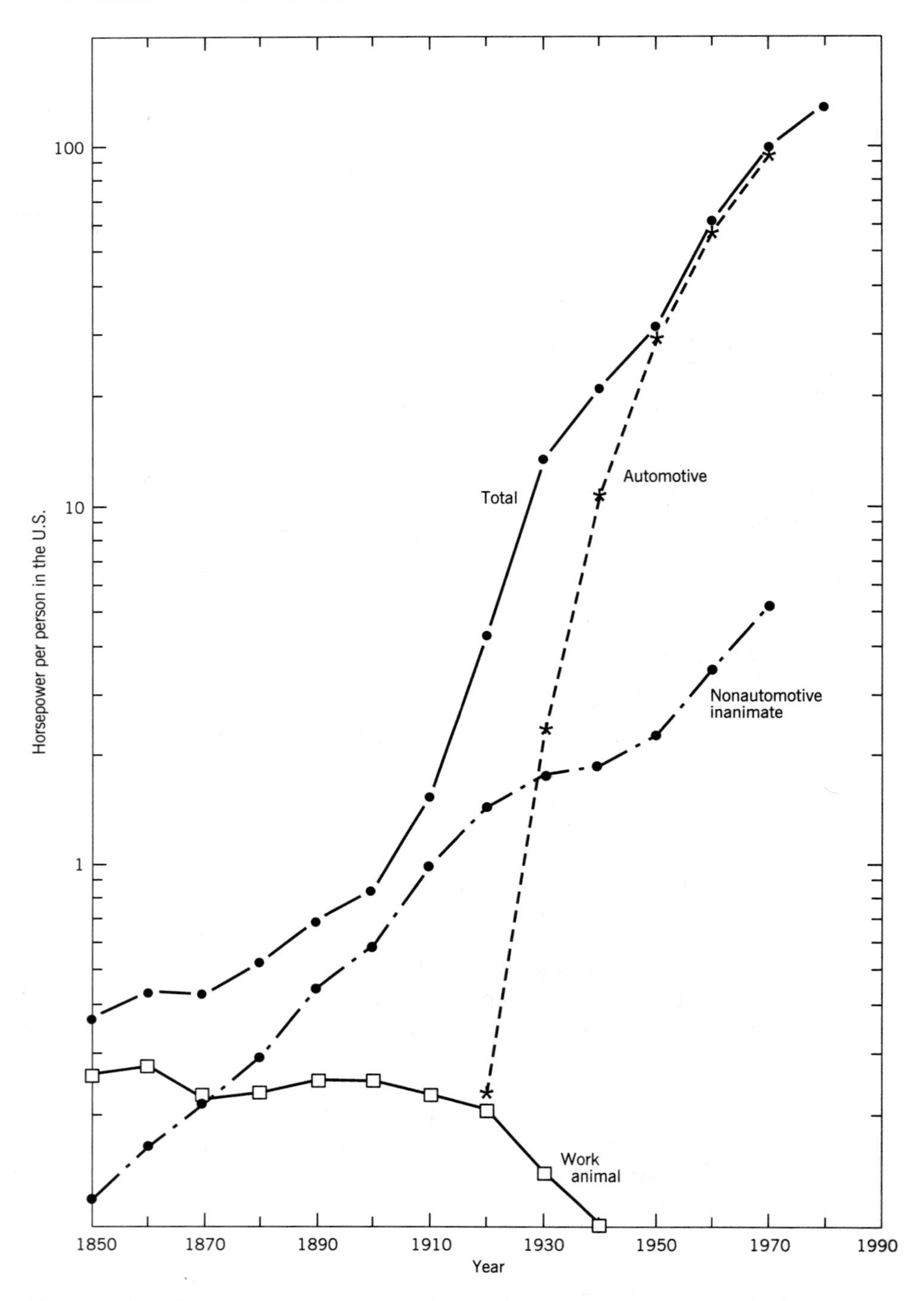

Figure 1.13 Total horsepower per capita of all prime movers in the United States since 1850 (Sources: *Historical Statistics of the United States Colonial Times to 1970* and *Statistical Abstracts of the United States 1990* Washington D.C.: U.S. Department of Commerce, Bureau of the Census.)

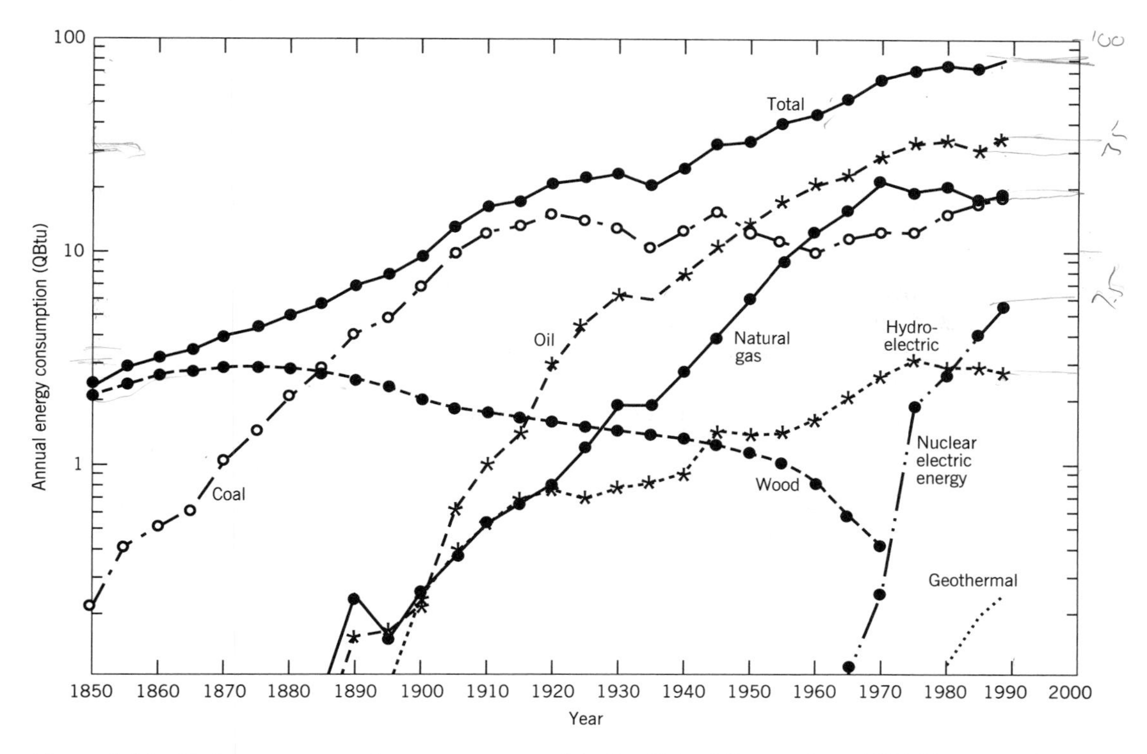

Figure 1.14 Various sources of energy consumed in the United States since 1850 (Sources: *Historical Statistics of the United States Colonial Times to 1970* Washington, D.C.: U.S. Department of Commerce, Bureau of the Census, 1975 and U.S. Energy Information Administration, *Annual Energy Review 1989*.)

United States and other industrialized nations. The total energy consumption per year in the United States since 1850, is shown in Figure 1.14. The total amount of energy used in the United States in 1850 was about 2.3×10^{15} Btu. (This is frequently written 2.3 QBtu or 2.3 Q, where Q stands for 10^{15}, or a quadrillion.) At that time there were about 20×10^6 people in the United States, so the energy used per person annually was about 1.2×10^8 Btu. In 1990, about 81 QBtu were used by a population of 250×10^6 persons, or 3.3×10^8 Btu per person. It is interesting to note that while the population has grown by a factor of 11, the energy consumption per person has increased only by a factor of 3 since 1850. This factor seems surprisingly low in a time of extensive use of jet airliners, automobiles, microwave ovens, and overheated buildings. It is clear that over the past 130 years, the overwhelming factor in the increased total energy consumption in the United States has been population growth. It is worth noting, also, that the primary fuel in 1850 was wood. If one omits wood and just considers the fossil fuels, then the per capita annual energy consumption has increased by a factor of about 30 between 1850 and 1990. The published figures, such as were used for the construction of Figure 1.14, do not take into account any solar energy use except that included in wood and hydropower. The solar energy component involved in growing food for people or feed for work animals, or the direct or indirect heating of houses by the sun, is not included. The use of the energy in the wind is also not included.

Figures 1.12, 1.13, and 1.14 are plotted on a semilogarithmic scale; that is, the abscissa is linear in time (years), but the ordinate is a logarithmic scale. We shall discuss such scales later. However, it is useful to note in Figure 1.14 that total annual energy consumption plotted in such a way can be rather closely approximated by a straight line, except for the dip that occurred around 1932, which was related to the great economic depression of the 1930s. A straight line on a semilog plot of this type is indicative of exponential growth. It is tempting to project the curve into the future based on its past straight-line behavior. Many estimates of the energy that the United States will consume in 2000 or 2020 are based on such projections. But there is no law of physics or economics that guarantees such behavior.

The energy consumed per year per person, stated as 3.3×10^8 Btu, does not make an immediate impression on most people. But both in an absolute and a relative sense, it is indeed a large amount of energy. To illustrate this point it is useful to convert this number of Btus to other equivalent energy terms that may be more familiar to the reader, such as barrels (bbl) of oil or tons of coal. Making use of energy equivalents we have

$$E = 3.3 \times 10^8 \frac{\text{Btu}}{\text{yr}} \times \frac{1 \text{ bbl of oil}}{5.80 \times 10^6 \text{ Btu}} = 57 \text{ bbl of oil/yr}$$

or

$$E = 3.3 \times 10^8 \frac{\text{Btu}}{\text{yr}} \times \frac{1 \text{ ton of coal}}{2.66 \times 10^7 \text{ Btu}} = 12.4 \text{ tons of coal/yr}$$

That each of us is annually using the equivalent of such prodigious amounts of energy, which for the most part is derived from nonrenewable fossil fuels, is a significant fact and is at the heart of the energy problem.

Another useful way to state the problem is to examine the rate at which each of us uses energy; that is, the average power.

$$P = \frac{E}{t} = \frac{3.3 \times 10^8 \text{ Btu/yr}}{3.15 \times 10^7 \text{ sec/yr}} \times \frac{778 \text{ ft} \cdot \text{lb}}{1 \text{ Btu}} \times \frac{1 \text{ hp}}{550 \text{ ft} \cdot \text{lb/sec}} = 14.8 \text{ hp}$$

or

$$P = \frac{E}{t} = \frac{3.3 \times 10^8 \text{ Btu/yr}}{3.15 \times 10^7 \text{ sec/yr}} \times \frac{1.06 \times 10^3 \text{ J}}{1 \text{ Btu}} = 1.11 \times 10^4 \text{ J/sec} = 11.1 \text{ kW}$$

Thus, our average power, or rate of energy consumption, is equivalent to 15 horses working for each of us continuously, or 111 100-watt lights burning day and night for each of us.

1.5 ENERGY CONSUMPTION AND GROSS NATIONAL PRODUCT

Figure 1.15 shows the average energy consumption per year per capita and the gross national product (GNP) in 1985 for a number of different countries. Although there have been some minor changes in the data since 1985, the same

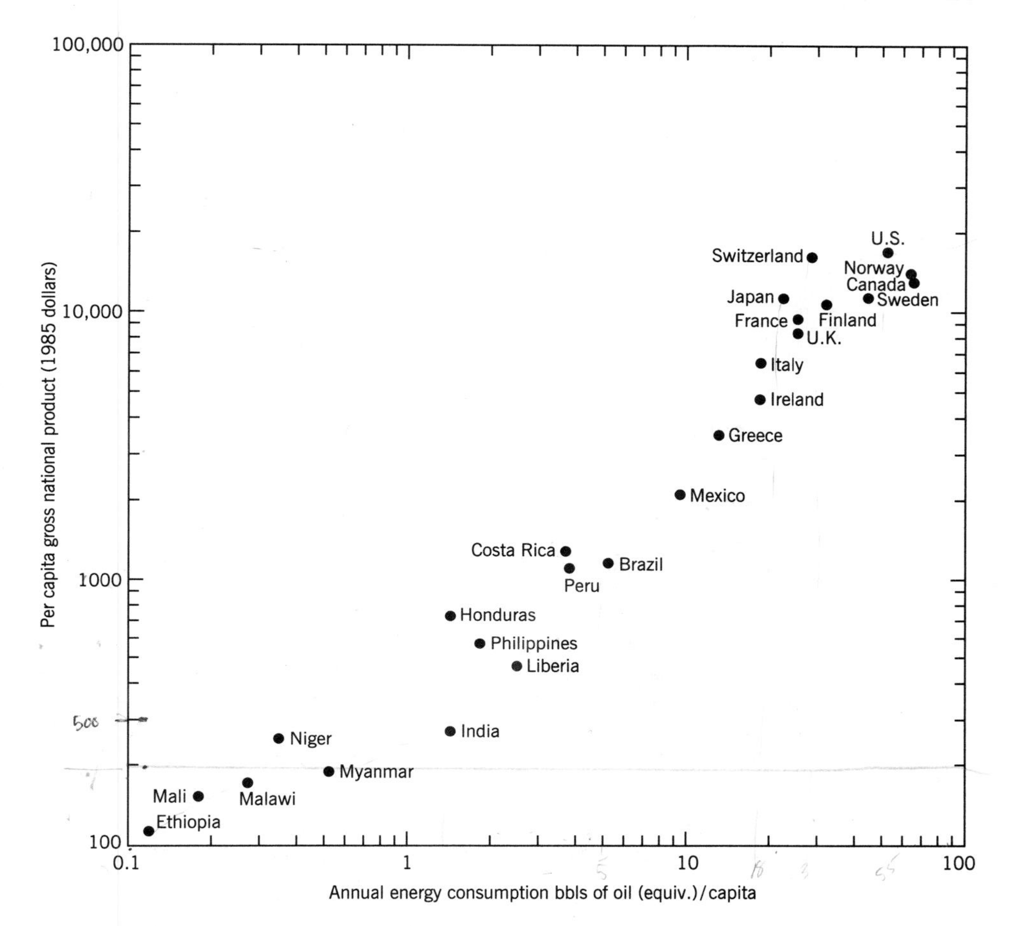

Figure 1.15 Per capita gross national product in 1985 dollars and per capita energy consumption per year in terms of the equivalent barrels of oil. (Source: World Bank [1987]. Adapted from E. S. Cassedy and P. Z. Grossman *Introduction to Energy, resources technology and society.* Cambridge: Cambridge University Press [1990].)

general picture prevails. U.S. per capita annual energy consumption (51.1 bbl of oil) is close to what was calculated in the preceding section, and this is considerably more than for any other country except Norway and Canada. It is considerably more than that of several of the industrialized countries of northern Europe, and 425 times as much as that of Ethiopia.

Many have noted the general correlation of per capita GNP with per capita energy consumed, which is shown in Figure 1.15. Indeed, if these variables are plotted for the United States as a function of time, as shown in Figure 1.16, there appears to be rather a close correlation. This relationship, however, cannot be held to be valid in great detail. For example, in recent years the per capita GNP of Sweden, Switzerland, and Germany have all equaled or exceeded that of the United States, with a per capita energy consumption less than that of the United States. There are certainly differences in industrial energy efficiency, such as Btu per ton of steel, that account for some of this difference, but it is also clear that the greater land area of the United States, the differences

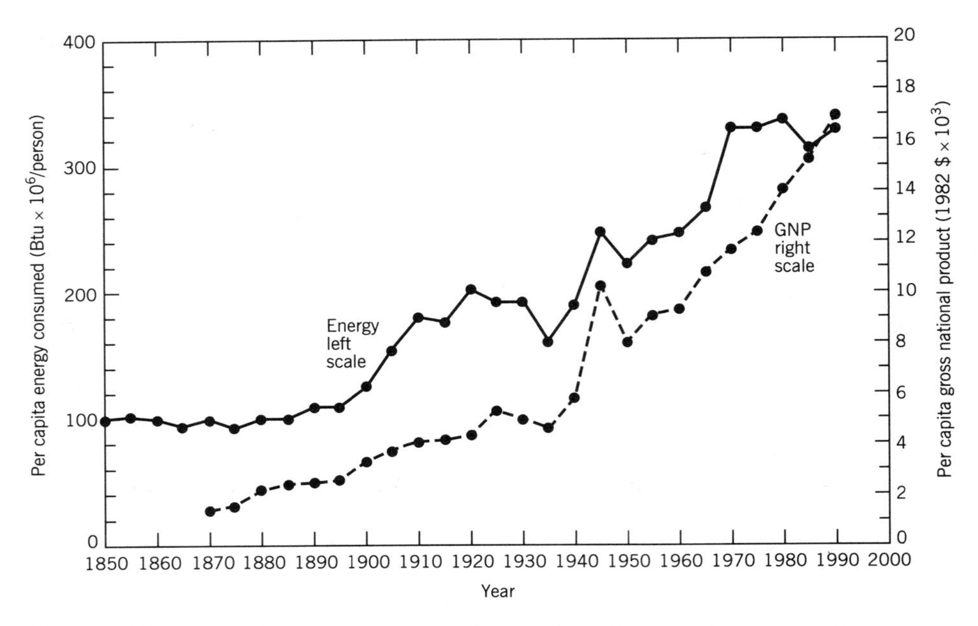

Figure 1.16 Per capita energy consumed annually and per capita gross national product since 1850 and 1870, respectively. The GNP has been adjusted for inflation using 1982 dollars. (Sources: *Historical Statistics of the United States Colonial Times to 1970* and *Statistical Abstracts of the United States 1990.* Washington, D.C.: U.S. Department of Commerce, Bureau of the Census.)

in life styles of the various countries, and the nature of the products produced by the nations' industries are also important factors.

1.6 CONCEPTS OF EXPONENTIAL GROWTH

We have noted that total U.S. energy consumption per year as a function of time appears to be an example of exponential growth. Not all quantities behave this way, however. For example, if we plot the Dow Jones stock average since 1960, as shown in Figure 1.17, it is difficult to project the stock prices into the future with any assurance because of their irregular behavior. On the other hand, if we plot the distance we are from a given city, if we are driving directly away from the city at a constant speed of 50 miles per hour, the plot would look like Figure 1.18. In this case, there is said to be a linear relationship between distance and time; that is, the distance, d, is proportional to the time, t. The relationship is

$$d = (\text{constant}) \times t = vt$$

where the constant in this case is the speed of the automobile, v. The amount of distance added in 1 hour is constant; that is, 50 miles.

Many interesting time dependencies do not behave in this linear fashion. There are numerous quantities that behave such that the amount of change, or

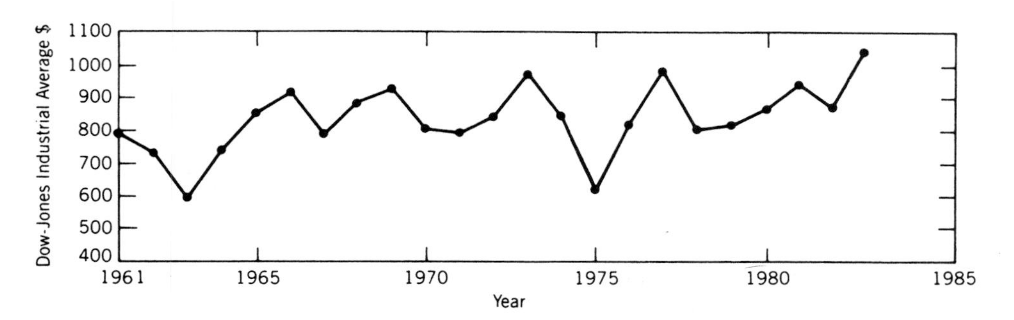

Figure 1.17 The Dow Jones industrial stock averages for the New York Stock Exchange from 1961 to 1983.

the amount added per unit of time, is not constant but is proportional to the quantity, N, at that time. The equation that expresses this is

$$\frac{\Delta N}{\Delta t} = (\text{constant} \times N) = \lambda N$$

where the Greek letter Δ (delta) indicates a change or increment of the variable, and λ (lambda) is a growth constant. It can be shown by methods of integral calculus that this equation can be transformed into the more useful form

$$N = N_o e^{\lambda t}$$

where N_0 is the amount of the variable N at the beginning of the time period under consideration, or at $t = 0$, and $e = 2.718$. A variable N that behaves in time according to this equation is said to grow exponentially. A plot of such a variable is shown in Figure 1.19.

With the relationships developed in Appendix B, we can, with the use of logarithms, write the preceding expression in a different way:

$$\ln N = \ln N_0 + \lambda t,$$

or

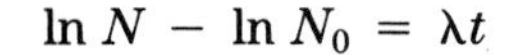

$$\ln N - \ln N_0 = \lambda t$$

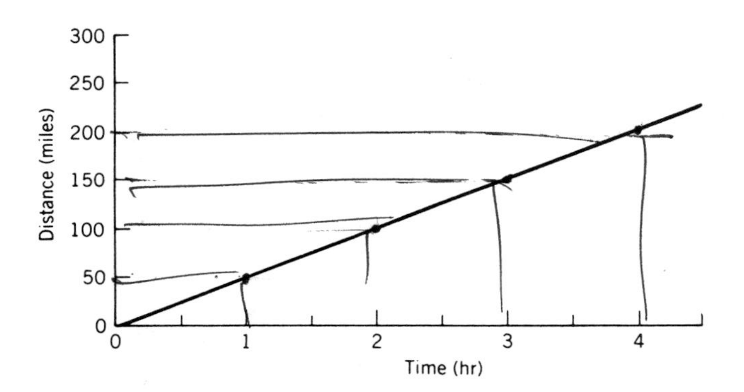

Figure 1.18 The distance traveled from some point as a function of the time traveled.

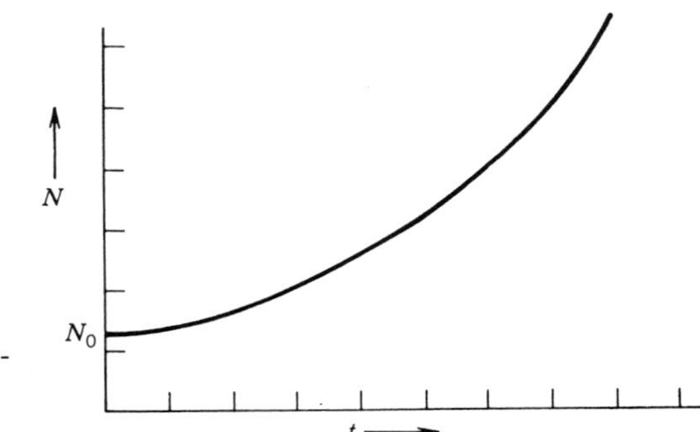

Figure 1.19 A growth pattern indicative of exponential growth, or $N = N_0 e^{\lambda t}$, plotted on a linear graph.

Thus, if we use logarithms, we have a linear dependency on time just as we did with the equation in the example involving a constant velocity. A plot of this equation is shown in Figure 1.20. Fortunately, we do not have to become involved with the computation of logarithms either to plot or to understand variables that behave exponentially. If we use semilogarithmic scales on our plot, the exponential equation appears as in Figure 1.21. It is not necessary to do anything but plot directly the quantities N at their corresponding times t, and if there is an exponential behavior a straight line will result. Such a dependency was shown for the total energy consumption in Figure 1.14.

There are some useful facts concerning exponential relationships and plots. For example, consider a time period t_D such that in this time a quantity doubles, or

$$e^{\lambda t_D} = e^{0.693} = 2$$

or

$$\lambda t_D = 0.693$$

or

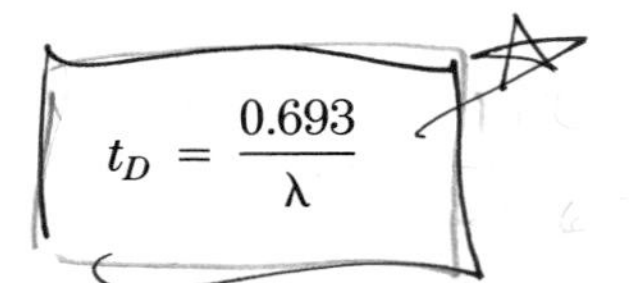

$$t_D = \frac{0.693}{\lambda}$$

In words, this says that when the exponent is equal to 0.693, the quantity N increases by a factor of 2. The time period t_D is thus called the doubling time.

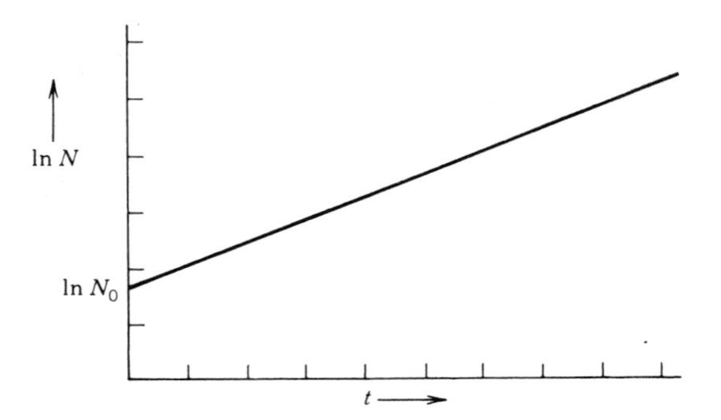

Figure 1.20 A plot of $\ln N$ where $\ln N = \ln N_0 + \lambda t$, plotted on a linear graph.

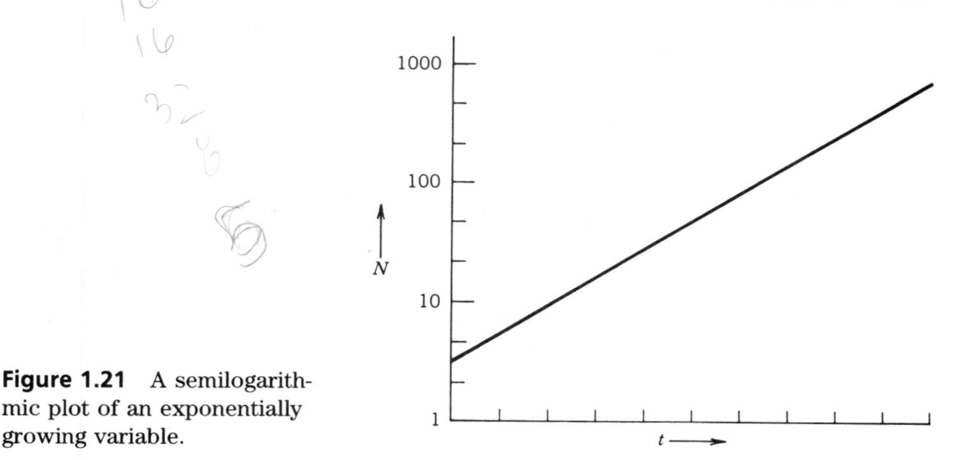

Figure 1.21 A semilogarithmic plot of an exponentially growing variable.

If the growth constant λ is 0.10 per year, that is, 10% per year, then

$$t_D = \frac{0.693}{0.10} = 6.93 \text{ years}$$

It is convenient to approximate this relationship as

$$\text{doubling time in years} \approx \frac{70}{\text{\% growth per year}}$$

or

$$\text{\% growth per year} \approx \frac{70}{\text{doubling time in years}}$$

It is not necessary for the times to be given in years; they could just as well be in minutes or decades or any other interval, as long as the time units are the same on both sides of the equation.

Some common situations exhibit exponential growth. In a bank account, the principal will increase every year by the interest rate times the principal if the interest is not withdrawn. This situation meets all the requirements for exponential growth. Under these conditions, an interest rate of 10% per year will double the money in just 6.9 years.

The increase in population is frequently cited as another example of exponential growth. For example, if at some time the population of the United States is 220 million and is increasing by 1% per year, for the first year the increase will be 0.01 times 220 million, or 2.20 million. The second year the increase will be 0.01 times 222.2 million, or 2.22 million, and so forth. There is no fundamental reason for the percentage growth or doubling time to be constant for extended periods, as the birth and death rates may change for various reasons. But, in fact, many population groups tend to grow exponentially with an approximately constant value for λ.

It is also possible for the exponent λt to be negative. In this case an ever-decreasing amount will be subtracted from the variable N for every successive unit time. This is an exponentially decreasing function.

1.7 SOURCES AND USES OF ENERGY IN A TECHNICAL SOCIETY

The sources of energy in a technical nation, such as the United States, are shown in Figure 1.14. After steam engines came into extensive use and as the population of the cities grew at the expense of the farm areas, wood was gradually replaced by coal as the primary source of energy. This shift took place during the last half of the 19th century. The widespread use of the internal combustion engine and the convenience of oil burning for home heating saw petroleum eventually replace coal to a large extent. In about 1950, oil and natural gas became the dominant sources of energy in the United States. The use of natural gas has increased rapidly and steadily as pipelines from Texas and Oklahoma have made fuel available to a large area of the country for home heating and cooking. Electric power plants are now largely fueled with coal, but a number of plants, particularly in the last 50 years, have used oil and to some extent natural gas because of the very low cost of these fuels. Hydroelectric power has been available and growing for over a century. However, because of the limited number of sites available, only 4.0% of our total energy comes from this source. Nuclear power has, until recently, been growing more rapidly than any other source of energy, but it furnishes only 7.0% of our total energy. Geothermal energy is 0.3% of the total. Table 1.2 lists the sources of energy in the United States in 1989, and Figure 1.22 shows the sources and

Table 1.2 U. S. ENERGY CONSUMPTION IN 1989

	Consumption			Conversion Factor
Energy Source	**Standard Units**	**QBtu**	**Percent**	**Values are Equivalent to 1 QBtu**
Coal[a]	837.7×10^6 tons	18.91	23.3	44.3×10^6 tons
Natural gas	18.95×10^{12} cubic feet	19.36	23.8	0.979×10^{12} cubic feet
Petroleum[b]	6192×10^6 barrels	34.21	42.1	181×10^6 barrels
Hydropower[c]	273.7 kilowatt hours	2.86	3.5	95.7×10^9 kilowatt hours
Nuclear power[c]	528.6 kilowatt hours	5.69	7.0	92.9×10^9 kilowatt hours
Other[c,d]		0.25	0.3	
Total		81.28	100.0	

[a] Includes bituminous coal, lignite, and anthracite.

[b] Includes natural gas plant liquids and crude oil burned as fuel, as well as refined products.

[c] The conversions from kilowatt hours to British thermal units are necessarily arbitrary for these conversion technologies. The hydropower thermal conversion rates are the prevailing heat-rate factors at fossil-steam electric power plants. Those for nuclear power and geothermal energy represent the thermal conversion equivalent of the uranium and geothermal steam consumed at power plants. The heat content of 1 kilowatt hour of electricity, regardless of the generation process, is 3413 Btu.

[d] Includes geothermal, wood, refuse, and other organic matter burned to generate electricity, but not solar energy.

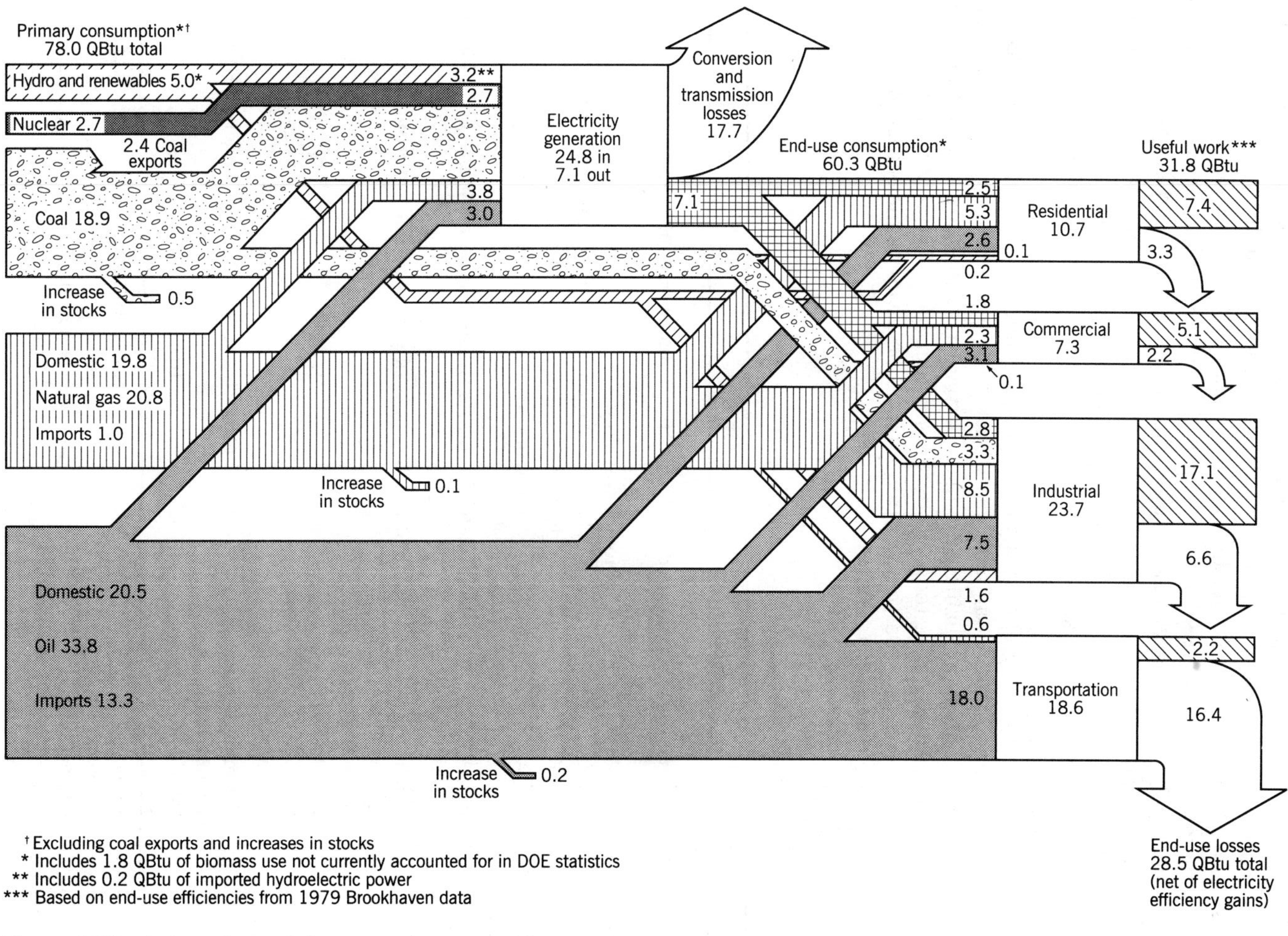

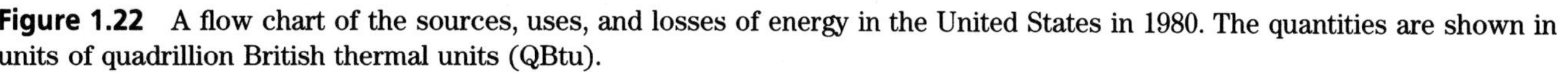

Figure 1.22 A flow chart of the sources, uses, and losses of energy in the United States in 1980. The quantities are shown in units of quadrillion British thermal units (QBtu).

uses of energy in the United States in 1980. The detailed energy pathways shown in this figure will become much more apparent after study of the material in the following chapters, particularly Chapter 2 and 3.

It is interesting to note that solar energy is not listed in statistical information of the kind displayed in Table 1.2 except for the hydroelectric component. This is due mainly to the difficulties encountered in trying to quantify the use of such a diffuse source of energy that is not subject to any accounting. Obviously, solar energy has a direct input into all our food crops as well as into the energy taken from firewood. The rapidly growing use of solar energy, both active and passive, in heating the living space of homes and heating water should be included in any discussion of energy use in the United States or any other country.

In the following chapters we discuss in some detail all of the sources of energy mentioned so far and a few that have not yet been mentioned. The amount of energy available from each particular source now and in the future will be of interest. So, too, will be the manner and efficiency with which we use the energy, and particular attention will be given to measures that would lead to greater conservation.

PROBLEMS

1. A mass of 10 kg is hoisted vertically up a distance of 5 m. Weight = mg (g = 9.8 m/sec^2).
   (a) How much work in joules was done?
   (b) If the mass is now quickly released and falls the 5 m to the ground, what is its kinetic energy as it hits the ground?

2. (a) An engine does 550 ft • lb of work in 10 sec. What is the average horsepower delivered by the engine?
   (b) A 1000-watt (1-kW) heater is run for 10 sec. How much energy in joules is delivered?

3. Which of the following quantities are units of energy as distinct from power?
   (a) watt
   (b) joule
   (c) ft • lb/sec
   (d) Btu
   (e) Newton-meter

4. Classify the following terms according to whether they represent energy (E), power (P), or neither (N).

calorie	______	electron volt	______
horsepower	______	(electron volt)/sec	______
joule/second	______	(electron volt) second	______
joule-second	______	foot-pound	______
kilowatt per hour	______	calorie per minute	______
watt	______	62.5 Btu	______
Btu/hr	______	MeV	______
kilowatt-hour	______	watt-year	______
Btu	______	(foot-pound) per day	______
horsepower per day	______	Btu per year	______

5. If the power consumption of a community is 100 MW and the annual growth rate is 3.5%, in how many years will the power consumpton be 200 MW?

6. A 3 kg mass falls from rest 8 meters in a gravitational field with $g = 9.8$ m/sec^2. What is its kinetic energy in joules?

7. A mass of 12 kg suspended at a height of 10 meters has what potential energy?

8. In 1 hour, 200×10^3 ft • lb of work is done. What is the average power in kilowatts?

9. A car has a mass of 1000 kg and is traveling at a speed of 30 m/sec. What is its kinetic energy in joules?

10. Assuming the engine converts gasoline to kinetic energy with 15% efficiency, how many gallons of gasoline must be burned to give the car the kinetic energy calculated in Problem 9? Ignore all energy-loss processes.

11. In one standard human lifetime (70 years) a city of 100,000 people which enjoys (?) a constant 5% annual growth rate will attain how large a population?

12. If a person lifts 10 lb upward a distance of 1 ft in 1 sec, at what rate is he/she working (i.e., power) in ft • lb/sec, kW, and hp? If she/he does this continuously for 200 sec, how much energy is expended in calories?

13. If an exponentially growing quantity increases by a factor of 8 in 1 year, what is the doubling time in months?

14. A household uses 3.0 kW • hr in a 24-hour period. What is its average rate of power consumption in watts?

15. If a quantity increases by 0.70% per year, in how many years will it double?

16. In a certain year, each person in the United States used the equivalent of 11.2 metric tons of coal (1000 kg). How many Btu per person a year is this?

17. An electric heater puts out 100,000 Btu in 24 hours. What is the average power in kilowatts?

18. If the population of a city grows steadily at 2% per year, in how many years will the population double?

19. (a) In a recent year about 340×10^6 Btu of energy in all forms was used by each person in the United States. What is the equivalent number of barrels of oil?
(b) How many of these Btus were accounted for by food?

20. Identify the type of energy involved for each of the samples below by indicating if it is chemical (c), kinetic (k), potential (p), heat (h), or electromagnetic radiation (r).

Flashlight battery ______
Speeding car ______
Car parked on top of a hill ______
Gasoline ______
Sunlight ______
Wind ______
Rotating flywheel ______
Firewood ______

21. If the mass of your pencil is 10 g, what is the equivalent mass energy in kW • hr?

22. A typical U.S. citizen consumes about how many times as much energy per year as does a typical Nigerian?

23. In the United States the average person eats about 3000 food calories per day. To what power in watts for the person to work and otherwise dissipate this energy does this correspond?

24. The following reaction represents the burning of sugar:

$$C_6H_{12}O_6 + 6O_2 \rightarrow 6CO_2 + 6H_2O + 690 \text{ cal/mole}$$

with C having an atomic mass of 12, H an atomic mass of 1, and O an atomic mass of 16. How many calories are released upon burning 360 grams of sugar?

25. How many grams of candy bars (pure sugar) does a 150-lb person need to eat to climb to the top of Green Mountain? Assume 100% efficiency and an elevation gain of 2000 ft.

26. Calculate how much energy you personally use each year in joules and then compute the average power for the year in kilowatts for the following categories.
    (a) Transportation—car, bus, airplane
    (b) Residential heating—natural gas, oil, or electricity
    (c) Water heating
    (d) Food
    (e) Electrical appliances, lights, and so forth
    (f) Manufactured goods*
    (Hint: Add up all inputs from cubic feet of gas, kilowatt hours, gallons of gasoline, and so forth in one year in some common unit such as joules and divide by number of seconds a year to obtain watts.)

SUGGESTED READING AND REFERENCES

1. Bartlett, A. A. "Forgotten Fundamentals of the Energy Crisis." In *Perspectives on Energy,* Lon C. Ruedisili and Morris W. Firebaugh, Eds. New York: Oxford University Press, 1982.
2. Christensen, John W. *Energy, Resources and Environment.* Dubuque: Kendall Hunt, 1981.
3. Crawley, Gerald M. *Energy.* New York: Macmillan, 1975.
4. Fowler, John M. *Energy and the Environment.* New York: McGraw-Hill, 1975.
5. Hubbert, M. King. "The Energy Sources of the Earth." *Scientific American,* **224** 3 (September 1971), pp. 60–70. (Also in *Energy and Power,* a Scientific American book. San Francisco: W. H. Freeman, 1971.)
6. Hubbert, M. King. "Energy Resources." *Resources and Man.* National Academy of Sciences—National Research Council. San Francisco: W. H. Freeman, 1969.
7. Krenz, Jerrold H. *Energy—Conversion and Utilization,* second edition. Boston: Allyn and Bacon, 1984.
8. Marion, Jerry B. *Energy in Perspective.* New York: Academic Press, 1974.
9. Priest, Joseph. *Problems of Our Physical Environment.* Reading, Mass.: Addison-Wesley, 1976.
10. Romer, Robert H. *Energy, an Introduction to Physics.* San Francisco: W. H. Freeman, 1976.
11. Saperstein, Alvin M. *Physics: Energy in the Environment.* Boston: Little, Brown, 1975.
12. Wilson, Richard, and Jones, William J. *Energy, Ecology and Environment.* New York: Academic Press, 1974.
13. Cassedy, Edward S., and Grossman, Peter Z. *An Introduction to Energy: Resources Technology and Society.* Cambridge: Cambridge University Press, 1990.
14. "Energy for the Planet Earth." *Scientific American* **263** 3 (September 1990), pp. 54–163.

*As a rough guide, assume that these commercial and industrial uses of energy are about equal to the sum of your personal use as in items (a) through (e).

CHAPTER TWO

ENERGY FROM FOSSIL FUELS

The combustion of a single pound of coal, supposing it to take place in a minute, is equivalent to the work of three hundred horses; and the force set free in the burning of 300 lbs. of coal is equivalent to the work of an able-bodied man for a lifetime.

J. Dorman Steele, PHYSICS, American Book Company, New York, 1878

2.1 INTRODUCTION

Our fossil fuel resources are of two general classifications: petroleum and coal. Each of these has subclassifications: petroleum includes a variety of compounds of hydrogen and carbon, ranging from gaseous methane through liquids like oil to the semiliquid tar sands and to the solid organic content of oil shale. Although natural gas, which is primarily methane (CH_4), is technically in the petroleum family, it is usually treated separately from liquid petroleum since it is processed, transported, and consumed in a quite different fashion. The natural gas liquids, propane and butane (bottled gases), on the other hand, are often classified with petroleum. What we call coal can be taken to include everything from peat through the hard anthracite coals of the eastern United States. These classifications are not rigorous; some authors consider oil shale to be a type of coal, with less carbon and more hydrogen than basic coal, and, as we shall learn, liquid and gaseous petroleum can be derived from most types of coal.

Each of the fossil fuels has a different story with regard to the method of formation, the extent of the resource, the role that it has played in meeting our need for energy, and the place that it will have in supplying energy in the future. The stories are important, for what is involved is not only the gasoline for our automobiles, the natural gas to heat our homes and cook our food, and the coal to provide electricity, but also international politics on a grand scale and perhaps the ingredients for world conflict.

2.2 GEOLOGY AND HISTORY OF PETROLEUM FORMATION

The oldest known petroleum is about 500 million years old. It dates back to a time when there was no life on the continents; plants and animals existed only in the saltwater oceans. These organisms were deposited on the floors of the ancient oceans and decomposed by bacteria. Fine sand and mud accumulated over the organic deposits. Eventually the weight of the sedimentary rock that formed over the deposits caused the temperature and pressure to increase sufficiently for the residues to be converted into oil and natural gas. The oil and gas could move about in the sedimentary rock and eventually become collected in some kind of trap. One such trap, called a structured trap, is shown in Figure 2.1. The oil and gas have risen to be trapped beneath a bulge in a layer of nonporous rock.

The locations of the ancient seas that were originally involved in the formation process have been changed by the movements of the earth's crust,

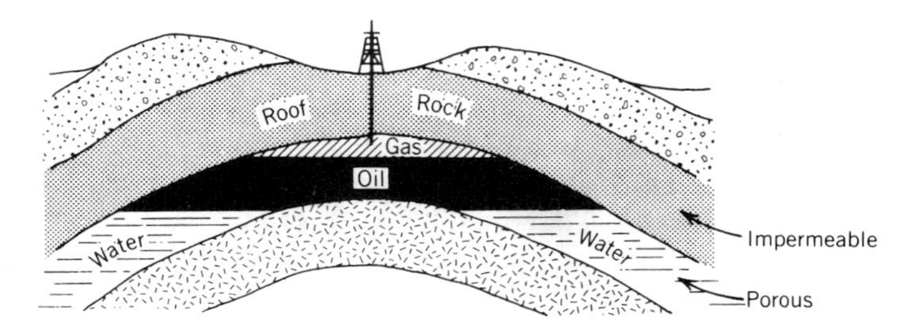

Figure 2.1 A typical geologic formation in which oil and natural gas are found. The gas is found above the oil, and the oil above water, all interspersed in a porous formation.

so that many petroleum deposits now lie under dry land. Offshore deposits, however, such as those found in the North Sea, are becoming a more significant source of petroleum as the deposits under land are exhausted.

Most commercially exploited deposits of petroleum are less than 250 million years old. Although the petroleum formation process may be continuing, there is no hope that appreciable amounts of petroleum can be formed in our lifetimes. The earliest recognizable stages of the processing of organic sediments into petroleum take some 3000 to 9000 years.

2.3 PETROLEUM PRODUCTION IN THE UNITED STATES

The petroleum industry in the United States began on August 27, 1859, when Edwin L. Drake drilled and struck oil near Titusville, Pennsylvania. Before that time oil was available in very small quanities where it naturally seeped to the surface in various locations. The combination of a prolific source of "rock oil," as it was called, and a refinery process that produced kerosene that could be used for lighting and cooking saw an explosion in the production of crude oil in the United States following Drake's discovery. Within 15 months there were 75 wells in Pennsylvania producing 450,000 barrels of oil per year, and by 1862 3 million barrels per year were being produced. The price dropped to 10 cents per barrel. In about 1870 John D. Rockefeller started a company for the refining and distribution of kerosene, which became the Standard Oil Company and which for many years dominated the industry.

Although the Pennsylvania oil fields were rather short-lived, new and larger oil discoveries were made in Ohio and Indiana. In the early 1900s major oil deposits were found in the San Fernando Valley of California and near Beaumont, Texas. By 1909, when the industry was just 50 years old, the United States was producing 500,000 barrels a day, which was more than was produced by all the other countries combined. Even though there were major oil-producing areas in Russia, Sumatra, Persia, and other places, petroleum was a major export item for the United States for many years. In fact, until 1950 the United States produced more than half the world's oil. Discoveries of oil in such places as Venezuela, Mexico, Saudi Arabia, Iraq, Kuwait, and Bahrain, however, meant that the cost of producing oil in the United States eventually became considerably larger than the cost of oil from these other abundant sources.

In 1882 Thomas Edison developed the incandescent lightbulb, which, coupled with electric generators and electric distribution systems, meant that by 1900 kerosene was used as an illuminant only in rural areas. Just as this occurred, horseless carriages started to make their appearance in great numbers. In a few short years the major users of petroleum were the millions of automobiles that needed a continuous source of gasoline and lubricants. The need for petroleum increased dramatically and the production of it followed. Figure 2.2 shows the petroleum produced annually in the United States. Aside from minor dips around 1930 and 1960, the production of oil increased systematically until 1970. This was accomplished by drilling many exploratory wells and by taking advantage of sophisticated location techniques developed by petroleum geologists and geophysicists. In 1970 the last major find, the deposits at Prudhoe Bay on the north coast of Alaska, were added to the proven reserves. Also in 1970, the production of oil in the United States peaked at 3.3×10^6

3 3,000,000

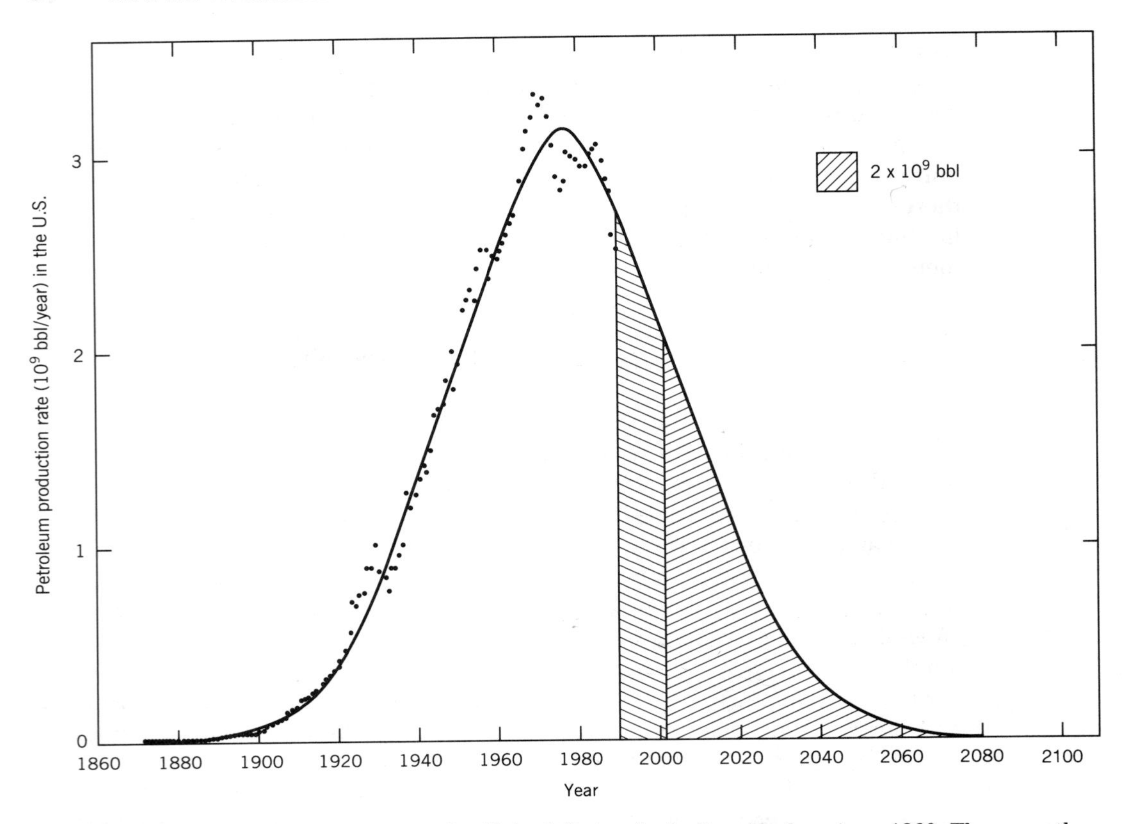

Figure 2.2 Petroleum production in the United States, including Alaska, since 1860. The smooth curve on the right side of the figure is drawn to reflect the shape of the actual production curve on the left. The significance of the shaded areas is discussed in the text. (Data from: *Basic Petroleum Data Book*, Vol. XII, Washington, D.C.: American Petroleum Institute, Jan. 1992.)

bbl/year. The use of oil in the United States, however, as we saw in Figure 1.14, has continued to increase. Since 1948 the United States has imported more oil every year than it has exported. At the present time we import nearly 50% of our needs.

2.4 PETROLEUM RESOURCES OF THE UNITED STATES

Fossil fuels clearly represent a finite, nonrenewable, depletable resource. As we burn petroleum and coal in our daily lives, we should retain our awareness that we are destroying ancient and precious entities in the process. The particular combinations of carbon and hydrogen atoms now serving us so well as fuels were assembled some 10 million to 500 million years ago, and they have been undisturbed since that time. Once we have oxidized them back into carbon dioxide and water (i.e., burned them), they are gone forever as far as humanity is concerned. But the continuance of our civilization may now depend on the understanding and wise use of fossil fuels. A necessary element of such an understanding is knowledge of the size of the fossil fuel resources available to us.

The estimation of the liquid petroleum resources of the United States provides a good case history that may apply also to other resources, both national and worldwide. In the following discussion of liquid petroleum, only the resources accessible through conventional drill hole methods, either by natural pressure or by pumping, are considered. Other techniques, such as those considered under the general description of advanced recovery techniques and oil mining, may increase the size of the domestic resource, possibly as much as doubling the amount considered in this discussion. However, the cost of oil obtained by these methods will be greater than that obtained by direct drilling. Such methods, coupled with increased market prices, can be expected to expand to some extent the magnitude of the economically recoverable resource.

The estimating process is dynamic; each year's experience adds new information, and the estimates from various sources of the size of the resource now seem to be coming into closer and closer agreement. This has not always been the case: 10 or 20 years ago the estimates were widely divergent. One estimate, published in 1962, predicted an eventual cumulative production of 590×10^9 bbl of oil in the United States, whereas other estimates were in the range of 150×10^9 bbl. The high estimate, made by A. D. Zapp of the U.S. Geological Survey (USGS), is based on what has become known as the Zapp hypothesis: "Oil discovery per foot of drilling will remain constant until we have one well per two square miles." Qualifying phrases include "in any given petroliferous region" and "to a maximum depth of 20,000 feet." This hypothesis was quickly and completely discredited. An examination of available data (see Figure 2.3) showed that oil discovery per foot of drilling was by no means constant, but was, in fact, a distinctly declining function of time. The Zapp hypothesis, therefore, produced a massive overestimate. The meaning of the declining discovery rate is, of course, that in spite of ever-more sophisticated exploration technology, the richest oil basins have almost certainly been discovered and

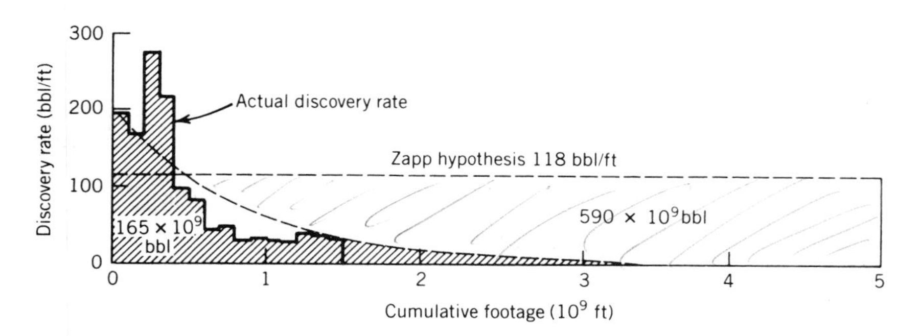

Figure 2.3 Actual discovery of petroleum in the United States compared to the predictions of the Zapp hypothesis. The discovery rates are given in barrels of oil per foot of drilling versus cumulative footage. Actual data extend to about 1.5×10^9 ft of cumulative footage. The area under the curve fit to the actual discovery rates represents a Q_∞ of 165×10^9 bbl, whereas the area under the horizontal line corresponding to a constant discovery rate represents a Q_∞ of 590×10^9 bbl. The large peak at a cumulative footage of 0.3×10^9 ft corresponds to the discoveries that took place in the 1930s. (Source: Reprinted with permission from M. K. Hubbert, *Resources and Man*, National Academy Press, Washington, D.C., 1969.)

drilled. Future oil discoveries are going to be more and more difficult and expensive.

An alternative approach, which now appears to be basically correct, was put forth (as early as 1956) by M. King Hubbert. His method is based on some straightforward and uncomplicated propositions. Hubbert has argued that for any finite resource consumed to depletion, one can plot a few curves tracing out the history of the exploitation of that resource.

Consider, for example, the graph shown in Figure 2.4. Let the vertical axis represent the total cumulative number of barrels of economically recoverable oil discovered in the United States as time progresses. A curve tracing the history of this discovery must begin at zero at the time of the initial discovery of oil in the United States (1859). As each unit of oil is discovered, the curve advances upward. It can never go down. As exploration techniques improve and as the incentive for discovery increases, the curve goes up ever-more steeply. There must come a time, however, when the easy discoveries have all been made, when the probability of unsuccessful drilling increases, and when the discovery rate, whether expressed in barrels per foot drilled or in barrels per year, must diminish. The curve will then advance more slowly; it will begin to level off. Occasional periods of success in exploration may produce minor fluctuations in the curve, but the overall trend is relentless; the curve must most certainly level off as new discoveries become less and less frequent. The cumulative discovery level that can never be exceeded we call Q_∞; even if one searched for an infinite time, no more economically recoverable oil could be discovered. A numerical estimate for Q_∞ (oil, United States) may be obtained by extrapolating the data of Figure 2.3. Hubbert derived a value of $Q_\infty = 165 \times 10^9$ bbl (exclusive of Alaska) in this way.

A second (dashed) curve is shown in Figure 2.4. This curve represents the cumulative production of oil in the United States as a function of time. Production, of course, must follow discovery in time and level off also at Q_∞. In the case of oil, production has been found to lag behind discovery by about 11 years.

Another interesting and informative graph (shown in Figure 2.5) can be derived from Figure 2.4. By taking the slopes of the two curves in Figure 2.4, the rates of discovery and production rather than the cumulative amount can be plotted. The discovery rate was low in the early history of the oil industry and then began to increase for reasons of technology and demand. It must attain some maximum value and then turn back toward zero as discovery becomes more difficult. As the cumulative discovery approaches Q_∞, the dis-

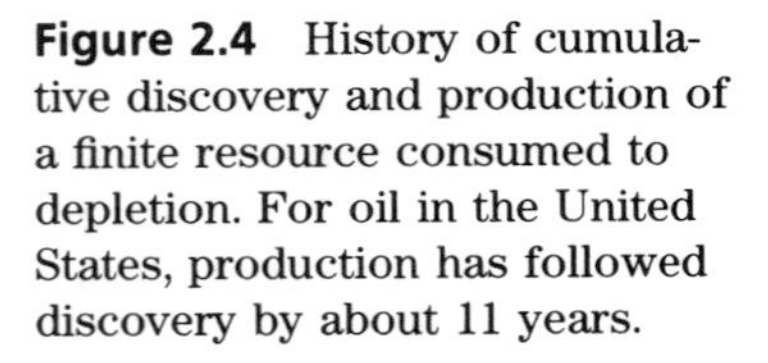

Figure 2.4 History of cumulative discovery and production of a finite resource consumed to depletion. For oil in the United States, production has followed discovery by about 11 years.

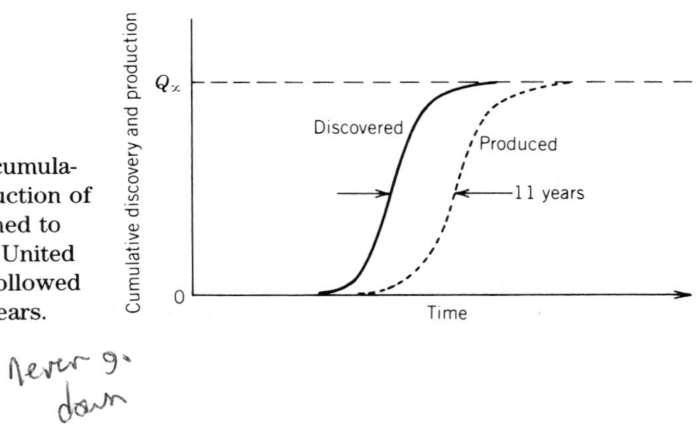

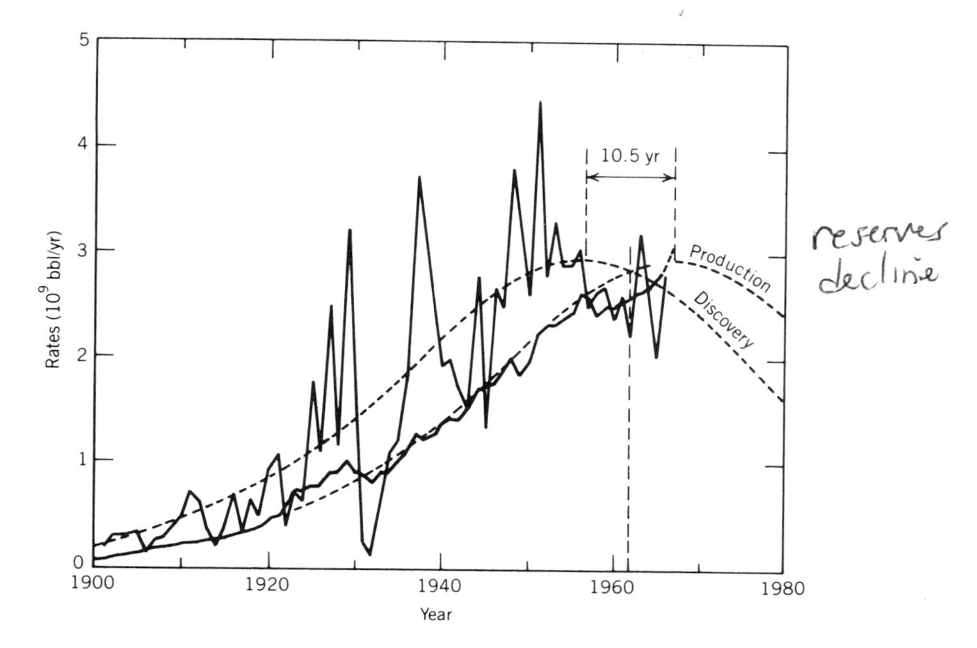

Figure 2.5 Actual discovery and production rates for oil in the United States are shown by solid lines. The smooth dashed lines represent average trends for these quantities. (Source: Reprinted with permission from M. K. Hubbert, *Resources and Man,* National Academy Press, Washington, D.C., 1969.)

covery rate must approach zero. A similar curve shown in Figure 2.5 for production follows the discovery curve by about 11 years. By projecting curves of this type beyond available data, Hubbert predicted *in 1956* that U.S. oil production would peak between 1966 and 1971. It actually peaked in November 1970. His analysis indicates that we have now produced well over 50% of our Q_∞ of oil and that trends from now on must be generally downward.

One point not often emphasized in public discussion of our national oil situation is that production will never exceed Q_∞, even though the exact value of Q_∞ might change with time. One might imagine that whatever value it now has, Q_∞ might be doubled by improved recovery techniques for in-ground oil. However, even such a circumstance would not alter the essential fact that Q_∞ is limited to some definite maximum value. If our oil companies should, through extraordinary measures, succeed in substantially increasing the domestic production rate, it would only hasten the end of the resource. A barrel for us now is one less for our grandchildren.

A fourth graph (shown in Figure 2.6) may be derived from Figure 2.4. This represents the *rate* at which our oil reserves grow and decline. It is evident that during times when discovery exceeds production, national oil reserves must increase, and when production exceeds discovery, reserves must decrease. The rate of this increase (or decrease) is given by the difference at any given time between the discovery and production curves. This difference is plotted in Figure 2.5, from which it can be seen that we are now in a period of declining reserves. Since about 1962, we have generally been pumping already discovered oil out of the ground faster than new discoveries are being made.

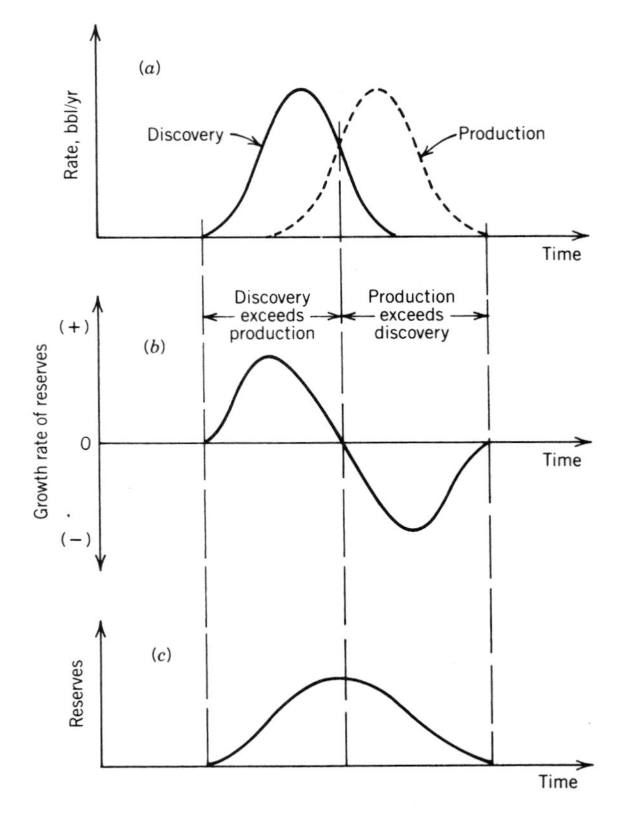

Figure 2.6 A generalized representation of the relationship between: (*a*) discovery and production rates for petroleum in the United States, (*b*) the growth rate of reserves, which had a maximum in the 1940s but is now negative, and (*c*) the reserves, which now are declining.

Figure 2.2 is constructed in the spirit of Hubbert's type of analysis and includes production data up through 1990, which include Alaskan and offshore contributions. It can be noted that there has been a dramatic decrease in the production rate since 1970. The right side of the production curve is a prediction, and there are many factors that could make the actual experience deviate significantly from this prediction.

The shaded area just to the right of the center of Figure 2.2 represents the proven reserves as of 1990. These are reserves that have been located in the ground and that are felt with a good degree of reliability to be available. The area of this section on the graph corresponds to proven 1990 reserves of 26.5×10^9 bbl.

The area under the entire curve corresponds to the Q_∞ for the United States. This corresponds to about 214×10^9 bbl. This estimate is not very much larger than Hubbert's early estimates of 165×10^9 bbl plus the later Alaskan discoveries. The section of the curve that is not shaded represents the 142×10^9 bbl of petroleum that has already been taken out of the ground. The two shaded areas on the right represent the remaining petroleum; the present estimate is 72×10^9 bbl or about 34% of Q_∞. The petroleum yet to be discovered is the shaded area furthest to the right; it corresponds to 46×10^9 bbl. This area is the most speculative of the three under the curve, as it represents yet undiscovered recoverable resources. These undiscovered resources have been the subject of many studies. Table 2.1 lists various estimates of this quantity, and there is not much agreement. It must be remembered, however, that an

Table 2.1 ESTIMATES OF UNDISCOVERED RECOVERABLE RESOURCES OF OIL AND NATURAL GAS IN THE UNITED STATES

	Oil and Natural Gas Liquids	Natural Gas
	Billion (10^9) Barrels	Trillion (10^{12}) Cubic Feet
Oil Companies		
Company A (Weeks 1960)	168	—
Company B (Hubbert 1967)	21–64[a]	280–500[a]
Company C (1973)	55[b]	—
Company D (1974)	89	450
Company E	90	—
U.S. Geological Survey		
Hendricks (1965)	346	1300
Theobald et al. (1972)	458	1980
McKelvey (1974)	200–400	990–2000
Hubbert (1974)	72	540
USGS (1975)	61–149	322–655
USGS (1989)	58	399
National Academy of Sciences		
National Research Council (1975)	113	530

[a] Exclusive of Alaska.

[b] Estimated discoverable between 1973 and 1985.

estimate made in 1974 must be reduced by the petroleum produced between 1974 and 1990 to be compared with the above estimate. Generally the more recent estimates are felt to be more reliable, as they are based on better data. They are smaller than earlier estimates, in most cases.

The uncertainties in estimating the amount of undiscovered oil are large, since it is impossible to predict when a major new find will occur. Most petroleum geologists are not optimistic on this score, since more than 1.7×10^6 wells have been drilled in the United States and offshore, and few surprises are expected. Another major uncertainty in Q_∞ is the effect of technical advancements in secondary and tertiary recovery of oil left in the ground at old fields. More is said about this later in the chapter, but the high costs of these processes have limited their usefulness so far.

If our present estimates of Q_∞ are too pessimistic, how much effect would a larger value have? This can be answered in part by reference to Figure 2.7, which depicts the world oil situation as determined by Hubbert. The two curves represent the extreme estimated values of Q_∞ (oil, world) $= 1.35 \times 10^{12}$ to 2.10×10^{12} bbl. Note that the upper estimate only changes the peak production year by 10 years compared to what it would be if the lower estimate were correct. A similarly minor effect would be expected in the United States for increases in Q_∞. For world production, the time period for producing 80% of the resource is about 65 years, as in the United States.

Could Hubbert be wrong by a large amount? Might normal technological advances render his derivation of Q_∞ from the data of Figure 2.3 entirely incorrect? What if we somehow learn to recover two thirds of the liquid petroleum known to be underground rather than the recovery ratio of one third

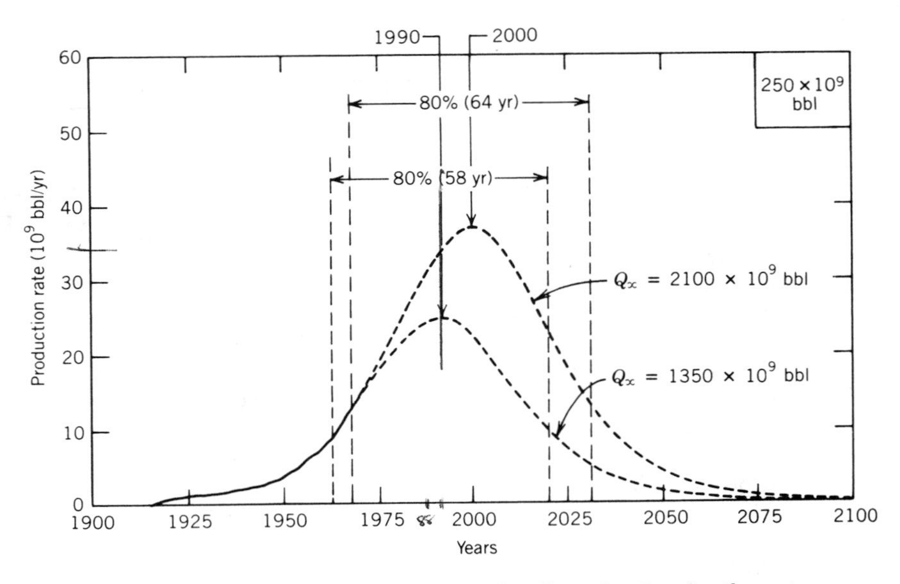

Figure 2.7 Projected cycles for world crude oil production for the extreme values of the estimated total resource. The years of estimated peak production are shown. (Source: Reprinted with permission from M. K. Hubbert, *Resources and Man*, National Academy Press, Washington, D.C., 1969.)

now possible? These three questions are similar in their intent and can all be addressed by some common answers. First, recall that the data of Figure 2.3, indicating an unquestionable decline in the discovery rate per foot of drilling, include the benefits of improvements in exploration technology. Without these improvements, the decline would have been even steeper. Second, even though Hubbert's estimates were at first regarded throughout the industry and by the USGS as far too pessimistic, after further study it is now generally accepted that Q_∞ (oil, United States) is indeed approximately the value originally put forth by Hubbert. Third, even if we were somehow able to increase Q_∞ by a large amount, say 100 or 200 $\times$ 10^9 bbl, at our present rate of consumption this amount represents only a few tens of years. The situation would be relieved temporarily, but there would still be a critical oil shortage impending within our lifetimes.

A completely different approach to estimating the lifetime of the petroleum resources of the United States was recently published by C. A. Hall and C. J. Cleveland. They calculated the cost in terms of energy (i.e., in terms of barrels of oil) needed to produce a new barrel of oil. The energy is expended in various exploration efforts, drilling the wells, and extracting the oil. As Figure 2.3 shows, the number of barrels of oil found per foot of drilling has been generally decreasing as the resource has been depleted. This is equivalent to saying that it costs more energy (barrels of oil) to find a new barrel of oil every year. Clearly, when it costs one barrel of petroleum fuel to find a new barrel of petroleum fuel, the process will come to a stop. By projecting the costs in terms of barrels per foot drilled and the finding rate (also in barrels per foot), Hall and Cleveland found the intersection of these two trends to lie between the years 2001 and 2011. After that time, the energy yield of the oil industry will be less than the energy cost, and U.S. oil production as an energy source will no longer exist. This is quite consistent with the information shown in Figure 2.2.

Now let us take a look at the present estimated remaining oil in the United States, including Alaska, on a per capita basis:

$$\frac{72 \times 10^9 \text{ bbl}}{250 \times 10^6 \text{ persons}} = 288 \text{ bbl/person}$$

Our present per capita consumption rate is approximately

$$\frac{6 \times 10^9 \text{ bbl/yr}}{250 \times 10^6 \text{ persons}} = 24 \text{ bbl/yr-person (or 1008 gal/yr-person)}$$

in the United States. This gives us a time until complete exhaustion of about

$$\frac{288 \text{ bbl/person}}{24 \text{ bbl/yr-person}} = 12 \text{ yr}$$

On a worldwide basis, the remaining oil is

$$\frac{2000 \times 10^9 \text{ bbl}}{5.0 \times 10^9 \text{ persons}} = 400 \text{ bbl per person}$$

So, as far as remaining oil alone is concerned, the per capita situation in the United States is not too different from that of the world at large. The United States has about 3.6% of the remaining oil and about 5% of the population. It is our prodigious rate of consumption that has created a particular problem in this country. Offshore oil wells as shown in Figure 2.8 have become an important source of oil for the United States.

Figure 2.8 An offshore drilling rig in the Gulf of Mexico.

There seems to be little doubt that the U.S. oil resource obtainable by conventional means is nearing its end. There is probably room for argument as to whether we have another 20 or another 40 years, but there is growing agreement that we must change our ways. We must learn to conserve, innovate, or do without. This is particularly true nationally, but it is also true worldwide. The time scales are only slightly different.

2.5 WORLD PRODUCTION OF PETROLEUM

Table 2.2 shows some data on the major oil-producing nations. It may be surprising that the United States is the second largest oil producer in spite of modest proven reserves. What is also noticeable is that the number of producing wells in the United States. (603,000) is very large compared to, say, Saudi Arabia (868), even though daily oil production is not very different. The reason for this is that the oil fields in the United States are very mature and the production per well is very small—only about 12 bbl/day on average. There are essentially no large reservoirs left in the United States where the oil flow is comparable to that in Saudi Arabia. There the average is about 7200 bbl/day per well. The countries with the largest proven reserves, such as Saudi Arabia, Iraq, and Iran, will be the sources of oil that the world will have to turn to as we head into the 21st century. The recent Gulf War with Iraq very much involved such considerations.

On a worldwide basis some 60.3×10^6 bbl of oil are produced every day from 910×10^3 wells. The world proven reserves are estimated to be 999×10^9 bbl. Of this proven reserve, about two thirds, or 663×10^9 bbl, is located in the Middle East. Figure 2.7 shows Hubbert's estimates for the world production cycle of petroleum with two extreme values for Q_∞.

2.6 PETROLEUM AND PETROLEUM REFINING

The crude petroleum taken from the earth contains hundreds of different chemical compounds called hydrocarbons. Hydrocarbons are compounds that contain only hydrogen and carbon. These two elements can be arranged in a

Table 2.2 MAJOR OIL-PRODUCING COUNTRIES OF THE WORLD

Country	Production ($\times 10^3$ bbl/day)	Estimated Proven Reserves ($\times 10^6$ bbl)	Number of Producing Wells
Former U.S.S.R.	11,500	57,000	145,000
U.S.	7,220	26,177	603,000
Saudi Arabia	6,215	257,504	858
Iran	3,120	92,850	361
China	2,755	24,000	43,700
Mexico	2,633	51,983	4,740
Venezuela	2,118	59,040	12,752
Iraq	2,083	100,000	820
United Kingdom	1,860	3,825	762
Nigeria	1,808	17,100	1,432

Source: *The Oil and Gas Journal,* February 1991.

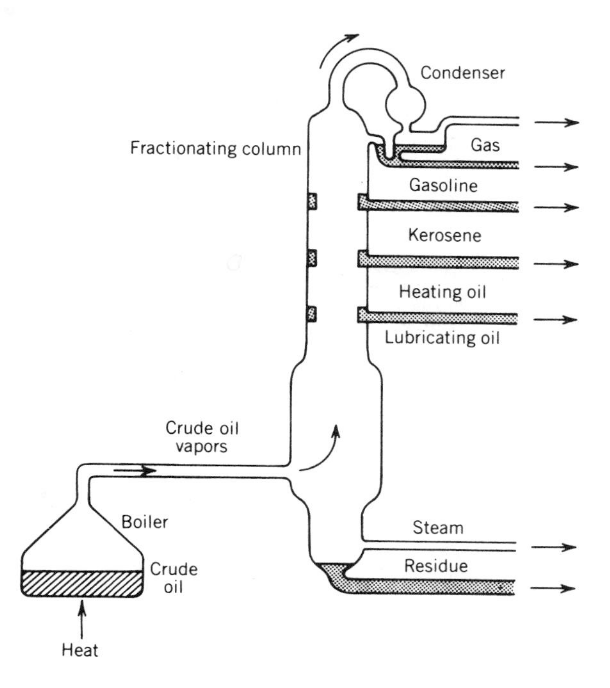

Figure 2.9 A fractionating column for the distillation of petroleum. The temperature of the column decreases from bottom to top so that the more volatile components are condensed toward the top of the column.

great variety of ways to form rather simple molecules, such as CH_4 (methane, the major constituent of natural gas), or very complicated molecules that contain more than 20 atoms of carbon or hydrogen. There are also trace elements in crude petroleum, the most important of which is sulfur, which is present in weight up to about 4%.

A variety of commercial products, essential for an industrialized society, come from petroleum. The process of obtaining these products, such as gasoline or heating oil, is called refining, and it is done in an interesting way.

The first step in the process is fractional distillation. The crude oil is vaporized by heating it to about 398°C, and the vapors are then introduced at the bottom of a fractionating column, or tower. The vapors will then condense at various collection points going up the tower, as the temperature for condensation decreases. Figure 2.9 shows the general arrangement schematically. Gasoline, with the lowest boiling point, condenses and is drawn off at the top, whereas heating oil, which condenses at higher temperatures, is drawn off lower down. The gases that do not condense leave from the top of the column. The residue that comes from the bottom of the column contains asphalt and heavy fuel oil. Table 2.3 shows some of the products of fractional distillation, their molecular size, boiling point range, and typical uses.

The mix of products from a distillation column is not normally the same as the mixture demanded by the marketplace. For this reason, it is desirable to alter the products of fractional distillation to make them more useful. The usual deficiency is in the lighter molecules that are usable in gasoline, such as octane (C_8H_{18}), and the excess is the heavy molecules of paraffin and tar. The oldest method of breaking up the heavy molecules into lighter ones is called thermal cracking, using high temperatures and pressures. A more efficient process for producing higher octane gasoline is catalytic conversion. In this

Table 2.3 PRODUCTS OF PETROLEUM DISTILLATION

Fraction	Molecular Size Range	Boiling Point Range Degrees Celsius	Typical Uses
Gas	C_1–C_5	−164 to 30	Gaseous fuel
Petroleum ether	C_5–C_7	30 to 90	Solvent, dry cleaning
Straight-run gasoline	C_5–C_{12}	30 to 200	Motor fuel
Kerosene	C_{12}–C_{16}	175 to 275	Fuel for stoves, diesel, and jet engines
Gas oil or fuel oil	C_{15}–C_{18}	Up to 375	Furnace oil
Lubricating oil	C_{16}–C_{20}	350 and up	Lubrication
Greases	C_{18}–up	Semisolid	Lubrication
Paraffin (wax)	C_{20}–up	Melts at 52–57	Candles
Pitch and tar	High	Residue in boiler	Roofing, paving

Source: Spencer L. Seager and H. Stephen Stoker, *Chemistry: A Science for Today* (San Francisco: Scott, Foresman, 1973), p. 299.

process petroleum vapor is passed over an alumina–silica mixture or certain types of clay that act as catalysts in bringing about the chemical changes. A photo of a typical refinery is shown in Figure 2.10.

The reverse process of joining together light hydrocarbon molecules and forming heavier ones is also carried out and is called polymerization. Natural gas and other gases from the cracking process are made into high-octane fluids in this way.

An interesting series of hydrocarbon molecules is the alkanes. They follow the sequence of C_nH_{2n+2}, where n is an integer starting at 1 and continuing to higher numbers. The first eight members of the sequence are shown in Table 2.4.

2.7 NATURAL GAS RESOURCES AND CONSUMPTION

Because of its cleanliness and convenience, the production and use of natural gas has grown more than that of any other form of energy in the United States in the last 40 years. Natural gas consists of approximately 85% methane and 15% ethane. Methane burns according to the formula

$$CH_4 + 2O_2 \rightarrow CO_2 + 2(H_2O)$$

with the release of about 1000 Btu for every cubic foot of gas at atmospheric pressure and normal temperature. Except for CO_2 and its possible effect on the temperature of the earth's atmosphere, the by-products of CH_4 burning are quite harmless. In addition to its advantage of cleanliness, natural gas can be delivered to its destination both conveniently and economically through pipelines. These reasons, as well as the relatively low cost of natural gas per Btu to the consumer, have made it a very desirable fuel for use in the home for space heating, water heating, and cooking. It is also an important ingredient in

Figure 2.10 A petroleum refinery is a complex array of modern chemical technology. Several fractional distillation towers are shown.

the manufacture of various organic chemicals and is used to some extent in the generation of steam for electric power production.

The formation of natural gas and of petroleum are closely related. Both resources are thought to have originated in the deposits of organic matter in the ancient oceans. Natural gas and petroleum are frequently found in the same deposit. In the early days of U.S. petroleum production, natural gas was not valued very highly and was often flared at the well to get rid of it. This, of course, is no longer the case in the United States; natural gas, whether associated with petroleum production, coming from separate gas wells, or occurring as a by-product of petroleum refining, is in great demand.

United States natural gas resources have been estimated by a number of people in the government (USGS), in oil companies, and in special committees set up, for example, by the National Academy of Sciences. One estimating technique takes advantage of the relatively constant ratio of gas discoveries to oil discoveries since 1940. The upper range of values is about 7500 ft^3/bbl and

Table 2.4 THE ALKANE SERIES OF HYDROCARBONS[a]

n	Molecule	Name	Primary Use
1	CH_4	Methane	Natural gas
2	C_2H_6	Ethane	Natural gas
3	C_3H_8	Propane	Bottled gas
4	C_4H_{10}	Butane	Bottled gas
5	C_5H_{12}	Pentane	Gasoline
6	C_6H_{14}	Hexane	Gasoline
7	C_7H_{16}	Heptane	Gasoline
8	C_8H_{18}	Octane	Gasoline

[a] The heat of combustion for these hydrocarbons ranges from about 53,000 Btu/kg for methane to 45,000 Btu/kg for octane.

the lower range about 6250 ft^3/bbl. With these values and with the value of Q_∞ (oil) = 200 × 10^9 bbl, Q_∞ (natural gas) would range from 1500 × 10^{12} ft^3 to 1250 × 10^{12} ft^3. Table 2.1 shows various estimates of the undiscovered recoverable natural gas resources in the United States. One can add to these estimates the cumulative production to 1990 (about 700 × 10^{12} ft^3) and the proven reserves of about 167 × 10^{12} ft^3 to obtain estimates of Q_∞. It is interesting to note in Table 2.1 that the estimates made recently, which are presumably based on better information, tend to be lower than some of the earlier estimates. If the cumulative production and proven reserves are added to the recent estimates of about 399 × 10^{12} ft^3, values of Q_∞ (natural gas) are obtained that are not inconsistent with those based on the Q_∞ for petroleum. Hubbert made an analysis of the complete cycle of U.S. natural gas production, similar to his earlier analysis for petroleum. The main predictions of this analysis were that the maximum production rate would be reached in 1980 and that 80% of the total natural gas resource would be used between 1950 and 2015, based on a Q_∞ of 1290 × 10^{12} ft^3, not including contributions from Alaska.

The annual U.S. consumption of natural gas is shown in Figure 1.14 for the years 1900 to 1990. The data, which can be approximately represented by a straight line on this semilog plot, are a fine example of exponential growth until about 1970. The growth constant until that time is about 6.3% a year. As we discuss later, the deficiencies in the supply of natural gas and, to some extent, the increase in price, have brought about a marked change in both the production and the consumption of natural gas since about 1973. Both production and consumption have been slowly decreasing since that date, or about seven years earlier than Hubbert predicted.

Proven reserves of natural gas increased at a rather constant rate until 1968. Since that time, proven reserves have decreased, with the exception of 1970, when the Prudhoe Bay, Alaska, reserves were added. The ratio of reserves to annual production has decreased steadily from a value of about 33 in 1945 to less than 10 at the present time.

To some extent, the deficiency in domestic supplies can be accounted for by the fact that natural gas is a finite resource that is rapidly being depleted. On the other hand, the control of natural gas prices by the federal government until recently made natural gas artificially inexpensive and did not provide much incentive for exploration and wildcat wells.

Because of the decrease in domestic production, imports of natural gas increased from about 1% of our total use in 1960, to 4% in 1974, to 5% in 1979, and to 7% in 1990, with Canada and Mexico being the primary suppliers. The completion of a pipeline that originates near Prudhoe Bay and comes down through Canada into the U.S. Midwest will provide some relief to the supply problem in a few years.

Because of the shortages of natural gas, tremendous investments have been made by some companies in liquefied natural gas (LNG) ships. These are specially constructed ships that can transport natural gas in its liquefied state at about 161°C below zero. Gas in its liquefied state takes up only 1/600 of the volume of room temperature natural gas. Libya, Algeria, Nigeria, and Venezuela are all possible sources of natural gas when LNG transports are used.

In addition to the conventional sources of natural gas, it is anticipated that coal gasification will with time become an expensive but possible source. Coal liquefaction and gasification is discussed later, as is methane production from biomass digestion.

2.8 COAL FORMATION

It is abundantly clear that the petroleum and natural gas resources of the United States are insufficient to provide energy very far into the 21st century. Coal, on the other hand, is a much larger resource, accounting for more than 97% of our fossil fuel energy reserve. About 20% of the world's coal resource lies within the United States, and we are the world's second largest coal producer. The importance of coal in the nation's energy future is assured. For this reason, it is crucial to have a good understanding of the nature and quantity of this resource.

About 440 million to 480 million years ago, plant life began on earth, and about 345 million years ago the great fern forests and swamps began to grow. The trees and algae of these swamps were the origin of our coal. The oldest coal, anthracite, is also the hardest and cleanest-burning. It has the highest carbon content, up to 95%. These almost pure carbon deposits evolved as organic matter in the form of peat or humus decayed, giving off carbon dioxide, methane, water, and other gases, leaving the carbon-rich coal behind. Bituminous coal, 50 to 80% carbon, is about 300 million years old. A third form of coal, lignite, is only about 150 million years old. Sometimes considered to be little more than a hard form of peat, lignite still shows wood fossils. In terms of quality, it is sometimes said to occupy the gap between bad coal and good dirt. The composition of some coal samples is given in Table 2.5.

Coal occurs in stratified deposits interlain with soil and rock in sedimentary basins. The average depth beneath the earth's surface is about 300 ft, and the average thickness of the coal strata (or seams) is about 2 to 8 ft. In parts of the West, however, seams 100 ft thick are being mined. These western coal resources are enormous, are accessible to surface mining, and have a sulfur content in the range of 0.7%, generally lower than for the eastern coals. The

Table 2.5 ANALYSIS OF SOME COALS OF THE UNITED STATES

Rank	Source	Percent Carbon	Percent Sulfur	Btu/lb
Anthracite	Pennsylvania	88	0.9	13,300
Semianthracite	Arkansas	79	1.7	13,700
Bituminous	Pennsylvania	57	1.4	13,870
	Illinois	49	0.9	11,930
	Michigan	50	1.2	11,780
	Kansas	53	4.3	12,930
	Utah	51	0.8	12,760
	Washington	50	0.5	12,250
Subbituminous	Wyoming	46	0.6	10,750
Lignite	North Dakota	27	0.6	6,700
	Texas	16	1.3	7,140

largest deposits are in Wyoming, North and South Dakota, Montana, and New Mexico.

2.9 COAL RESOURCES AND CONSUMPTION

Coal resources are somewhat easier to evaluate than petroleum or natural gas resources, since coal tends to reside in large beds relatively close to the surface. There are three major regions in the United States where coal is deposited: the Appalachian Basin (parts of West Virginia, Pennsylvania, Ohio, and eastern Kentucky), the Illinois Basin (Illinois, western Kentucky, and Indiana), and the northern Great Plains and Rocky Mountain regions (Montana, Wyoming, Colorado, North Dakota, Arizona, and New Mexico).

P. Averitt of the USGS estimated in 1971 that the United States had 1486 $\times 10^9$ metric tons of minable coal. This includes beds 14 in. or greater in thickness to a depth of about 4000 ft. Table 2.6 shows the distribution of coal throughout the world according to Averitt.

In 1974, the USGS and the Atomic Energy Commission estimated the proven recoverable reserves at 430 $\times 10^9$ tons and additional recoverable resources at 1070 $\times 10^9$ tons for a total of 1500 $\times 10^9$ tons, which is consistent with Averitt's estimates. More recently, the U.S. Bureau of Mines put proven recoverable reserves at 438 $\times 10^9$ tons. There is a great deal more coal in the ground than 1500 $\times 10^9$ tons, but it is either not on accessible land or it is not economically possible to extract. Minable coal is defined as 50% of the coal present, as much of it occurs in very narrow seams or at depths that make it too costly to extract. Of the minable coal in the United States, 71% is bituminous, 28% is lignite, and only 1% is anthracite. The anthracite deposits are essentially all in Pennsylvania and involve underground mines. The bituminous coal is mostly east of the Mississippi in Illinois and West Virginia. The subbituminous and lignite deposits are in the West, largely in North Dakota, Wyoming, and

Table 2.6 ESTIMATED WORLDWIDE COAL RESERVES[a]

	Amount (tonne[b])	Percentage of Total	Equivalent Energy[c] (kW • hr)
Former Soviet Union	4.3×10^{12}	56	30.2×10^{15}
United States	1.5×10^{12}	20	10.5×10^{15}
Asia (not including former Soviet Union)	0.68×10^{12}	9	4.8×10^{15}
North America (not including United States)	0.60×10^{12}	8	4.2×10^{15}
Western Europe	0.38×10^{12}	5	2.7×10^{15}
Africa	0.11×10^{12}	1	0.7×10^{15}
Australasia	0.06×10^{12}	1	0.4×10^{15}
South America	0.01×10^{12}	—	0.1×10^{15}
	7.64×10^{12}	100	53.6×10^{15}

[a] According to USGS, 1971.

[b] 1 tonne = 1000 kg = 2205 lb = 1.1 ton.

[c] Based on an average of 6400 kW • hr per ton or 7020 kW • hr per tonne.

Figure 2.11 A surface coal mine near Gillette, Wyoming. Vast areas of thick coal seams are being mined in this area.

Montana. The western coal is largely extractable by surface mining. The sulfur content of western coal is less per ton, but it is generally comparable on a per-Btu basis with that from West Virginia. Surface mining of Wyoming coal is shown in Figure 2.11.

As Figure 1.14 shows, coal was the major source of energy in the United States between about 1885, when it replaced wood, and 1950, when it was replaced by oil and natural gas. The annual production of coal since 1850 is shown in Figure 2.12. It is interesting to note the sharp rise in coal production between 1870 and 1920. The industrialization of the United States was largely based on this energy source.

About 70% of the coal mined in 1977 provided fuel for electric power plants that produced more than 45% of U.S. electricity. Exports accounted for about 10% of the production and coke production about 17%. The small remainder went to steel mills, cement plants, and retail deliveries. Coal is used primarily for producing steam, and to a lesser degree, coke. Coke is produced for the steel industry, where it is used in the reduction of iron oxides in furnaces to obtain free iron. The process of making coke involves heating bituminous coal to a high temperature (about 1200°C) in the absence of air. Volatile materials are driven off, and rather pure carbon is left. The coke, when used in a blast furnace, not only provides the fuel for heating the iron, but also forms carbon monoxide, which reduces the iron oxides to iron. In the process of producing coke, a number of useful by-products are formed, such as coal tar and coal gas. Coal tar is used for producing valuable organic compounds, such as benzene, toluene, and naphthalene. Coal gas is useful as a fuel in steel production or, with purification, for home heating and cooking.

It is obvious that if we continue to use coal at the rate of about 1000×10^6 tons a year, our resource of 1.50×10^{12} tons will last a very long time—about 1500 years. Many articles and advertisements have appeared in recent years stressing the size of our coal resources and stating that we have hundreds of years of energy available. A. A. Bartlett in particular has emphasized the dangers of such reasoning. Although such statements concerning the lifetime of our coal resource are true *if there is no growth*, most people ignore the fact that the rate of consumption will not remain constant for 500 years or probably

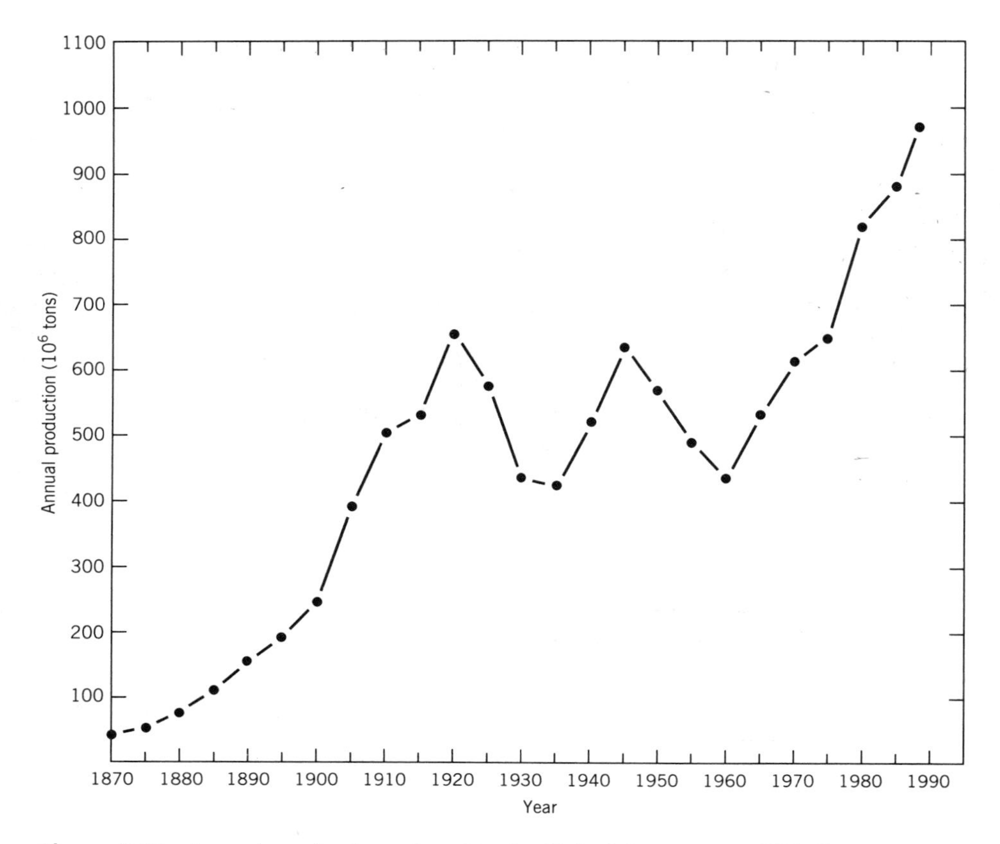

Figure 2.12 Annual production of coal in the United States since 1870 (Source: *Historical Statistics of the United States Colonial Times to 1970* Washington, D.C.: U.S. Department of Commerce, Bureau of the Census, 1975 and U.S. Energy Information Administration, *Annual Energy Review, 1989.*)

even 5 years. It is currently the policy of the U.S. government to increase the use of coal and to decrease the use of natural gas and petroleum because, as we have seen, these fuels are far less abundant than coal. Major programs are being initiated to convert coal to gas and to oil, as well as to replace these fuels where they are still used for electric power production and even to export more coal. It would not be surprising in the light of these policies to have the use of coal increase at the rate of 3 to 5% a year. Under these conditions, the lifetime of our coal resources becomes remarkably less than the projections based on present rates of consumption. With a growth rate of 5% and an initial consumption rate of 1000×10^6 tons, the 1.5×10^{12} ton resource would last only 86 years. On the other hand, if we assume that coal is going to furnish all our energy needs, the lifetime is reduced to 66 years. Exponential growth makes the difference. Many people, including those in positions of influence, do not seem to appreciate the consequence of exponential growth. The necessary relationships for calculating the lifetime of a finite resource are developed in Appendix B.

2.10 SUMMARY OF CONVENTIONAL FOSSIL FUELS, RECENT DEVELOPMENTS, AND THE ENERGY CRISIS

The years since 1970 have witnessed some abrupt changes in the patterns of fossil fuel production and consumption that were established in the United States in the preceding 50 years. Many feel that the formation of OPEC (Organization of Petroleum Exporting Countries) and the Arab oil embargo of November 1973 were the cause of the "energy crisis," but, as we have seen, the stage was set long ago for the production of oil and natural gas in the United States to peak out in the early 1970s. The reality of the fossil fuel dilemma, however, was made real to the average American by the long lines at service stations in 1973 and the shortages of natural gas that followed a few years later. The average American need not feel badly about lack of foresight, as few professional economists and financial advisers correctly predicted the end of the age of cheap fossil fuel energy. Whatever the future may hold, the events of the 1970s brought about fundamental changes in a number of ways.

1. The domestic production of crude oil peaked in 1970, and proven reserves decreased from 39×10^9 bbl in 1970 to 26×10^9 bbl at the end of 1990. Figure 2.13 shows the basic data. Except for the addition of the resources found at Prudhoe Bay, Alaska, in 1970 and small increases in 1984 and 1987, the United States has found less oil than it produced every year since 1967.

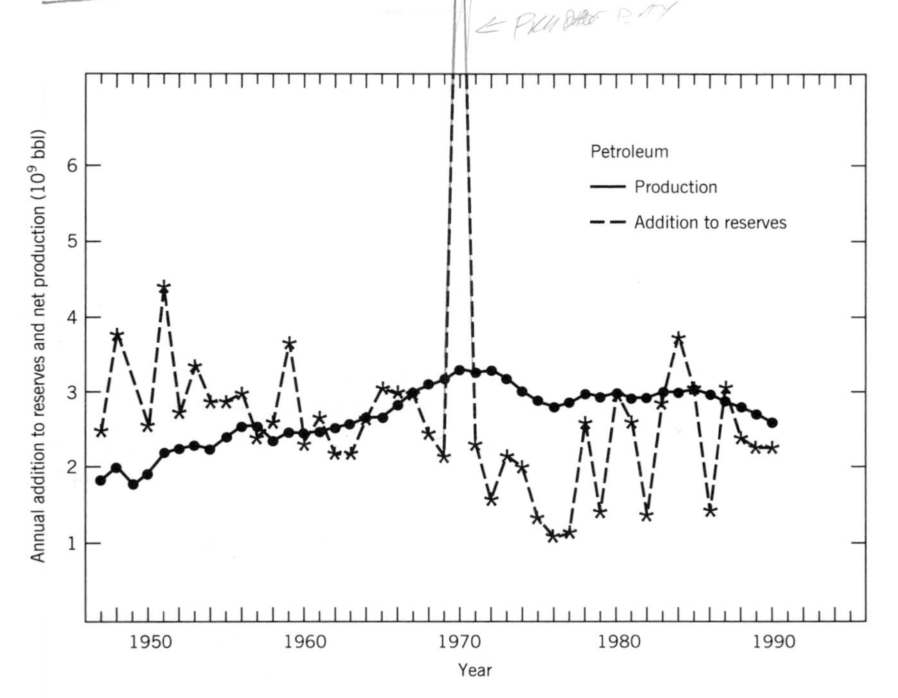

Figure 2.13 Annual addition to the reserves and net production of petroleum in the United States from 1947 to 1990. (Source: *Basic Petroleum Data Book*, Vol. XII, Washington, D.C.: American Petroleum Institute, Jan. 1992.)

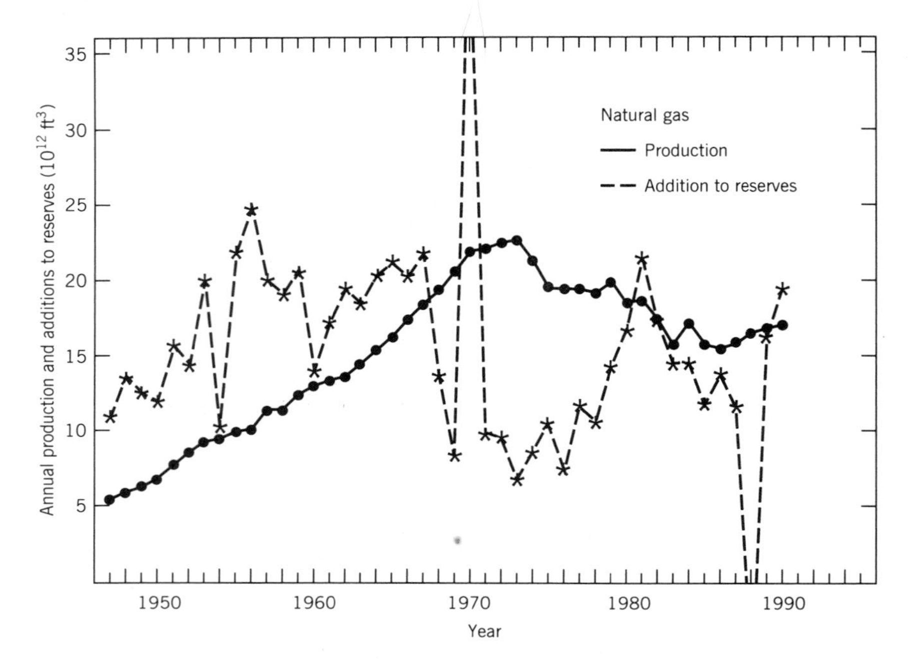

Figure 2.14 Annual addition to reserves and production of natural gas in the United States from 1947 to 1990. (Source: *Basic Petroleum Data Book*, Vol. XII, Washington, D.C.: American Petroleum Institute, Jan. 1992.)

2. As Figure 2.14 shows, the production of natural gas peaked in 1973 and proven reserves decreased from 291×10^{12} ft^3 in 1970 to 167×10^{12} ft^3 in 1990. As with oil, we have produced more natural gas than has been discovered every year since 1968, with the exceptions of 1970 when Prudhoe Bay was added to the reserves and small increases in 1981 and 1990.
3. Consumption of oil increased rather steadily, in keeping with historic trends, until about 1977 when it began to level off and decrease slowly for the first time since the depression years of the 1930s. This decrease in consumption was directly related to the steep increase in the cost of oil and to many programs directed toward energy conservation. Following the oil price collapse in 1986, consumption has now started to increase again.
4. Consumption of natural gas peaked in the early 1970s and has decreased slightly since that time. Since very little natural gas is imported, production is very closely related to consumption.
5. In 1970 26% of the total petroleum used in the United States was imported. In 1977 this rose to 52%. The amount of petroleum imported then decreased as our consumption went down; consumption then began to rise again in 1985 for the reasons noted in item 3. In 1989 we imported 2.9×10^9 bbl of oil, or 46% of our needs. At $20/bbl this corresponds to over $58 billion worth of oil, which is a significant portion of the U.S. trade deficit.
6. The cost of a barrel of oil from the Middle East was about $2 in 1970. Then, mainly through control of prices by OPEC, the cost of OPEC oil rose to about $36 per barrel in 1981. These high prices brought about conser-

vation measures such as fuel-efficient cars, and stimulated greater domestic production. As demand fell, so did the price of oil, and OPEC revenues dropped to the point where the OPEC nations could no longer afford to control the price through production cutbacks. Soon the market was flooded with oil, and the price fell to about $8 per barrel. With lower prices, demand has again increased and the cycle continues, with oil at about $20 per barrel in 1991.

7. Beginning in the 1960s, the number of feet drilled each year in exploring for oil in the United States began to decrease because the price of crude oil was low and the finding rate (barrels per foot drilled) was declining. The trend of decreasing exploration was abruptly reversed in 1977 as the price of oil began to escalate. In 1980, 65,000 wells were drilled in the United States—the most since 1956. In 1982 a record high of 4530 drill rigs were in place. Since then there has been an easing of the price of oil, and the finding rate has been found to be low. Consequently, there has been a dramatic decrease in exploration since 1982. The number of drill rigs working in 1987 was at a postwar low of 663, and in 1991, was 1100.

8. The energy policy of the Bush administration gave more emphasis to increased petroleum production rather than conservation. In particular, it has been recommended that the Arctic National Wilflife Refuge in Alaska be opened up for exploration and that the requirements for permits for offshore exploration on the Pacific Coast be eased.

9. One of the surprising aspects of the petroleum problem is that the real cost of gasoline in the United States is now less than it was in 1947. Figure 2.15 shows the cost per gallon in terms of 1990 dollars since 1947. The cost of gasoline in the United States is far less than in other industrialized countries, mainly due to the reluctance of the government to add taxes.

Table 2.7 summarizes the conventional fossil fuel resources of the United States as of about 1990. Most (about 97%) of this remaining resource is in the form of coal. In the last 70 years, the growth rates in the use of petroleum and natural gas have been about 5.8 and 6.3%, respectively. It can certainly be expected that the use of coal will become much more significant, replacing petroleum and natural gas in many applications. It is interesting to see how long our total conventional fossil resources would last under the assumption of some reasonable growth rate in their use. The increased cost of oil and natural gas as well as conservation measures can be expected to reduce the traditional growth rates. Possibly, our annual use of energy, which was about 81 QBtu in 1990, will be reduced. However, if one assumes a growth rate in our use of total fossil fuels of about one half of that noted for petroleum and natural gas, namely, 3.5% (a doubling time of about 20 years), one can compute, using the relationships developed in Appendix B, the lifetime of our total conventional fossil fuel resources. The answer is 84 years, or depletion by the year 2077. This lifetime is comparable to that obtained earlier for coal alone.

This rather surprisingly short lifetime is not the full story. As has been discussed, the growth in the use of petroleum and natural gas has been brought about by their suitability for our needs in transportation and home heating. To make use of coal for these same needs, it must be put through a rather expensive conversion process.

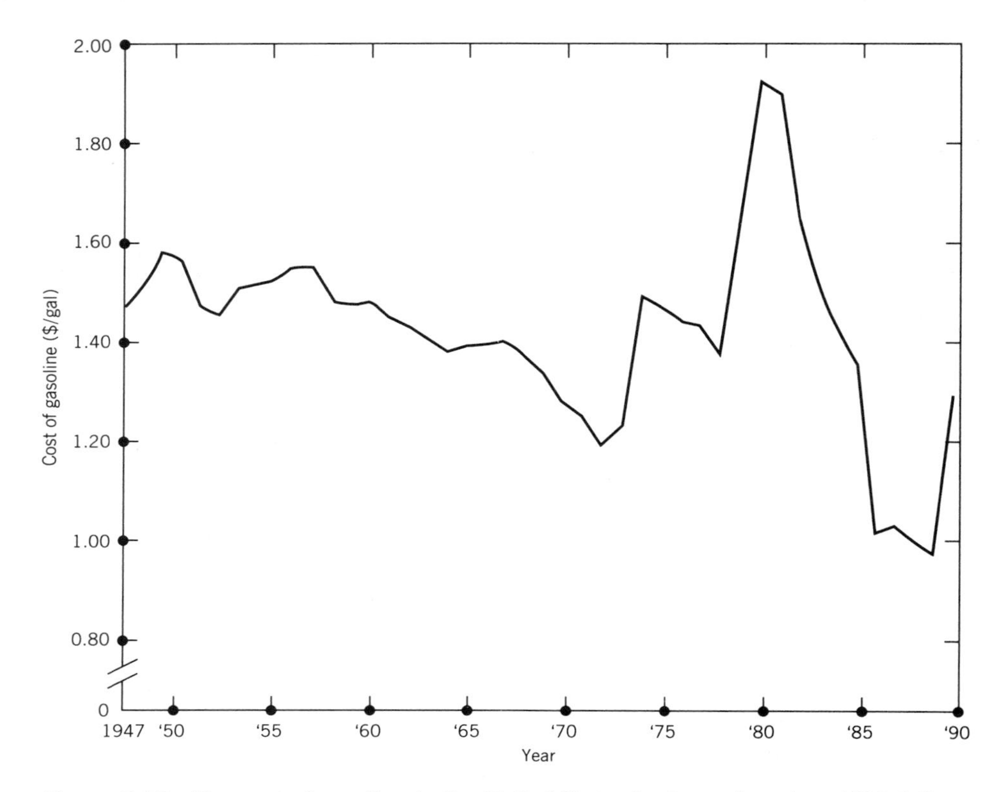

Figure 2.15 The cost of gasoline in the United States is shown based on 1990 dollars for the years 1947 to 1990. For 1990 only the months of January to September are included. (Source: Cambridge Energy Research Associates. As shown in Yergin, Daniel, *The Prize*. New York: Simon and Schuster, 1991. Copyright © 1991, 1992 by Daniel Yergin. Reprinted by permission of Simon and Schuster, Inc.)

In the next sections we examine fossil fuel resources other than the conventional resources discussed above. These alternative resources are oil shale, tar sands, and petroleum recovered by tertiary methods and by mining. In addition, we describe the processes of liquefaction and gasification of coal.

2.11 SHALE OIL

In 1882, Mike Callahan, a pioneer in Rio Blanco County in western Colorado, invited friends over for a housewarming. In keeping with the occasion, he started a fire in his new fireplace, which he had recently built out of attractive gray rocks streaked with blue that he had found in great abundance in the area. Unfortunately, the housewarming turned into a house burning, as the fireplace and then the house went up in flames. Oil shale has a number of uses, but fireplace building is not one of them. Although he did not realize it, Mr. Callahan may have made one of the greatest mineral strikes of all time.

The story of the formation of the petroleum content of oil shale provides an interesting contrast to that of liquid petroleum. About 50 million years ago, the Green River Formation, the most important source of oil shale in the United

Table 2.7 CONVENTIONAL FOSSIL FUEL RESOURCES IN THE UNITED STATES

Kind of Fuel	Q_∞	Used to 1990	Remaining	Remaining Energy Content*(Btu)	Percent
Petroleum	214×10^9 bbl	142×10^9 bbl	72×10^9 bbl	4.0×10^{17}	0.95
Natural Gas	1300×10^{12} ft^3	700×10^{12} ft^3	600×10^{12} ft^3	6.3×10^{17}	1.50
Coal	1486×10^9 tons	68×10^9 tons	1418×10^9 tons	4.1×10^{19}	97.55
Total				4.20×10^{19}	100.00

* Calculated on the basis of 5.51×10^6 Btu/bbl, 1.05×10^3 Btu/ft^3, and 2.9×10^7 Btu per ton.

States, was covered by two large tropical lakes. Organic matter from the surrounding hills was deposited along with sedimentary material on the bottom of the lakes to a depth of 3000 ft over a period of many millions of years. The combined materials from plants and possibly from aquatic animals at the lake bottom finally formed a mudstone, or marlstone, which is the source of shale oil. Through changes in the earth's crust, these deposits were brought up to their present elevations. Then erosion by rivers cut into the formations, leaving much of the oil shale exposed.

Oil shale is a marlstone that contains an organic substance known as kerogen. Kerogen is a solid, waxlike substance that vaporizes when heated sufficiently. The kerogen can then be converted into useful fuel products. A vast amount of oil shale is widely distributed in the United States, but it is only the rich deposits in the Green River Formation in Colorado, Utah, and Wyoming that have a potential in the near future for providing a useful source of fossil fuel energy. Figure 2.16 shows the regions of the highest yield oil shale in the Green River Formation.

The production of liquid fuel from oil shale is not an entirely new development. The February 1918 issue of *National Geographic* magazine featured an article entitled "Billions of Barrels of Oil Locked Up in Rocks" by Guy Elliot Marshall of the USGS. The article discussed the 70-year-old (in 1918) oil shale industry in Scotland and included a photo of an "oil shale distillery near Juab, Utah used by Mormons a generation ago." Within the context of what at that time was perceived to be an impending national shortage of liquid petroleum, the following figure caption appeared:

> New Experimental Oil Still Near Debeque, Colorado. No man who owns a motor-car will fail to rejoice that the United States Geological Survey is pointing the way to supplies of gasoline which can meet any demand that even his children's children for generations to come may make of them. The horseless vehicle's threatened dethronement has been definitely averted and the uninviting prospect of a motorless age has ceased to be a ghost stalking in the vista of the future.

This article also pointed out that

> Oil-shale distillation is not new in the United States; yet it is doubtful if there are many people alive who remember anything about the earlier industry. Before petroleum was discovered in Pennsylvania, about 50 small companies in the eastern United States were crudely distilling oil from shales; but after subterranean pools were discovered the companies went out of business.

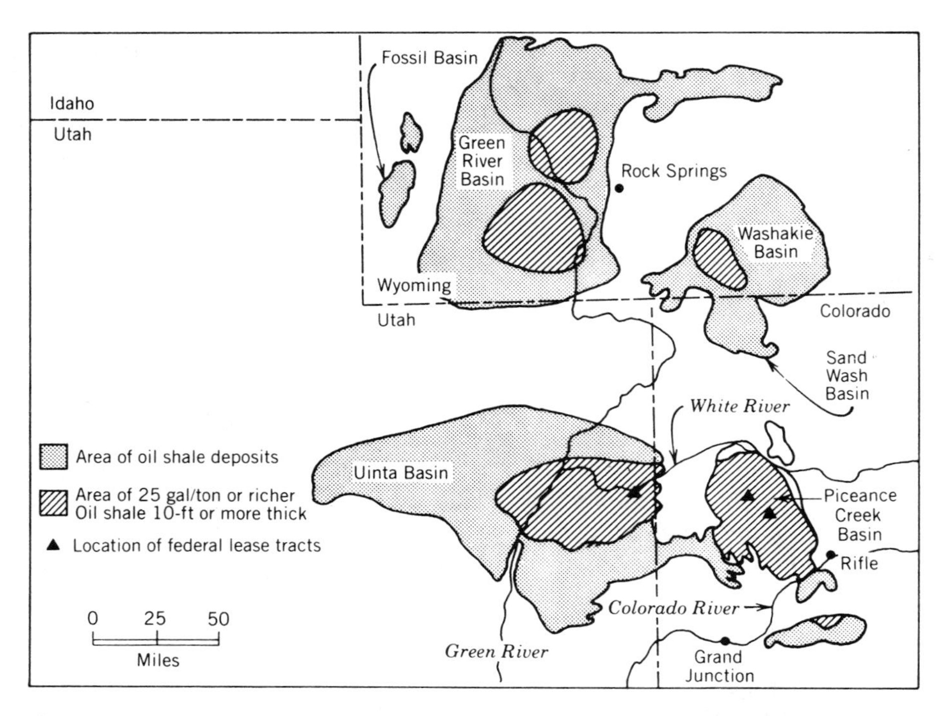

Figure 2.16 Oil shale deposits in the Green River formation of the western United States.

The molecules of kerogen and petroleum differ in that petroleum molecules are more linear whereas the kerogen structure exhibits extensive cross-linking. When kerogen is heated, the cross-linkings break, and the molecules become more linear, like petroleum molecules. Shale oil has a considerable sulfur and nitrogen content and has to be further refined to be useful as petroleum.

Oil shale deposits are classified by the number of gallons of oil that can be obtained per ton of shale. It is currently felt that shale that yields 25 gallons of oil or more per ton may be economically minable. The energy content per ton of oil shale, even for the richer deposits, such as that shown in Figure 2.17,

Table 2.8 SHALE OIL DEPOSITS IN THE GREEN RIVER FORMATION

	Billions of Barrels of Oil in Place			
	Colorado	**Utah**	**Wyoming**	**Total**
Intervals of 10 ft or more thick averaging 25 gallons or more of oil per ton	480	90	30	600
Intervals 10 ft or more thick averaging 10 to 25 gallons of oil per ton	800	230	400	1430
Total: Intervals 10 ft or more thick averaging more than 10 gallons of oil per ton	1280	320	430	2030

Figure 2.17 A rich sample of oil shale found in western Colorado.

is, however, several times less than that of the better coal deposits. Table 2.8 shows the estimated deposits in the Green River Formation.

It is clear that the magnitude of the resource is vast compared, for example, to Q_∞ for petroleum of 214×10^9 bbl. Enthusiasm, however, must be tempered by several facts. First, it has been estimated that of the total oil shale resource with 25 gal per ton or more (600×10^9 bbl), only about 80×10^9 bbl appear to be recoverable under present conditions because of the depth of the overburden and other factors. A second problem is that extracting shale oil is an expensive and technically complicated procedure. Efforts have been made for more than 50 years, but no production beyond that from pilot plants has been achieved in the United States. Significant production on the order of 10^5 bbl/day may not be possible for many years. Problems of water supplies, disposition of the spent shale, and environmental damage must also be solved.

Example 2.1

How many tons a day need to be excavated for an oil shale industry that produces 200,000 bbl a day if the shale has an oil content of 20 gal/ton?

Solution

$$\frac{2 \times 10^5 \text{ bbl/day} \times 42 \text{ gal/bbl}}{20 \text{ gal/ton}} = 4.2 \times 10^5 \text{ tons/day}$$

This is in addition to any overburden that must be removed.

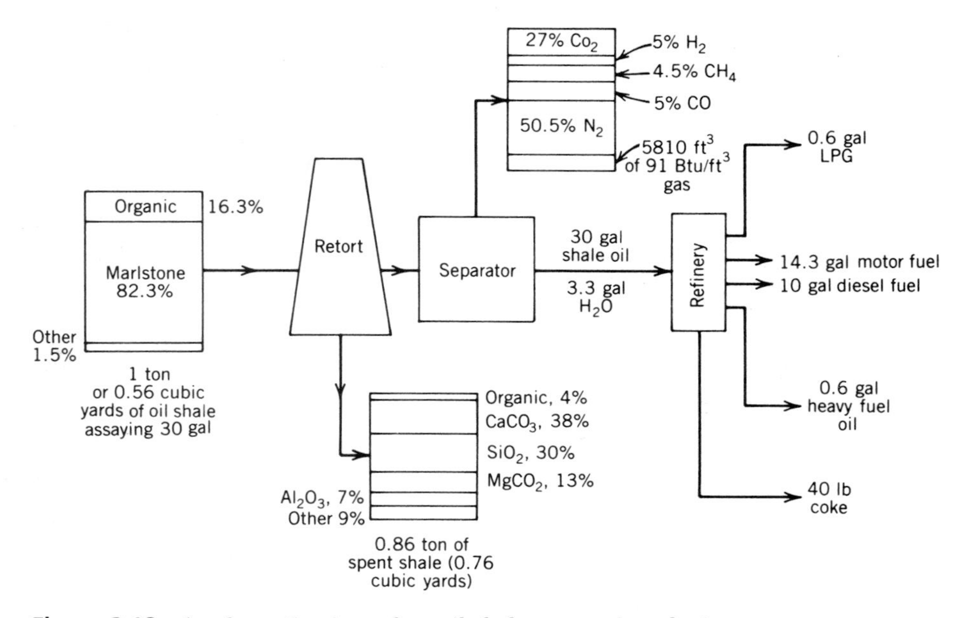

Figure 2.18 A schematic view of an oil shale processing plant.

A schematic diagram of an oil shale processing plant is shown in Figure 2.18. The retort can take on a variety of forms. Basically, the shale must be heated to about 900°F (482°C) to vaporize the kerogen. In some types of retorts, raw shale is introduced at the top and the spent or cooked shale is drawn off the bottom. Air, as well as burnable gas (frequently from the oil shale), must be injected near the middle of the retort to sustain combustion of the gas and heat the shale. In such an arrangement, the oil vapors would be extracted from the top of the retort. A number of different types of retorts have been developed by various companies in pilot shale oil programs.

One interesting feature shown in Figure 2.18 is that the volume of the shale increases by about 36%, or, in the example given, from 0.56 cubic yards for the raw shale to 0.76 cubic yards for the spent shale. A large part of the expense of shale oil is involved in moving so much material. The raw shale is usually obtained from surface mines that require extensive earth-moving equipment. The shale must then be hauled to a crushing facility and then into storage prior to retorting. After retorting, the spent shale must again be hauled away and placed where the environmental damage will be minimal. Whenever possible, revegetation is planned to restore the landscape.

Another problem faced by the shale oil industry is the water supply needed for the process. For the type of process shown in Figure 2.18, it is estimated that 3 bbl of water are needed for 1 bbl of oil. This is equivalent to 14,000 acre-ft/yr for 10^5 bbl of oil a day. Unfortunately, the Green River Formation is located in a very arid part of the country. The Green River is the major source of water in the area, and it has an average flow of 5×10^6 acre-ft/yr. This is about 45% of the flow of the Colorado river after the two rivers join. River water is extremely valuable for agricultural irrigation, and water rights to the Green

River are almost entirely allocated. The available surface water in the area appears adequate for shale oil production of roughly 1×10^6 bbl a day, which is about 5% of our national oil consumption.

Because of the problems in disposing of spent shale and the shortage of water resources, a different method of retorting is actively under study by several companies. It is called in situ retorting, and essentially involves retorting the shale by heating it in the ground before it is mined. The general scheme is shown in Figure 2.19. The shale formation must be fractured by hydraulic or explosive techniques. Air and gas are injected as shown and then ignited to heat the shale in a localized area. The oil vapors condense adjacent to the heated shale, and the liquid oil is pumped to the surface from a second well. Even though the in situ method is not going to realize the yield per ton that surface retorting does, it is estimated that a 60% cost reduction per barrel can be achieved. As long as normal petroleum can be obtained by the usual methods, it has been very difficult for oil shale to become an economically attractive alternative. There has been some activity to develop efficient production methods, but no full-scale plant is in operation. A production level of 1×10^5 bbl a day, 0.5% of U.S. consumption, is estimated to involve about $1 billion and 10 years.

2.12 TAR SANDS

Tar sands represent an appreciable fraction of the world's fossil fuel resources. They are deposits of sand impregnated with a very viscous heavy crude oil (bitumen). Because of the high viscosity of the bitumen, the normal process of letting the petroleum flow into wells from which it can be pumped to the surface does not work. The tar sands must instead be mined and taken to a processing plant where the bitumen is extracted, usually by steam or hot water. The bitumen must then be refined in the same fashion by which normal heavy crude oil is refined.

There are several known tar sand deposits around the world (in Venezuela

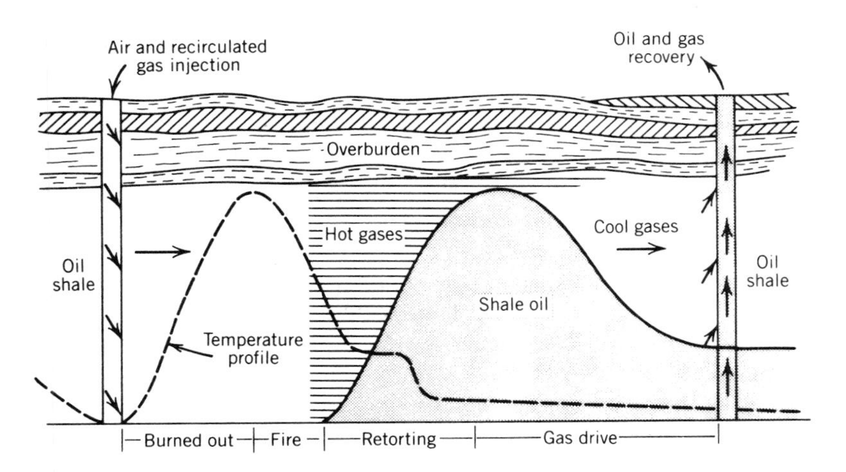

Figure 2.19 A schematic view of an in situ oil shale retorting operation.

and Colombia, for example), but the major deposit is in North America, in the Athabasca deposit near Fort McMurray in the northeastern part of the province of Alberta, Canada. In the United States, there are lesser deposits in Utah that contain about 20×10^9 bbl of recoverable petroleum. The Athabasca deposit contains some 300 to 700×10^9 bbl of bitumen under a surface area of more than 11,000 square miles. The overburden ranges from 0 to 2000 ft; roughly 10% of the tar sands is bitumen. Each grain of sand is covered by a rather thick layer of the material. In the winter, the bitumen binds the sand into a solid mass that is difficult to excavate, and in warmer weather the tar sands become sticky and tend to cling to everything they touch.

Although earlier efforts failed from an economic standpoint, there has been a large mining and extraction effort near Fort McMurray since 1966, and in 1976 it began to become profitable. A few years ago this plant produced 45,000 bbl of oil a day, and this involved moving about 100,000 tons of sand daily. It can be expected that production in the Athabasca area will increase with time, since production methods have been worked out even for the rather severe climate of northern Alberta. The process in use employs bucket-wheel excavators to strip the tar sands and conveyor belts to take the material to a hot-water extraction plant. A schematic diagram of the processing plant is shown in Figure 2.20.

Although the developments in processing tar sands are encouraging, the process is clearly expensive. With the increased costs of finding normal petroleum, however, tar sands have become competitive. There have been estimates that 1.25×10^6 bbl/day could be produced in Alberta in the near future. Although this increased Canadian production does not immediately help the U.S. domestic production of petroleum, there is ample reason to believe that the technique developed in Canada can also be applied to the deposits in Utah.

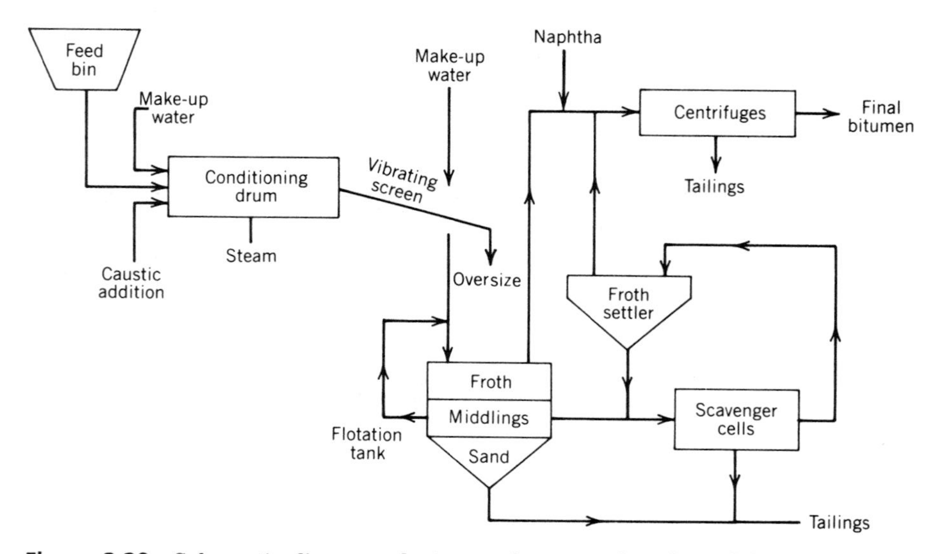

Figure 2.20 Schematic diagram of a tar sands processing plant. Oil rises to the surface in the flotation tank. Water and fine sand are removed in the centrifuge.

2.13 ENHANCED RECOVERY OF OIL

Geologists believe that after a well stops producing by the normal processes, about 70% of the petroleum is left behind in the ground. In the primary extraction process, oil either comes to the surface under natural geologic pressures or is pumped out. About 15% of the total petroleum underground can be extracted in this way. When the primary process ceases to yield a reasonable flow, secondary processes are initiated that mainly involve flooding with water to force out more oil. Secondary recovery processes can produce another 15%, leaving about 70% of the petroleum still underground.

The tertiary, or enhanced recovery, techniques now being tested at a number of U.S. oil fields are based on the two primary effects that lock the oil in: interfacial tension and viscosity. Oil is not found in underground pools of free oil but in layers of rock or sand of varying porosity. If the high viscosity of the oil is the major problem, gas (carbon dioxide) can be injected under pressure. When the gas dissolves in the petroleum, the viscosity will be reduced, and oil will be able to flow more readily to the well. Steam injection also can be used to heat the oil and reduce the viscosity. By injecting air or oxygen, a fraction of the petroleum in the ground can be burned in place. This heats the general area and reduces the viscosity of the remaining oil. The heated petroleum is pushed in front of the combustion front of the air–oil mixture until the heated oil reaches the shaft where the well is being pumped.

If the oil is locked in the pores of the rock, efforts are made to reduce the surface tension of the oil by injecting detergent material, followed by flooding with water containing polymers.

The exact technique that is best for a particular oil field must be determined experimentally. The processes are expensive and generally require long periods of time. Enhanced recovery certainly provides hope for tapping the vast resource of oil that is left underground.

Various estimates have been made of the number of barrels of oil that can be extracted using enhanced recovery. If 60% of the oil could be recovered, Q_∞ (United States) for oil could be increased by 126×10^9 bbl. A more moderate estimate is that the total yield attributable to enhanced recovery could range from 20×10^9 to 75×10^9 bbl. The Harvard Business School study *Energy Future* states that it is unlikely that as much as 1×10^6 bbl of oil a day will be available through enhanced recovery through the early 1990s. This is about 5% of U.S. consumption.

2.14 OIL MINING

One of the alternative methods of extracting oil from the ground is to mine it rather than pump it to the surface. As with enhanced recovery, the objective would be to extract oil left behind after the primary and secondary processes were completed. There are three general schemes for mining oil. The first, called block caving, is applicable if the oil is trapped in weak rock. Tunnels are dug in strong rock beneath the oil-bearing rock, and the oil-bearing rock is then caved in and collected in chutes and transported to the surface by a series of conveyor belts. At the surface, the oil can be separated from the rock in a fashion similar to that described for tar sands.

Another scheme involves gravity drainage of an oil-bearing formation. If there is strong rock underneath the oil, a series of access tunnels can be constructed. A number of holes are then drilled up from the access tunnels into the oil-bearing region, and gravity causes the oil to flow down through these holes to the tunnels. The oil is collected by a series of pipes leading to a storage tank from which the oil is pumped to the surface.

A third variation in the mining of oil can be utilized if there is no strong rock formation beneath the oil-bearing zone. Again, a number of access tunnels are excavated, but above the oil; then short holes are drilled down into the oil-bearing rock. Each hole requires its own pump.

Although appreciable amounts of oil are not now being mined in the United States, it can be expected that oil mining, along with other enhanced-recovery schemes, will become more feasible economically as the cost of oil increases. Oil mining, of course, has the handicaps of the cost and complexities of any underground mine.

2.15 COAL GASIFICATION AND LIQUEFACTION

It would be useful to be able to convert coal into a liquid fuel that could be used in automobiles and into a gaseous fuel that could be used for the home, because coal is far more abundant in the United States than are the conventional resources of oil and natural gas. Processes that will accomplish these goals have, in fact, been known for many years. Much of the liquid fuel used during World War II by Germany's war machine came from coal liquefied by the Fischer–Tropsch process. The SASOL plant in South Africa is currently using the same process to convert about 3500 tons of coal a day into synthetic fuels. Gas manufactured from coal was quite common in the United States prior to 1950. Much of the gas then used for cooking and lighting was from coal.

The fact that liquid and gaseous fuels from coal are not used extensively in the United States today is due primarily to the still relatively low cost of these fuels when they are extracted directly from the earth. There are some technical and environmental problems, but the big barrier to coal conversion is economics, as is the case for shale oil, tertiary recovery of oil, and other nonconventional sources of fossil fuels. With the rapid depletion of our domestic sources of petroleum and natural gas, one can expect that a point will be reached where fuels from coal conversion will again play an important role. To that end, it is important to understand the most promising of the several conversion processes that are being explored by a number of private companies and by the U.S. government. The government established the Synthetic Fuels Corporation to encourage private participation in coal conversion during the transitional period.

SYNTHETIC GAS FROM COAL

A number of different processes and chemical reactions have been used for coal gasification, but all of them involve finely crushing the coal and then

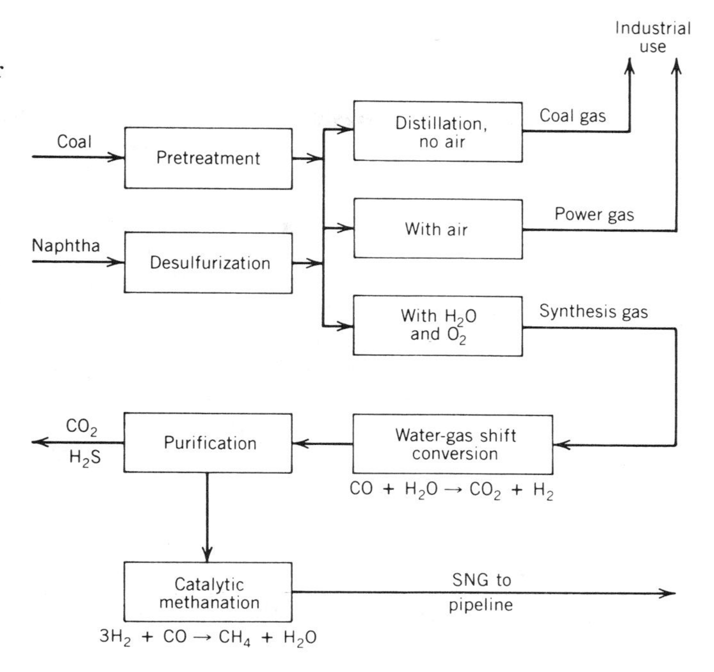

Figure 2.21 Steps in the coal (or naphtha) gasification process. Coal gas has a very low heating value, being about 50% H_2 + CO and the remainder CO_2, CH_4, and others. Power gas has intermediate heating value useful for on-site power production but not for pipeline transport. It is about 14% CH_4, 11% CO, 16% H_2, 11% CO_2, 28% H_2O, and 30% N_2. Synthesis gas is high-quality gas useful for producing pipeline gas: it has about 10% CH_4, 21% CO, 40% H_2, 28% CO_2, and 1% other. SNG is usually 90% CH_4. (Source: D. W. Devins, *Energy: Its Physical Impact on the Environment,* New York: John Wiley, 1982. Copyright © John Wiley & Sons Inc. Reprinted by permission.)

having it react at high temperatures with steam and air or oxygen. Some of these reactions are shown in Figure 2.21. The products are different combinations of hydrogen, carbon monoxide, carbon dioxide, and some methane. The first stage, shown in Figure 2.21, produces a low (150 to 300 Btu/ft^3) or medium (300 to 500 Btu/ft^3) Btu gas, depending largely on the amount of methane produced. The low Btu gas, because of its low heating value, cannot be distributed through pipelines as a substitute for natural gas, which has about 1000 Btu/ft^3. It can be used for electric power generation and industrial processes, however. If oxygen is used instead of air, the methane production is increased; this can lead to medium Btu gas (300 to 500 Btu/ft^3). As Figure 2.21 shows, a purification process is usually needed to remove the hydrogen sulfide (H_2S) before the product is used. The Lurgi, Winkler, and Koppers–Totzek processes differ in detail, but all use oxygen and steam to produce what is usually a medium Btu gas.

To increase the heating value of the gas to about 1000 Btu/ft^3 so that it can be used as a direct substitute for natural gas, medium Btu gas is further treated by a process such as catalytic methanation. Hydrogen and carbon monoxide in the gas are combined in this process to increase the methane concentration. Straightforward application of these techniques leads to a product that is quite usable, but considerably more expensive than natural gas.

A number of different approaches are being tested in small pilot plants to reduce the cost; however, no single process that is clearly superior to the others has emerged. In situ production of synthetic gas is also being tested, particularly for western coal. It is hoped that not only could economies be realized as a result of the reduced handling of the coal, but also that environmental problems could be diminished.

LIQUEFACTION OF COAL

Of the four methods of coal liquefaction that are generally considered—indirect liquefaction, pyrolysis, solvent extraction, and catalytic liquefaction—only the first is in commercial use (at SASOL in South Africa) at the present time. In this process, the coal is first gasified, and the carbon monoxide and hydrogen are synthesized to a liquid after the gases are cleaned of impurities. The final products have a wide range of uses depending on the details of the process and the catalyst. A different type of process involving solvent extraction is shown in Figure 2.22. A number of different processes are being explored in the United States by private organizations (mainly oil companies) with the help of government funds. Three pilot projects are being constructed that have a capacity of several hundred tons of coal per day. It is difficult to make very precise estimates of the production costs per barrel from the information available. Larger scale production facilities will be needed before realistic costs can be calculated.

The production of synthetic liquid and gaseous fuels from coal is not going to happen quickly. Because the investment in a full-scale facility is so large, there must be some assurance that the process being used is technically and economically sound. Federal subsidies or price guarantees would appear to be needed to attract private investment for the construction of large facilities because of risks involved in coal conversion at this time.

The combustion of the products of coal gasification and liquefaction will probably be less polluting than the direct combustion of coal. What is not known very well is the extent to which the conversion process itself will contribute to pollution problems, because there has been such limited experience with large-scale conversion plants. Many fear that if the conversion plants are located near the sources of western coal, there could be an extensive environmental impact on these regions.

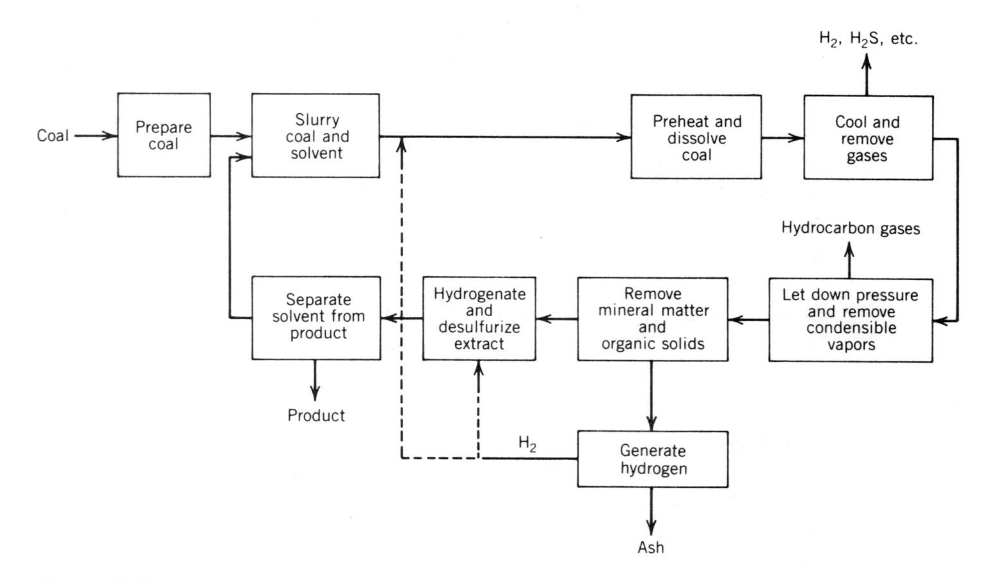

Figure 2.22 Flow diagram of the liquefaction of coal.

PROBLEMS

1. One cord of firewood delivered and stacked costs about \$100. It will release about 25×10^6 Btu when burned. Compare the cost in dollars of heating your home with this firewood on a cold winter day to the cost of heating it with natural gas. Assume that natural gas costs \$5 per 10^3 ft^3 and that your home is the typical home that requires 1×10^6 Btu on a cold winter day. Consider the efficiencies of the stove and the furnace to be equal at 60%.

2. Each U.S. citizen, on the average, utilizes energy (food, heat, light, etc.) at the steady rate of 10^4 W (10 kW).
   (a) *Assume* that this energy is derived solely from oil at 100% efficiency and calculate the number of gallons (or barrels) of oil that each such person consumes daily.
   (b) Based on Hubbert's estimate of remaining U.S. oil, calculate the number of years before the country runs out of oil at the rate of consumption found in part (a) above. Assume no oil imports.

3. (a) What is oil shale and how can it be converted into a usable fuel for cars or heating?
   (b) Discuss briefly the role you feel that shale oil, tar sands, and enhanced recovery of petroleum can play in meeting the energy needs of the United States.

4. Describe, being as brief and quantitative as possible:
   (a) The history of the use of petroleum in the United States.
   (b) About how much oil is used now in the United States and its uses.
   (c) What you see as the future for petroleum use (not production) in the country.

5. Shale oil has received much attention. Answer *briefly* the following questions concerning this resource.
   (a) How can usable oil be obtained from the oil shale by above-ground retorting?
   (b) By in situ retorting?
   (c) About how much shale oil can one expect to have produced per day in 10 years? Compare this to how much total oil per day the United States uses at the present time.
   (d) What are the major environmental problems of shale oil development?

6. It has been proposed by some that a 50-cents-a-gallon tax should be put on every gallon of gasoline or diesel oil used in the United States to encourage conservation.
   (a) Estimate how much money would be collected in 1 year by this tax.
   (b) The President of the United States just phoned and said that the tax is a great idea, but an energy program is needed on which to spend this collected revenue to bring an end to all imported petroleum and natural gas by 2000. Briefly outline a specific program that is realistic and directed toward this goal. Give details of the research and development work, tax incentives, subsidies, and so forth that you would propose and estimate the contribution these would make toward the solution. Consider the depletion of the resources in the United States during this period and the environmental consequences of your program.

7. As discussed, coal is the most abundant fossil fuel the United States has left. Answer as concisely as possible the following questions about this resource:
   (a) How and when was the coal formed?
   (b) About what fraction of our total energy needs are supplied by coal? About how many tons is this per year?

(c) Why has the use of coal in the United States not increased very much since 1910?
(d) About how large is the U.S. coal resource and where is it located?

8. The following data represent the annual energy, in QBtu, in the natural gas used in the United States from 1900 to 1970.

Year	QBtu/yr
1900	0.25
1910	0.51
1920	0.80
1930	2.0
1940	2.6
1950	6.3
1960	11.0
1970	22.0

(a) Plot these data on semilog paper so that the years 1900 to 2020 are covered.
(b) Draw a "best" straight line through the data points and determine the average doubling time and the average annual percentage growth.
(c) Project the line to estimate how much will be used in 2000 and 2020. Assume for this purpose that there will be no problem with resource depletion.

9. The oil consumed in the United States in the past was as follows:

Year	QBtu/yr
1920	2.5
1930	5.1
1940	7.3
1950	12.5
1955	16.0
1960	18.0
1965	21.0
1970	29.0

(a) Plot these data on semilog graph paper.
(b) Draw a "best" straight line through the data points and determine the doubling time and the average annual percentage growth.
(c) Estimate the QBtu/yr to be used in 2000.
(d) What are these values in barrels of oil per year?

10. It has been estimated that a person can perform continuous manual labor at a power of 50 watts for an entire working day.
(a) How many pounds of coal, oil, or wood contain the energy equivalent of the useful physical work a person can perform in an eight-hour day?
(b) Estimate the dollar cost of this amount of fuel.
(c) Make reasonable assumptions and calculate answers as in (a) and (b) above, but for the useful work one can perform in a lifetime.

11. If there are 30×10^9 bbl of recoverable oil in Alaska, for how many years could the United States supply its present demand for oil from that source?

12. Considering the annual U.S. domestic production of oil and natural gas, the energy contained in the natural gas is approximately what fraction of the energy contained in the oil?

13. What is the energy content of a ton of oil shale relative to a ton of coal, if the oil shale contains 25 gallons of oil per ton with an energy content of 5.8×10^6 Btu/bbl, while the energy content of the coal is 26.6×10^6 Btu/ton?

14. How many tons of oil shale are required each day to supply a 10,000 bbl/day retort, if the shale yields 25 gallons per ton?

15. One estimate of the coal remaining in the United States is 1.5×10^{12} tonnes. If this coal were uniformly spread over the lower 48 states (estimate the area as a rectangle 1000 miles by 3000 miles), and each ton occupied 1 yd^3 (27 ft^3), what would be its thickness in feet?

16. One gram molecular weight of octane (C_8H_{18}) weighs how many grams?

17. How much energy, in foot-pounds, is needed to raise one barrel of oil 25,000 ft? How much is this energy in terms of the equivalent barrels of oil?

18. The U.S. coal reserves are estimated at 1.5×10^{12} tonnes. It is shown in Appendix B that this coal would be consumed in 97 years if our coal consumption rate increased by 5% each year. Calculate how many years it would last if we could hold our energy consumption to the present 81×10^{15} Btu/year and coal were our sole source of energy. Assume it is all bituminous coal.

19. The methane in natural gas burns as shown:

$$CH_4 + 2O_2 \rightarrow CO_2 + 2H_2O$$

How many tons of CO_2 are produced for each ton of methane that is burned?

SUGGESTED READING AND REFERENCES

1. Dick, Richard A., and Wimpfen, Sheldon P. "Oil Mining." *Scientific American,* **243** 4 (October 1980), pp. 182–188.
2. Fowler, John M. *Energy and the Environment.* New York: McGraw-Hill, 1975.
3. Hall, C. A. S., and Cleveland, C. J. "Petroleum Drilling and Production in the United States: Yield Per Effort and Net Energy Analysis." *Science,* **211** (February 6, 1981), p. 576.
4. Hubbert, M. King. "Energy Resources." *Resources and Man.* National Academy of Sciences—National Research Council. San Francisco: W. H. Freeman, 1969.
5. Hubbert, M. King. "The Energy Sources of the Earth." *Scientific American,* **224** 3 (September 1971), pp. 60–70. (Also in *Energy and Power,* a Scientific American book. San Francisco: W. H. Freeman, 1971.)
6. Jensen, M. L., and Bateman, A. M. *Economical Mineral Deposits.* New York: John Wiley, 1979.
7. Krenz, Jerrold H. *Energy—Conversion and Utilization,* Second Editon. Boston: Allyn and Bacon, 1984.
8. Lovins, Amory B. *Soft Energy Paths—Towards a Durable Peace.* New York: Harper Colophon, 1977.
9. "Amory Lovins and His Critics." Hugh Nash, Ed. *The Energy Controversy.* San Francisco: The Friends of the Earth, 1979.
10. Marion, Jerry B. *Energy in Perspective.* New York: Academic Press, 1974.
11. Meadows, Donella H.; Meadows, Dennis L.; Randers, Jorgen; and Behrens, William W., III. *The Limits to Growth.* New York: Universe, 1972.
12. Menard, H. William. "Towards a Rational Strategy for Oil Exploration." *Scientific American,* **244** 1 (January 1981), pp. 55–65.
13. Penner, S. S., and Icerman, L. *Energy.* Vol. 1. Reading, Mass.: Addison-Wesley, 1981.

14. Priest, Joseph. *Problems of Our Physical Environment.* Reading, Mass.: Addison-Wesley, 1973.

15. Ruedisili, Lon C., and Firebaugh, Morris W. *Perspectives on Energy.* New York: Oxford University Press, 1978.

16. Stobaugh, Robert, and Yergin, Daniel, Eds. *Energy Future: Report of the Energy Project at the Harvard Business School.* New York: Random House, 1983.

17. Stocker, H. Stephen; Seager, Spencer L.; and Capener, Robert L. *Energy from Source to Use.* Glenview, Ill.: Scott Foresman, 1975.

18. Wilson, Richard, and Jones, William J. *Energy, Ecology, and the Environment.* New York: Academic Press, 1974.

19. Final Report of the Committee on Nuclear and Alternative Energy Systems (CONAES). *Energy in Transition 1985–2010.* San Francisco: W. H. Freeman, 1979.

20. Hirsch, Robert, L. *Impending United States Energy Crisis,* Science **235** 4795 (March 20, 1987), pp. 1467–1473.

21. Yergin, Daniel. *The Prize.* New York: Simon and Schuster, 1991.

22. Cassedy, Edwards, and Grossman, Peter Z. *An Introduction to Energy Resources, Technology, and Society.* Cambridge: Cambridge University Press, 1990.

69

CHAPTER THREE

HEAT ENGINES AND ELECTRIC POWER

3.1 THE ENERGY CONTENT OF FOSSIL FUELS

We burn fossil fuels, as well as renewable fuels, for only two basic purposes: to provide direct heating and lighting and to power heat engines. The general definition of *heat engine* covers a very broad range of possibilities, but for our purposes we use the term to mean a man-made device, or system of devices, that extracts mechanical energy from the heat energy given off by burning fuels, for example.

The general pathways by which we utilize energy from fossil fuels are presented in Figure 3.1.

Much more will be said later about heat engines and the thermodynamic principles that govern their behavior; but let us first look at how heat is derived from fuels. In simplified form, the burning of fuel is merely the combining of carbon from the fuel with oxygen from air. It is the reverse of photosynthesis and proceeds according to the chemical reaction

$$C + O_2 \rightarrow CO_2 + \text{heat energy}$$

Consider, for example, a more realistic situation, the 100% burning of heptane, a colorless liquid constituent of gasoline. This burning process is

$$C_7H_{16} + 11\ O_2 \rightarrow 7\ CO_2 + 8\ H_2O + 1.15 \times 10^6 \text{ calories per gram molecular weight of } C_7H_{16} \text{ (heptane)}$$

The number at the right, 1.15×10^6 calories, is the *heat of combustion*. Every fuel has a tabulated value for this quantity, usually given in calories per gram molecular weight (cal/g mol wt). If 1 g mol wt of a fuel is burned completely in air at atmospheric pressure and at 20°C, the amount of energy given off is equal to the heat of combustion. This is the definite maximum amount of energy available from a fuel, which cannot be exceeded by clever techniques of combustion or carburetion. For a fossil fuel, it can be considered a measure of the solar energy stored in the fuel since ancient times. The quantity *gram molecular weight* is just an amount in grams equal to the sum of the atomic weights of one molecule of the fuel. This quantity is also known as the mole, as discussed in Section 1.3. One mole, or 1 gram molecular weight, of any substance always contains 6.02×10^{23} (Avogadro's number) of atoms or molecules. For the case at hand, C_7H_{16}, 1 g mol wt is 7×12 plus 16×1, which is equal to 100 g. The

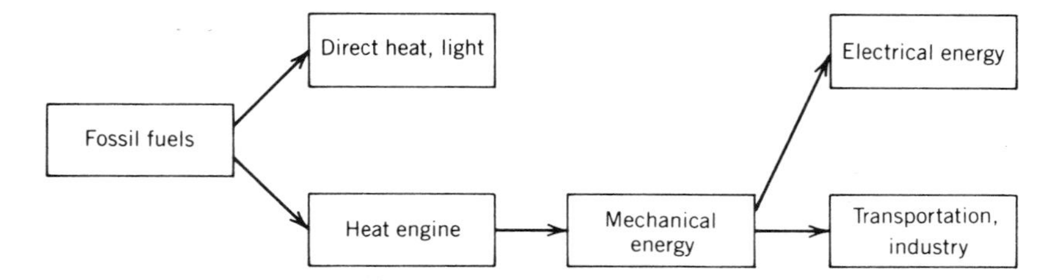

Figure 3.1 The general pathways by which we utilize energy from fossil fuels.

number 12 is the atomic weight of carbon, and hydrogen has an atomic weight of approximately 1. For each 100 g of heptane burned, 1.15×10^6 cal of heat energy is liberated. From this we can derive the energy liberated by burning 1 kg of heptane:

$$\frac{1000 \text{ g/kg}}{100 \text{ g/g mol wt}} \times 1.15 \times 10^6 \text{ cal/g mol wt} = 1.15 \times 10^7 \text{ cal/kg}$$

or

$$4.18 \text{ J/cal} \times 1.15 \times 10^7 \text{ cal/kg} = 4.8 \times 10^7 \text{ J/kg}$$

Given this information we can work a sample problem.

Example 3.1

How many kilograms per hour of heptane must be burned to power a 100% efficient 25 horsepower (hp) engine? (Such a thing doesn't exist.)

Solution

$$1 \text{ hp} = 746 \text{ W} = 746 \text{ J/sec}$$

or

$$1 \text{ hp} = 746 \text{ J/sec} \times 3600 \text{ sec/hr} = 2.69 \times 10^6 \text{ J/hr}$$

A 25-hp engine would require

$$25 \text{ hp} \times 2.69 \times 10^6 \text{ J/hr-hp} = 6.7 \times 10^7 \text{ J/hr}$$

and we have available 4.8×10^7 J/kg. Therefore

$$\frac{6.7 \times 10^7 \text{ J/hr}}{4.8 \times 10^7 \text{ J/kg}} = 1.4 \text{ kg/hr}$$

is the rate at which the heptane fuel would be consumed. This is roughly equivalent to 3 lb or 2 liters, or $\frac{1}{2}$ gal/hr. If the engine were 50% efficient, fuel would be consumed at twice this rate; at 25% efficiency, fuel would be burned at four times the calculated rate, and so forth.

3.2 THE THERMODYNAMICS OF HEAT ENGINES

Whenever we observe a heat engine in operation, it is evident that not all the heat energy released from the burning of the fuel is used to perform useful work. Our automobiles, motorcycles, and mopeds all dissipate waste heat through their radiators, cooling fins, and exhausts; coal-burning power plants for electricity generation are all connected to rivers, lakes, cooling towers, or have some other means of carrying off the waste heat. Why is this? Have the designers carelessly permitted energy to become waste heat rather than useful work? Not entirely. There is a fundamental requirement that some heat energy must be rejected. The discussion of this point is best expressed in terms of the laws of thermodynamics, which like all other laws of physics are merely statements of human observation that have so far been found to hold without exception.

In this discussion we shall need to employ, among other factors, three basic concepts: pressure, volume, and temperature in absolute units. The first, pressure, is defined as force per unit area, or the total force divided by the area over which it is exerted. The units of pressure are pounds per square foot, or pounds per square inch, or newtons per square meter, or some similar ratio of force to area. The second concept, volume, is relatively obvious. The usual units are cubic feet, cubic meters, or cubic centimeters. Cubic inches have been commonly used for the volume swept out by the pistons of a reciprocating engine per revolution of the crankshaft. Liters and quarts are also units of volume. For thermodynamic discussions, temperature must be in absolute units, that is, units such that the zero of the temperature scale is at absolute zero, a concept we mention here without going into its precise definition. An appropriate temperature scale that has come into common scientific use is the Kelvin scale, in which temperatures are expressed in degrees Kelvin (°K). This scale is similar to the Celsius scale in that there are 100 units between the freezing and boiling points of water. In the Kelvin scale, absolute zero is at 0°K, water freezes at 273°K, and water boils at 373°K at atmospheric pressure. Normal room temperature is just a few degrees below 300°K, and the human body temperature is about 310°K.

The simple laboratory exercise shown in Figure 3.2 allows us to illustrate the absolute temperature scale and the determination of absolute zero. It is called the constant volume ideal gas thermometer. An ideal gas is one in which the molecules behave independently; one molecule does not interact with another. For this purpose, ordinary gases, such as nitrogen, are reasonable approximations to ideal gases at atmospheric pressure and lower pressures.

A sealed volume of gas is connected to a pressure gauge having a small volume compared to the main volume of the gas. The gas container is immersed in boiling water, which we know to be (by definition) at 100°C at a pressure of 1 standard atmosphere (atm). The boiling temperature, of course, will be less at lower barometric pressures, but by a known amount. This gives us one point on the graph of pressure gauge reading versus temperature in degrees Celsius. A second point is obtained by immersing the gas container in ice water at 0°C (again by definition). These data points are then extrapolated linearly down to the point where the pressure would be equal to zero. The temperature spacing between this zero pressure point and the 0°C point is found to be equal to 273°C, setting absolute zero at −273°C. Absolute zero, then, is the temperature at which a confined gas would be expected to exert no pressure on the walls of its container. In practice, this is not quite true; all gases condense to liquids at temperatures above absolute zero. The Kelvin temperature scale is a convenient scale for all calculations involving the relationships between pressure and temperature of a gas, as well as for thermodynamic calculations in general.

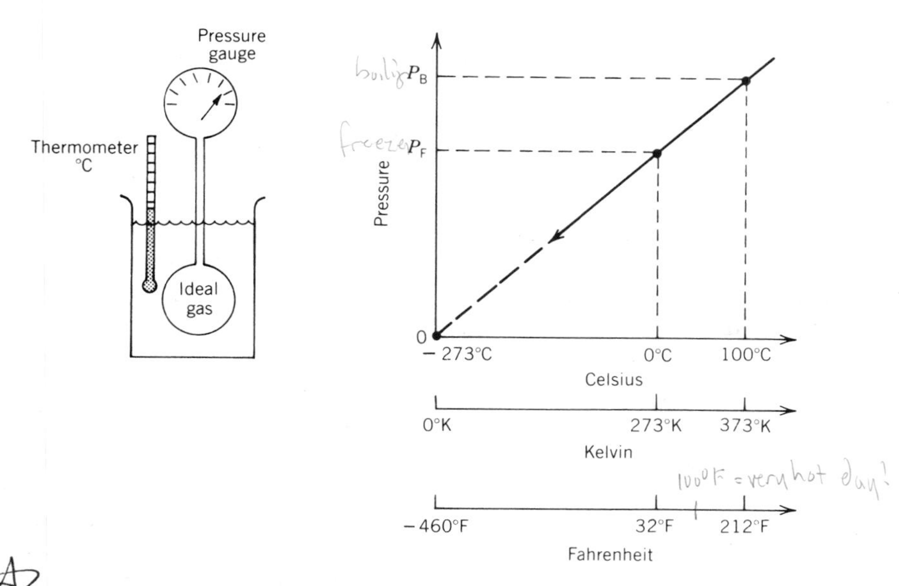

Figure 3.2 A constant volume ideal gas thermometer. The pressures at which water boils (P_B) and freezes (P_F) are plotted against their corresponding temperatures. By extrapolation to zero pressure, absolute zero on the temperature scale is determined.

Now, just what are heat engines? Consistent with our earlier definition, we shall consider a heat engine to be any device that exploits a temperature difference to perform work. There are numerous examples of mechanical heat engines. They range from intercontinental ballistic missiles to rocks being split by alternate cycles of freezing and thawing to nails working out of a tin roof as a result of the day–night and summer–winter cycles of expansion and contraction of the roofing material. People have been absolutely ingenious in devising heat engines to do work wherever a temperature difference can be found (geothermal or solar) or created, usually by burning some fuel. Figure 3.3 shows one example of a heat engine that we all use. This is a typical gasoline engine used in most automobiles. Engines of this type operate at an efficiency of 10 to 20%. Whatever the form of the heat engine, its operation corresponds to the diagram shown in Figure 3.4. All of the elements of Figure 3.4 are present in any heat engine. An important point is that Q_{cold}, the heat energy rejected to the low temperature reservoir, can never be zero.

There are only four laws of thermodynamics that govern all thermal behavior. They are usually numbered in the following way:

(0) There exists the concept of temperature. Systems are in equilibrium when they are at the same temperature.

(1) Energy is conserved within any closed system.

(2) This law is sometimes expressed in terms of entropy, but for our purposes the second law may be stated as: It is not possible to extract heat energy from a reservoir and perform work without rejecting some heat energy to another reservoir at lower temperature or to transfer heat from one

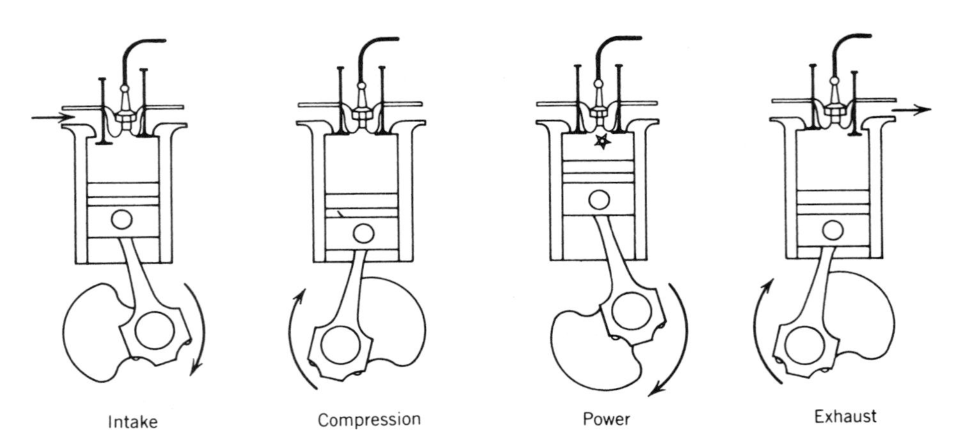

Figure 3.3 Operation of a common heat engine, the four-stroke internal combustion engine. The heat source is provided by burning gasoline above a piston in a cylinder; the heat sink is provided by the surrounding air. During each half-revolution of the crankshaft (bottom), the piston moves either from the top of the cylinder to the bottom or vice versa, giving four strokes during two revolutions of the shaft. Only the power stroke delivers work. A fuel-air mixture is admitted during the intake stroke, and the exhaust gases are expelled during the exhaust stroke. The valve openings and closings are controlled by a camshaft mechanism not shown in this drawing.

reservoir to another at a greater temperature without doing external work on the system.

(3) It is impossible to attain absolute zero.

In general, it can be said that we can extract work from the universe around us only because there exist high temperature points such as stars, as well as low temperature regions in space. If the time ever comes when the entire universe is in equilibrium at the same temperature, the energy content of the universe will be exactly what it is today, but the energy will be unavailable for doing work. This fate, although by no means certain, is sometimes called the "heat death of the universe."

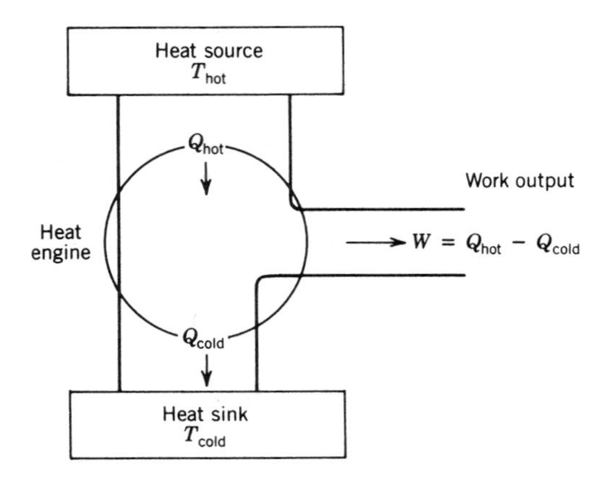

Figure 3.4 A thermodynamic diagram of a heat engine operating between a heat source and a heat sink at a lower temperature. The work output, because of conservation of energy, must equal the difference between the heat energy extracted from the source and that rejected to the sink.

Referring to Figure 3.4, we can develop an expression for the thermodynamic efficiency for any heat engine. By the usual definition of efficiency

$$\text{efficiency} = \frac{\text{work done}}{\text{energy put into the system}}$$

For the heat engine this becomes

$$\text{efficiency} = \frac{Q_{\text{hot}} - Q_{\text{cold}}}{Q_{\text{hot}}}$$

where the numerator is a result of the first law of thermodynamics. To express efficiency in terms of percent, the above expression should be multiplied by 100. The symbol η (Greek letter eta) is often used for efficiency. With a minor algebraic rearrangement of terms, the above expression can be written as

$$\eta = \left(1 - \frac{Q_{\text{cold}}}{Q_{\text{hot}}}\right) \times 100\%$$

It was shown convincingly in 1824 by Sadi Carnot, a French engineer, that for an ideal engine the ratio of Q_{cold} to Q_{hot} must be the same as the ratio of the temperatures of the reservoirs between which the heat engine is operating. This ideal engine has never existed and never will. It would require no friction, perfect insulation, and so forth, but the concept of an ideal engine does lead to an expression for the maximum efficiency of a heat engine if all practical engineering problems could be overcome. This, of course, is an ideal efficiency never quite attainable. It is known as the Carnot efficiency. If we take the Carnot statement

$$\frac{Q_{\text{cold}}}{Q_{\text{hot}}} = \frac{T_{\text{cold}}}{T_{\text{hot}}}$$

where the T is measured on the absolute (Kelvin) temperature scale, and substitute it in the above expression for efficiency, we arrive at the Carnot efficiency:

$$\eta_c = \left(1 - \frac{T_{\text{cold}}}{T_{\text{hot}}}\right) \times 100\%$$

as the maximum efficiency of a heat engine. Note that this efficiency depends only on the temperatures of the reservoirs between which the heat engine operates. Let us consider an example. For a typical steam turbine electricity-generating system powered by a coal-burning boiler, T_{hot} (the boiler temperature) would be about 825°K, and T_{cold} (the cooling tower) would be about 300°K. This leads to

$$\eta_c = \left(1 - \frac{300}{825}\right) \times 100\% = (1 - 0.36) \times 100\% = 64\%$$

as the Carnot efficiency. A maximum of 64% of the energy in the fuel can go to turning the dynamo, and no less than 36% of the fuel's energy must be rejected as waste heat by the system. The waste heat may still be used for other purposes, space heating, for instance, but it cannot be used for producing mechanical work unless there is available another reservoir at a temperature lower than 300°K. To make the Carnot efficiency as high as possible, one would like to increase T_{hot} and decrease T_{cold}. In practice, the limits are imposed by the materials from which the boilers can be constructed and the availability in nature of large heat reservoirs.

3.3 ENTROPY

In evaluating any thermodynamic system or any system for the recovery of resources from waste, the physical scientist often employs the concept of entropy. Entropy has many definitions, ranging from "time's arrow" to highly analytical thermodynamic expressions. It is relevant to social organizations, information theory, and progress toward the heat death of the universe. In general, entropy is a quantity that is low for highly ordered systems and that increases as any system tends toward randomness or disorder. In any closed system, including the universe, entropy will increase with time; to effect a local decrease of entropy, work must be brought in from the outside. This is, of course, impossible for the universe. Just after you have cleaned and arranged your room, it is in a state of low entropy. As time passes, it will tend toward a state of randomness; restoring it will require work, or the expenditure of energy obtained through your food and electricity for the vacuum cleaner. Entropy increases as things decay, as organisms age, as rich deposits of ores are washed downstream. Localized iron ore deposits are of low entropy; as time passes, the iron becomes distributed randomly throughout the junkyards of the nation. To reclaim it into a neat, orderly pile requires work. All rich fossil fuel deposits are of low entropy. Once their atoms are randomly distributed, where will we obtain the energy needed to restore the local order so necessary in our society? Local order includes words in books, data on magnetic tapes, and legislators sitting in session. Also schools, bridges, clocks, the human memory, and solar collectors are far from random combinations.

Two important points were expressed by Clausius, in 1865, as the first and second laws of thermodynamics: (1) The energy of the universe is constant. (2) The entropy of the universe tends toward a maximum.

To this we can add that without sources of energy to maintain locally ordered systems, our world must tend toward disorder of all sorts. This is inescapably required by the second law of thermodynamics. We have not yet observed a single exception to this law.

3.4 GENERATION OF ELECTRICITY

Now that we have seen how heat energy is derived from fossil fuels and have gained an understanding of thermodynamic efficiency, let us examine how these ideas are integrated into the design of an electric power plant. Still missing from our discussion are the actual mechanical details of how heat energy is converted to rotational motion of machinery and how the rotational motion ultimately results in electrical energy. Let us consider the use of a steam turbine to utilize a flow of high-temperature, high-pressure steam against the fanlike turbine blades to produce rotational motion. The rotating turbine shaft is connected directly to the shaft of a dynamo, or generator, causing rapidly moving coils of wire to sweep through a magnetic field. It is well known in physics that whenever an electrical conductor is moved across a magnetic field, a voltage will be generated across the ends of the conductor, causing current to flow through any load connected to the system. This is the Faraday law of electromagnetic induction, discovered in 1831. Some generators have stationary coils and moving magnetic fields, but the effect is the same.

The main components of a typical electricity-generating plant are shown in Figure 3.5. Plants of this type have been in use for a long time, certainly more than 90 years. Engineering improvements have resulted in steadily increasing efficiency over the years, as Figure 3.6 shows. We see that actual efficiencies today are only about one third, meaning that two thirds of the energy in the fuel ends up as waste heat. This is about double the waste that one would expect from the calculated Carnot efficiency. The difference is due to the realities of engineering. There can be no such thing as an ideal Carnot engine. Bearings do have friction, heat does escape up the exhaust stack, insulation does leak heat, pumps are not perfectly efficient, and so forth. At this point it is interesting to look at our former expression for the Carnot efficiency with the terms slightly rearranged.

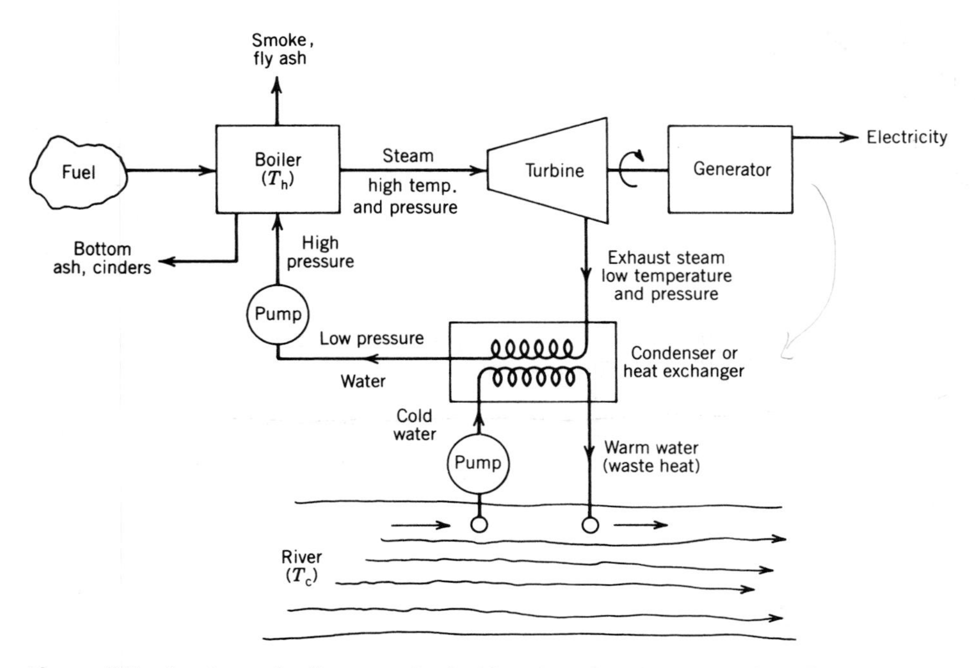

Figure 3.5 A schematic diagram of a fuel-burning electric power plant. Here a river provides cooling water to the condenser, but lake water or a cooling tower could serve the same purpose.

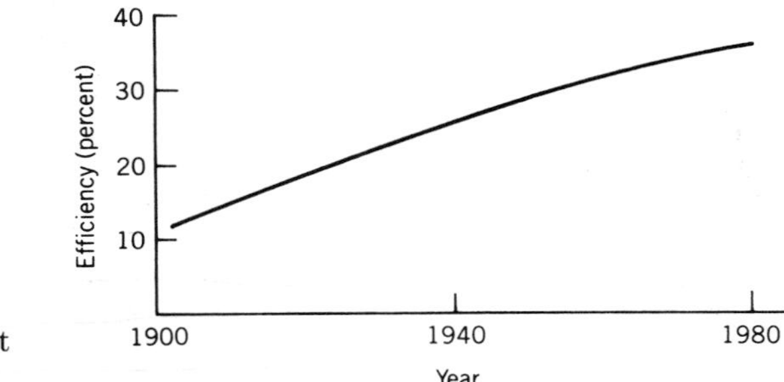

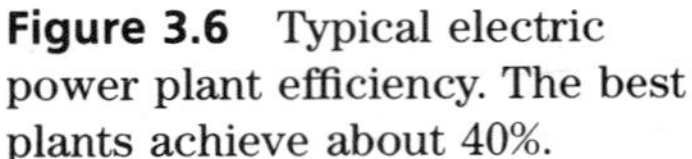

Figure 3.6 Typical electric power plant efficiency. The best plants achieve about 40%.

$$\eta_c = \left(1 - \frac{T_c}{T_h}\right) \times 100\% = \frac{T_h - T_c}{T_h} \times 100\% = \frac{\Delta T}{T_h} \times 100\%$$

The term ΔT represents the difference between the hot and cold temperatures between which the heat engine operates. The entire power plant is the heat engine. The temperatures are measured in degrees Kelvin. It is apparent from the above expression that the efficiency will be improved by any measures that increase the temperature difference without increasing T_h proportionately, or by any measures to decrease T_h without reducing the temperature difference. This leads to some difficult choices; usually the maximum possible T_h and the minimum possible T_c are sought.

It is important to realize why, in the system shown in Figure 3.5, the pump that returns the water to the boiler does not consume the entire power produced by the turbine and generator. On a pound-for-pound basis, it does handle as much water per unit time as came out of the boiler as steam, and the pressure difference against which it must pump is the same as the pressure drop across the turbine and condenser. The answer is that water, the working fluid in this system, is reduced in volume by a factor of about 1000 as it condenses from steam to liquid water. The work done by a pump (or on a turbine) is given by the product of pressure difference times the volume pumped. In this case, the pump handling the liquid water has its work reduced by a factor of about 1000 compared to what it would have to do if it were handling steam. Thus, the power consumed by the pump is just a tiny fraction of that produced by the turbine.

In summary, a power plant that provides 1000 MW of electricity typically releases 2000 MW of power to pollute the environment thermally and requires a 3000 MW boiler. This means that with electric resistance heat, three units of fuel must be burned to put one unit of heat into a home. This problem can be offset to some degree by using electrically powered heat pumps or by directly piping the warm exhaust water or low-pressure steam from power plants to buildings needing heat. Figure 3.7 shows a large power plant with cooling ponds to carry off the waste heat.

3.5 HEAT PUMPS

In the discussion of heat engines illustrated by Figure 3.4, we saw that it is possible to devise a system such that the heat energy is extracted from some source, with part of this energy being converted to work and the remainder

Figure 3.7 The Four Corners power plant near Farmington, New Mexico. This heat engine converts chemical energy from coal into electrical energy that is supplied through transmission lines to the Los Angeles region at a distance of 500 miles. The waste heat is discharged to the environment through the cooling ponds.

being rejected to a heat sink at a temperature lower than that of the source. The diagram of Figure 3.4 would still be consistent with all the laws of thermodynamics if the directions of all three arrows indicating energy flow were reversed. This new diagram is shown in Figure 3.8; it corresponds to a device that uses an energy input to cause the transfer of heat energy from one reservoir to another at a higher temperature. Heat energy is thus pumped uphill in a temperature sense, and the quantity of heat energy transferred can far exceed the work input. This can make heat pumps extremely attractive for space heating purposes.

In Figure 3.8, the first law of thermodynamics requires that $Q_h = W + Q_c$; the second law requires that W not be equal to zero. The effectiveness of a heat pump is often expressed in terms of a coefficient of performance (C.O.P.), which is the ratio of Q_h to W, or C.O.P. $= Q_h/W$. By algebraic rearrangement and by use of the Carnot relationship, $Q_c/Q_h = T_c/T_h$, the C.O.P. for an ideal heat pump is

$$\text{C.O.P.} = \frac{Q_h}{W} = \frac{Q_h}{Q_h - Q_c} = \frac{1}{1 - Q_c/Q_h} = \frac{1}{1 - T_c/T_h} = \frac{T_h}{T_h - T_c}$$

This ideal C.O.P. is thus determined entirely by the temperatures (in °K) of the two reservoirs. As with the Carnot efficiency of the heat engines, this coefficient of performance applies only to an ideal heat pump; it is thus a theoretical maximum value never attained in practice.

When a heat pump is used to transfer heat energy out of an enclosure, thus maintaining the enclosure at some desired T_c less than the surrounding T_h, it is known either as a refrigerator or as an air conditioner. It is usually designated simply as a heat pump when used to transfer heat energy into an enclosure,

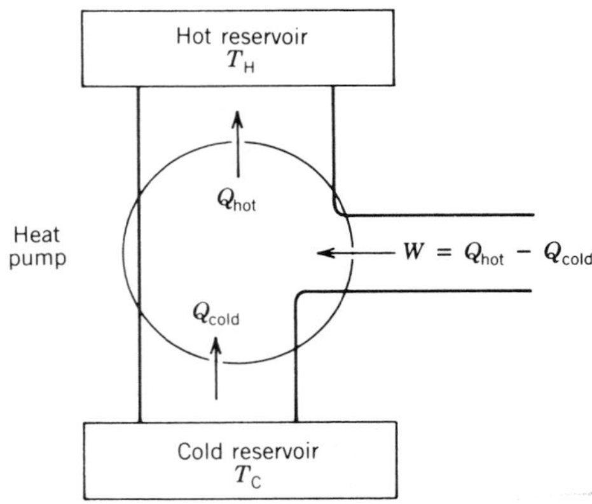

Figure 3.8 A thermodynamic diagram of a heat pump. A work input, W, is required to transfer an amount of energy, Q_c, out of a cold reservoir and a larger amount, Q_h, into a hot reservoir. Because energy is conserved, Q_h must equal $W + Q_c$.

such as a house, in order to maintain the house at some T_h warmer than the surrounding T_c. The same device can often be operated in either direction, providing warmth in winter and cooling in summer. In space-heating applications, the cold reservoir can be the outside air, ground water, lake or river water, or a solar-heated swimming pool. In residential heating applications, heat is usually pumped from the outside air (thus refrigerating the outdoors) by a heat pump powered by an electric motor. The C.O.P. calculated in Example 3.2 suggests that 13.3 W of heating can be provided for every watt consumed by the heat pump. In practice, a C.O.P. in the range of 2 to 6 is typical, depending on outside temperature. As the outside temperature drops (or as $T_h - T_c$ increases), both the ideal and actual coefficient of performance diminish. Because of this, electrically driven air-to-air heat pumps are most useful in moderate climates, losing their advantages over resistance heating when the outside temperature falls below 15°F. Air conditioners and refrigerators sometimes carry the specification of energy efficiency ratio (EER). This is the rate at which heat energy is removed in Btu/hr divided by the rate at which energy is consumed by the appliance in watts.

Example 3.2

Calculate the ideal coefficient of performance for an air-to-air heat pump used to maintain the temperature of a house at 70°F when the outdoor temperature is 30°F. This example is not realistic for two reasons: Ideal mechanical devices cannot be achieved, and to maintain a house interior at 70°F, the heating air delivered must be somewhat above that temperature.

Solution

$$T_h = 70°\text{F} = 21°\text{C} = 294°\text{K}$$

$$T_c = 30°\text{F} = -1°\text{C} = 272°\text{K}$$

$$\text{C.O.P.} = \frac{T_h}{T_h - T_c} = \frac{294°\text{K}}{294°\text{K} - 272°\text{K}} = \frac{294}{22} = 13.3$$

Thus, for every watt of power used to drive this ideal heat pump, 13.3 W is delivered to the hot reservoir (the interior of the house), and 12.3 W is extracted from the cold reservoir (the outside air). In practice, the C.O.P. for such a situation would be much less favorable. It would probably have a value of about 3.

The actual mechanism by which a heat pump operates can vary from device to device, but a typical system is schematically illustrated in Figure 3.9. In this system, a compressor driven by an electric motor compresses Freon gas to raise its temperature and pressure. The gas then flows through a radiator-type heat exchanger where it is cooled by a flow of room-temperature air and condensed to a liquid, still at high pressure, thus giving up heat to the room. On passage through a small orifice in an expansion valve, the liquid expands into a gas, becoming much colder in the process. This cold gas then passes into a second heat exchanger located outside, where it is warmed to the temperature of the ambient outside air, thus extracting heat from the outside air. The gas then passes into the compressor to repeat the cycle.

The overall effect of heat pumps is to make it possible to use electrical energy for space heating with an overall (heat pump plus power plant) efficiency as great as if the fuel burned in the power plant had been burned directly

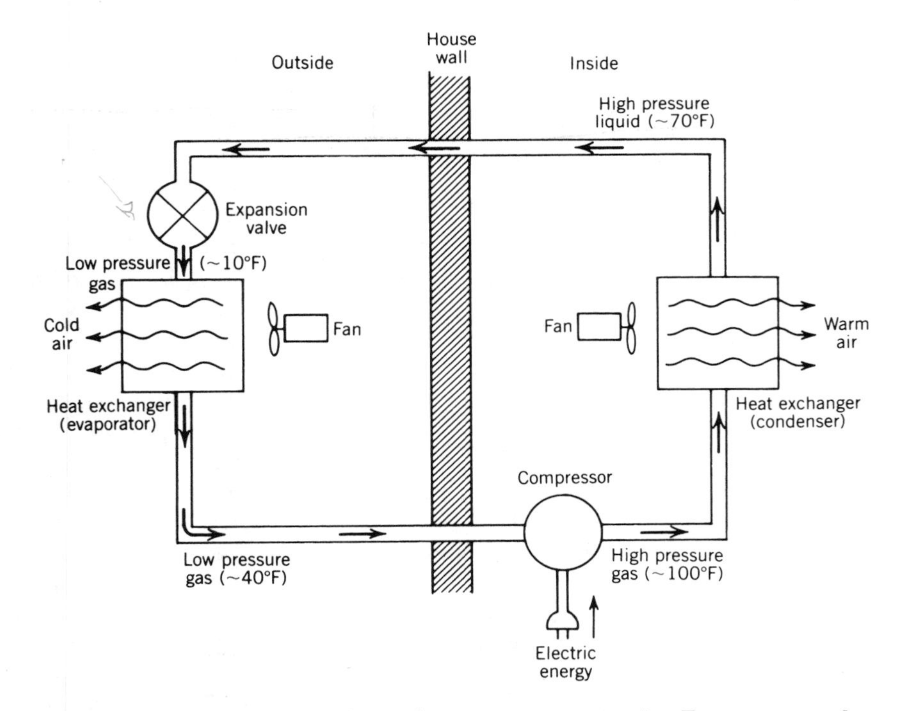

Figure 3.9 An electrically driven heat pump system using Freon as a working fluid. In principle, the system becomes an air conditioner if the fluid flow direction is merely reversed, although in practice the reversal of function is more complex.

in a furnace in a home. The drawbacks to heat pumps center around their initial cost and lack of long-term reliability as compared to electric resistance heating, as well as their reduced coefficient of performance under conditions of extremely cold weather.

3.6 COGENERATION

The operation of a heat engine for any purpose necessarily involves the rejection of large amounts of heat. Generally, this rejected heat is dissipated in the atmosphere through the use of a cooling tower or into an adjacent body of water, such as a cooling pond, a river, or an ocean. Not only is there expensive equipment involved in the process of throwing the heat away, but frequently there are also environmental concerns about the effect of the added heat to the body of water or the atmosphere. By far the greatest concern from an energy standpoint, however, is the waste of heat energy. A coal-fired electric power plant will typically have an efficiency of about 38%, which means that 62% of the heat derived from the coal is rejected to the environment.

There is no basic reason why the rejected heat from a heat engine cannot be used in some beneficial way. The rejected heat is said to be of lower quality; it is at a lower temperature and is more limited in its application than higher quality heat. There are, however, many uses for such low-quality heat, ranging from space and water heating in homes, institutions, and factories to industrial process heat used directly in manufacturing. In principle, there can be several steps in the use of rejected heat—a cascading of applications until the final temperature is equal to that of the atmosphere. The general term given to this process of waste heat utilization is cogeneration.

The obvious example of a large, central, coal-fired, steam-turbine electric power plant, owned by a public utility, piping heated water or steam for space or water heating or industrial process steam, may not be an optimum application of cogeneration. Central power plants of this kind have been constructed only for electricity production, and there could be large capital expenditures in restructuring such plants for cogeneration. Because of the heat losses in transporting steam or hot water large distances, the industries or housing units would have to be located near the power plant. This frequently is not the existing situation. There is also the problem that often the demand for heat does not coincide in time with the demand for electricity. Because of these complications, some feel that a better application of cogeneration would be in smaller, decentralized electricity-generating units primarily in industry. These units would be located at the point of use. They would be designed for the generation of industrial process steam and of electricity that would either be used by the industry involved or be put back into the public utilities' transmission lines for use elsewhere. Ross and Williams have estimated that up to 2×10^6 bbl of oil a day (6% of all energy used in the United States) could be saved if all the industrial users of steam cogenerated electricity.

There are other situations in which cogeneration of electricity and steam may be economically practical, such as in large apartment house complexes or in institutions such as universities. The idea of cogeneration can be extended to homes in which the hot air from clothes driers could be used for space heating or the cooling water from a diesel engine–heat pump combination could be used directly for space heating and to increase the efficiency of the heat pump.

PROBLEMS

1. A typical room temperature is 68°F. What are the corresponding Celsius and Kelvin temperatures?
2. The basic equation of carbon burning is $C + O_2 \rightarrow CO_2 + 95$ kilocalories per mole (1 mole of carbon is 12 g). From this information calculate the number of Btus from burning 1 ton of coal assuming it is pure carbon.
3. (a) Starting from the principle of energy conservation, explain the reasoning that gives us the expression for thermodynamic efficiency, η, in terms of the temperatures T_c and T_h. You may assume that $Q_c/Q_h = T_c/T_h$ for an ideal heat engine.
   (b) What is a heat engine?
4. An inventor claims to have developed a wonderful new heat engine that operates with a relatively cool (and therefore nonpolluting) flame at 150°C and discharges waste heat to the environment at 20°C. His promotional literature advertises that 45% of the fuel energy is converted into useful work. Calculate the maximum efficiency that can be expected for such an engine and compare it to the inventor's claim.
5. (a) The other morning a 1-kW electric heater was switched on for 1 hour. Trace back as far as you can the various forms that this electric energy had before it was used for heat. Be explicit, give details, and utilize the basic principles of physics and of the transformation of one kind of energy into another. What is the ultimate source of the energy? Assume a coal-fired electric power plant.
   (b) Now trace the path the electric energy will take after it has been used for heat as described above. What will be its final destiny?
6. Why are fossil fuel–burning electric power generating plants always situated near lakes or rivers or provided with cooling towers? Support your answers by arguments derived from fundamental laws of physics.
7. Power plants may have a maximum thermodynamic efficiency of perhaps 64% calculated from temperatures of boiler and heat sink, but in practice, the best power plants convert into electricity only 40% of the energy released by the burning of fuels. Why?
8. (a) A steam engine receives steam from a boiler at 300°C and at the end of the heat cycle exhausts it to the air at 100°C. What is the maximum possible efficiency of the engine?
   (b) State in your own words the meaning of the second law of thermodynamics and explain its significance for the efficiency of heat engines.
9. An electric heat pump can deliver more energy than it draws from the power line without violating the principle of energy conservation. Explain how this can happen.
10. How many tons of CO_2 are produced for each ton of methane burned?
11. What is the efficiency of a Carnot engine operating between a heat source at 100°C and a heat sink at 0°C?
12. A heat engine in each cycle extracts 50,000 Btu of thermal energy and rejects (or releases) 20,000 Btu of thermal energy.
    (a) How many Btus of work are done every cycle?
    (b) What is the efficiency of the engine?

13. The Four Corners Power Plant (Units 1, 2, and 3) consumes 7200 metric tons of coal a day. The average heat value of this coal is 2.0×10^4 Btu/kg. The overall thermal efficiency of the plant is 30%.
(a) What is the averge electric power output of the plant in kilowatts?
(b) What is the amount of heat energy in kilowatt hours that is dumped into the atmosphere in one day from the waste heat from this plant?

14. Because individual natural gas–fueled water heaters in homes do not have air pollution control devices, it has been proposed that water in homes be heated with electrical energy generated at a distant power plant that is equipped with an air pollution control device. The power plant burns natural gas and produces electrical energy with 40% efficiency; the electrical transmission line between the plant and the home loses 10% of the electrical energy as heat. All that remains goes into heating the water. In contrast, the gas water heater gets only 60% of the fuel energy into the water. Which scheme uses more natural gas to heat a given quantity of water? Explain your reasoning in detail.

15. An electric power plant can deliver 1000 MW (10^9 W) of power continuously.
(a) Assuming the overall thermal efficiency of the plant is 33%, how many tons of coal need to be put into the boilers every 24 hours?
(b) How many tons of CO_2 are released by the above power plant to the atmosphere every 24 hours? Use the approximation that coal is 100% carbon.
(c) How many Btus of thermal energy must be extracted by the heat exchanger and dissipated every 24 hours?

16. About 3.0×10^{12} kW • hr of electricity could be used in one year in the United States. Using the parameters of the above power plant:
(a) How many tons of coal would be needed a year?
(b) How many tons of CO_2 would be given off? Use the approximation that coal is 100% carbon.

17. An electric power plant operates with a boiler temperature of 900°C and a condenser temperature of 100°C. What is the absolute thermodynamic upper limit to the number of kilowatt-hours that it can obtain from a ton of bituminous coal?

18. A nuclear power reactor produces steam at 288°C and has an overall efficiency of 34%. If the condenser temperature is 36°C, what fraction of its theoretical maximum efficiency does it achieve?

19. A heat engine operating between a geothermal heat source at 210°C and a river at 20°C achieves an efficiency of 20%. What percentage of its theoretical maximum efficiency is it achieving?

20. An ideal heat pump delivers 8 Btu of heat energy to a house for each 6 Btu of heat that it draws from its low temperature reservoir. What is its coefficient of performance?

21. Show that the combination of a 40% efficient power plant with a heat pump having a coefficient of performance of 4.0 would actually deliver 60% more heat than if the fuel were used directly to heat the house with 100% efficiency.

22. A power plant produces electricity from bituminous coal with 40% efficiency. Electricity is transmitted to a house with 90% efficiency. A house with a heat pump that has a coefficient of performance of 5 heats air to 100°F. How many Btus are delivered to the house for each ton of coal burned for that purpose?

23. If a refrigerator has an energy efficiency ratio of 10, for each unit of input energy (W) drawn from the electric power company, how many units of heat energy (Q_c) are removed from the cold box, and how many are delivered to the room (Q_h)?

24. The consumption of electric energy in the United States from 1910 to 1979 is listed below.

Year	kW • hr Consumed/yr
1910	1.97×10^{10}
1920	5.66×10^{10}
1930	1.15×10^{11}
1940	1.80×10^{11}
1950	3.89×10^{11}
1960	8.44×10^{11}
1970	1.64×10^{12}
1973	1.95×10^{12}
1979	2.32×10^{12}

(a) Plot these data on semilog paper with the time on the horizontal linear scale.
(b) Draw the "best" straight line through the data points and determine the average doubling time.
(c) From the doubling time, calculate the average percentage growth per year.
(d) Extrapolate the curve to estimate the electric energy consumed in 1990.
(e) Discuss briefly how you think the electric energy consumption in the United States will *actually* grow in the next 50 years.

SUGGESTED READING AND REFERENCES

1. Crawley, Gerald M. *Energy.* New York: Macmillan, 1975.
2. Devins, D. W. *Energy, Its Physical Impact on the Environment.* New York: John Wiley, 1982.
3. *Energy and Power.* A Scientific American Book. San Francisco: W. H. Freeman, 1971.
4. Fowler, John M. *Energy and the Environment.* New York: McGraw-Hill, 1970.
5. Krenz, Jerrold H. *Energy—Conversion and Utilization,* Second Edition. Boston: Allyn and Bacon, 1984.
6. Priest, Joseph. *Problems of Our Physical Environment.* Reading, Mass.: Addison-Wesley, 1973.
7. Romer, Robert H. *Energy, an Introduction to Physics.* San Francisco: W. H. Freeman, 1976.
8. Ross, Marc H., and Williams, Robert H. *Our Energy, Regaining Control.* New York: McGraw-Hill, 1981.
9. Wilson, Richard, and Jones, William J. *Energy, Ecology, and the Environment.* New York: Academic Press, 1974.

CHAPTER FOUR

NUCLEAR ENERGY

4.1 INTRODUCTION

Following the discovery by Hahn, Strassman, Meitner, and Frisch, in 1938 and 1939, that uranium nuclei would fission under neutron bombardment, it became reasonable to consider a new form of energy to meet the world's needs. The developments during World War II that led eventually to nuclear weapons provided the scientific and engineering background for the design and construction of a nuclear reactor that could power an electrical generator. At the very beginning of these developments, in 1942, Enrico Fermi and his coworkers constructed a nuclear reactor in which, using natural uranium and many blocks of graphite, a self-sustaining chain reaction was demonstrated for the first time. The neutrons from the fissioning of one uranium nucleus were slowed down by the graphite moderator so that they could then induce fission in other uranium nuclei. Each time a uranium nucleus fissioned, energy was released, since the mass of the fission product nuclei was less than the mass of the uranium nucleus. In the early reactors, the energy released was not used for any purpose, and it was only 15 years later at Shippingport, Pennsylvania, that a commercial power reactor was first put into use. This reactor generated steam that was used to power turbines connected to electric generators.

As we saw briefly in Chapter 1, the equivalence of mass and energy can be understood in terms of Einstein's special theory of relativity. The now famous equation $E = mc^2$ describes the basis for energy generation in a nuclear reactor. The magnitude of the energy that can be obtained from direct conversion of nuclear mass to energy is quite impressive compared to the energy obtained by burning a fuel such as coal.

Considering the burning of 1 kg of coal. The energy released would be

$$E_c = 2.8 \times 10^7 \frac{\text{J}}{\text{kg}}$$

Now consider the energy that would be released if we could convert all of the mass in 1 kg in coal (or 1 kg of anything) into energy:

$$E_M = mc^2 = 1 \text{ kg} \times (3 \times 10^8 \text{ m/sec})^2 = 9 \times 10^{16} \text{ J}$$

Although there is no way at the present time to produce energy from all of the mass of coal, the ratio of the mass energy to the chemical energy is impressive. This ratio is

$$\text{R} = \frac{E_M}{E_c} = \frac{9 \times 10^{16} \text{ J}}{2.8 \times 10^7 \text{ J}} \approx 3.2 \times 10^9$$

Instead of having 100 years of energy from our coal resource, we would, in this hypothetical case, have energy for 320×10^9 years, or more than 10 times the age of our universe.

Around 1950, there was tremendous optimism regarding nuclear energy. Many scientists and engineers working in the field felt that a vast new source of energy that was both inexpensive and nonpolluting was about to be made available. Some individuals even predicted that the cost of electricity would be so low that there would be no need for electric meters to monitor how much was used by various residential and industrial customers. More recently (around 1970), it was widely believed that by the year 2000 half of our electric energy would come from fossil fuels and half from nuclear fission. The first prediction appears to be clearly incorrect, as the cost of electric energy from coal and from the nucleus is about the same (although there is continuing debate over the relative costs), and both costs are increasing faster than the inflation rate. The second prediction will not be fulfilled; no new nuclear reactors are on order by power companies in the United States. In 1990, 22% of the U.S. domestic electricity generation was derived from nuclear fission; in France it was about 75%. In addition, France exports some nuclear electric power to neighboring countries.

The reasons that the original promise of nuclear energy is not being fulfilled are interesting. They involve nuclear physics, engineering, health and safety factors, psychology, and politics. Before attempting to delve into the problems of nuclear energy, however, one needs to have some background in the basic nuclear physics underlying this new form of energy.

4.2 THE STRUCTURE OF NUCLEI

The fundamental constituents of nuclei are protons and neutrons. The proton is the same as the nucleus of the most common hydrogen atom. The neutron is very similar to a proton in its mass and other properties, but it has no electric charge; it is electrically neutral. All known nuclei can be constructed from combinations of protons and neutrons. The atomic number (Z) of a nucleus is equal to the number of protons in the nucleus; the atomic mass number (A) is the sum of the number of neutrons (N) and the number of protons.

Atoms each consist of a nucleus, which has almost all of the mass of the atom, and electrons orbiting around the nucleus. For a neutral atom, these electrons are equal in number to the protons, since the electron has a charge of the same magnitude as that of the proton, but of opposite sign. The diameter of the atom is about 10^{-10} meters, whereas the nucleus is very much smaller (10^{-15}m). The masses of the three basic building blocks of atoms can be expressed either in conventional units of mass, kilograms (kg), or in units of energy, million electron volts (MeV), according to $E = mc^2$.

$$m_{\mathrm{p}}\ (\text{proton}) = 1.672 \times 10^{-27}\ \text{kg} = 938.26\ \text{MeV}$$

$$m_{\mathrm{n}}\ (\text{neutron}) = 1.675 \times 10^{-27}\ \text{kg} = 939.55\ \text{MeV}$$

$$m_{\mathrm{e}}\ (\text{electron}) = 9.108 \times 10^{-31}\ \text{kg} = 0.511\ \text{MeV}$$

There are some important points to note concerning these masses. First, the proton and neutron are each about 1836 times as massive as the electron. The neutron and proton have about the same mass, with the neutron being somewhat heavier. The unit million electron volts is used for energy measurements in nuclear physics because it is more convenient than the joule. An electron volt is the energy that an electron would gain if it were accelerated through an electrical potential difference of 1 volt. The metric unit of charge is the coulomb (C), and the electron has a charge of 1.60×10^{-19} C. Now, since the volt is defined as 1 J/1 C, the product of the charge of the electron times 1 volt is 1.60×10^{-19} J. Hence

$$1 \text{ eV} = 1.60 \times 10^{-19} \text{ J}$$

$$1 \text{ MeV} = 1.60 \times 10^{-13} \text{ J}$$

The other common units of energy in nuclear physics are keV (10^3 eV) and GeV (10^9 eV).

Figure 4.1 shows the beginning section of the Chart of the Nuclides. The complete chart shows all of the known stable and radioactive nuclei. The chart is arranged such that the atomic number, *Z*, increases vertically (0 to 107) and the neutron number, *N*, horizontally (0 to 158). An individual nucleus occupies one box. If it is radioactive, its half-life is stated, and if it is stable, the percentage of the element represented by that one nucleus or isotope is shown. The horizontal lines correspond to nuclei with a particular value of *Z;* hence, they are all the same chemical element with the same chemical behavior but with different atomic masses (or numbers of neutrons). They are called isotopes of the element. The usual way of designating a particular isotope (in this case, oxygen-18) is

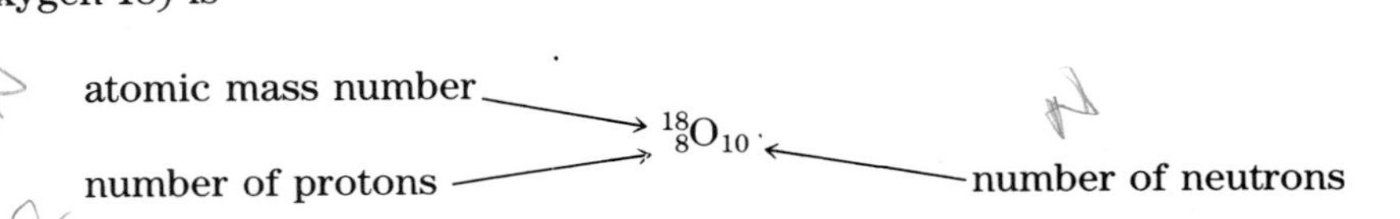

The first stable nuclei in the chart shown in Figure 4.1 are $^{1}_{1}H_0$ and $^{2}_{1}H_1$, the proton and the deuteron. Although the deuteron is present in hydrogen as found in nature to only 0.015% abundance, we shall see later that it plays a crucial role in the possibility of obtaining useful amounts of energy from the fusion reaction. For $Z = 2$ there are two isotopes of helium: $^{3}_{2}He_1$ and $^{4}_{2}He_2$. The nucleus $^{4}_{2}He_2$ is known as an alpha particle, and is one of the particles emitted in the decay of radioactive nuclei such as $^{235}_{92}U_{143}$.

The stable isotopes tend to follow the line formed by equal numbers of protons and neutrons ($Z = N$) until about ^{40}Ca. At this point, more neutrons than protons are needed for stability. One of the heaviest stable isotopes is $^{208}_{82}Pb_{126}$. It has 44 more neutrons than protons. Beyond lead and bismuth, all of the heavier isotopes are unstable, that is, radioactive.

The question of what forces hold nuclei together is one that has occupied the attention of nuclear physicists for many years. There are two basic forces we know about from classical physics: the gravitational force and the electromagnetic force. Both forces vary in strength according to the inverse square law ($1/r^2$, where r is the distance between particles). The gravitational force is always attractive; the electromagnetic force is attractive for particles having charges of different sign and repulsive for particles having charges of the same sign.

To keep the two protons in $^{4}_{2}He_2$, for example, from flying apart under the strong repulsive electrostatic or Coulomb force, a stronger attractive force is needed. The gravitational force acts in the correct direction, but it is too weak by a factor of 10^{36}. A new type of force is needed to explain the stability of

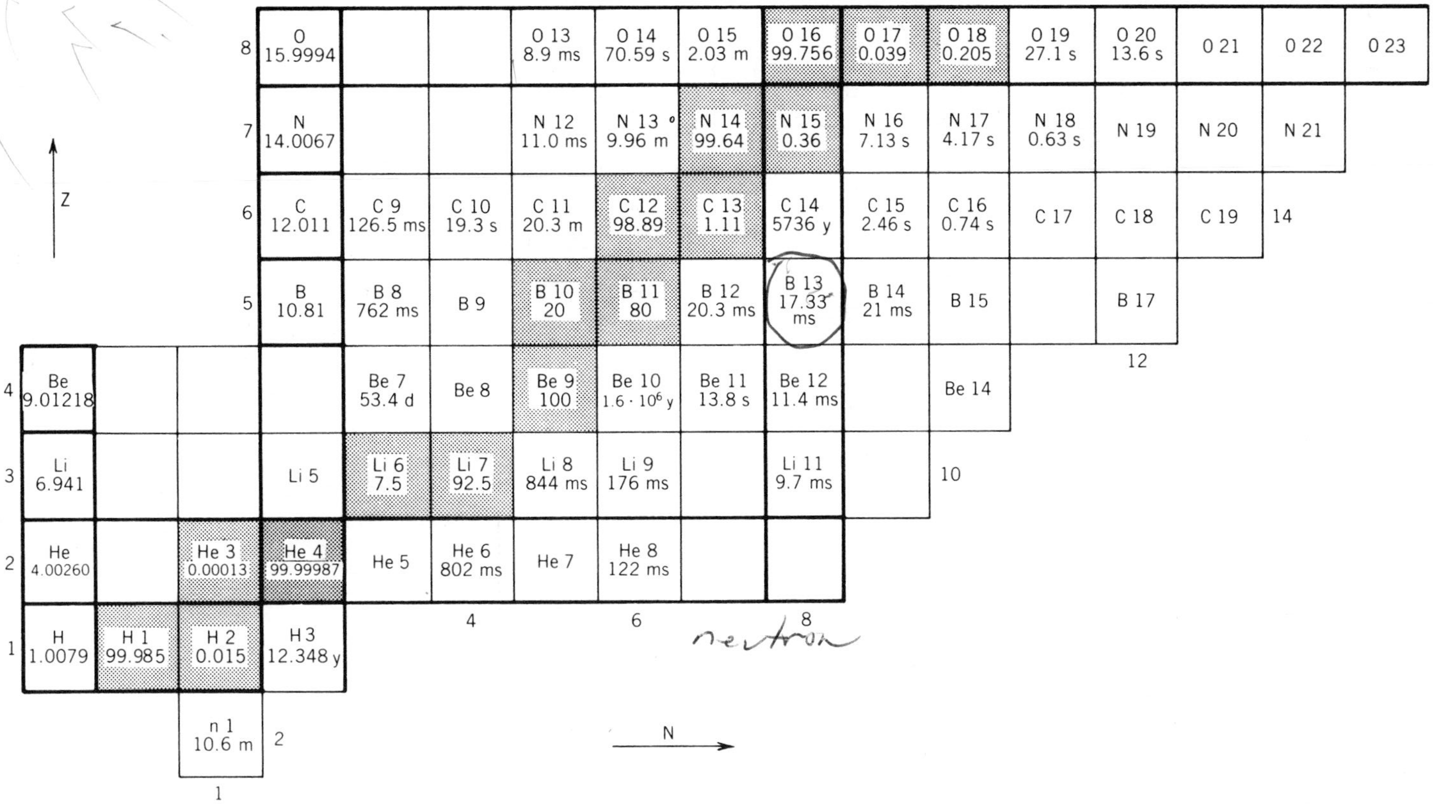

Z \ N	Element	0	1	2	3	4	5	6	7	8	9	10	11	12	13	14	15
8	O 15.9994						O 13 8.9 ms	O 14 70.59 s	O 15 2.03 m	O 16 99.756	O 17 0.039	O 18 0.205	O 19 27.1 s	O 20 13.6 s	O 21	O 22	O 23
7	N 14.0067						N 12 11.0 ms	N 13 9.96 m	N 14 99.64	N 15 0.36	N 16 7.13 s	N 17 4.17 s	N 18 0.63 s	N 19	N 20	N 21	
6	C 12.011				C 9 126.5 ms	C 10 19.3 s	C 11 20.3 m	C 12 98.89	C 13 1.11	C 14 5736 y	C 15 2.46 s	C 16 0.74 s	C 17	C 18	C 19		
5	B 10.81				B 8 762 ms	B 9	B 10 20	B 11 80	B 12 20.3 ms	B 13 17.33 ms	B 14 21 ms	B 15		B 17			
4	Be 9.01218				Be 7 53.4 d	Be 8	Be 9 100	Be 10 $1.6 \cdot 10^6$ y	Be 11 13.8 s	Be 12 11.4 ms		Be 14					
3	Li 6.941			Li 5	Li 6 7.5	Li 7 92.5	Li 8 844 ms	Li 9 176 ms		Li 11 9.7 ms							
2	He 4.00260		He 3 0.00013	He 4 99.99987	He 5	He 6 802 ms	He 7	He 8 122 ms									
1	H 1.0079	H 1 99.985	H 2 0.015	H 3 12.348 y													
0			n 1 10.6 m														

Z ↑ N →

Figure 4.1 A portion of the chart of the nuclides. The proton number is plotted in the vertical direction, the neutron number in the horizontal direction. Each square is identified by its chemical symbol and mass number. The solid squares represent nuclei that are stable against radioactive decay; the radioactive half-lives are shown, if known, for the others. Each solid square is labeled with the percentage of the element in nature made up by that particular isotope.

nuclei. It is called either the strong force or the nuclear force and is a short-range attractive force considerably more complicated than the gravitational or electrostatic force. The same attractive nuclear force is present between two protons, two neutrons, or between a neutron and a proton. As the number of protons in a nucleus increases, more and more neutrons are needed to provide the attractive nuclear forces needed to overcome the repulsive electrostatic forces, since the neutrons add to the binding forces but not to the repulsive electrostatic forces.

The stable nuclei in Figure 4.1 are shaded. For elements with an odd number of protons, there are often only one or two stable isotopes. However, elements with an even number of protons can have as many as 10 stable isotopes. When the number of protons is either too large relative to the number of neutrons or when the number of neutrons is too large relative to the number of protons (although this is not obvious), the nucleus will be unstable (radioactive) and will undergo a transformation until it becomes one of the stable isotopes. All of the nonshaded isotopes in Figure 4.1 are radioactive and have been produced by various reactions using energetic particles from accelerators, or by neutron-induced reactions in a nuclear reactor. A number of nuclei (about 80) are naturally radioactive; many of them have lifetimes so long that they are still present in nature, remaining from the birth of the universe. Others are found in the decay of these very long-lived radioactive nuclei or are produced by the ever-present cosmic radiation.

4.3 RADIOACTIVITY

An example of radioactive decay is provided by the decay of a free neutron, a neutron not bound into an atomic nucleus. Because the neutron has somewhat more mass than the proton, it is energetically possible for a neutron to decay into a proton. This process, although infrequent, can be observed under laboratory conditions. It occurs by the emission of an electron, which for historical reasons is called a beta particle, but which is identical to an atomic electron. From the masses given for the proton and neutron, one can see that the neutron is heavier by 1.29 MeV. In the decay of the neutron, some of the mass energy will be needed to provide the mass of the electron that is created in the process, 0.51 MeV. One might expect, because of energy conservation, that the excess energy ($1.29 - 0.51 = 0.78$ MeV) would appear as the kinetic energy of the emitted electron. In this case the proton has negligible kinetic energy because it is so much heavier than the electron. However, when physicists first measured the energies of electrons coming from radioactive decay (of sources other than the neutron), they were surprised to find that there was a continuous spectrum of electron energies ranging from zero up to the maximum possible energy. Many electrons apparently came off with less than the maximum energy, and hence must have shared the energy available with some other undetected particle. This particle, now called the neutrino (or, in some instances, the antineutrino), was predicted to exist by Pauli in 1931 and has since been detected. It is an electrically neutral particle with either no mass or a very, very small mass that has not yet been reliably measured. The question of the neutrino mass is extremely interesting on theoretical grounds, having implications for the evolution of the universe. The equation that summarizes the beta decay of the free neutron is

$$^{1}_{0}\text{n}_1 \rightarrow {}^{1}_{1}\text{H}_0 + \beta^- + \bar{\nu}, \qquad T_{1/2} = 10.6 \text{ min}$$

where the symbol $\bar{\nu}$ stands for an antineutrino, β^- is the negative beta particle, and $T_{1/2}$ is the half-life. Beta decay of this type can also take place for neutrons inside many nuclei, generally those with a neutron excess relative to the stable nuclei, but the energetics and half-life differ from those of the free neutron decay. Nuclei created by nuclear fission generally have a neutron excess, so they are radioactive and decay by β^- emission.

It is also possible to have a nucleus with too many protons for stability. Essentially all of the nuclei in Figure 4.1 to the left of the stable nuclei are of this kind, and they tend to decay by the emission of a β^+ particle (a positively charged electron, often called a positron). An example of such a process is

$$^{13}_{7}\text{N}_6 \rightarrow {}^{13}_{6}\text{C}_7 + \beta^+ + \nu, \qquad T_{1/2} = 9.96 \text{ min.}$$

where again the β^+ and ν share the energy available. In this process, a proton is converted to a neutron. The two kinds of neutrinos involved in beta decay are the antineutrino, $\bar{\nu}$, emitted in β^- decay and the neutrino, ν, emitted in β^+ decay. The positive electron, β^+, created in the process, is the antiparticle of the negative electron, β^-. There is an additional process, also considered as beta decay because it results in a proton being converted to a neutron, that can compete with β^+ emission. It is called electron capture. In this process, one of the orbiting atomic electrons is captured by the nucleus and a neutrino is emitted. In some instances there is not enough energy available for β^+ emission, and electron capture is then the only way beta decay can proceed.

The emission of an alpha particle (the nucleus of a helium atom) is a very common way for the unstable nuclei heavier than lead to decay. An example of such a decay process involves a common isotope of plutonium:

$$^{239}_{94}\text{Pu}_{145} \rightarrow {}^{235}_{92}\text{U}_{143} + {}^{4}_{2}\text{He}_2(\alpha \text{ particle}), \qquad T_{1/2} = 24{,}200 \text{ yr}$$

Since there is no third particle sharing the decay energy, the alpha particles will have an explicit energy, not a continuous energy spectrum as do the beta particles. Actually, in the decay of ^{239}Pu a number of different alpha groups (alphas with an explicit energy) are emitted. These correspond to alpha particle decay leading to the population of various states of ^{235}U at different excitation energies.

The population of excited states of daughter nuclei by beta or alpha emission is a very common occurrence. These excited states usually decay to a lower excited state by the emission of a gamma ray, an electromagnetic quantum of energy similar to an x-ray. This is the primary source of gamma radiation from radioactive sources.

Energy is conserved in all nuclear processes, including radioactive decay, according to the *principle of energy conservation.* Other fundamental quantities, such as the total electric charge and the total number of nucleons (neutrons or protons), are also conserved in these processes.

4.4 RADIOACTIVE HALF-LIVES

Every radioactive decay occurs with a characteristic time called a half-life. It is the time during which one half of the radioactive nuclei in a given sample will decay. If some particle or gamma ray associated with the decay process from a given source is being measured, the number of these emitted per unit of time will also decrease by a factor of 2 in one half-life. The mathematical expression for radioactive decay is identical to that discussed in Chapter 1 for exponential growth, except that the exponent is negative instead of positive:

$$N = N_0 e^{-\lambda t}$$

where λ is called the decay constant, t is the time, N_0 is the number of radioactive nuclei at $t = 0$, and N is the number at any later time t. Since $e^{-0.693} = ½$, when $N/N_0 = ½$, we must have

$$T_{1/2} = \frac{0.693}{\lambda}$$

where $T_{1/2}$ is the half-life. If the emission rate from radioactive decay or the number of radioactive nuclei present is plotted on semilog graph paper with time on the linear scale, a straight-line dependence will be noted. The behavior of radioactivity with time is analogous to exponential growth, with the exception, of course, that exponential growth increases with time, whereas radioactivity decreases with time.

Table 4.1 lists a number of radionuclides (radioactive nuclei) with their half-lives. A number of these isotopes will be of interest when nuclear reactors and nuclear weapons are considered. The natural occurrence of some radionuclides, such as ^{40}K and ^{232}Th, is the direct result of their half-lives being comparable to the age of the universe.

Radiocarbon dating is based on the continuous production of ^{14}C in the atmosphere by neutrons that result from cosmic rays. The photosynthetic process responsible for the formation of organic matter, of course, requires atmospheric CO_2 as a source of carbon. Any organic material will have a carbon content consisting of radioactive ^{14}C and stable carbon (^{12}C and ^{13}C). At any time after the death of an organism, the ratio of ^{14}C to stable carbon will be decreased by radioactive decay from the ratio in the atmosphere and will be an indication of the number of years the ^{14}C has had to decay.

4.5 NUCLEAR REACTIONS

A number of nuclear reactions can take place using neutrons from nuclear reactors or beams of nuclear particles (protons, deuterons, etc.) that have been accelerated to an energy in the million electron volts range in a device such as a cyclotron. Another source of energetic charged particles is a plasma at a very high temperature (about 10^{8}°K), such as exists in the interior of the sun or in thermonuclear plasma experiments such as the Tokamaks built in various

Table 4.1 SOME RADIONUCLIDES OF INTEREST FROM THE POINT OF VIEW OF OUR ENVIRONMENT AND THE GENERATION OF NUCLEAR POWER

Radionuclide		Decay Particle	$T_{1/2}$	Source
$^{14}_{6}C_8$	(carbon-14)	β^-	5568 yr	Naturally occurring, $^{14}N(n,p)^{14}C$ in atmosphere
$^{40}_{19}K_{21}$	(potassium-40)	β^-	1.28×10^9 yr	Naturally occurring
$^{232}_{90}Th_{142}$	(thorium-232)	α	1.39×10^{10} yr	Naturally occurring
$^{237}_{93}Np_{144}$	(neptunium-237)	α	2.20×10^6 yr	Naturally occurring
$^{235}_{92}U_{143}$	(uranium-235)	α	7.13×10^8 yr	Naturally occurring
$^{238}_{92}U_{146}$	(uranium-238)	α	4.51×10^9 yr	Naturally occurring
$^{232}_{92}U_{141}$	(uranium-232)	α	1.59×10^9 yr	Reactors
$^{239}_{94}Pu_{145}$	(plutonium-239)	α	2.41×10^4 yr	Reactors
$^{3}_{1}H_2$	(hydrogen-3, tritium)	β^-	12.35 yr	Reactors
$^{90}_{38}Sr_{52}$	(strontium-90)	β^-	29 yr	Fission product
$^{131}_{53}I_{78}$	(iodine-131)	β^-	8.04 days	Fission product
$^{137}_{55}Cs_{82}$	(cesium-137)	β^-	30.17 yr	Fission product
$^{85}_{36}Kr_{49}$	(krypton-85)	β^-	10.72 yr	Fission product

laboratories. There are two aspects of nuclear reactions that interest us at the moment: the conservation laws and the energetics of the reactions. The conservation laws are exactly the same as those listed for radioactive decay; that is, the conservation of the number of nucleons and total charge. The energetics of nuclear reactions follow directly from the equivalence of mass and energy. For example, suppose that the following reaction takes place:

$$a + A \rightarrow b + B + Q$$

where a is the light bombarding particle, b is a particle emitted by the reaction, and A and B are the initial and final nuclei. An alternative way to write this reaction is $A(a,b)B$. The Q in the reaction is the amount of energy that is either given off (exoergic) or taken up (endoergic), and it can be calculated directly from the masses of the four nuclei.

$$Q = m_a + m_A - m_b - m_B$$

If the Q value is positive, energy will be given off in the reaction; that is, the kinetic energy of the outgoing particles will exceed by Q MeV the kinetic energy of the initial particles. On the other hand, if the Q value is negative, that amount of energy must be put into the reaction for it to be energetically possible. An example of a nuclear reaction is

$$^{4}_{2}\text{He}_2 + {}^{9}_{4}\text{Be}_5 \rightarrow {}^{12}_{6}\text{C}_6 + {}^{1}_{0}\text{n}_1 + Q$$

One can see in this $^{9}\text{Be}(\alpha,\text{n})^{12}\text{C}$ reaction that there are 13 nucleons (6 protons and 7 neutrons) on both sides of the reaction equation. The neutral atomic masses of the interacting nuclei are, in atomic mass units (amu),

^{4}He 4.002603 amu	^{12}C 12.000000 amu
^{9}Be 9.012186	n 1.008665
13.014789	13.008665

There is thus more mass initially than finally; so the Q value will be positive and equal to the mass difference of 0.006124 amu. The Q value is usually stated in MeV. The relationship between the units is 1 amu = 931 MeV. The Q value for this reaction is 6.124×10^{-3} amu $\times$ 931 MeV/amu = +5.70 MeV.

4.6 NUCLEAR BINDING ENERGIES—FUSION AND FISSION

The binding energy of a nucleus is defined as the energy equivalent of the sum of the masses of the individual neutrons and protons contained in the nucleus minus the mass of the nucleus. For example, some nucleus $^{A}_{Z}X_{N}$ will have a binding energy in MeV where the masses are in atomic mass units:

$$\text{B.E. (MeV)} = 931 \quad (Zm_{\text{H}} + Nm_{\text{n}} - m_{\text{x}})$$

The simplest nucleus beyond the proton is the deuteron, ^{2}H, composed of one neutron plus one proton bound together by the attractive nuclear force. When those two particles come together, the excess energy, 2.2 MeV in this case, is given off as a gamma ray. To separate the deuteron into its two separate particles, the amount of energy must be provided from some external source. In other words, the binding energy is the amount of energy that would be needed to separate the nucleus into its individual constituents.

A useful quantity to consider is the binding energy per nucleon, or how tightly bound the nucleus is per particle.

Figure 4.2 shows a plot of this quantity (binding energy per nucleon) in MeV per nucleon as a function of the atomic mass number A. Note that the binding energy is zero for the free nucleon, as shown for the proton, ^{1}H. The deuteron, ^{2}H, is bound by 2.2 MeV, or 1.1 MeV per nucleon as shown on this graph. The curve rises initially very rapidly as A increases. There are some particular high values for ^{4}He, ^{12}C, and ^{16}O of 7.07, 7.68, and 7.95 MeV per nucleon, because these particular nuclei involve a very stable configuration or a tightly bound number of both protons and neutrons, similar to the atomic electron structure of He, Ne, A, and so forth. The curve in Figure 4.2 can be seen to reach a broad maximum at around 8.7 MeV per nucleon for values of A between 50 and 100 and then decrease slowly to values of 7.6 MeV per nucleon for the very heavy nuclei.

Example 4.1

Calculate the binding energy in MeV per nucleon for ${}^{235}_{92}U_{143}$. In order to account for the electrons, neutral atomic masses must be used instead of the bare nuclear masses. The relevant atomic masses in amu are:

$$\begin{aligned} n &- 1.008665 \\ {}^{1}_{1}H_0(p) &- 1.007825 \\ {}^{235}U &- 235.043943 \end{aligned}$$

Solution

The summed masses of the constituents of ${}^{235}U$ are

$$(92 \times 1.007825) + (143 \times 1.008665) = \\ 92.719900 + 144.239095 = 236.958995 \text{ amu}$$

The total nuclear binding energy is

$$236.958995 - 235.043943 = 1.915052 \text{ amu}$$

which is equal to

$$1.915052 \text{ amu} \times 931 \text{ MeV/amu} = 1782.9 \text{ MeV}$$

Per nucleon, that is

$$\frac{1}{235} \times 1782.9 \text{ MeV} = 7.59 \text{ MeV}$$

Figure 4.2 provides a means for understanding how energy can be released from the two types of nuclear reactions that provide energy. The understanding and exploitation of this single curve, the curve of binding energy, have increasingly dominated global politics for the past decades, initially because of the weapons implications, but now equally because of our need for new energy sources. In the fusion reaction, two light nuclei come together and become one larger nucleus. As the two nuclei "fuse" together, the value of binding energy per nucleon increases from the initial value of no more than about 1 or 2 MeV to a final value of perhaps 7 MeV. The difference of 5 or 6 MeV per nucleon times the number of nucleons, say four, is equal to the energy released in the reaction. That is, because the binding energy per nucleon is greater for the final configuration, there must be a reduction in the total mass, and it is just the

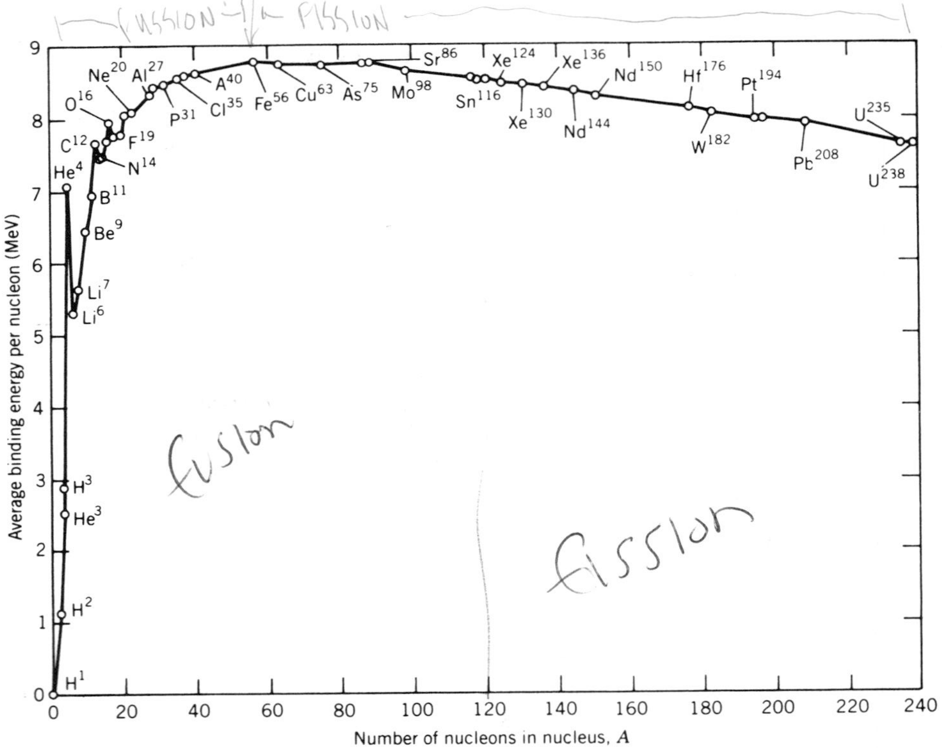

Figure 4.2 The curve of binding energy. The curve reaches a maximum near $A = 60$ and drops significantly by $A = 235$. The drop between $A = 120$ and $A = 235$ provides a measure of the energy release in fission. The energy release in fusion is related to the steeply rising left portion of this curve.

mass energy corresponding to this reduction that is released. Three very important fusion reactions involve the deuteron (${}^2_1H_1$) and the triton (${}^3_1H_2$). These exoergic reactions can be written as

$$ {}^2_1H_1 + {}^3_1H_2 \rightarrow {}^4_2He_2 + n + 17.6 \text{ MeV} $$

$$ {}^2_1H_1 + {}^2_1H_1 \rightarrow {}^3_2He_1 + n + 3.2 \text{ MeV} $$

$$ {}^2_1H_1 + {}^2_1H_1 \rightarrow {}^3_1H_2 + {}^1_1H_0 + 4.0 \text{ MeV} $$

These D–T or D–D reactions, as they are usually called, form the basis of thermonuclear energy.

On the other end of the binding energy curve, it is also possible to have a release of energy by a completely different type of nuclear reaction. If a very slow neutron is absorbed by ${}^{235}U$, the resulting ${}^{236}U$ nucleus is unstable against breaking into two pieces. The nucleus behaves somewhat like a liquid drop that is jostled. The two pieces, known as fission fragments, generally have different masses. There is usually a lighter fragment that will have a mass number in the range of about 85 to 104, and a heavier fragment with a mass number in the range of 130 to 149, in addition to a few neutrons. Of course,

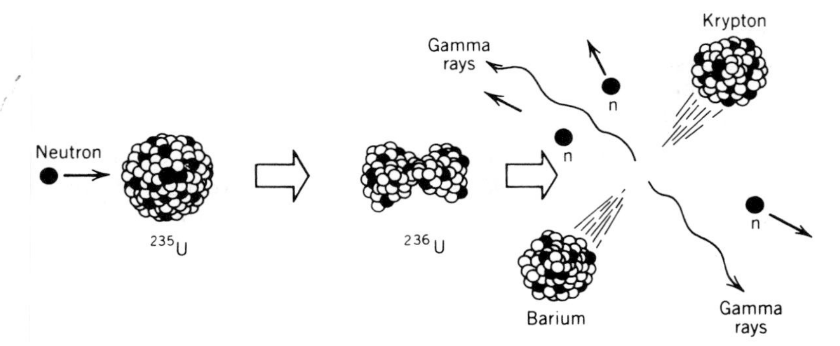

Figure 4.3 Neutron-induced fission of ^{235}U. The combination of a neutron and ^{235}U forms ^{236}U in a highly excited state, which promptly fissions into two lighter nuclei, emitting neutrons and gamma rays in the process.

the total number of nucleons cannot change in the reaction. One example of the many possible neutron–induced fissions of ^{235}U, shown in Figure 4.3, is

$$n + {}^{235}_{92}U_{143} \rightarrow {}^{236}_{92}U_{144} \rightarrow {}^{144}_{56}Ba_{88} + {}^{89}_{36}Kr_{53} + 3n + 177 \text{ MeV}$$

The intermediate nucleus, ^{236}U, is in a highly excited state, and it fissions spontaneously. The binding energy per nucleon is about 7.6 MeV for nuclei around ^{235}U and roughly 8.5 MeV for the fission fragment nuclei.

In this fission process, the 235 nucleons originally in the ^{235}U nucleus are rearranged such that 233 are in fission fragment nuclei and two are free neutrons. The fission fragments, or product nuclei, each have more binding energy per nucleon than does ^{235}U; the neutrons have no binding energy per nucleon. Because energy is released in this reaction, the mass of ^{235}U plus one neutron is greater than the sum of the masses of the two product nuclei plus three neutrons. Because the change in binding energy per nucleon is about +0.9 MeV in going from ^{235}U to the product nuclei and about −7.6 MeV in going from ^{235}U to free neutrons, the energy release in this fission process can be approximately calculated as $(233 \times 0.9 - 2 \times 7.6) = 195$ MeV. Just as on page 95, where it is shown that the energy release in a nuclear reaction is given by the initial mass minus the final mass, here it is seen that the energy release is also given by the change in total binding energy. This is the final binding energy minus the initial binding energy, which in this case can be put in the form $(233 \times 8.5) - (235 \times 7.6) = 195$ MeV. This value differs somewhat from the explicit example given below because the 0.9 MeV binding energy difference used here is an average for all ^{235}U fissions. On average, considering all possible combinations of fission fragments, it is known that 198 MeV is released by neutron-induced fission of ^{235}U. (The explosion of a TNT molecule releases only about 30 eV.) This energy appears mainly as kinetic energy of the fission fragments. Since both fission fragments are positively charged nuclei, they are accelerated away from each other by their mutual repulsion, in this way gaining their kinetic energy. The fission fragments will vary in type from one fission event to another.

The curve shown in Figure 4.2 is important for understanding nuclear energy. Nuclear reactors using the fissioning of ^{235}U, ^{239}Pu, or certain other isotopes can certainly provide a source of energy for many years. Eventually, however, the earth's resource of fissionable fuels will be depleted, and the only source of energy, with the exception of geothermal and tides, will be the fusion reaction, either as it takes place in the sun or in combination with thermonuclear fusion reactors yet to be developed. We explore the fission and fusion aspects of nuclear energy in the following sections.

Example 4.2

Starting from the known mass values, calculate in MeV the energy released in the thermal neutron fissioning of ^{235}U for the following reaction:

$$n + {}^{235}_{92}U_{143} \rightarrow {}^{236}U \rightarrow {}^{144}_{56}Ba_{88} + {}^{89}_{36}Kr_{53} + 3n$$

You can assume the following masses in atomic mass units (1 amu = 931 MeV):

^{235}U – 235.04394	^{89}Kr – 88.91660
n – 1.00867	^{144}Ba – 143.92000

Solution

$$\text{initial mass} = 235.04394 + 1.00867 = 236.05261 \text{ amu}$$
$$\text{final mass} = 143.92000 + 88.91660 + (3 \times 1.00867) = 235.86261$$
$$\text{difference} = 0.19000 \text{ amu}$$
$$\times\ 931 \text{ MeV/amu}$$
$$\text{energy released} = 176.9 \text{ MeV}$$

4.7 THERMAL NEUTRON REACTORS

To produce usable amounts of electrical energy, it is necessary to have a controlled, self-sustaining chain reaction. This means, in effect, that there must be at least one neutron that comes from each fission reaction and that on average can then induce another fission reaction. By some circumstance (it is not clear yet whether it is fortunate or unfortunate), nature provides 2.4 and 2.9 neutrons on the average from the slow neutron fission of ^{235}U and ^{239}Pu, respectively. A chain reaction is, therefore, quite possible.

The commercial reactors developed in the United States are largely based on the use of slow or thermal neutrons, enriched uranium fuels, and normal water as a neutron energy moderator. Other varieties of reactors are possible and, in fact, have been constructed both in the United States and elsewhere, but the light (or normal) water thermal reactors are by far the most common, and we shall focus our attention on them. There are two major types of such reactors: the boiling water reactor (BWR) and the pressurized water reactor (PWR), and they differ mainly in the heat-transfer mechanisms used.

The term slow, or thermal, neutrons refers to the energy that neutrons have after they have scattered in hydrogenous material such as water a sufficient number of times. They share their energy at each scattering, so that their energy eventually becomes equal to the energy of the atoms of the material, which is just the thermal energy of motion. If a material is at room temperature (T = 20°C = 293°K), the thermal energy of the atoms and the minimum energies of the multiply scattered neutrons will be about 1/40 eV (0.025 eV). The material in which neutrons have their energy reduced by multiple scattering is known as the moderator. The neutrons at this reduced energy are designated thermal neutrons and are the basis for BWR and PWR reactors. Figure 4.4 shows the basic reason for the choice of neutron energy. It can be seen that the probability of having the ^{235}U nucleus fission increases very rapidly as the neutron energy is reduced. Thus, it is advantageous to use very low-energy neutrons; the arrow indicates the lowest energy neutrons that can conveniently be obtained; namely, thermal neutrons at 0.025 eV. Figure 4.4 also shows that the probability of having ^{238}U fission is negligible unless neutrons above 1 MeV in energy are used. Of the two isotopes of uranium that are present to any extent in nature, ^{235}U (0.72% abundant) is of direct use in thermal reactors, whereas ^{238}U (99.27% abundant) plays an indirect role that is discussed later. A thermal reactor using water as a coolant and moderator must use enriched fuel in which the ^{235}U, instead of being 0.7% of the total uranium present, has been increased to about 3%.

The neutrons that come directly from the fissioning of ^{235}U have an average energy of 2 MeV. To reduce their energy to 0.025 eV, a moderator is positioned such that it is likely that a neutron will become thermalized before it is absorbed by a nonfissionable nucleus. In the BWR and PWR reactors, water, which is both the moderator and the coolant, surrounds the rods containing the uranium fuel. To slow neutrons down quickly, it is advantageous to have them collide with a nucleus that has about the same mass as the neutron. As any billiards player knows, when a billiard ball collides with another of the same mass head-

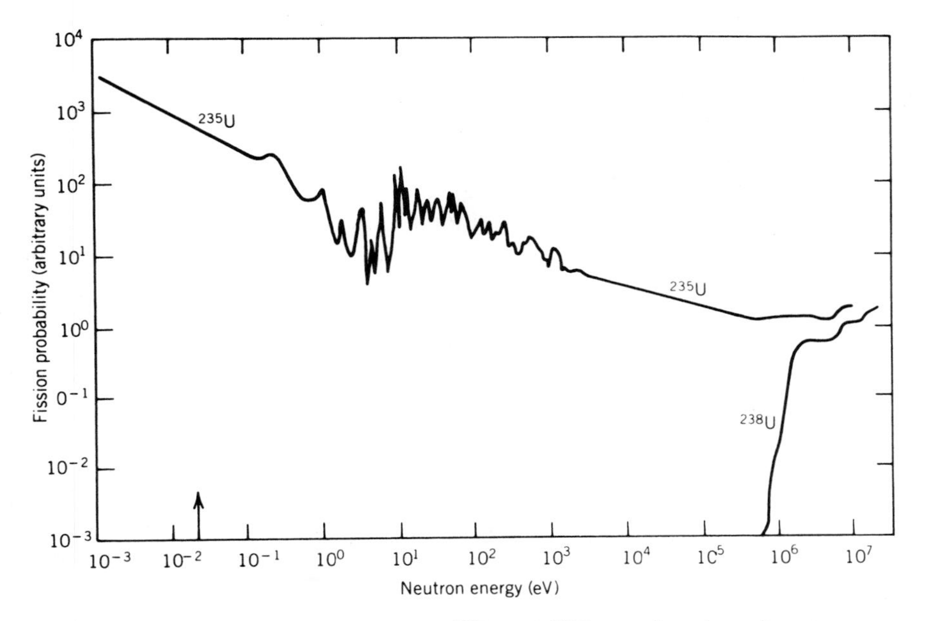

Figure 4.4 The fission probability for ^{235}U and ^{238}U as a function of neutron energy. The arrow at 0.025 eV indicates the energy of thermalized neutrons. For ^{238}U, the fission probability becomes appreciable only above 1 MeV neutron energy.

on, the first ball will stop dead while essentially all of the initial kinetic energy is transferred to the second ball. The hydrogen atom in the water molecule makes water a useful moderator material. The average distance a neutron has to go in water to slow down is only 5.7 cm, whereas in graphite (carbon) it is 18.7 cm. There is an appreciable possibility that a neutron and a proton (the nucleus of the hydrogen atom) will interact to form a deuteron, and thus the neutron is lost for further fissioning. For this reason, some reactors, such as the Canadian CANDU reactor, use heavy water (water with deuterium as the nucleus of the hydrogen atom) because the deuteron does not absorb as many neutrons as does the proton.

Of the 2.5 neutrons that on average come from the fissioning of ^{235}U, not all will survive to cause other fission events. Some of the neutrons may be captured by ^{235}U or ^{238}U with the emission of gamma rays rather than resulting in the fission process. This type of radiative neutron capture can also take place in other materials in the reactor, such as the fuel cladding or structural parts. Neutrons can also escape from the core of the reactor into the surrounding shielding or containment vessel. There is one effect that offsets some of these losses: A fast neutron can cause a fission reaction in ^{238}U, although the probability is not very great. Such an event will provide heat energy as well as contribute neutrons. Another function that ^{238}U serves in a reactor is a consequence of its capturing a thermal neutron and becoming ^{239}U. The following processes occur:

$$^{238}\mathrm{U} + \mathrm{n} \rightarrow {}^{239}\mathrm{U} \xrightarrow[(T_{1/2} = 24 \text{ min})]{\beta^-} {}^{239}\mathrm{Np} \xrightarrow[(T_{1/2} = 2.3 \text{ days})]{\beta^-} {}^{239}\mathrm{Pu}$$

The end result is ^{239}Pu that lives long enough ($T_{1/2} = 2.4 \times 10^4$ yr) to play an important role in a reactor. The thermal neutron fission probability of ^{239}Pu is even greater than that of ^{235}U. Thus, as ^{239}Pu builds up in the reactor, it can fission and contribute to the fuel of the reactor. About one third of the energy of the entire uranium-fueled reactor in one fuel cycle comes from ^{239}Pu, and at the end of the fuel cycle, 60% of the fissionings are due to ^{239}Pu.

A reactor must be designed so that, given the neutron losses noted above and steady operating conditions, exactly one neutron, on average, will be left from every fission reaction that can be moderated and finally find its way to induce a second fission. The geometrical arrangement of the reactor is clearly important. In addition, there is a certain critical size below which the reaction will not be self-sustaining because the neutron losses through the surface of the reactor are too great compared to the volume of the reactor where the reactions take place.

To control a reactor so that neutron production and fission events do not increase in an uncontrolled fashion to power levels that would damage the reactor, a set of control rods made of materials that readily absorb neutrons is inserted into the core of the reactor. If the control rods are fully inserted, the reactor will be completely shut down. The control rods also allow the reactor to be run at any desired power level. Boron has a very high probability for absorbing neutrons, and so the control rods are often made of a boron compound.

In addition to the prompt neutrons that come from a fission event, there are delayed neutrons that have their origin in the radioactive decay of some of the fission products. A few of the radioactive fission products decay by emitting a neutron rather than a beta particle. These neutron-emitting nuclei are formed by beta decay from nuclei with half-lives that range from a fraction of a second to many seconds. Of all of the neutrons in a uranium-fueled thermal reactor, about 0.5% of them are delayed. The presence of these delayed neutrons is a great stabilizing force in the control of the reactor because they do not permit the neutron population to be changed instantaneously. Were it not for these delayed neutrons, the control rods could not be moved in and out with sufficient speed to control the reactor power level.

When a reactor is to be shut down, the control rods are inserted and the prompt neutrons are sufficiently absorbed so that the reactor goes subcritical and will no longer generate power. Because of the large amount of radioactivity built up in the fuel rods, however, an appreciable amount of heat is still generated. Immediately after shutdown, after an extended operating period, somewhat over 7% of the pre-shutdown power level will remain because of the heat from the radioactivity. During normal operation, 7% of the power level is due to this radioactivity. In about 1 hour this heat production will be reduced to 1% of the normal power level. One of the necessary features of a power reactor is some means of carrying away this heat in the event that there is a loss of the primary coolant. Emergency core cooling systems have been devised for such contingencies and are discussed later.

4.8 THE BOILING WATER REACTOR

The BWR is used here to illustrate how a commercial reactor is constructed and functions. The pressurized water reactor, however, has also played an important role in both power plants and ship propulsion. In the PWR, the water in the core is maintained at sufficiently high pressure that it remains liquid. The hot pressurized water is then put through a heat exchanger to produce steam to power a turbine and generate electricity. It is a PWR design that has been used so successfully by the U.S. Navy for propulsion in more than 150 vessels, many of them submarines. In this latter application, the PWR can go 15 years without refueling because the initial fuel loading is enriched beyond the 3% common in power reactors.

Table 4.2 lists many of the specifications for a 1220-MW_e (or 1220 megawatts of electrical power generated) BWR. The reactor heats water as it flows through the core, causing the water, which is under about 71 atmospheres of pressure, to boil as it reaches the top of the core. Dry steam is then separated out and piped to the turbines for the generation of electrical power. A schematic view of the overall way that the reactor functions is shown in Figure 4.5. Figure 4.6 shows the pressure vessel and core of the reactor in more detail. The water pumps on each side circulate the water so that it can rise up through the core and become heated in the process. The fuel rods, of which there are about 46,000, have an active length of about 12 ft and are somewhat less than 0.5 in. in diameter. The fuel rods are made of zirconium alloy tubes and are filled with UO_2 pellets. The uranium, when it is first put into the reactor, is enriched to 2.8% ^{235}U and is removed when the enrichment is reduced to 0.8%, which takes a year or more. There are roughly 155 tonnes of fuel in the reactor. Approximately 54 fuel rods are assembled into a bundle; there are four bundles to a

Table 4.2 REPRESENTATIVE CHARACTERISTICS OF A BOILING WATER REACTOR

Plant electrical output	1220 MW_e
Plant efficiency	34%
Core diameter	193 in.
Core (or fuel rod) active length	150 in.
Core power density	54 kW per liter
Cladding material for fuel rods	Zircaloy-2
Fuel material	UO_2
Fuel pellet size	0.41 in. diameter × 0.41 in. long
Number of fuel rods	46,376
Control rod type	"Cruciform" control rods inserted from the bottom
Number of control rods	177
Amount of fuel (UO_2)	342,000 lb
Coolant material	water
Coolant pressure	1040 lb/in^2
Coolant temperature	551°F (288°C)
Fresh fuel enrichment	2.8% ^{235}U
Spent fuel assay	0.8% ^{235}U, 0.6% $^{239,241}Pu$
Refueling sequence	About ¼ of fuel per year to ⅓ per 18 months
Vessel wall thickness	5.7 to 6.5 in.
Vessel wall material	Manganese–molybdenum–nickel steel
Vessel diameter	19 ft 10 in.
Vessel height	71 ft

Source: Adapted from General Electric specifications.

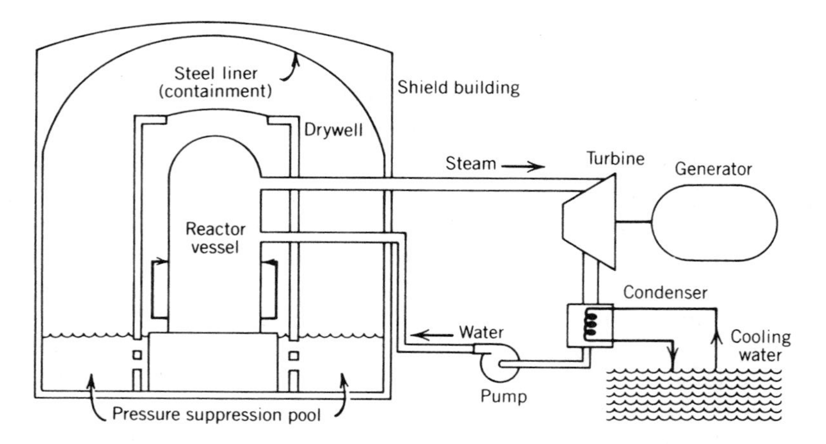

Figure 4.5 A diagram of a boiling water reactor power plant. Steam is produced in the reactor vessel and flows at high pressure to the turbine. After condensation at the low-pressure side of the turbine, it is recirculated to the reactor core. The components of the reactor containment building are discussed in the text. (Source: A. V. Nero, Jr., *A Guidebook to Nuclear Reactors*, Berkeley: University of California Press, 1979.)

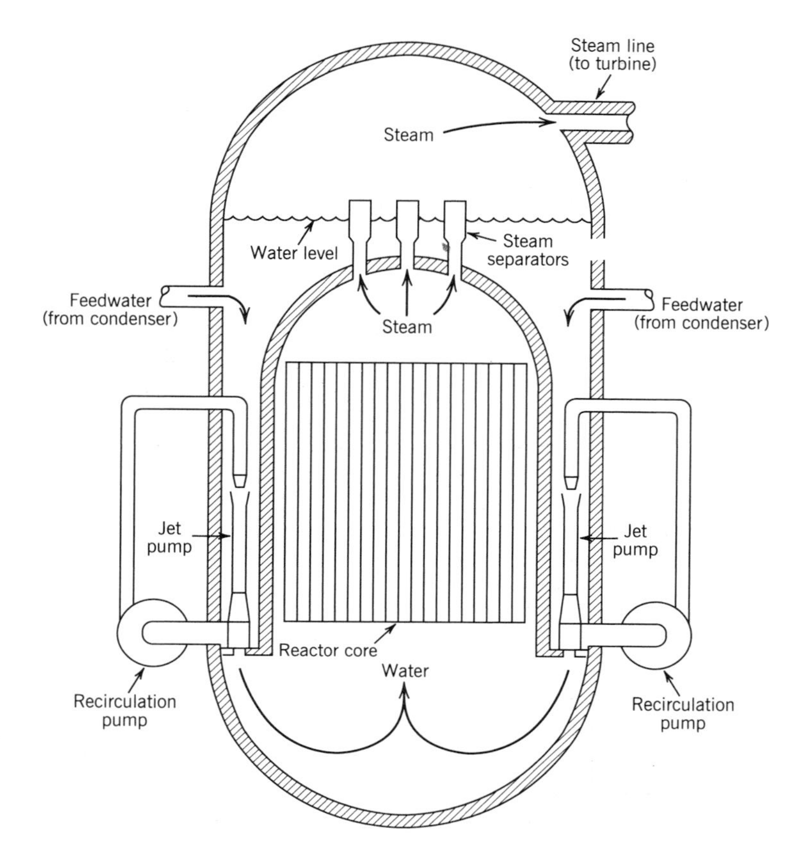

Figure 4.6 A detailed view of a boiling water reactor core and surrounding components. (Based on WASH-1250.)

module. A control rod can be inserted from the base of the reactor so that it comes up in the middle of the assembly. There are water passageways between the fuel rods, as water is not only the coolant but also the moderator. Figure 4.7 shows fuel being loaded into a reactor core.

There are, of course, a number of safety features built into the basic design of the reactor. The drywell shown in Figure 4.5 separates the reactor pressure vessel from the rest of the building. A pressure suppression pool is at the bottom of the drywell so that any coolant released as steam will be condensed in the pool, thus reducing the pressure. The steel liner shown in Figure 4.5 is provided to withstand the temperatures and pressures that would be experienced if there were a loss-of-coolant accident. The building surrounding the liner is also protection against release of radioactivity. In fact, the space between the liner and building is below atmospheric pressure so that any released material would collect there.

In the event that the control rods do not function, it is possible to add boron to the coolant water to shut down the reactor promptly. In the event that no feedwater can enter the reactor vessel, auxiliary systems are provided to keep the water level in the reactor up to normal. There are, in addition, several core spray inlets where water from the suppression pool can be sprayed into the reactor to prevent a meltdown.

Figure 4.7 Loading the first fuel bundles into the core of the nuclear reactor at the Duane Arnold Energy Center near Palo, Iowa. This 550-MW reactor uses about 19 tons of nuclear fuel per year.

4.9 URANIUM RESOURCES

In order for the U.S. nuclear program to furnish a significant fraction of the electric power generated in the country, there must, of course, be an ample supply of uranium. The lifetime of a reactor is usually taken to be 30 years, during which period about 5000 tons of uranium (U_3O_8) will be committed to a 1000-MW_e reactor. Actually, this will range between 4000 and 6000 tons, depending on the reactor design and to what extent fuel recycing is carried out.

Table 4.3 indicates the number of nuclear power plants either operating or under construction in the United States. These plants supply about 22% of the electricity in the United States. If all of these reactors come into operation and continue to operate for 30 years, about 120 times 5000 tons, or 600,000 tons, of U_3O_8, will be required. How much uranium can one expect to find in the United States?

The answer to the question depends on how many dollars a pound one is willing to spend. Yellowcake, the U_3O_8 concentrate, that in 1980 cost $30/lb or less, and was derived from ore that ranged from 0.05% to 0.5% uranium, was considered economically usable. Various estimates put the number of tons of unenriched U_3O_8 in the United States in the range of 2.4 to 2.9 million, if ore classified as speculative is not included. There is a much larger resource of lower grade ore that is less than 0.01% uranium, but at the present time the cost of processing makes it too expensive for nonbreeder reactors.

Assuming a resource of 2.5×10^6 tons and 5000 tons for the lifetime of a 1000-MW_e reactor, it is clear that fuel for 500 such reactors is available within the United States. Whether 500 reactors will ever be in operation is somewhat uncertain, because of delays in licensing and construction being encountered now. If somehow 500 large power reactors were in operation by 2000, it would not appear likely that U.S. prime uranium resources would be adequate beyond 2030.

Table 4.3 NUCLEAR POWER PLANTS IN THE UNITED STATES AS OF MARCH 1990

Reactors operating or licensed	112	100,000 MW_e
Reactors with construction permits	9	14,000 MW_e
Reactors on order	0	—
Total	121	114,000 MW_e

The power reactors presently in use in the United States use enriched ^{235}U as the primary fuel. The much more abundant isotope ^{238}U is only incidentally of any value in these reactors. In the next section, we see what kind of reactor is needed to utilize this more abundant isotope of uranium and, in so doing, extend our uranium resources many times over.

4.10 FAST BREEDER REACTORS

Many people feel that the national investment in research and development needed for a large nuclear reactor program would be poorly spent if the expiration of our ^{235}U resources brought the whole program to a halt by 2030. As we have seen, however, it is possible to convert ^{238}U into ^{239}Pu by neutron capture. Can a sufficient number of neutrons be produced in a reactor such that from each fission one neutron proceeds to induce another fission, more than one neutron is captured by ^{238}U to form ^{239}Pu, and some neutrons are by necessity lost by other processes? If such a series of events could take place, more fuel could be produced than is used. That is, for every ^{235}U or ^{239}Pu nucleus that fissions, more than one nucleus of ^{238}U is converted into ^{239}Pu.

The answer to this question depends on how many neutrons are emitted per neutron absorbed by a particular nucleus. Table 4.4 shows the number of neutrons emitted for various reactions. The interesting quantity is the number of neutrons given off per neutron absorbed. As Table 4.4 shows, ^{233}U and ^{239}Pu fuels with fast neutrons offer the greatest hope. Fast neutrons are those with kinetic energy in the MeV range. A major effort is under way in the United States and other countries to develop a fast breeder reactor with ^{239}Pu as the fissionable fuel. The ^{239}Pu would initially come from a thermal ^{235}U-fueled reactor, but later the breeder reactor would produce its own ^{239}Pu from ^{238}U. The breeding ratio is the number of ^{239}Pu atoms produced divided by the number consumed. A breeding ratio of 1 would represent a break-even point. The liquid metal fast breeder reactor (LMFBR) proposed at Clinch River, Tennessee, is designed to have a breeding ratio of about 1.2; that is, 20% more

Table 4.4 NEUTRONS LIBERATED PER NEUTRON CAPTURE

	Neutrons per Thermal Neutron Induced Fission	Neutrons per Thermal Neutron Absorbed	Neutrons per Fast Neutron Absorbed
^{233}U	2.50	2.27	2.60
^{235}U	2.43	2.06	2.18
^{239}Pu	2.90	2.10	2.74

^{239}Pu atoms are produced from ^{238}U than are burned up. The overall effect is that the ^{238}U, which is now not a primary fuel component in thermal reactors, will become a source of fuel. Instead of just 0.7% of the uranium being useful, essentially all of it would be. Our resource lifetime, which we earlier estimated at about 30 years, would be more than 140 times greater. Because the anticipated fuel costs of a breeder reactor are so small, it is expected that much lower grade uranium ore will become useful. Estimates of about 50,000 years have been made for the time it would take to deplete the uranium resources in the United States with a full-scale breeder reactor program.

Although reasonably large fast breeder reactors have been built in France, Great Britain, and Russia, construction of the Clinch River LMFBR has been suspended. Fast breeder reactors differ in a number of ways from the light water thermal reactors that have been described. One of the major differences is that since the reactors operate primarily with fast neutrons, neutron moderators play a reduced role. However, although the fission of ^{239}Pu is induced in these reactors primarily by fast neutrons, breeding occurs as a result of neutron capture in ^{238}U (or possibly ^{232}Th) at much lower neutron energies, certainly below 1 keV. The necessary moderation of the neutron energies results from neutron scattering within the reactor, even though no moderating material is deliberately included in the design. The concept of the breeder reactor is made possible by the natural presence on earth of ^{232}Th and ^{238}U, the only known fertile nuclei; that is, nuclei that can capture a neutron to form fissionable nuclei.

Because of the very high power densities in the reactor core, a coolant with superior heat-transfer properties is essential. Liquid sodium metal was selected for that reason as well as a number of others: It can be run at high temperatures at atmospheric pressure, and it will not quickly moderate the neutron energies because of its higher mass number. Liquid sodium, however, is chemically very reactive, and great care must be taken to see that it is never exposed to the atmosphere or to water. It also becomes intensely radioactive. Because of its radioactivity, two heat exchangers are used in tandem to generate steam for the turbines. Figure 4.8 shows schematically the arrrangement of the heat exchangers.

The fission probabilty is so low for fast neutrons in ^{239}Pu that it is necessary for the fuel in a fast breeder reactor to be considerably more enriched than in a thermal reactor. The Clinch River reactor design has about 15% of the contents of the fuel rod fissionable, compared to about 3% in a thermal light water reactor. The fuel, in the form of plutonium oxide, is packed into the center section of a very thin fuel rod (0.6 cm diameter) with ^{238}U on both the upper and lower ends of the fuel rod. The ^{238}U thus forms a blanket surrounding the plutonium fuel. The other components of the LMFBR are similar in function to the BWR discussed above, but the details are quite different.

4.11 NUCLEAR FUSION AS AN ENERGY SOURCE

We have just seen that nuclear fission, which rearranges nucleons from one large cluster (a heavy nucleus such as uranium) into two lighter clusters (a pair of medium-mass nuclei such as a barium and krypton), results in the release of energy. This energy release comes about as the nucleons in effect

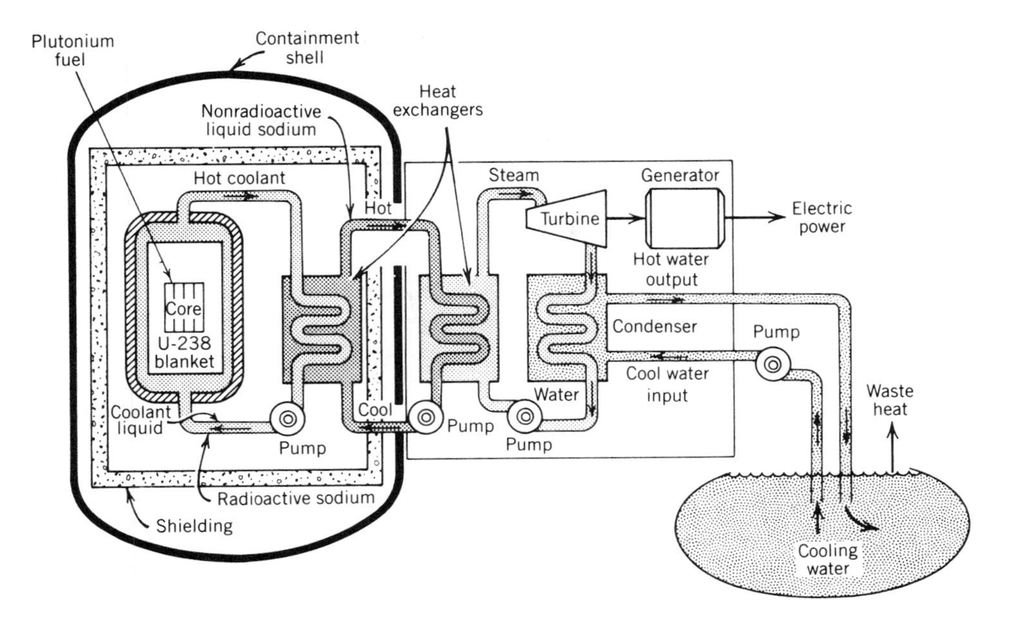

Figure 4.8 A liquid metal fast breeder reactor (LMFBR). Both the primary and secondary heat-transfer loops use circulating liquid sodium. (Based on ERDA 76-107).

move from one position along the curve of binding energy to another position, with their final position corresponding to greater binding energy per nucleon than did their initial position. There is a second obvious possibility for rearranging nucleon configurations such that greater binding energy per nucleon is achieved, with an attendant release of energy. This can occur if one starts with a pair of light nuclei, to the left of the maximum in the curve of binding energy, and combines them into one heavier nucleus at a position on this curve corresponding to greater binding energy per nucleon. This combining, or fusing, of two light nuclei into a heavier one is known as nuclear fusion.

The nuclear energy production processes in the sun and other stars provide a good example of the conversion of nuclear mass-energy into other forms of energy. There are a number of nuclear fusion reactions occurring continuously within the sun. The proton–proton cyce is one of the most fundamental and is used here to illustrate how hydrogen nuclei can be fused into helium nuclei, with an accompanying release of energy.

This reaction proceeds by first fusing two hydrogen nuclei (protons) into a deuterium nucleus:

$$ {}^{1}_{1}\mathrm{H}_0 + {}^{1}_{1}\mathrm{H}_0 \rightarrow {}^{2}_{1}\mathrm{H}_1 + \beta^+ + \nu + \text{energy} \qquad (1)$$

In this process one proton becomes a neutron within the deuterium nucleus; this fusion process requires that $p \rightarrow n + \beta^+ + \nu$. The positive electron, or positron (β^+), and the neutrino (ν) must be emitted in the reaction to conserve detailed properties of nuclear reactions such as charge, energy, and angular momentum. These concepts are not discussed here. At this point it is only necessary to observe that two nucleons have been moved upward on the curve of binding energy such that energy is released. This energy appears as kinetic energy of the three reaction products ($^2_1H_1$, β^+, ν). In order for the reaction (1) to occur at a reasonable rate, it is necessary to have protons present in the sun at sufficient density so that one moving proton is likely to encounter another. Also, the velocity with which the protons move about in random directions and with which they eventually collide must be great enough to overcome the repulsion that the two positively charged protons have for each other. The requisite dense concentration of protons occurs because of gravitational confinement due to the large mass of the sun. The velocities with which the protons collide are due to the temperature of 2×10^{7}°K in the solar interior. This thermal component of reaction (1) is the reason it is known as a thermonuclear reaction. The temperature is also responsible for the electrons being stripped from otherwise neutral hydrogen atoms, leaving the bare nuclei, or protons, free to interact with each other. The electrically neutral mixture of dissociated electrons and ions is known as a plasma.

Once reaction (1) is accomplished, there are both protons and dueterons moving about in the solar interior. This leads to another reaction:

$$^1_1H_0 + {}^2_1H_1 \rightarrow {}^3_2He_1 + \text{energy} \tag{2}$$

in which a mass-3 nucleus of helium is formed. Two of these helium nuclei can then interact to form a mass-4 helium nucleus plus two protons:

$$^3_2He_1 + {}^3_2He_1 \rightarrow {}^4_2He_2 + 2\,{}^1_1H_0 + \text{energy} \tag{3}$$

If we now let reactions (1) and (2) occur twice and reaction (3) occur once, we have an interesting reaction sequence, or cycle:

$$2\,{}^1_1H_0 + 2\,{}^1_1H_0 \rightarrow 2\,{}^2_1H_1 + 2\beta^+ + 2\nu \tag{1'}$$

$$2\,{}^1_1H_0 + 2\,{}^2_1H_1 \rightarrow 2\,{}^3_2He_1 \tag{2'}$$

$$2\,{}^3_2He_1 \rightarrow {}^4_2He_2 + 2\,{}^1_1H_0 \tag{3'}$$

Looking at all three of these reactions, we have on the left $6\,{}^1_1H_0$, $2\,{}^2_1H_1$, and $2\,{}^3_2He_1$. On the right we have $2\,{}^1_1H_0$, $2\,{}^2_1H_1$, $2\,{}^3_2He_1$, and one $^4_2He_2$, plus $2\beta^+$ and 2ν. As a *net* result, four protons are put into this reaction cycle and one helium nucleus results. The cycle can be summarized as follows:

$$4\,{}^1_1H_0 \rightarrow {}^4_2He_2 + 2\beta^+ + 2\nu + \text{energy}$$

or, more simply, as

$$4p \rightarrow \alpha + 2\beta^+ + 2\nu + \text{energy}$$

In effect, hydrogen nuclei are burned into helium nuclei, releasing energy during the process. There are no radioactive reaction products.

From the curve of binding energy, it can be seen that in going from protons to alpha particles, the binding energy per nucleon increases by 7.1 MeV. Taken as a fraction of a nuclear mass (1 amu = 931 MeV), this is 7.1 MeV/931 MeV

= 0.76%; one part in 132 of the mass of each proton participating in this reaction is converted to energy. This fraction is about eight times greater than that achieved in nuclear fission and about 10^8 times greater than in typical chemical reactions. Thus, there is increasing destructiveness as we go from chemical to fission to thermonuclear bombs, and there is an increasing ratio of energy provided per unit mass of fuel as we go from fossil fuel burning to uranium fission to (possibly) hydrogen fusion power plants.

Each day our sun burns 5.3×10^{16} kg of hydrogen nuclei into alpha particles. Since only 0.76% of the total participating mass is converted into energy, we can calculate the mass converted into energy to be

$$5.3 \times 10^{16} \text{ kg} \times 7.6 \times 10^{-3} = 4.03 \times 10^{14} \text{ kg}$$

The energy released daily by this process, then, is

$$\begin{aligned} E = mc^2 &= (4.03 \times 10^{14})(3 \times 10^8)^2 \\ &= 3.6 \times 10^{31} \text{ J/day} \end{aligned}$$

4.12 CONTROLLED THERMONUCLEAR REACTIONS

Our use of energy from thermonuclear reactions occurs daily as we enjoy the many benefits of solar energy. But we would profit further if we could somehow produce fusion reactions at a controlled rate, so that we had available an even flow of energy day and night, summer and winter. We can routinely produce fusion reactions in any nuclear physics laboratory equipped with an accelerator, such as a cyclotron. This laboratory process is hopelessly slow and inefficient; the accelerator operation alone consumes more energy than is released by the relatively infrequent reactions. We can also routinely release thermonuclear energy by detonating hydrogen bombs. However, it has not yet been possible to harness this energy release in a carefully controlled manner. A scheme has been proposed (Project Pacer) in which hydrogen bombs would be periodically exploded in a deep underground cavern with the heat energy being used to generate electricity. This scheme has not gone beyond the discussion stage.

Our hopes for controlled thermonuclear reactions (CTRs) now hinge on a small number of possible reactions involving the fusion of hydrogen nuclei. The solar proton–proton cycle is far too slow an energy producer for interest on earth. The energy production rate per gram of fuel, although impressive in a body as large as a star, is only about 1% that of the metabolic heat production of the human body. On earth we hope to use the deuterium–tritium (D–T) reaction, written as follows:

$$^{2}_{1}H_1 + ^{3}_{1}H_2 \rightarrow ^{4}_{2}He_2 + n + 17.6 \text{ MeV}$$

or

$$D + T \rightarrow \alpha + n + 17.6 \text{ MeV}$$

The energy released (17.6 MeV) is divided between the kinetic energy of the alpha particle and the neutron, with the neutron carrying off 14.1 MeV of the 17.6 MeV released. The D–T reaction has a minimum ignition temperature of $40 \times 10^{6\circ}$K. The corresponding kinetic energies of the deuteron or triton can be computed according to an equation derived from the equipartition of energy, a result of thermodynamic considerations. For any temperature, each particle present has an average kinetic energy given by

$$E_{\text{kin}} = \frac{3}{2}kT$$

where k is Boltzmann's constant

$$k = 1.38 \times 10^{-23} \text{ J/}^\circ\text{K}$$

Thus, for a deuteron or triton at $40 \times 10^{6\circ}$K we have

$$E_{\text{kin}} = \frac{3}{2} \times (1.38 \times 10^{-23} \text{ J/}^\circ\text{K}) \times (40 \times 10^{6\circ}\text{K})$$

$$= 8.3 \times 10^{-16} \text{ J}$$

or

$$E_{\text{kin}} = 8.3 \times 10^{-16} \text{ J} \times \frac{1}{1.6 \times 10^{-19} \text{ J/eV}} = 5000 \text{ eV}$$

$$= 5 \text{ keV}$$

or a modest energy by laboratory standards.

Another reaction of interest is the D–D reaction, which has two branches of almost equal probability:

$$\text{D} + \text{D} \rightarrow {}^{3}_{2}\text{He}_{1} + \text{n} + 3.3 \text{ MeV}$$

and

$$\text{D} + \text{D} \rightarrow {}^{3}_{1}\text{H}_{2} + \text{p} + 4.0 \text{ MeV}$$

For this reaction, the ignition temperature is very high, in excess of $100 \times 10^{6\circ}$K. The D–D reaction has the attractive feature that the only fuel needed is deuterium, an abundantly available stable isotope of hydrogen. Of course, once the D–D reaction starts, the tritons produced mean that the D–T reaction will also be present. In normal water, such as the world's oceans, there is one deuterium atom per 6500 normal hydrogen atoms. A simple calculation shows that a mere 1% of the deuterium in the world's oceans has an energy equivalent, through the D–D reaction, of 500,000 times the Q_∞ for the world's fossil fuels. If D–D CTRs could be achieved, our energy needs could be met for millions of years. However, because of the high ignition temperature required, this reaction is presently far from feasible as a means of energy production.

Example 4.3

Calculate the amount of energy in joules that could be obtained if all of the deuterons ($^2_1H_1$) in the oceans were used in the D–D fusion reaction to generate electricity. Assume the following:

1. Area of oceans $= 1.2 \times 10^8$ km^2.
2. Average depth $= 1.2$ km.
3. Water has density of 1 g/cc $= 10^3$ kg/m^3.
4. One out of 7000 hydrogen atoms is deuterium.
5. There are 6×10^{23} molecules in 18 g of H_2O.
6. Four MeV is released for every D–D reaction.

Solution

$$\begin{aligned}(\text{total energy}) = {} & (\text{number of deuterons reacting}) \\ & \times (\text{energy released per reaction}) \\ & \times \left(\frac{1\ \text{reaction}}{2\ \text{deuterons}}\right)\end{aligned}$$

$$(\text{number of deuterons reacting}) = 1.2 \times 10^8\ \text{km}^2 \times 1.2\ \text{km} \times 10^9\ \frac{\text{m}^3}{\text{km}^3}$$

$$\times\ 10^3\,\frac{\text{kg}}{\text{m}^3} \times 10^3\,\frac{\text{g}}{\text{kg}} \times \frac{6 \times 10^{23}\ \text{molecules}}{18\ \text{g}} \times 2\frac{\text{hydrogen atoms}}{\text{molecule}}$$

$$\times\ \frac{1\ \text{deuterium atom}}{7000\ \text{hydrogen atoms}} = 1.37 \times 10^{42}\ \text{deuterons}$$

$$\begin{aligned}\text{total energy} &= (1.37 \times 10^{42}) \times (4\ \text{MeV}) \times (1.6 \times 10^{-13}\ \text{J/MeV}) \times {}^1\!/_2 \\ &= 4.38 \times 10^{29}\ \text{J}\end{aligned}$$

Note that this is equal to 4×10^{11} QBtu or 5×10^9 times the total U.S. annual energy consumption.

This brings us back to the D–T reaction as the focus of current efforts in controlled fusion. In order for our energy needs to be met through the D–T fusion reaction, we must have abundant supplies of both deuterium and tritium. As we have just seen, the deuterium supply is no problem, but the tritium for the initial D–T fusion reactors must be obtained from fission reactors that can produce large amounts of tritium by a neutron-induced reaction on lithium. The tritium-producing reaction in the fission reactor is

$$n + {}^6\text{Li} \rightarrow \alpha + T$$

Once both deuterium and tritium are available, the D–T reaction can be supplemented by the presence of lithium-6, a stable isotope found in nature, so that the tritium supply will be self-sustaining. We then have

$$\text{D} + \text{T} \rightarrow \alpha + \text{n} + 17.6 \text{ MeV}$$

and

$$\text{n} + {}^{6}_{3}\text{Li}_3 \rightarrow \alpha + \text{T} + 4.8 \text{ MeV}$$

The first reaction provides neutrons for the second, and the second reaction provides tritons for the first. This process is schematically illustrated in Figure 4.9.

The reaction sequence can be summarized as follows:

$$\text{D} + {}^{6}_{3}\text{Li}_3 \rightarrow 2\alpha + 22.4 \text{ MeV}$$

where alpha particles are the only reaction products. The fuels for this reaction are deuterium and lithium, so the Q_∞ for this reaction depends on whether we first run out of deuterium or lithium.

The world's oceans are known to contain about 10^{42} atoms of deuterium. Lithium-6 constitutes about 6.5% of natural lithium, which is present at a level of about 7 parts per billion in seawater, about 2 parts per million in many rock formations, and about 4 parts per million in the Great Salt Lake in Utah. The U.S. reserves represent 10^7 tons, including about 6.5×10^{34} atoms of ^{6}Li. Since the D–T reaction requires equal numbers of deuterium and lithium atoms, it is apparent by reference to the number of deuterons given above that we shall run out of lithium before we have used even a tiny fraction of the deuterium.

From each D–T reaction cycle, 22.4 MeV (3.6×10^{-12} J) are released. When all the ^{6}Li atoms in the U.S. reserves are used up, the energy release will be $6.5 \times 10^{34} \times 3.6 \times 10^{-12}$ J $= 2.3 \times 10^{23}$ J for the United States alone. This compares to the world Q_∞ for fossil fuels of about 2.7×10^{23} J; so it can be seen that the D–T reaction is not a permanent solution. United States lithium

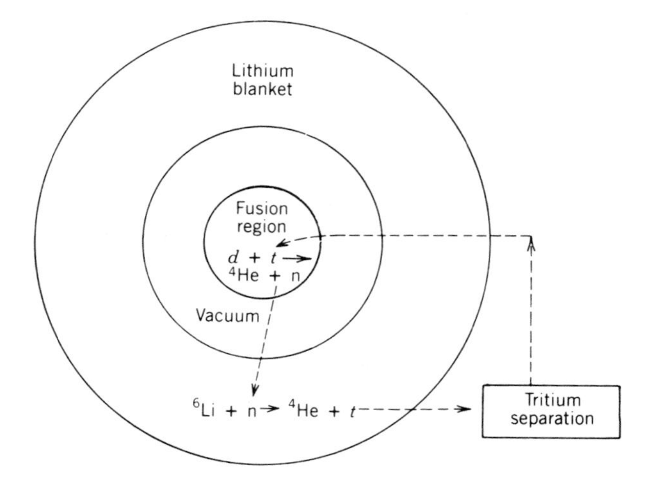

Figure 4.9 A schematic D–T fusion reactor. Deuterium and tritium react in the high-temperature fusion region, producing 14 MeV neutrons that penetrate to a surrounding blanket of liquid lithium. The neutrons deposit their kinetic energy in the lithium and induce reactions that provide a source of tritium for the fusion reaction. Heat is then removed from the lithium blanket by a circulating coolant and used to drive a steam turbine.

resources would probably last for several hundred years; after that the world's oceans could probably provide lithium fuel for several thousands of years.

4.13 A FUSION REACTOR

Current laboratory efforts are focused on producing a plasma at sufficient temperature and particle density, held together long enough, that the fusion reaction is a net producer of energy. At the temperature encountered in a fusion reactor, the energy of the ions is so high that the atoms of deuterium and tritium have lost all of their electrons. This electrically neutral mixture of dissociated electrons and nuclei at high temperature is known as a plasma. The binding energies of the electrons are only 13.6 eV, whereas the energies of the ions are some thousand times as great. It has been demonstrated that for the D–T reaction to be a useful producer of energy, the product of particle density and the time period during which the particles are confined at this density must exceed a certain minimum value. This finding is known as the Lawson criterion, which specifies that $n\tau > 10^{14}$ sec/cm^3, where n is the particle density in number per cubic centimeter and τ is the confinement time in seconds. This criterion could be met by having 10^{14} particles/cm^3 held together for 1 sec, 10^{15} particles/cm^3 confined for 0.1 sec, or by any other combination of density and time giving a product equal to or greater than 10^{14} sec/cm^3. For comparison, normal air has a density of about 10^{19} molecules/cm^3, which is several orders of magnitude greater than the density anticipated in fusion reactors. Of course, the necessary high temperature must also be achieved in addition to the Lawson criterion.

If and when the fusion reaction can be demonstrated as a useful energy producer on a laboratory scale, a later step will be the construction of a fusion reactor system for producing electricity for our distribution system. A diagram of such a system is shown in Figure 4.10. The fusion plasma region shown in this figure is the region where the plasma is confined by magnetic fields or some other means. The high-temperature plasma region must be isolated from contact with the walls of any container, not so much because the container would be destroyed if brought up to the temperature of the hot plasma, but because the plasma would be cooled on contact and the reaction extinguished. The energy density in the fusion plasma is not as high as might be supposed; it is one or two orders of magnitude less than in fission reactor and inadequate to destroy materials.

One common approach to confinement and isolation from container walls is to use magnetic fields. Charged particles, as in a plasma, cannot move across strong magnetic fields. If the fields are properly shaped, the plasma can be confined within the desired region. One such confinement scheme, shown in Figures 4.11 and 4.12, is the Tokamak (from the Russian *to*—toroidal, *ka*—chambers, *mak*—magnetic), which was initially developed in the Soviet Union and is currently the most promising magnetic device.

Other proposed confinement schemes, classified as inertial confinement, include laser-induced fusion and heavy ion-induced fusion. The laser method is illustrated in Figure 4.13. The heavy-ion method is similar, but beams of particles from accelerators, rather than laser beams, converge on a D–T pellet. These inertial schemes are still in the early development stages.

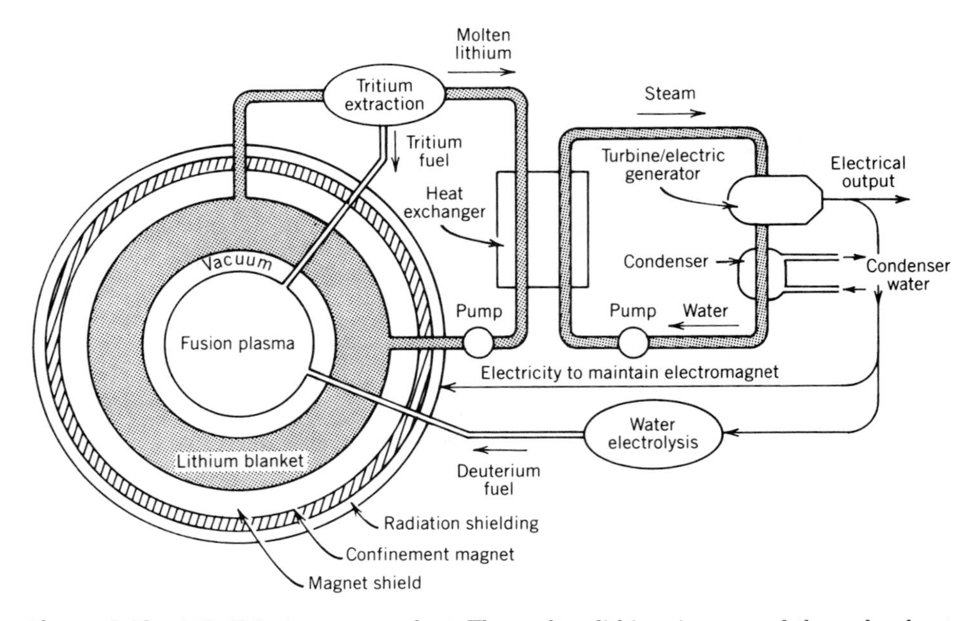

Figure 4.10 A D–T fusion power plant. The molten lithium is pumped through a heat exchanger that produces high-pressure steam to drive a turbine. The fusion plasma and lithium blanket regions are in the shape of a torus, a cross section of which is shown.

Fusion reactors will produce neither the problematic emissions now experienced from fossil fuel–burning plants nor the long-lived fission fragments and transuranic elements resulting from fission reactors. They will, however, contain a large inventory of radioactive tritium and, during operation, the reactors' structural components will become radioactive to varying degrees. Neither of these factors, however, is considered to present anything near the hazards associated with fission reactors. Studies that have considered lithium fires, failure of magnets, and other possibilities have concluded that there would be no severe damage outside the plant itself and little potential for damage to the surrounding community. Such findings indicate further possible economies in the use of waste heat and in power distribution, as the reactors could be located within cities, near the point of need of both heat and electricity.

At this time, it is by no means evident that we shall ever achieve controlled fusion. The optimists see progress over the past decades, but others point out that there still remain at least four barriers. These are proof of scientific feasibility of the basic reactions, economics of electric power generation, materials damage to the reactor due to the intense neutron flux, and materials availability for the reactor construction. The last point arises primarily with regard to the metals: beryllium, manganese, chromium, nickel, tin, niobium, and vanadium, some of which are needed for superconducting magnet construction. Reliance on foreign markets is expected for these materials, and it is possible that not even the world's resources would be adequate for large-scale fusion power generation. The other points are undergoing intense engineering and scientific effort, and several ideas are emerging. The overall prospects for the

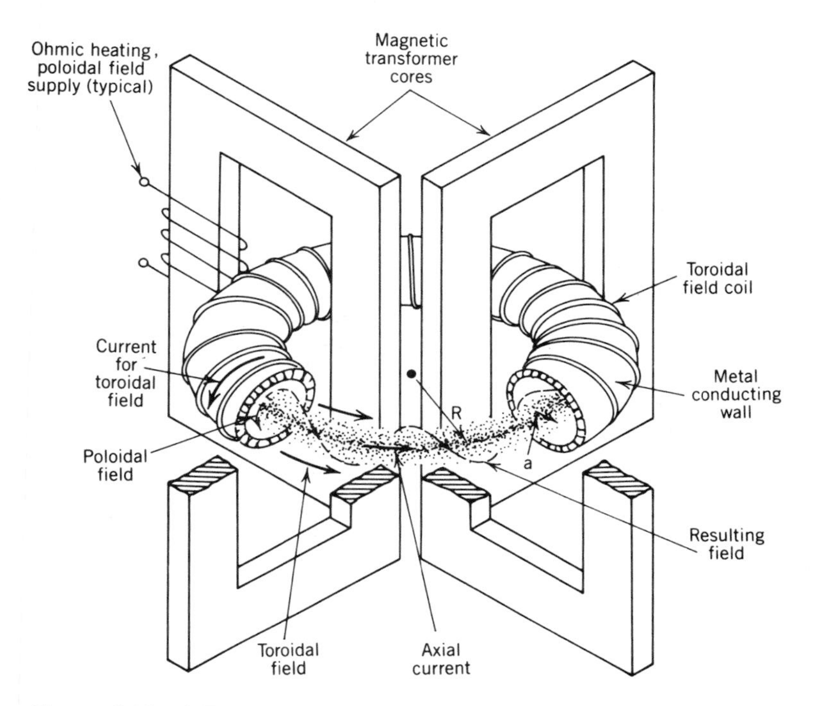

Figure 4.11 A Tokamak magnetic confinement fusion system. A magnetic field confines the plasma within a torus.

fusion power program are summarized in Figure 4.14, a schedule put forth in 1976. It is now apparent that this figure presents a too optimistic picture of our prospects. As yet there is no sign that the plasma test reactor experiments are leading to an experimental power reactor.

Figure 4.12 The Tokamak fusion test reactor recently brought into operation at the Princeton University Plasma Physics Laboratory.

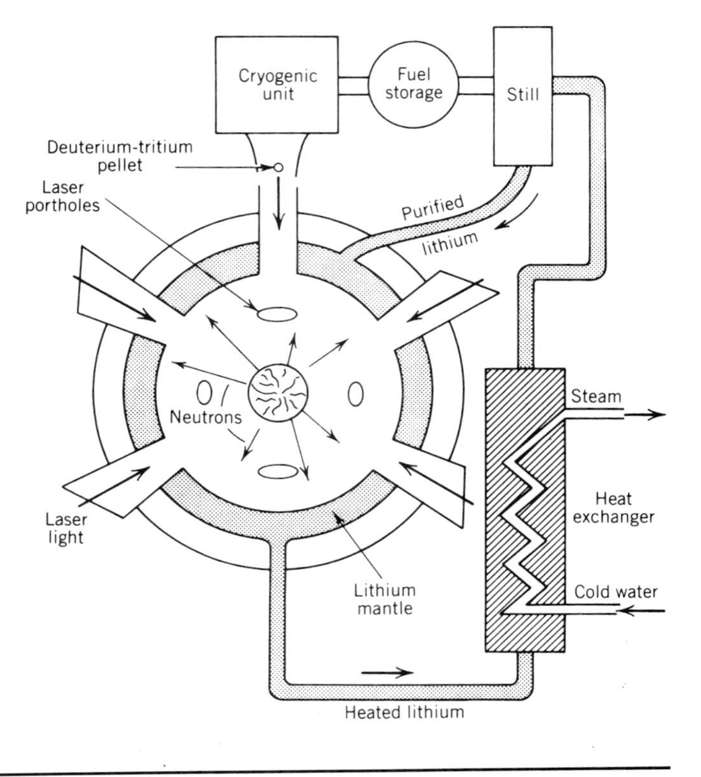

Figure 4.13 A cutaway view of a proposed laser fusion reactor. A microscopic deuterium–tritium pellet is injected into the reaction chamber several times per second. Precisely timed and sharply focused pulsed laser beams of high intensity converge in the chamber, thereby creating the requisite conditions of temperature and density to initiate fusion. As in the magnetic confinement schemes, the resulting energetic neutrons are absorbed in a surrounding mantle of molten lithium, which is circulated to remove the heat energy. The tritium produced in the lithium mantle is separated out and incorporated into new D–T pellets.

PROBLEMS

1. Calculate the number of joules that can be obtained by the fissioning of 1 kg of ^{235}U assuming 198 MeV average energy released per fission. How much energy can be obtained from the ^{235}U in 1 kg of natural uranium, which is only 0.72% ^{235}U?
2. Calculate how much energy, in joules, can be obtained from burning 1 kg of coal.
3. Compare the total energy in the U.S. resources of coal and ^{235}U.
4. What is the mass energy of your pencil in kW • hr?
5. Fission fragment X with an activity of 4.0 megacuries has contaminated a room, exposing the workers in the room to 1.0 rem per year. The half-life of X is 10 years. In how many years will the dose rate be down to 250 mrem per year, if we rely only on its radioactive decay to diminish its activity? (Exact definitions of rem and curie are presented in Chapters 5 and 11 but are not needed here.)
6. Calculate the binding energy of the deuteron in units of MeV given the following information on masses in atomic mass units:

$$\begin{aligned} \text{neutron} &- 1.008665 \\ {}^1_1H_0(\text{proton}) &- 1.007825 \\ {}^2_1H_1(\text{deuteron}) &- 2.014102 \end{aligned}$$

7. In your own words, what is a:
   (a) proton?
   (b) neutron?
   (c) electron?
   (d) beta particle?
   (e) gamma ray?
   (f) alpha particle?

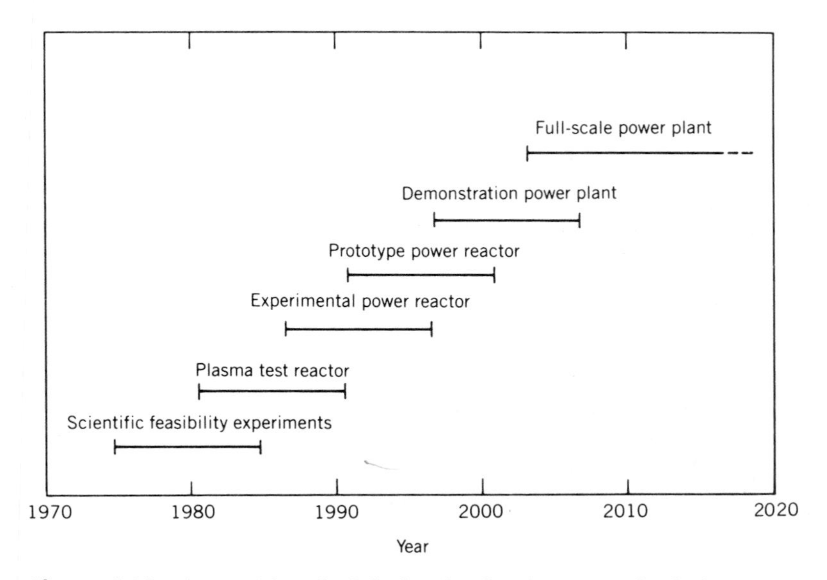

Figure 4.14 A possible schedule for the development of a fusion power plant.

8. The following reaction takes place:

$$^{9}_{4}\mathrm{Be}_{5} + ^{4}_{2}\mathrm{He}_{2} \rightarrow ^{12}_{5}\mathrm{B}_{7} + X$$

What is X?

9. In the following reaction for the spontaneous fission of $^{238}\mathrm{U}$, apply an appropriate conservation law to determine the number of neutrons emitted.

$$^{238}_{92}\mathrm{U}_{146} \rightarrow ^{140}_{55}\mathrm{Cs}_{85} + ^{92}_{37}\mathrm{Rb}_{55} + (?)\ \mathrm{n}$$

10. (a) Approximately how many MeV *per nucleon* are liberated in the typical uranium fission event?
    (b) What fraction of the total mass-energy of the uranium nucleus is thus liberated?

11. Draw the relevant portion of the chart of the nuclides (a plot of proton number versus neutron number) near uranium, labeling both axes carefully. Indicate with carefully drawn and labeled arrows the transitions which each of the three nuclei below would make on the diagram if it underwent the specified nuclear reaction. Be sure to label both the initial and final nuclei in each case. Some relevant atomic numbers are Th = 90, Pa = 91, U = 92, Np = 93 and Pu = 94.
    (a) $^{232}\mathrm{Th}$ captures a neutron.
    (b) $^{237}\mathrm{U}$ undergoes β^- decay.
    (c) $^{238}\mathrm{U}$ emits an alpha particle.

12. It has been estimated that in the year 2010, the United States will need 10^{12} W of electrical power produced by one thousand 1000-MW$_e$ power plants. These would most likely be powered either by coal or by nuclear reactors.
    (a) If the energy is provided by $^{235}\mathrm{U}$-burning reactors, how many metric tons of the isotope $^{235}\mathrm{U}$ will be needed annually? Assume an efficiency of 33%. One metric ton equals 10^6 g.

(b) How many total metric tons of uranium must be mined annually?

(c) If the energy is provided by coal, how many metric tons of coal will be needed annually? Again assume an efficiency of 33%.

(d) If the coal is obtained from land that provides 170,000 metric tons per acre (as would be true for a 100-ft thick seam), how many square miles of land would be mined annually? One square mile equals 640 acres.

13. Calculate the binding energy per nucleon in MeV for a typical fission product nucleus ${}^{140}_{55}Cs_{85}$.

$$\text{Mass of } {}^{140}Cs - 139.917110 \text{ amu}$$

14. If both fission product nuclei had the same binding energy per nucleon as ${}^{140}Cs$, as calculated above, and if three neutrons were released, how much energy would be released on the thermal neutron-induced fissioning of ${}^{235}U$?

15. The following D–T reaction is the basis for obtaining energy from fusion at the present time. Calculate from the masses given in atomic mass units how much energy is released in MeV.

$${}^{2}_{1}H_1 + {}^{3}_{1}H_2 \rightarrow {}^{4}_{2}He_2 + {}^{1}_{0}n_1 + Q$$

$${}^{2}_{1}H_1 - 2.014102 \qquad {}^{4}_{2}He_2 - 4.002603$$

$${}^{3}_{1}H_2 - 3.016066 \qquad {}^{1}_{0}n_1 - 1.008665$$

16. Calculate the energy released in MeV in the D–D reaction (${}^{2}_{1}H_1 + {}^{2}_{1}H_1 \rightarrow {}^{3}_{1}H_2 + {}^{1}_{1}H_0$) with the following masses in amu:

$${}^{2}_{1}H_1 : 2.014102$$

$${}^{3}_{1}H_2 : 3.016050$$

$${}^{1}_{1}H_0 : 1.007825$$

17. Sketch a boiling water reactor, label the important components within the pressure vessel, and write a sentence explaining the major function(s) of each.

18. Using the complete chart of the nuclides, explain:

(a) Why are neutrons, and not protons, emitted by the fission process?

(b) Why are the fission product nuclei usually radioactive?

19. In your own words, describe the curve of binding energy and its meaning.

20. What conditions are necessary for operation of a fission reactor? Include description of concepts such as critical mass, enrichment, moderation, etc.

SUGGESTED READING AND REFERENCES

1. Choppin, G. R., and Rydberg, J. *Nuclear Chemistry—Theory and Applications.* New York: Pergamon Press, 1980.
2. Crawley, Gerald M. *Energy.* New York: Macmillan, 1975.
3. Devins, D. W. *Energy, Its Physical Impact on the Environment.* New York: John Wiley, 1982.
4. Fowler, John M. *Energy and the Environment.* New York: McGraw-Hill, 1975.

5. Hottel, H. C., and Howard, J. B. *New Energy Technology—Some Facts and Assessments.* Cambridge, Mass.: M.I.T. Press, 1971.

6. Inglis, David Rittenhouse. *Nuclear Energy—Its Physics and Its Social Challenge.* Reading, Mass.: Addison-Wesley, 1973.

7. Krenz, Jerrold H. *Energy—Conversion and Utilization*, Second Edition. Boston: Allyn and Bacon, 1984.

8. Lovins, Amory B. *Soft Energy Paths—Towards a Durable Peace.* New York: Harper Colophon, 1977.

9. Marion, Jerry B. *Energy in Perspective.* New York: Academic Press, 1974.

10. Nero, Anthony V., Jr. *A Guidebook to Nuclear Reactors.* Berkeley: University of California Press, 1979.

11. Priest, Joseph. *Problems of Our Physical Environment.* Reading, Mass.: Addison-Wesley, 1973.

12. Romer, Robert H. *Energy and Introduction to Physics.* San Francisco: W. H. Freeman, 1976.

13. Ruedisili, Lon C., and Firebaugh, Morris W. *Perspectives on Energy.* New York: Oxford University Press, 1978.

14. Saperstein, Alvin M. *Physics: Energy in the Environment.* Boston: Little, Brown, 1975.

15. Stobaugh, Robert, and Yergin, Daniel, Eds. *Energy Future—Report of the Energy Project at the Harvard Business School.* New York: Random House, 1979.

16. Stoker, H. Stephen; Seager, Spencer L.; and Capener, Robert L. *Energy from Source to Use.* Glenview, Ill.: Scott, Foresman, 1975.

17. Wilson, Richard, and Jones, William J. *Energy, Ecology, and Environment.* New York: Academic Press, 1974.

18. The Nuclear Energy Policy Study Group. *Nuclear Power Issues and Choices.* Cambridge, Mass.: Ballinger Publishing, 1977.

19. "Amory Lovins and His Critics." Hugh Nash, Ed. *The Energy Controversy.* San Francisco: The Friends of the Earth, 1979.

20. Final Report of the Committee on Nuclear and Alternative Energy Systems (CONAES). *Energy in Transition 1985–2010.* San Francisco: W. H. Freeman, 1979.

CHAPTER FIVE

ENVIRONMENTAL AND SAFETY ASPECTS OF NUCLEAR ENERGY

5.1 INTRODUCTION

We saw in Chapter 4 that by 1990 there were no new power reactors on order for the generation of electricity in the United States. Whatever nuclear power we will have at the turn of the century will in all likelihood be provided by reactors now operating or under construction.

Projections for the amount of installed nuclear reactor electric generating capacity in the United States in the year 2000 have decreased drastically since 1970. Official predictions made in 1970 called for about 1200 GW_e (or 1200 gigawatts of electrical generating capacity) by the year 2000. Ten years later, the predictions were for about 200 GW_e installed by the year 2000. In 1990 there were 112 reactors in operation and nine under construction. These 121 reactors have a total capacity of 114 GW_e. By the year 2000 the nuclear reactor generating capacity will possibly be even less than 114 GW_e, as some reactors are being shut down and there are no new reactors on order.

The promise made years ago of large amounts of inexpensive clean energy from the nucleus is not becoming a reality. What has caused this rather abrupt halt in the growth of a new source of electrical energy, just at a point when liquid and gaseous fossil fuel resources are becoming limited? The answer is complex, involving real and imaginary problems, loss of public support, cost overruns in the installations, weapons connections, and some technical problems. Nuclear energy is the only technologically new source of energy developed since the beginning of the industrial revolution, and it is interesting and important to understand the reaction of society to its full-scale use.

The various aspects of nuclear energy that are of concern to many will be examined in the next sections in some detail in an effort to evaluate objectively the actual problems and risks involved. It is difficult if not impossible to make a valid judgment on the worthiness of one particular form of energy without considering the alternative forms of energy that would have to be used in its stead. In the next several chapters we discuss a number of alternative forms of energy that will certainly play an important role in the years ahead. At the moment, however, it seems quite clear that the choice of an electric utility for a new generating station is generally either nuclear (^{235}U-burning thermal reactor) or coal. Hydroelectric, solar, and geothermal power are attractive alternatives, but except for rather special situations, they are not options now open to the average U.S. electric power company. Thus, we focus the following discussions on a comparison between nuclear and coal power stations. Consideration of many of the important details related to the hazards of ionizing radiation and the hazards of air pollution from burning coal must be postponed until later, but it is important to deal now with an overview of the relative risks and problems, even without having studied all of the technical details.

A viewpoint frequently expressed is that since both coal and nuclear power plants present some risks to the public, the best policy would be not to increase the use of either but to stop the growth in the use of electric energy through conservation. The annual growth rate in the use of electric energy has, in fact, decreased in the last several years from its historical values of about 8% since 1920 to less than 3%. The average rate of increase from 1984 to 1989 was 2.8%. Can it go to 0% or actually decrease and still allow us to maintain a reasonable living standard with some increase in the gross national product? The U.S. population is increasing at the rate of 1% per year, and arguments could be made that even with reasonable attempts to conserve, we shall experience a 1% annual growth rate in electric energy consumption. The replacement of existing power plants that become obsolete is another factor. There was, in 1982, 515 GW_e of generating capacity, or the equivalent of 515 1000-MW_e plants. If one assumes a lifetime for each plant of 30 years, about 17 new plants would have to be constructed on average each year just to maintain the present generating capacity. In addition to the replacement plants, there are still a number of power plants that use either natural gas or liquid petroleum as a fuel. Because of the expense and shortage of these fuels, one can anticipate the conversion of these power plants to some other energy source. Likewise, one can anticipate an increase in the use of electricity for transportation as petroleum prices rise.

In summary, it appears that new electric generating stations will be needed in the years ahead, and the great majority of them will be either coal-fired or ^{235}U thermal nuclear reactors in the immediate future.

5.2 RADIOACTIVE EMISSIONS DURING THE NORMAL OPERATION OF A NUCLEAR REACTOR

The risks associated with nuclear reactors are centered about the effects of ionizing radiation on the health of the public. Ionizing radiation can lead, in some cases, to cancer and genetic defects. A unit that is used as a measure of the biological damage done to human beings by such radiation is the rem. The mrem (10^{-3} rem) is also used. This unit takes into account the more destructive nature of alpha particles as compared to x-rays. A complete discussion of these units appears in Chapter 11. The government has established guidelines for the maximum amount of radiation to which various population groups can be exposed. Those persons who are subject to exposures because of their occupations are limited to 5 rem/yr; individual members of the general public are limited to 0.5 rem/yr, and the public in general (as a group) to 170 mrem/yr. There are, of course, many more detailed considerations and qualifications to these limits, but these are considered in a later chapter. It should be pointed out that if an individual receives the above dose, ill effects will not necessarily be observed. These limits are far lower than the levels from which direct consequences can be expected. The effects could only be observed over a period of many years with a statistical study involving a large group of people. For example, if 100,000 people received a single brief exposure of 10 rem each, 790 additional cancer deaths would be experienced during their lifetime.

A nuclear reactor is designed to contain the radioactive fission products within the fuel cladding until the fuel is reprocessed at some later date. There are some radioactive releases, however, that come about in normal operation because gases, such as krypton and xenon, can diffuse through metal, and occasionally there can be a small crack in a fuel rod. These radioactive gases, such as ^{85}Kr ($T_{1/2}$ = 10.7 yr), are released into the atmosphere in amounts that range from 300 to 50,000 curies a year (the curie is defined later in this chapter) for a 1000-MW_e reactor. About 100 to 1000 curies of radioactive tritium are also released a year, mainly in the discharged water. Various filters and holdup systems have been incorporated into reactor designs to be certain that the emissions do not exceed limits designated by the Nuclear Regulatory Commission (NRC). The maximum dose of radiation to be received by anyone outside the reactor site has been specified by the NRC as 3 mrem/yr from liquid effluents and 5 mrem/yr from gaseous effluents. These limits are far lower than the total radiation limits previously mentioned.

It turns out that the maximum dose received by individuals, even those living in the vicinity of a typical reactor site, is about 1 mrem/yr. This dose is very small compared to the general natural background radiation of approximately 100 mrem/yr. The average dose received by members of the general public at the present time from operating reactors is estimated to be something like 0.005 mrem/yr. If there were a vigorous reactor construction program this dose might increase to about 0.5 mrem/yr by the year 2000. These are indeed very small doses. They are far less than the difference between the dose received from living in a brick or wooden house or the difference between living at 5000-feet altitude and sea level. It seems quite clear that the radioactive emissions from the normal operations of a nuclear reactor present a negligible risk to the public.

Coal, as commonly used as a fuel, contains naturally occurring radioisotopes such as ^{238}U and ^{232}Th. When the coal is burned some of the radioactive decay products of these two isotopes are released to the environment along with some other pollutants. Calculations have been carried out that indicate that the radiation dose received by the general public from the burning of coal in a power plant is equal to or greater than the dose from a nuclear power plant of the same generating capacity.

5.3 THE CHINA SYNDROME

What has raised the fears of the general public in regard to nuclear power more than anything else is the image of a reactor somehow exploding or melting down and releasing large amounts of radioactivity onto the surrounding communities. The consequences of this type of accident are perceived to be the possible deaths of thousands of people followed by the greatly increased incidence of cancer and genetic defects. The film *The China Syndrome* and the 1979 accident at Three Mile Island and the more recent and severe 1986 accident at Chernobyl have made this image very real.

A study of the worst credible accident that could be experienced with a nuclear reactor was undertaken at Brookhaven National Laboratory in 1957. The predictions in that study (WASH 740) of thousands of fatalities and tens of thousands of cases of serious illness for a 150-MW_e reactor have been frequently quoted and have fanned the fires of public fear of nuclear energy. Of course, the overall risk factor must take into account the probability of such an event occurring. There appears to be little public understanding of risk evaluation, and the picture of thousands dying in some catastrophic accident is the basis for many opinions about the safety of nuclear power.

A nuclear reactor cannot explode in the way that a nuclear weapon explodes. The structure of the reactor and the enrichment of the uranium make such an event impossible. What can happen is that the core of the reactor can overheat if the cooling is lost or if the reactor gets out of control. Overheating could be due to the residual radioactivity after a reactor is shut down. Chapter 4 discusses the elaborate precautions built into every reactor in the form of an emergency core cooling (ECC) system and the containment vessel. The "China

Syndrome" refers to the possibility that a reactor might lose its primary coolant and then its ECC system and backup systems for some reason might also not function, while the temperature of the core continues to increase. In the usual scenario, the core becomes molten and then melts its way down through the bottom of the containment vessel, continuing on into the ground where presumably it eventually cools off. Radioactivity would then be released to the atmosphere and perhaps also to the ground water. This event could indeed be disastrous and could lead to many deaths.

What must be evaluated is the probability of such events taking place. This involves knowledge of the reliability of various interlocks, controls, automatic valves, pumps, and, to some extent, the reliability of the human beings who are at the controls of the reactor. To evaluate the probability of a major accident at a nuclear reactor, an extensive study was made for PWR and BWR reactors for the Nuclear Regulatory Commission. The final report (WASH-1400) of this study was issued in 1975 and is known as the Rasmussen Report after its director, Norman Rasmussen, of the Massachusetts Institute of Technology. The task of evaluating such probabilities is difficult and involves the identification of possible sequences of failure that could lead to an accident. The next step involves the calculation of the probability of a particular event in a sequence taking place, and then the probability that the whole sequence will occur. Finally, the consequences in terms of radioactive releases and damage to the population and property surrounding the nuclear reactor must be calculated. The main concern of the Rasmussen Report was a loss-of-coolant accident (LOCA), and this involved the evaluation of many failure sequences known as event trees. Not only did the study involve, for example, the probability of some automatic valves not closing or opening, but it also involved assumptions about the effects of ionizing radiation on people, meteorological conditions at the time of release, and earthquake probabilities.

The results of the Rasmussen Report can be presented in a number of different ways. The report estimates that a LOCA will have a probability of 1 in 2000 per reactor year. Failure of the ECC system is estimated as 1 per 10 accidents, or a meltdown frequency of 1 in 20,000 per reactor year. Such a meltdown would lead to a serious release of radioactivity in only 1 case of 100. The chances of an accident with severe consequences are, thus, about 1 in 2×10^6 per reactor year. This can be restated to say that if we had 200 operating nuclear reactors (about twice as many as we have now in the United States), there would be a serious release of radioactivity from one reactor every 10,000 years. Table 5.1 casts these probabilities in another way by comparing the risks of our present reactors and 100 times as many reactors as we had in 1972, with common life-shortening causes. Figure 5.1 shows a plot of frequency (events per year) versus number of fatalities. As the number of fatalities in any one type of event increases, the likelihood for that event decreases. The curve for 100 operating nuclear plants lies below the other man-made causes by about a factor of 1000. An often cited prediction of the Rasmussen Report is that the chance of an accident that kills 100 or more people occurring with 100 operating nuclear reactors is the same as that of a meteor impact that has the same consequence, namely, once in 100,000 years. No matter how one portrays the statistics of the Rasmussen Report, the conclusion is that the chance of a major accident with nuclear reactors is very small compared to many other man-made and natural events.

The publication of the report brought forth a variety of criticisms and

Table 5.1 COMMON RISKS COMPARED TO RISKS FROM NUCLEAR POWER

Factors Tending to Decrease Average Lifetime	Decrease of Average Lifetime
Overweight by 25%	3.6 years
Male rather than female	3.0 years
Smoking	
1 pack a day	7.0 years
2 packs a day	10.0 years
City rather than country living	5.0 years
Actual radiation from nuclear power plants in 1970	Less than 1 minute
Estimate for the year 2000 assuming 100-fold increase in nuclear power production	Less than 30 minutes

Source: Advisory Committee on the Biological Effects of Radiation, "The Effects on Populations of Exposure to Low Levels of Ionizing Radiation," Division of Medical Science, National Academy of Sciences, National Research Council (November 1972).

comments from individuals and organizations. Some maintained that the common-mode type of failure was given too high a probability. Others faulted the study for not using the mean rather than the median in discussing frequency of accidents. Many felt that the contribution of latent cancer fatalities had been seriously underestimated. It is difficult to respond to these criticisms by assigning some number that can be used to correct the various probabilities in the Rasmussen Report without a complete reanalysis of the problem. Factors of 10 and as high as 500 have been mentioned as appropriate multipliers. The correction by such factors still has the accident probability for reactors far lower than that for the other man-made causes of deaths shown in Figure 5.1. A general conclusion regarding the report, which is shared by a number of groups, is that there was no consistent bias shown in the study, but that the uncertainties to be attached to the results are considerably larger than the factor of 5 stated in WASH-1400.

The accident at Three Mile Island is sometimes cited as proof that nuclear reactors are not safe and that the Rasmussen Report is not valid. Although the reactor experienced a partial meltdown, the containment vessel was not breached and significant amounts of radioactivity were not released to the environment. It is clear, however, that it was a major accident with serious financial implications. The causes of the accident seem rather complex, but it certainly involved poor design of some reactor elements, malfunction of some safety features, and poor judgment on the part of some operators. The design of future reactors will benefit from the various studies of the accident initiated by the Nuclear Regulatory Commission.

5.4 THE CHERNOBYL DISASTER

On April 26, 1986, an explosion and fire destroyed a reactor of the Chernobyl power plant in the Soviet Ukraine. This was clearly the most disastrous accident in the history of nuclear reactors, and it is important to understand its causes and consequences for nuclear development in the United States and elsewhere.

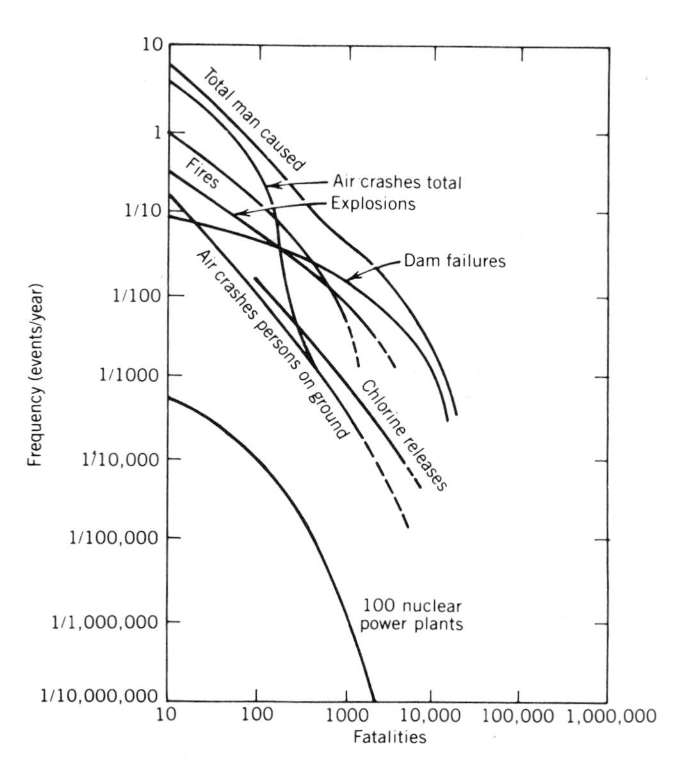

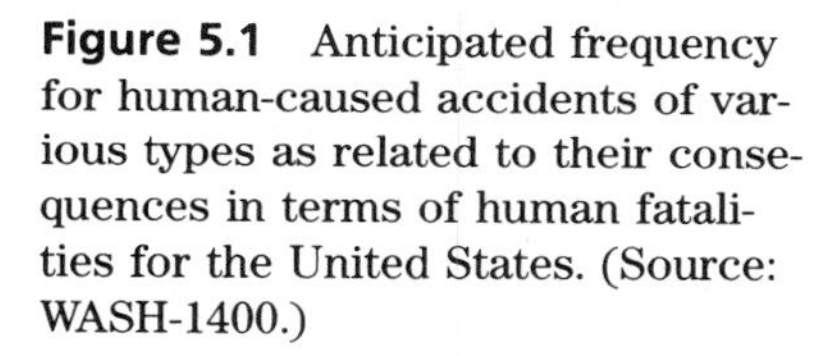
Figure 5.1 Anticipated frequency for human-caused accidents of various types as related to their consequences in terms of human fatalities for the United States. (Source: WASH-1400.)

The 1000-MW_e reactor was a boiling-water graphite-moderated reactor, unlike any power reactor in the United States. The RMBK reactor, as reactors of that type are called, is designed to produce ^{239}Pu for nuclear weapons as well as electric power. In order to maximize the production of ^{239}Pu from the ^{238}U in the fuel and to minimize production of ^{240}Pu, which is unwanted for nuclear weapons, it is important to have a graphite-moderated reactor where the fuel rods can be taken out after only 30 days without shutting down the reactor. Removing the fuel rods during operation requires a large open space above the reactor, which in turn precludes the use of a steel shell or reinforced concrete containment vessel common to power reactors in the United States. The RMBK reactor is also particularly unstable and hazardous in that as the water in the reactor is boiled and turned to steam, there is reduced neutron absorption. This causes the power level to increase and induce more water to boil. This positive feedback condition is compensated at high power levels by a negative temperature coefficient, but at low power levels it is a cause of instability and concern. The positive feedback can in principle be controlled by insertion of control rods, but this takes time. As discussed in Chapter 4, the delayed neutrons normally provide this necessary safety factor of time for control rod insertion. The design of the RMBK reactor, however, is such that the reactivity (basically the number of neutrons) can rise to the point where the reactor is prompt critical. This means that the reactor is critical without the contribution from the delayed neutrons; hence the safety factor of time no longer exists. Once a reactor goes into a condition of prompt criticality the power level can very rapidly increase to the point of meltdown and destruction.

On the night of April 25, 1986, the operators of unit 4 had planned to conduct an experiment at low thermal power levels near 200 MW before a shutdown for routine maintenance. In order to perform the measurements, certain safety systems, such as the emergency cooling system, were shut off in violation of safety rules. To control the reactor at this low power level, the operators had to withdraw the control rods and turn off the generator that powered the pumps providing cooling water. At this point the reactor went out of control; the thermal power surged, the fuel started to disintegrate, the cooling channels ruptured, and an explosion ripped open the reactor, exposing the core and starting many fires. The graphite moderator began to burn.

Some 50 megacuries of radioactivity, including a large fraction of the cesium and iodine in the fuel rods, was released to the environment. The radioactivity was deposited most heavily within a radius of 30 kilometers, and 135,000 people in that area were evacuated. Significant levels of radioactivity were experienced by much of Europe. Estimates for the number of excess cancer deaths due to the released radioactivity have been in the range of 10,000 to 40,000 people over the next 50 years in Europe and the former Soviet republics. These additional cancers will be extremely hard to detect against the background of the 600×10^6 cancers normally expected in the same population group over that period. In addition, 203 plant personnel and firefighters experienced acute radiation sickness, and 31 of these individuals died from radiation exposure.

The causes of this tragic accident can be laid at least partially to design flaws in the RMBK reactor and to an insufficient containment system. There were also obvious errors made by the operators. While power reactors in the United States do not share the particular design flaws of the RMBK reactors, and presumably have better containment vessels, the Chernobyl disaster presented a sobering demonstration of what can be expected from catastrophic reactor failure. It has been a major blow to public confidence in nuclear power.

Example 5.1

Calculate the combined hypothetical probability that if a plane crashes, it will be on a day that is rainy ($P_1 = 0.12$), that it will be a Thursday ($P_2 = 1/7$), and that it will be in October ($P_3 = 1/12$).

Solution

For compounding individual, causally unrelated probabilities, it is correct to compute the product of the individual probabilities. Thus, the probability of a given plane crash occurring on a rainy Thursday in October would be as follows:

$$P_T = P_1 \times P_2 \times P_3$$

$$= 0.12 \times 1/7 \times 1/12 = 1.4 \times 10^{-3}, \text{ or roughly 1 in 700}$$

5.5 NUCLEAR WEAPONS

Although a nuclear reactor cannot explode like a bomb, there is, unfortunately, a relationship between nuclear weapons and nuclear reactors designed for electric power. The technical aspects of nuclear bombs are presented in some detail in Chapter 16, but there are two essential facts needed to discuss the relationship between reactors and bombs. First, any fissionable material can be used for a bomb, whether it is ^{233}U, ^{235}U, or ^{239}Pu, but it must be highly enriched, usually to about 90% or better. Second, the amount of fissionable material needed for a critical mass (the minimum size assembly that will explode) depends strongly on the shape of the material and the surroundings, but is about 20 kg for ^{235}U and 10 kg for ^{239}Pu.

Now, since normal light-water reactor fuel is only enriched to about 3% ^{235}U, it cannot be used for bomb material without further enrichment. Thus, if some country or group of individuals wanted to make a uranium-based nuclear bomb, it would have to enrich either natural uranium or uranium fuel very extensively to bring it up to 90%. During World War II, the United States tried a variety of schemes to enrich uranium, but the process that has proven most economical is gaseous diffusion. This is a very complex and expensive process, and it would be difficult for any country to set up a gaseous diffusion plant without its being detected; it would be virtually impossible for some criminal gang to build such a plant. Laser isotope separation may be a simpler process for uranium enrichment.

A far more likely scenario would be for a country to construct a light-water nuclear reactor and run it either for military purposes or as an electric power reactor and generate ^{239}Pu from neutron capture on ^{238}U. The amount of plutonium in a spent uranium fuel rod is about 0.6% of the fuel rod loading. When the spent fuel rods are chemically processed, the plutonium can be rather easily separated from the uranium and accumulated until sufficient quantities are present to make a nuclear weapon. One of the weapons that the United States exploded in Japan was made in this way, as were weapons subsequently made in Great Britain, the Soviet Union, France, China, and India. Except in the case of India, all of these countries have special military reactors for plutonium production. Apparently, every country that has tried to explode a nuclear weapon has succeeded.

To control the spread of weapons, a Nuclear Non-Proliferation Treaty was signed in 1957–58 by more than 100 countries; the primary aim was control over the processing of spent fuel rods through inspections by the International Atomic Energy Agency. Under the treaty, the United States, for example, has control over the reprocessing of fuel rods manufactured from uranium produced or enriched in the United States. Table 5.2 shows many of the countries and their party or nonparty status. Also listed in the table are the present and projected nuclear power capacities and the annual bomb equivalent. Unfortunately, many of the nations that are not party to the treaty have the industrial and scientific capability to produce nuclear weapons. Countries such as Israel, South Africa, Brazil, Spain, Agentina, and Pakistan are probably in such a category. Thus, it is to be expected that as more and more countries construct nuclear reactors, some will divert sufficient plutonium to start making nuclear weapons. It would be surprising to many if Israel and South Africa have not already done so, and there is evidence that Iraq is trying.

The connection between the United States embarking on a vigorous nuclear power reactor program and the proliferation of nuclear weapons is somewhat indirect, and it is also complicated. One of the complications is the fact, discussed in the previous chapter, that the continuation of a 500-reactor nuclear power program beyond roughly 2030 depends on the development and use of breeder reactors, in which ^{238}U is converted to fissionable ^{239}Pu. Under a full-

Table 5.2 NUCLEAR POWER AND NUCLEAR PROLIFERATION CAPABILITIES

Country	Non-proliferation Treaty Status	Operational Nuclear Power Capacity[a]	Forecast Nuclear Power Capacity Mid-1980s[a]	Annual Bomb Equivalent[b]
Nuclear Weapons States				
United States	Party	37,600	208,400	4,168
Soviet Union	Party	4,600	14,400	288
United Kingdom	Party	5,300	11,800	236
China	Nonparty	?	?	?
France	Nonparty	2,800	21,300	426
Insecure States				
Israel	Nonparty	0	?	?
South Africa	Nonparty	0	?	?
South Korea	Party	0	1,800	36
Taiwan	Party	0	4,900	98
Yugoslavia	Party	0	600	12
Status-seeking States				
Brazil	Nonparty	0	3,200	64
India	Nonparty	600	1,700	34
Iran	Party	0	4,200	84
Spain	Nonparty	1,100	8,300	166
Rivals to States in Preceding Categories				
Argentina	Nonparty	300′	900	18
Egypt	Signatory	0	?	?
North Korea	Nonparty	0	?	?
Pakistan	Nonparty	100	100	2
Politically Constrained Major States				
Czechoslovakia	Party	100	1,900	38
East Germany	Party	900	2,700	54
Italy	Party	500	5,200	104
Japan	Party	5,100	15,500	310
Poland	Party	0	400	8
West Germany	Party	3,300	23,300	466
Other Developed Countries				
Australia	Party	0	?	?
Austria	Party	0	700	14
Belgium	Party	1,600	5,400	108
Bulgaria	Party	900	1,800	36
Canada	Party	2,500	11,800	236
Finland	Party	0	2,200	44
Hungary	Party	0	1,800	36
Luxembourg	Party	0	1,300	26
Netherlands	Party	500	500	10
Romania	Party	0	400	8
Sweden	Party	3,200	8,400	168
Switzerland	Signatory	1,000	5,800	116
Other Developing Countries				
Chile	Nonparty	0	0	0
Greece	Party	0	0	0
Indonesia	Signatory	0	0	0

Table 5.2 (con't.) NUCLEAR POWER AND NUCLEAR PROLIFERATION CAPABILITIES

Country	Non-proliferation Treaty Status	Operational Nuclear Power Capacity[a]	Forecast Nuclear Power Capacity Mid-1980s[a]	Annual Bomb Equivalent[b]
Mexico	Party	0	1,300	26
Philippines	Party	0	1,400	28
Thailand	Party	0	?	?
Turkey	Signatory	0	0	0

[a] Capacity in megawatts of electricity derived from "World List of Nuclear Power Plants, December 31, 1975," *Nuclear News*, February 1976. Many countries' programs have undergone change since then, but the overall picture is reflected here.

[b] "Annual bomb equivalent" is a rough approximation that assumes that each 1000 MW_e of forecast capacity is operated in such a way as to produce 2000 kg of plutonium annually as a by-product and that 10 kg of plutonium is required for one bomb.

scale breeder program, ^{239}Pu will be the basic fuel for reactors. This means that all fuel rods must be processed to remove the ^{239}Pu and, of necessity, includes the handling, transportation, and storage of large amounts of plutonium. The rather complicated fuel processing scheme is shown in Figure 5.2. The chances that some of this bomb-grade material may fall into the hands of a terrorist group, a criminal gang with intentions for blackmail, or a demented person will be very much larger than they are today. A country with a nuclear power reactor and that is intent on making a plutonium bomb could probably do so and escape detection. It was consideration of this possibility that led President Carter to discontinue funding for the Clinch River breeder reactor and to suspend the reprocessing of spent fuel rods.

Arguments can be made that other countries, such as France, because of their lack of fossil fuels, must be firmly committed to nuclear power, and what the United States does unilaterally will not have a major impact on nuclear proliferation. It has also been argued that with new simpler techniques for uranium enrichment based on the atomic excitation by lasers of particular uranium isotopes, or the use of new very-high-speed centrifuges, nuclear weapons will become more available whether or not we have more nuclear power reactors.

On the other hand, many people in the United States feel that since nuclear reactors and nuclear weapons originated here, we have a moral obligation to set an example for the world by renouncing entirely the use of nuclear energy. Such a step, it is argued, would be a moral deterrent to any nation further developing a nuclear reactor program. A less drastic step would be for all nations to follow the Carter initiatives and to forego the processing of all fuel rods and, of course, the use of breeder reactors. Under such a program, fuel would be used on a once-through basis. The spent fuel rods would eventually be put into permanent storage after removal from the reactor. Obviously, this option precludes the long-term use of nuclear energy, as we would be dependent on our ^{235}U resources and relatively high-grade uranium ores. This option would give the world about 50 years of nuclear fission energy, and might permit a transition to renewable energy sources or perhaps to nuclear fusion energy.

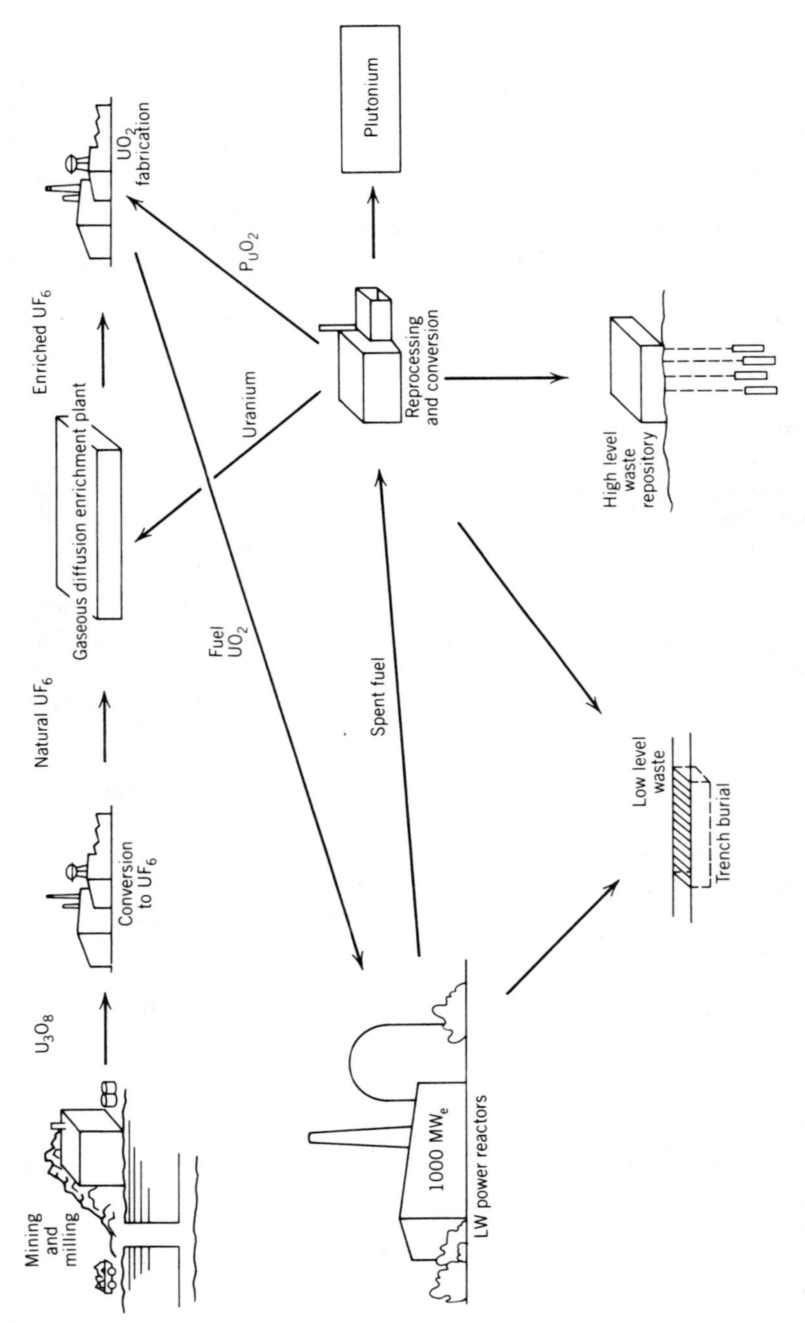

Figure 5.2 The fuel cycle for light-water reactors.

The weapons aspect of a nuclear reactor policy is the most difficult one to assess and about which to draw absolute judgments. The questions raised involve national security as viewed by many different nations and other international policy matters that are difficult to fold into some optimization calculation. It does not appear that there is any quick technological solution to the weapons dilemma. The fact that the United States, the former Soviet Union, and some other countries have already stockpiled sufficient nuclear weapons to wipe out essentially all civilization as we know it is a concern that to many completely dwarfs the problem of future proliferation.

5.6 THE STORAGE OF HIGH-LEVEL RADIOACTIVE WASTE

The basic source of energy in a nuclear reactor is a fission reaction of a nucleus, such as ^{235}U, that leads to two lighter fission fragments that are, in general, radioactive. Most of the radioactive fission products have reasonably short half-lives in the range of seconds to days, so that storage of the fuel rods for periods of one or two years will render them relatively harmless.

Unfortunately, there are fission products such as ^{90}Sr and ^{137}Cs with half-lives of 28.8 and 30.0 years, respectively, that are present in great abundance and that require isolation for hundreds of years. Even more troublesome, however, are isotopes such as ^{129}I, ^{99}Tc, and ^{135}Cs with half-lives in the range of 10^5 to 10^7 years. In addition to the fission products, there are various isotopes of uranium, plutonium, thorium, americium, neptunium, curium, and protactinium (called the actinides) that are built up by neutron capture and radioactive decay processes in the fuel rods as power is generated. Two of the more significant isotopes are ^{237}Np and ^{239}Pu, with half-lives of 2.14×10^6 years and 24.3×10^3 years, respectively.

Example 5.2

Calculate the number of years needed for the following radioisotopes to decay to one thousandth of their initial activity. (Note that $1/1000 \approx 1/2^{10} = 1/1024$.)

$$
\begin{aligned}
{}^{239}\text{Pu} \quad & (T_{1/2} = 24{,}000\ \text{yr}) \\
{}^{137}\text{Cs} \quad & (T_{1/2} = 30.1\ \text{yr}) \\
{}^{3}\text{H} \quad & (T_{1/2} = 12.35\ \text{yr}) \\
{}^{89}\text{Sr} \quad & (T_{1/2} = 50.5\ \text{days})
\end{aligned}
$$

Solution

Each radioisotope will decay to one thousandth of its initial activity in about 10 half-lives. For the listed radioisotopes this will require

^{239}Pu	240,000 yr
^{137}Cs	301 yr
^{3}H	124 yr
^{89}Sr	1.38 yr

To a lesser extent, the elements such as zirconium, nickel, iron, and cobalt, which make up the fuel cladding, become radioactive and add to the storage

Table 5.3 RADIOACTIVITY IN THE INVENTORY AND IN THE FUEL DISCHARGED AFTER ONE FUEL CYCLE FROM A 1000-MW_e URANIUM-FUELED LIGHT-WATER REACTOR

	Reactor Inventory (10^6 curies)	In Discharged Fuel (10^6 curies/yr)		
		At Discharge	After 150 Days	After 10 Years
Fission products	11,970	3,970	130	9.98
Total actinides	3,614	1,198	4.45	2.55
Fuel cladding	12.9	4.3	1.0	0.1
Total	15,600	5,150	135	12.6

Source: Adapted from A. V. Nero, Jr., *A Guidebook to Nuclear Reactors* (Berkeley: University of California Press, 1979).

problem. Table 5.3 shows the radioactivity from a 1000-MW_e uranium-fueled light-water reactor. After 10 years there are about 12.6 megacuries of radioactivity left, most of it from the fission products. The common unit of radioactive source activity is the curie; 1 curie is equal to 3.7×10^{10} radioactive decays per second. One gram of radium has an activity of 1 curie. As time goes on, the activity of the longer-lived actinides will dominate that of the fission products.

The present policy of the United States is not to recycle the spent fuel from power reactors. This policy is based on fears of nuclear weapon proliferation as well as on some of the difficulties encountered in earlier recycling attempts. The spent fuel rods from commerical power reactors are stored in tanks of water near the reactor to allow the radioactivity and heat to subside and to allow time for preparation of a site for long-term isolation by burial in a geologically stable repository. Before such storage could take place, the spent fuel material would be encapsulated in a heat- and corrosion-resistant glass. Surrounding the glass would be several different protective cylinders. These cylinders would be inserted into holes in the floors of tunnels about 400 m below the surface of the earth. A test storage facility of this kind is shown in Figure 5.3. Special claylike material would be backfilled around the deposited cylinders. It has been estimated that the scheme would keep the radioactivity out of the environment for at least 10^5 years. The U.S. Congress in 1987 selected Yucca Mountain in the southern part of Nevada as the first site for such a repository. Tests are being carried out and measurements of ground water made in preparation for the first use in 2010. The state of Nevada has however, resisted the use of Yucca Mountain as a national repository for high-level waste. Based upon fears that it is a region of uncertain hydrology, the state has taken the issue to court. It is not clear at this point that radioactive storage will begin in 2010 at that or any other location. In the meantime, the facilities for storing the spent fuel rods at the various reactors are being saturated.

Other accumulations of high-level waste exist at Hanford, Washington, and Savannah River, South Carolina, where plutonium and tritium have been produced for a number of years for nuclear weapons. These defense-related wastes have the same problem with long-lived fission products and actinides as do the spent fuel rods from power reactors, and there is no accepted solution at this time. There is also a second category of waste that is far less radioactive. These

Figure 5.3 A spent reactor fuel storage test facility in Nevada. The storage test assesses the effects of heat and radiation on the behavior of granitic rock.

low-level wastes have resulted mainly from the nuclear weapons programs and will be treated separately from the high-level wastes. The Waste Isolation Pilot Plant (WIPP) has been established near Carlsbad, New Mexico, for this type of waste, but again there have been many delays. Eventually the site is scheduled to accommodate 300,000 55-gallon steel drums of transuranic waste buried in salt beds 2000 feet below the surface. This waste consists mainly of tools, clothes, laboratory equipment, etc. that are contaminated with plutonium. Before permanent disposal can take place, a five-year test will be made with a small number of drums.

Because of the lack of credibility on the part of the general public and the political problems that have surfaced, alternative procedures have been examined by the Department of Energy national laboratories. The general schemes being discussed involve partitioning the radioactive waste and then transmutation of the longer lived fission products and actinides. For example, if 99.9% of the actinides were removed from the spent fuel, the remaining radioisotopes would be primarily ^{137}Cs and ^{90}Sr, and the Environmental Protection Agency standards could be met after only 300 years of burial. It is also desirable to remove the fission products ^{99}Tc and ^{129}I. According to one proposal, the removed ^{239}Pu and uranium could be used in new reactor fuel and the remaining radioisotopes subjected to transmutation under intense neutron bombardment. Some of the nuclei would undergo fission and produce useful energy, and others would undergo neutron capture to shorter lived isotopes. Both special reactors and a linear proton accelerator have been put forth for the source of the neutrons for transmutation. The net effect of these processes would be that the length of time necessary for geologic storage of high-level radioactive waste would be reduced from many thousands of years to 300 years. The shorter time span would make the waste isolation project far more acceptable. There are, however, many practical problems to overcome with the partitioning and transmutation scheme, and as yet no decisions have been made to initiate such a program.

The public image of the radioactive waste disposal problem is rather far removed from actual fact. The volume of the liquid high-level waste from a 1000-MW_e reactor operating for 1 year is about 20,000 liters. This corresponds to a cubic volume of 2.7 m on a side. For all the nonmilitary reactors now operating in the United States, this corresponds to a cubic volume each year of about 11 m on a side. The entire waste from a full-scale reactor program to the year 2000 could easily be accommodated in a repository with a surface area of a few square miles.

B. L. Cohen has made some interesting observations concerning the relative hazards of nuclear wastes compared to other poisonous materials widely used in the United States. Materials such as barium, arsenic, chlorine, phosgene, and ammonia are quite poisonous and are liberally distributed around the country with little concern for the fact that the number of lethal doses from these substances exceeds the number of lethal doses from nuclear wastes (assuming 100 operating reactors) by factors of 10^3 to 10^7. Under any system of storage, the nuclear wastes will be well isolated from the environment using procedures similar to those just described, but arsenic and other poisons are spread onto agricultural fields with little worry about the fact that they are completely stable and will never decay.

In summary, it appears that the storage of nuclear wastes is a solvable problem, and one hopes that the government can by prompt action demonstrate to the concerned public that this is the case.

5.7 THE COSTS OF NUCLEAR POWER

One of the major concerns about nuclear power is related not to environmental problems or hazards to the public but rather to costs. It is obvious that utilities are not going to invest in nuclear reactors if less expensive alternatives are available. If nuclear energy is not economically competitive it will not be used. The cost of nuclear electric energy has been compared to that from coal-fired plants for many years, with coal winning in one study and nuclear in another. The calculations are subject to a variety of different assumptions concerning the future cost of uranium, recycling of the spent fuel, reclamation of mined land, the pollution control equipment required of the coal-fired plant, and so forth. The high rate of inflation in recent years has added greatly to the uncertainty of such calculations. Generally speaking, coal-fired plants involve a smaller capital investment but a higher fuel cost. Until about 1980, cost estimates were comparable for coal and nuclear at about 42 mills/kW • hr in 1980 dollars (1000 mills = $1). Since that time a number of factors have escalated the costs of nuclear power plants to a point where, as we have seen, nuclear reactors are no longer competitive. What has brought this change about?

The experience in the United States with nuclear reactor costs has been far less favorable than in Europe and Japan. In the United States, private utilities have in the past ordered nuclear reactors on a customized basis dependent on the utilities' needs and particular site requirements. There has been little standardization by the manufacturers. Each reactor has had to have separate approval by the Nuclear Regulatory Commission (NRC) and detailed inspection before an operating license was granted. After the Three Mile Island accident, concern for safety led the NRC to require many changes in the mechanical and electrical details of the reactors being designed or constructed. It is very expensive to make changes in a complex device such as a reactor when it is under construction, and these changes, along with legal delays, further lengthened the construction time. Mainly because of changes required by the NRC, the average construction time almost doubled between 1971 and 1980, and this resulted in about a doubling of the final cost of the power plant. The increases were due largely to costs of financing and refitting at the site. Some power plants experienced a million-dollar cost increase from every day of delay. In addition to these costs, inflation added another factor of 2, so the final construction costs were frequently four times the original estimates.

In addition to higher construction costs, power reactors in the United States have not fared as well operationally as those elsewhere. The average availability of U.S. reactors is about 60% compared to 75 to 85% in some countries, and the cost of operation and maintenance in the United States is about double what it is elsewhere.

While no new reactors are being ordered in the United States because of these high costs, the Department of Energy and reactor manufacturers have been working on a next generation of nuclear power plants that will overcome many of the present difficulties. Design work is being concentrated on smaller 600,000 kW_e light-water plants that can be standardized and receive NRC approval prior to construction. These smaller reactors are designed to have passive stability, simplified controls, greater ruggedness, ease of operation, and greater standardization. The term "passive stability" refers to the reactor's inherent property of having the reaction rate go down as the temperature of the coolant or fuel goes up, without relying on the operator or external control devices. The AP-600 is an example of this new breed of reactor. Developed by Westinghouse and the Department of Energy, it is a 600-MW_e pressurized water reactor that has the above features and has reduced the number of valves, pumps, heat exchanges, ducting, and control cables by 35 to 80% compared to a conventional reactor of that size. This work is being submitted to the NRC for review and comments so that the final licensing will be greatly expedited. Other companies are developing other types of reactors with similar properties to compete with the AP-600, so several new reactors will be available for the utilities to select when and if the nuclear option again appears favorable.

5.8 OTHER CONCERNS ABOUT NUCLEAR POWER

The question of thermal pollution is sometimes raised in connection with nuclear reactors. As we stressed earlier in the discussion of heat engines, all heat engines must reject heat to the environment. The efficiency of the light-water reactor power plants used in the United States today is about 32 to 34%. Coal-burning power plants have efficiencies that are somewhat better (about 38%). In either case, cooling towers can be used to avoid locally heating rivers or the ocean, but these towers will add heat and humidity to the local atmosphere. Example 5.3 shows that for the same electrical output capacity, a power plant operating at 32% efficiency rejects 30% more waste heat than does a plant operating at an efficiency of 38%. Thus, the plant with the lower thermal efficiency requires a more extensive cooling system and contributes more heavily to all problems arising from thermal pollution. There is increasing interest in making the waste heat from either type of power plant available for space heating or industrial processes.

Example 5.3

Compare the relative amounts of waste heat energy rejected by two 1000-MW_e power plants, one operating at 32% efficiency ($\eta = 0.32$) and the other operating at 38% efficiency ($\eta = 0.38$).

Solution

In general

$$(\text{electrical energy produced}) = (\text{efficiency}) \times (\text{energy input})$$

and

$$\begin{aligned}(\text{waste heat energy}) &= (\text{energy input}) - (\text{electrical energy produced}) \\ &= (\text{energy input})\,(1 - \eta)\end{aligned}$$

Also

$$(\text{energy input}) = \frac{1}{\eta}\,(\text{electrical energy produced})$$

so that

$$(\text{waste heat energy}) = (\text{electrical energy produced})\,\frac{(1 - \eta)}{\eta}$$

For the $\eta = 0.32$ plant, the waste power is

$$\begin{aligned}(\text{waste power})_{0.32} &= 1000\ \text{MW}_e \frac{(1 - 0.32)}{0.32} \\ &= 2125\ \text{MW}\end{aligned}$$

and for the other plant

$$(\text{waste power})_{0.38} = 1000\ \text{MW}_e \frac{(1 - 0.38)}{0.38} = 1632\ \text{MW}$$

Thus, the $\eta = 0.32$ plant rejects 2125/1632 = 1.30 times as much power, or 30% more, than does the other plant.

5.9 RELATIVE RISKS

It is impossible to calculate exactly how many deaths will occur per gigawatt-year for nuclear power reactors and coal-fired plants, but it is important to try to do so. Such a calculation will certainly point up where reductions in health hazards can be achieved with either type of electric energy production; more important, it serves as some basis for a rational discussion of the relative hazards. Without some considerations and calculations, we can be reduced to emotional arguments that add little to the discussion. It is unfortunate that so much of the nuclear versus coal discussion in the past has been conducted on an emotional basis.

We have discussed the concerns about nuclear power in some detail because people are generally less familiar with this relatively new form of energy. Coal, as one might guess, also has its problems, even if they are not of such an exotic nature. Coal mining has been and continues to be a hazardous occupation. There have been 88,000 coal miners killed in the United States in underground mine accidents since 1905. Black lung disease had taken an additional toll. Because of the volume of the coal needed to fuel a power plant, many trainloads of coal must be shipped, sometimes long distances. There are accidents in handling coal, and there are a surprising number of fatalities in auto collisions with coal trains at grade crossings. A summary of accidental deaths during routine operation is listed in Table 5.4 for four common energy sources per gigawatt-year. It really should not be a surprise to see coal 13 to 20 times worse than nuclear since so much more material must be handled. Uranium mines have taken their toll in lung cancer from the radon in the air in the mines, but because there are far fewer uranium miners than coal miners the numbers are smaller. Radon is a radioactive gas that includes 56-sec ^{220}Rn from the decay of thorium, and 3.82-day ^{222}Rn from the decay of uranium. There is also the possibility of exposure of the general public to radon from mines

Table 5.4 ACCIDENTAL DEATHS DURING ROUTINE OPERATION, BY ENERGY SOURCE PER GIGAWATT-YEAR

Energy Source and Quantity Required	Extraction	Processing	Transport	Power Station	Total[a]
Coal (3×10^6 tons)		0.02	2.3[b]	0.01	
Deep	1.7				4.0
Surface	0.3				2.6
Oil, onshore and offshore (12×10^6 bbl)	0.2	0.08	0.05	0.01	0.4
Natural gas (67×10^9 ft^3)	0.16	0.01	0.02	0.01	0.2
Uranium oxide[c] (150 tons from 75,000 tons of ore)	0.2	0.001	0.01	0.01	0.2

[a] Totals do not add up due to rounding.

[b] The estimates are not based on coal trains per se, but on the overall rate of train accidents. Furthermore, many accidents with trains are not the fault of cargo nor of the carrier, and the responsibility for them may be incorrectly charged. For meaningful statistics, the matter needs further study. A forthcoming review cites figures based on the exclusive use of unit trains that scale to 0.5 deaths per gigawatt-plant-year, less than one fourth of the entry in the table. Carl W. Gehrs, David S. Shriner, Steven E. Herbes, Harry Perry, and Eli Salmon, "Environmental, Health, and Safety Implications of Increased Coal Utilization," in *Chemistry of Coal Utilization*, M. A. Elliot, Tech. Ed. Chap. II, Suppl. Vol. 2 (New York: Wiley Interscience, in press).

[c] With reprocessing, the uranium oxide requirement could be reduced to 1.4 tons. Presumably, the extraction risk would be reduced proprotionately and the processing risk increased. The net result could be lower total risk.

Source: Adapted from *Energy in Transition, 1985–2010*, Committee on Nuclear and Alternative Energy Systems (1979).

and piles of ore tailings. About 80% of the original radioactivity of uranium ore remains in the tailings. Among the remaining radioisotopes, there are very long-lived isotopes and their daughters (for instance, 77,000-year ^{230}Th and 1600-year ^{226}Ra). Radon gas can diffuse into the surrounding air from exposed tailing piles, although a few feet of earth on the surface of the pile is an effective seal against escape, because the gas has only a 3.82-day half-life and diffuses slowly.

The emissions from coal-burning plants will be discussed in some detail in chapter 13. Of particular concern are the tons of sulfur dioxides and particulates that are emitted even with reasonable pollution controls. Various epidemiological studies have estimated that tens of thousands of people in the United States die prematurely each year because of emissions from all the coal-burning plants. The carbon dioxide produced, which is not generally classified as a pollutant, may have very important and deleterious effects on the earth's climate in 20 to 30 years.

H. A. Bethe has summarized the relative risks of coal and nuclear power plants per gigawatt-year as follows: coal burning without scrubbers near cities, 74 fatalities per year; coal burning without scrubbers far from cities, 28 fatalities per year; coal burning with scrubbers near cities, 11 fatalities per year; coal burning with scrubbers far from cities, 7; and nuclear reactors, about 1. For coal, air pollution is generally the biggest problem.

H. Inhaber has completed a rather extensive study of relative risks from various conventional and nonconventional energy sources. Table 5.5 summarizes his findings for the total man-days lost per megawatt year for both occupational groups and members of the public. It may seem surprising to see solar energy rated as being more hazardous than nuclear, but it should be remembered that solar energy has a very low power density and hence much more material must be manufactured, transported, maintained, and so forth for an equivalent amount of power.

In summary, all energy forms have their associated risks. Although nuclear energy has risks that are somewhat more unusual than some of the other forms

Table 5.5 TOTAL MAN-DAYS LOST PER MEGAWATT-YEAR FOR BOTH OCCUPATIONAL WORKERS AND THE GENERAL PUBLIC

Source	Man-Days Lost
Coal	3000
Oil	2000
Wind	1000
Solar thermal electric	700
Solar photovoltaic	700
Methanol	300
Hydroelectric	45
Ocean thermal	30
Nuclear	10
Natural gas	6

Source: Herbert Inhaber, "Risk Evaluation," *Science* ***203*** (1979), p. 718.

of energy, the overall hazards presented by nuclear energy appear to be less than for almost all other forms of energy. In normal operation, there is strong evidence that nuclear power results in appreciably fewer prompt deaths than does almost any other form of power generation, but some questions still linger with regard to long-term latent cancers due to radioactive emissions, either from the nuclear power industry or from coal-burning plants. Many would say that the choice of coal over nuclear is an acceptance of a certain known fatality rate over the rate known to be much lower for nuclear in steady operation, coupled with the unknown, but small, likelihood of a catastrophic nuclear accident. It appears that both nuclear and coal-fired electric power plants will be needed at least during the next 25 to 50 years. Conservation should reduce appreciably the number of new power plants needed compared to historical trends. The weapons proliferation question is serious and can only be met through international agreements; it is not obvious how it affects the use of nuclear energy in the United States as a single nation.

PROBLEMS

1. Calculate the frequency of serious accidents involving releases of radioactivity if all the electric power in the United States were generated by nuclear reactors. Use the probability of serious accidents predicted by the Rasmussen Report.

2. How many megawatt-years would a reactor have to run to produce enough ^{239}Pu for the critical mass for one bomb? See Table 4.2 and the footnote to Table 5.2 and neglect the amount of ^{241}Pu produced.

3. (a) How would the high-level radioactive waste problem be changed if the United States went from a complete dependence on thermal LWR reactors to fast breeder reactors?

(b) What changes would be brought about in the radioactive waste problem if all of the electric energy in the United States were produced in fusion reactors?

4. Discuss briefly the reasons for the sharp curtailment of new nuclear reactors in the past few years in the United States.

5. You buy a lottery ticket that gives you a 10% chance of winning a prize and you also go to a bingo game where there is a 5% chance of winning a prize. What is the chance that you will win both the lottery and bingo?

6. The Three Mile Island accident is estimated to have released into the environment 20 curies of ^{131}I, which has a half-life of 8 days. How many curies remained after 48 days?

7. The Rasmussen Report estimates the probability of having a loss of cooling accident followed by failure of the emergency cooling system followed by a serious release of radioactivity as one chance in 2×10^6 reactor-years. If we have 200 nuclear reactors operating in the United States, how frequently should we expect such accidents?

8. Describe briefly the major differences between the reactor that was involved in the accident at Chernobyl and a typical U.S. light-water reactor.

9. For a new nuclear power plant, the designers estimate the following probabilities: the probability of a loss-of-coolant accident is 10^{-3} per reactor year. The probability that the emergency core cooling system will work is 80%. The probability that a meltdown (failure of the emergency core cooling system) will result in a dangerous release of radioactivity is 3%. What is the probability per reactor-year that the plant will have a loss-of-coolant accident and be saved by the ECC system?

10. In 1983 the rest of the world (non-U.S.) produced 6.2×10^{11} kW•hr of electricity with nuclear reactors. If each 1000-MW_e power plant produced enough plutonium for 20 nuclear bombs each year, how many bombs could be produced from the power reactor plutonium produced in 1983 by the rest of the world?

SUGGESTED READING AND REFERENCES

1. Bethe, H. A. "The Necessity of Fission Power." *Scientific American*, **234** 1 (January 1976), pp. 21–31.
2. Cohen, B. L. "The Disposal of Radioative Wastes from Fission Reactors." *Scientific American*, **236** 6 (June 1977), pp. 21–31.
3. Inhaber, H. "Risk Evaluation." *Science*, **203** 4384 (Feb. 4, 1979), pp. 718–723.
4. Lewis, H. W. "The Safety of Fission Reactors." *Scientific American*, **242** 3 (March 1980), pp. 53–65.
5. Nero, A. V., Jr. *A Guidebook to Nuclear Reactors.* Berkeley: University of California Press, 1979.
6. Ross, M. H., and Williams, R. H. *Our Energy: Regaining Control.* New York: McGraw-Hill, 1981.
7. Report of Ford Foundation Study Group. *Energy: The Next Twenty Years.* Cambridge, Mass.: Ballinger, 1979.
8. Final Report of the Committee on Nuclear and Alternative Energy Systems (CONAES). *Energy in Transition 1985–2010.* San Francisco: W. H. Freeman, 1979.
9. U.S. Nuclear Regulatory Commission. *Reactor Safety Study: An Assessment of Accident Risks in U.S. Commercial Nuclear Poweer Plants.* Washington, D.C.: U.S. Nuclear Regulatory Commission (WASH-1400), 1975.
10. "Report to the American Physical Society by the Study Group on Nuclear Fuel Cycles and Waste Management." *Reviews of Modern Physics*, **50** 1 Part 2 (January 1978), pp. S1–S183.
11. Wilson, Richard. "A Visit to Chernobyl." *Science*, **236** 4809 (1987), pp. 1636–1640.
12. Cohen, B. L. *The Nuclear Energy Option—An Alternative for the 90's*. New York: Plenum Press, 1990.
13. "Postmortem on Three Mile Island" *Science* **238** 4832 (December 4, 1987), pp. 1342–1345.
14. Upton, Arthur C. "Health Effects of Low-level Ionizing Radiation." *Physics Today*, **44** 8, (August 1991), p. 34.
15. Cohen, Bernard L. "The Nuclear Reactor Accident at Chernobyl, USSR." *American Journal of Physics*, **55** 12 (December 1987), pp. 1076–1083.
16. Taylor, John J. "Improved and Safe Nuclear Power." *Science*, **244** 4902 (April 21, 1989), pp. 318–325.

CHAPTER SIX

THE USES OF SOLAR ENERGY

6.1 INTRODUCTION

It is clear that from the earliest times, people in search of warmth have oriented their dwellings toward the sun. This sensible tendency is expressed in a quotation from more than 2000 years ago:

> He approached the problem thus: "When one means to have the right sort of house, must he contrive to make it as pleasant to live in and as useful as can be?"
>
> And this being admitted, "Is it pleasant," he asked, "to have it cool in summer and warm in winter?"
>
> And when they agreed with this also, "Now in houses with a south aspect, the sun's rays penetrate into the porticos in winter, but in summer the path of the sun is right over our heads and above the roof, so that there is shade. If, then, this is the best arrangement, we should build the south side loftier to get the winter sun and the north side lower to keep out the cold winds. To put it shortly, the house in which the owner can find a pleasant retreat at all seasons and store his belongings safely is presumably at once the pleasantest and the most beautiful."
>
> Xenophon
> *Memorabilia Socratis* III:viii

We have another example in the 1000-year-old structures of Mesa Verde shown in Figure 6.1. In recent centuries, however, this approach to gaining energy from the sun has been largely ignored, certainly in part because of the block-by-block arrangement of our cities and the easy availability of fuel, which has made the sun's warmth seem unimportant. But the utility of the sun beyond growing crops and providing daylight and suntans has never been completely forgotten. Two quotations from a publication of a century ago describe the direct use of solar energy for space heating, the distillation of wine into brandy, and the pumping of water:

> Since May, last year, M. Mouchot has been carrying on experiments near Algiers with his solar receivers. The smaller mirrors (0.80 m. diameter) have been used

Figure 6.1 The Cliff Palace, a notable structure in Mesa Verde National Park, Colorado. The sun's direction is indicated by the shadows.

successfully for various operations in glass, not requiring more than 400° to 500°. Among these are the fusion and calcination of alum, preparations of benzoic acid, purification of linseed of oil, concentration of syrups, sublimination of sulphur, distillation of sulphuric acid, and carbonization of wood in closed vessels. The large solar receiver (with mirror of 3.80 m.) has been improved by addition of a sufficient vapor chamber and of an interior arrangement which keeps the liquid to be vaporized constantly in contact with the whole heating surface. This apparatus on November 18, last year, raised 35 litres of cold water to the boiling point in 80 minutes, and an hour and a half later showed a pressure of eight atmospheres. On December 24, M. Mouchot with it distilled directly 25 litres of wine in 80 minutes, producing four litres of brandy. Steam distillation was also successfully done, but perhaps the most interesting results are those relating to mechanical utilization of solar heat. Since March the receiver has been working a horizontal engine (without expansion or condensation) at a rate of 120 revolutions a minute, under a constant pressure of 3.5 atmospheres. The disposable work has been utilized in driving a pump which yields six litres a minute at 3.50 m. or 1,200 litres an hour at 1 m., and in throwing a water-jet 12 m. This result, which M. Mouchot says could be easily improved, is obtained in a constant manner from 8 A.M. to 4 P.M., neither strong winds nor passing clouds sensibly affecting it.

"Progress in Utilization of Solar Heat," *Science,* Vol. 1 (first series), July–December 1880, Aug. 7, 1880, p. 69.

Mr. Morse drew attention to this device a year ago, before the National Academy of Sciences. At that time he was able to offer only crude computations as to the operations of the heater, derived from its use at the museum of Salem, Mass.

The device consists mainly of a slaty surface painted black, standing vertically upon a wall, outside the building, with flues to conduct warmed air to the inside. The slates are inserted in a groove, much as one might place glass in a frame. One made within the last year was three feet wide and eight long. It was placed where it received the sun's rays as directly as practicable. Its service was to warm a room used for a library. During an entire winter the room was thus made comfortable, except on a few of the coldest days. The current of air passing through it, when the sun's rays impinged directly upon it, was raised about 30°; it discharged 3,206 feet of warmed air in an hour. This was in the morning. At 11:45 the air of the apartment was raised 29°, with 3,326 cubic feet of air discharged; at 12:45, 29° and 4,119 feet; at 1:55, 24° and 3,062 feet; at 2:45, 20° and 1,299 feet. The room measured 20 × 14, and was ten feet high.

The apparatus works to most advantage in a room that is ventilated by an open chimney. But some very good results have been obtained in closed rooms. One was cited, where the air in a public building was raised by such means to nearly 40° above the outside temperature. In general, a difference of 30° to 35° can thus be secured during four or five working hours of the day. . . .

E. S. Morse, "The Utilization of the Sun's Rays for Warming and Ventilating Apartments," *Science,* Vol. 2 (old series), July–December 1883, Aug. 31, 1883, p. 283.

In spite of their promise, these early efforts gave way to the burning of fuels in place of direct use of solar energy. The solar installations were too expensive for the energy they produced. The sun's energy is too diffusely distributed on earth, its direction and intensity vary throughout the day, and it is only as reliable as the weather. In the far northern and southern reaches of the globe, there are long, completely sunless periods of extreme cold. However, in recent years, the scarcity and increasing cost of conventional fuels have forced a reexamination of the potential of solar energy. In some ways, the products of modern technology have made new approaches possible; we can now speak of orbiting collector satellites, photovoltaic cells, plastic films, selective surfaces, hydrogen–oxygen fuel cells, and so forth. We now have effective means for storing solar energy for use during sunless periods. And most of the world's population and, therefore, most of our energy demand, is found in regions experiencing fairly predictable solar influx.

In this chapter we examine some of the basic characteristics of solar radiation and the methods and devices likely to be useful for capturing solar energy and putting it to beneficial use.

The indirect uses of solar energy such as wind turbines, hydroelectric power, and ocean thermal energy are examined in the next chapter. Biomass is treated in Chapter 10. Both the direct and indirect forms of solar energy are termed "renewable energy" to contrast them with fossil fuel and nuclear energy. It is interesting to note that energy schemes that rely on the kinetic energy of molecules, such as almost all forms of solar energy, do not produce waste. Energy conversions based on chemical or nuclear reactions change the molecules or atoms involved and thus produce waste that can be environmentally harmful.

6.2 BASIC CONCEPTS

Of the total flow of energy available on the earth's surface, more than 99.9% is due to incoming solar radiation. The remaining fraction is from geothermal energy, gravitational (tidal) energy, and nuclear energy. In Chapter 4 we discussed the nuclear origins of solar energy. The solar radiation of interest to us is electromagnetic radiation, which includes radio waves, x-rays, and ultraviolet, infrared, and visible light. This radiation is transmitted at a constant speed, the speed of light, which is independent of wavelength. The frequency (f), wavelength (λ), and speed (c) (a constant quantity) are related by $f\lambda = c$, requiring that electromagnetic radiation of high frequency has short wavelength and the converse. In considering the interaction of radiation with matter, it is often useful to describe the radiation as arriving in quanta, discrete bundles of electromagnetic energy, each having an energy related to the frequency by $E = hf$, where h is a constant of nature known as Planck's constant. This constant has a value of $h = 6.626 \times 10^{-34}$ joule seconds (J • sec).

The solar spectrum is usually presented as in Figure 6.2. This figure shows the wavelength distribution at the top of the atmosphere, or in other words, at 93×10^6 miles (1.48×10^8 m) from the sun with no intervening atmosphere or clouds. Both the temperature of the sun's surface (5800°K) and the radiated power (64 megawatts per square meter of solar surface) can be deduced from this spectrum. The power in this spectrum at 93×10^6 miles from the sun is 1.97 calories/min • cm^2, where the intercepting area is perpendicular to a radius vector having its origin at the sun's location. This number, rounded off to 2 cal/min • cm^2 and measured at the top of the earth's atmosphere, is known as the solar constant. It varies only slightly in time, being 3% stronger in the winter and 3% weaker in the summer, because of the varying sun–earth distance.

To compute the average power available per square centimeter of horizon-

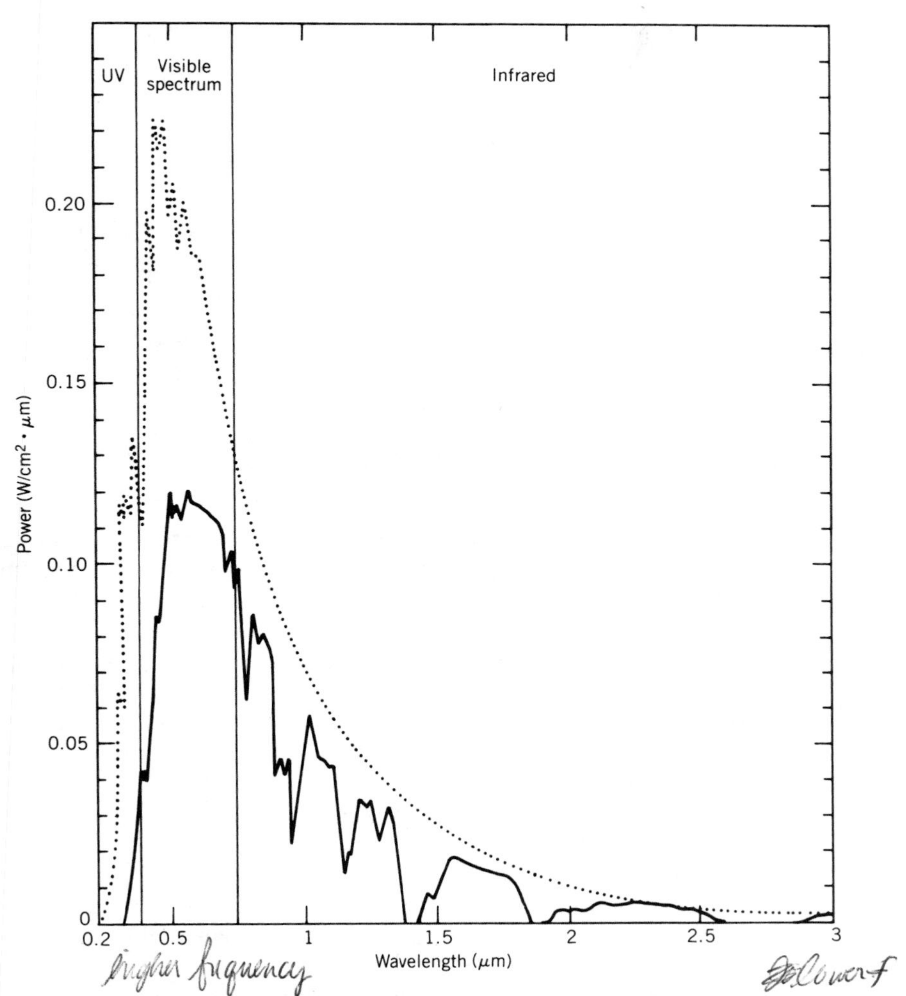

Figure 6.2 The wavelength distribution of solar radiation above the atmosphere (dotted line) and at the earth's surface (solid line). The solar constant is given by the area under the dotted curve.

tal surface at the top of our atmosphere, we must correct for the fact that only one face of the earth is illuminated at any one time. As seen by the sun, the earth appears to be a disk of radius r_e and area πr_e^2. But the earth's total surface area is $4\pi r_e^2$, so the average power per unit horizontal area at the top of the atmosphere is the total incident power divided by the total area, or

$$\frac{2 \text{ cal/min} \bullet \text{cm}^2 \times \pi r_e^2}{4\pi r_e^2} = \tfrac{1}{2} \text{ cal/min} \bullet \text{cm}^2$$

This average value is over day and night and over all latitudes.

The solar spectrum on the earth's surface, beneath its protective atmosphere, is quite different from the incident spectrum at the top of the atmosphere because of the absorption properties of the atmospheric gases. This situation is shown in Figure 6.2.

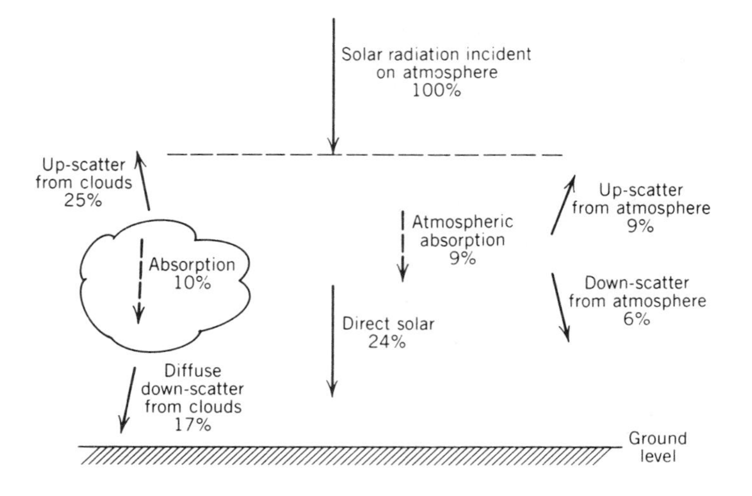

Figure 6.3 Absorption and scattering of solar radiation in the atmosphere. The values shown are for average weather, and are averaged over all seasons and latitudes.

In addition to comparing the wavelength distribution at ground level to that at the top of the atmosphere, it is also interesting to examine the intensity of solar radiation at ground level relative to the upper-atmosphere intensity, considering all wavelengths that reach the earth. This can be illustrated by a simple figure. From Figure 6.3 we see that on average, 47% (17% + 24% + 6%) of all incident solar energy reaches the earth, 19% is absorbed by the atmosphere, and 34% is reflected away from the earth. Of the incident radiation at the earth's surface (on average 47% of all solar energy incident on the upper atmosphere), about 51% is direct and unscattered; about 49% is scattered and diffuse. On a relatively clear day, perhaps 75% would be direct and unscattered. Of course, different regions of the globe may differ either upward or downward, by a factor of perhaps 2, from these average figures because of typical local weather conditions. A photo of the earth, shown in Figure 6.4, strikingly illustrates how clearly cloud patterns can be seen owing to the solar radiation scattered upward.

With this information, the average value (again, over all weather conditions, times of day, latitudes, seasons, etc.) can now be calculated in familiar units for the solar radiation reaching the earth's surface.

Figure 6.4 A view of earth from an Apollo spacecraft. Almost the entire western hemisphere is included in this photograph. South America, in the lower half, is almost completely covered with clouds except for a portion of the high Andes along the western coast.

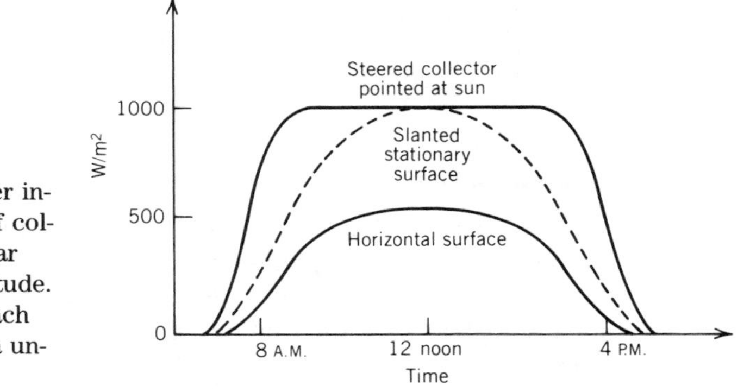

Figure 6.5 Solar power incident on three types of collectors for a typical clear winter day at 40° N latitude. The energy collected each day is given by the area under each curve.

$$\tfrac{1}{2}\ \text{cal/min}\cdot\text{cm}^2 \times \tfrac{1}{60}\ \text{min/sec} \times 4.184\ \text{J/cal} \times 10^4\ \text{cm}^2/\text{m}^2 \times 0.47\ (\text{transmitted}) \times 1\ \text{W}\cdot\text{sec/J} = 164\ \text{W/m}^2 \text{ of horizontal surface area}$$

This, of course, is only an average value, including all hours of the 24-hour day. If we consider only the 8-hour day centered about noon, at 40° N latitude, this 8-hour average is substantially larger, 600 W/m^2 of horizontal surface area, averaged over a typical year. For each such average 8-hour day, the energy incident on a 1 m^2 horizontal surface would be

$$8\ \text{hr} \times 600\ \text{W/m}^2 = 4.8\ \text{kW}\cdot\text{hr/m}^2$$

This is equivalent to the energy content of 0.13 gallons of gasoline per square meter each day, or 1 gallon of gasoline a day for each 7.7 m^2 horizontal surface. For a typical single-family residence of 93 m^2 (1000 ft^2) horizontal roof surface, this would amount to an equivalent of 12 gallons of gasoline a day. The annual 8-hour average of 600 W/m^2 for a city at this latitude is reduced to 300 W/m^2 for the coldest months of winter and is about 1000 W/m^2 for the summer months. The winter 8-hour average is about 600 W/m^2 for a properly slanted collector. On June 20, at 40° N latitude, the noontime sun is incident at an angle about 17.5° south of vertical; on December 20, the noontime sun angle is about 27.5° above the horizontal. The general description of how horizontal, slanted, and steerable collectors compare is shown in Figure 6.5.

6.3 HEAT TRANSFER

Much of our current and projected use of solar energy relies on converting the solar radiation to heat energy and then using this heat energy for space heating, driving heat engines, generating electricity, and so forth. Exceptions to this use of the sun's energy as a source of heat include the direct photovoltaic conversion of solar radiation to electricity and the fixing of atmospheric carbon into biomass (photosynthesis). To proceed with a serious discussion of heat energy, we must first understand something of the mechanisms by which this energy may be transferred from one location to another, and from one medium to another. Heat transfer is accomplished by three quite distinct modes: conduction, convection, and radiation, acting either alone or in concert. They can be considered one at a time.

CONDUCTION

Thermal conduction will take place through any material medium between any two points at differing temperatures. Heat energy will flow by conduction from regions of higher temperature to regions of lower temperature. There is no heat transfer by conduction through empty space, such as in a vacuum chamber. Thermal conduction proceeds by transferring the energy of vibrating atoms, molecules, and electrons to their less energetic neighbors. The ability to conduct heat in this way varies enormously from one material to another, depending on, among other factors, the density of free electrons in the material. Every material has an associated thermal conductivity, k, which is the rate of heat flow across a unit thickness per unit cross-sectional area per unit temperature gradient. Appropriate units for thermal conductivity are: Btu per hour per square foot per degree Fahrenheit across a 1-inch thickness (Btu • in./hr • ft^2 • °F) or, in metric terms, joules per second per square centimeter per degree Celsius across a 1-centimeter thickness (J • cm/sec • cm^2 • °C). In the latter case, the thermal conductivity is equal numerically to the number of watts (or joules per second) conducted from one face of a cubic centimeter of some material to the opposite face when the two faces differ in temperature by 1 degree Celsius. Thermal conductivities for several substances are given in Table 6.1. Note that the mechanisms for thermal conduction can be especially complex for porous materials, gases, and liquids because conduction, convection, and radiation processes are simultaneously present in the space occupied by the medium. In the strictest sense, the term "thermal conductivity' can apply only to solid media. Nevertheless, the given values, which result from measurement under laboratory conditions, are useful in heat-transfer calculations. They do represent how well heat is transmitted through the bulk of each material.

A calculation of how much heat is conducted per unit time, Q/t (Btu/hr), through some material with a thickness of l (inches), an area A (ft^2), and a thermal gradient of $T_2 - T_1$ (°F), can be carried out with the use of the following expression:

$$Q/t = \frac{kA(T_2 - T_1)}{l}$$

It is frequently useful to rewrite the above expression in terms of the conductance, $U = k/l$, so that the above expression is

$$Q/t = UA(T_2 - T_1)$$

Example 6.1

How many Btu/hr are conducted through a 5-in.-thick wall of concrete if the temperature on one face is 70°F and the temperature on the other face is 15°F and the area of the wall is 80 ft^2?

Solution

It is convenient to calculate first the conductance of the concrete. Obtain a value of k from Table 6.1.

$$U = \frac{k}{l} = 12.0 \frac{\text{Btu} \cdot \text{in}}{\text{hr} \cdot \text{ft}^2 \cdot {}^\circ\text{F}} \times \frac{1}{5 \text{ in.}} = 2.4 \frac{\text{Btu}}{\text{hr} \cdot \text{ft}^2 \cdot {}^\circ\text{F}}$$

The heat-transfer rate is then

$$\begin{aligned} Q/t &= UA(T_2 - T_1) \\ &= 2.4 \frac{\text{Btu}}{\text{hr} \cdot \text{ft}^2 \cdot {}^\circ\text{F}} \times 80 \text{ ft}^2 \times (70^\circ\text{F} - 15^\circ\text{F}) \\ &= 10{,}560 \frac{\text{Btu}}{\text{hr}} \end{aligned}$$

In practical situations, there are frequently several layers of different material that make up any insulating wall. This type of problem is treated in more detail in a later chapter when we discuss energy conservation. For the moment, it is sufficient to recognize that the effect of multiple layers of insulation will

Table 6.1 TYPICAL VALUES OF THERMAL CONDUCTIVITY k NEAR 20°C

Substance	$\frac{\text{J} \cdot \text{cm}}{\text{sec} \cdot \text{cm}^2 \cdot {}^\circ\text{C}}$	$\frac{\text{Btu} \cdot \text{in.}}{\text{hr} \cdot \text{ft}^2 \cdot {}^\circ\text{F}}$
Metals		
Silver	4.23	2930
Copper	3.85	2680
Gold	2.93	2030
Brass	1.09	750
Iron and steel	0.46	320
Aluminum	2.01	1390
Mercury, liquid	0.063	44
Nonmetallic solids		
Brick, common	7.1×10^{-3}	5.0
Concrete	1.7×10^{-2}	12.0
Wood (across grain)	1.3×10^{-3}	0.9
Glass	5.9×10^{-3}	4.0
Ice	2.2×10^{-2}	15.4
Porous materials		
Fiber-blanket insulation	3.8×10^{-4}	0.27
Glass wool or mineral wool	3.8×10^{-4}	0.27
Sawdust	5.9×10^{-4}	0.41
Corkboard	4.2×10^{-4}	0.30
Liquids		
Water	5.99×10^{-3}	4.15
Ethyl alcohol	1.76×10^{-3}	1.23
Gases		
Air	2.34×10^{-4}	0.16
Hydrogen	1.7×10^{-3}	1.16

be to reduce the heat flow through any one layer. It is convenient to define a quantity for each layer that can be directly added to those of the other layers to assess the effect of the composite wall. For this purpose the R-factor is defined as

$$R_1 = \frac{1}{U_1} = \frac{l_1}{k_1}$$

If a wall is made up of layers of brick, fiber blanket insulation, and wood, as shown in Figure 6.6, the individual conductances U_b, U_i, and U_w can be calculated, and then the values of R_b, R_i, and R_w can be obtained. The effect of the entire wall can be calculated by simply adding the R-factors:

$$R_{\mathrm{T}} = R_{\mathrm{b}} + R_{\mathrm{i}} + R_{\mathrm{w}}$$

The heat conducted per unit time for the entire wall is readily found from

$$Q/t = \frac{A}{R_{\mathrm{T}}}(T_2 - T_1)$$

R-values for various materials used in the construction of houses or solar panels are available so that heat conduction losses can be calculated.

In the previous example of heat conducted through a wall, the inner and outer wall temperatures were specified. In practical calculations of heat losses through walls and windows, it is important to realize that the inner and outer surfaces may not be at the same temperature as the air at some distance from the surface. Each surface has associated with it a thin layer of relatively immobile air that acts as an additional insulator. The approximate effect is as follows: On the interior surfaces an additional R-factor of about 0.7 (hr • ft^2 • °F/Btu) is provided by this boundary layer of air, and on exterior surfaces the added R-factor is typically 0.17 (hr • ft^2 • °F/Btu), depending on wind velocity. For single-pane glass windows, these boundary layers are the dominant part of the total thermal resistance. The temperature drop across the glass itself is typically quite small, as $\frac{1}{8}$-in. thick glass has an R-value of only about $R = 0.03$ (hr • ft^2 • °F/Btu). The main role of the glass is to prevent direct movement of air between inside and outside and immobilize millimeter-thick layers of air on

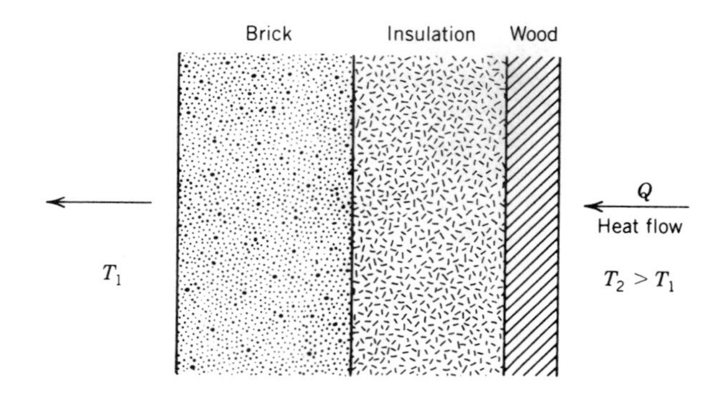

Figure 6.6 Heat flow through a layered wall.

both inner and outer surfaces. A $\frac{1}{8}$-in. sheet of aluminum or a thin plastic sheet would do approximately as well as glass.

CONVECTION

Liquids and gases transfer heat principally by convection, which is motion of a medium, such as air, between regions at different temperatures, which thus transfers heat energy. Convection may be forced, as by a blower for air or a pump for liquids, or it may be natural, driven by gravity. When air warmed near the surface of a stove rises and cooler air near the inner surface of a window moves downward, natural convection takes place. Convective heat transfer is difficult to analyze exactly; semiempirical rules are usually used. The medium's viscosity, specific heat, density, and thermal conductivity all influence the efficiency of heat transfer by this means. Most of our space heating systems operate through convective heat transfer.

THERMAL RADIATION

All materials are constantly emitting and absorbing thermal radiation. The sun's light, the redness of a very hot stove, and the direct heat we feel from a campfire are all examples of thermal radiation. The radiation is electromagnetic, in the same general class as radiowaves, microwaves, x-rays, and gamma radiation. Thermal radiation has its origin in the motion of electrons. These charged particles, in random motion, with abrupt changes in direction, radiate electromagnetic energy. All electromagnetic radiation arises from the acceleration of electric charges. The properties of thermal radiation are of particular importance to us because this is the means by which we receive the sun's energy.

The intensity and wavelength distribution of the thermal radiation emitted by any body depend on the surface temperature and on a property of the surface known as the emissivity. The power radiated per unit area is given by Stefan's law:

$$\frac{P}{A} = \epsilon\sigma T^4$$

where P/A is the power in watts emitted per square meter, ϵ is the surface emissivity, $\sigma = 5.67 \times 10^{-8}$ W/m$^2 \cdot$ °K^4 is the Stefan–Boltzmann constant, and T is the surface temperature in degrees Kelvin. Note that the radiated power increases dramatically with increasing temperature because of the fourth-power dependence. A doubling of the absolute temperature produces a 16-fold increase in the radiated power because $2^4 = 16$.

The *emissivity*, ϵ, is a dimensionless factor that is related to the rate of thermal radiation from a particular surface. It varies from a maximum of 1.0 down to very small values, close to zero. It varies with surface temperature, roughness, color, and degree of oxidation. Several values are shown in Table 6.2. It is interesting to note that most building materials have emissivities near 0.9, but that the aluminum foil commonly used in thin layers on insulating panels has an emissivity of less than 0.10. The low value obviously leads to reduced heat transfer by thermal radiation. To perform a complete radiant heat-transfer calculation, it is also necessary to have another factor, *absorptivity*

Table 6.2 EMISSIVITY OF SURFACES NEAR T = 300°K

Material	$\epsilon(T \approx 300°K)$
Aluminum	
Polished	0.04
Rough plate	0.06
Oxidized	0.15
Cast iron	0.50
Sheet steel	0.70
Wood, black lacquer, white enamel, plaster, roofing paper	0.90
Porcelain, marble, brick, glass, rubber, water	0.94

(α)—the fraction of the radiation impinging on a surface that is directly absorbed as heat. *Reflectivity* (r) is the complement ($r = 1 - \alpha$) of absorptivity for opaque surfaces.

Fortunately, for ease of calculation of radiant heat transfer between two facing surfaces, it can be shown that the emissivity is numerically equal to the absorptivity when the absolute temperatures of the two surfaces in question are not very different, as in cases involving building insulation. In other words, a surface that is a good reflector of incident radiation from a source at its own temperature is also a poor radiator of thermal energy. In contrast to the common cases involving building insulation, it is not usually true that the absorptance of a solar collector surface for the sun's radiation is numerically similar to the emittance of the collector surface. This difference is due to the large difference between temperatures of the collector surface ($T \approx 300°K$) and the sun's surface ($T \approx 6000°K$), and it leads to the concept of a selective surface, which will be discussed in greater detail later.

A common situation requiring the computation of radiant heat transfer is that between two parallel flat surfaces. It can be shown by computations too detailed to go into here that the rate of radiant heat transfer between two surfaces, each having area A, is

$$\text{power} = \frac{\sigma A(T_1^4 - T_2^4)}{1/\epsilon_1 + 1/\epsilon_2 - 1}$$

where subscript 1 refers to one surface and subscript 2 refers to the facing surface. If $\epsilon_1 = \epsilon_2$ (as with two facing aluminum foil surfaces), this expression reduces to

$$\text{power} = \frac{\epsilon}{2 - \epsilon}\sigma A(T_1^4 - T_2^4)$$

This equation tells us that the radiant heat transfer varies with the difference between the fourth powers of the absolute temperatures of each surface [not to be confused with $(T_1 - T_2)^4$], and as a factor that depends on the emissivity.

As a practical example, let us calculate the radiant heat-transfer rate in watts per square meter between two plane surfaces, one at −20°C (253°K) and the other at 25°C (298°K), where the emissivity of both surfaces is 0.9, as in common building materials.

$$\frac{\text{power}}{\text{area}} (\text{W/m}^2) = \frac{\epsilon}{2 - \epsilon} \sigma(T_1^4 - T_2^4)$$

$$= \frac{0.9}{2.0 - 0.9} \times (5.67 \times 10^{-8} \text{ W/m}^2) \times (298^4 - 253^4)$$

$$= 175.8 \text{ W/m}^2 \text{ (or 55.5 Btu/hr} \cdot \text{ft}^2)$$

Now, what would the radiant heat-transfer rate be if both surfaces were faced with aluminum foil ($\epsilon = 0.1$)?

$$\frac{\text{power}}{\text{area}} (\text{W/m}^2) = \frac{0.1}{2.0 - 0.1} \times (5.67 \times 10^{-8} \text{ W/m}^2) \times (298^4 - 253^4)$$

$$= 11.3 \text{ W/m}^2 \text{ (or 3.58 Btu/hr} \cdot \text{ft}^2)$$

These calculations show that the heat lost by radiation is about 15.5 times less when aluminum foil is added. To gauge the importance of radiation heat loss, the earlier figure, 55.5 Btu/hr • ft^2, can be compared to the heat lost by conduction and convection across an air gap having an effective R-1 thermal resistance factor. Then the heat loss rate by conduction and convection would be

$$\text{Power (Btu/hr} \cdot \text{ft}^2) = \frac{\Delta T(°\text{F})}{R} = \frac{81}{1} = 81 \text{ Btu/hr} \cdot \text{ft}^2$$

where $\Delta T = 81°$F is the same as the 45°C temperature difference.

So in the case at hand, we have the radiation heat loss equal to 55.5/(55.5 + 81) = 41% of the total heat transfer without the aluminum foil surfaces; it is reduced to 3.58/(3.58 + 81) = 4.2% of the total heat loss when the facing aluminum foil surfaces are present.

Earlier in this chapter, we mentioned that the surface temperature of the sun could be deduced from the solar spectrum. Figure 6.7 shows the spectra of the emitted power for several temperatures for blackbody radiation (meaning $\epsilon = 1$). These spectra display maxima at different wavelengths depending on temperature. The relationship between the wavelength location of these maxima and the temperature of the radiating surface is given by the Wien displacement law:

$$\lambda_{max}(\mu\text{m}) \times T(°\text{K}) = 2897.8 \ \mu\text{m } °\text{K}$$

Thus, if we know (by measurement) that $\lambda_{max} = 3.5 \ \mu$m for some particular blackbody, then we can deduce the temperature of the radiating surface to be

$$T = \frac{2897.8}{3.5} = 828°\text{K}$$

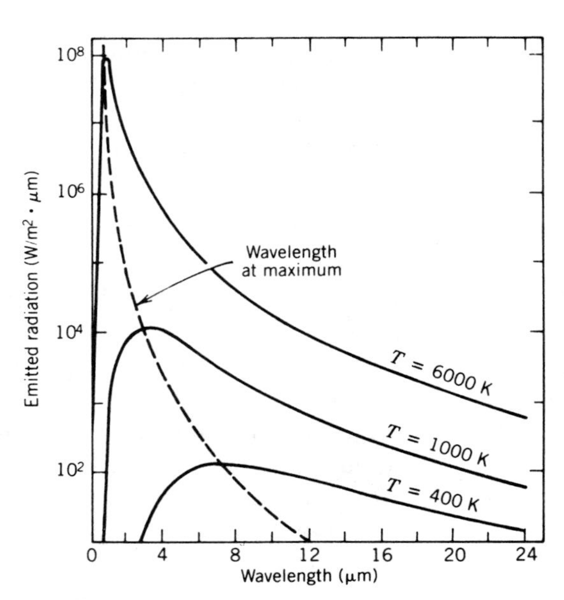

Figure 6.7 Wavelength distribution for emitted radiation from a blackbody.

Note that whereas the Stefan's law relates the radiated power to the surface temperature, the Wien displacement law relates the wavelength distribution (which appears as color) to the surface temperature. The Wien displacement law describes the systematic progression toward longer wavelengths of emitted radiation as the surface temperature becomes cooler. It is interesting to note that the surface temperature of the sun corresponds to maximum radiated intensity at wavelengths to which the human eye is most sensitive.

6.4 HEAT STORAGE

There are a few applications of solar energy, such as water pumping, crop drying, and solar stills, where the intermittent nature of the incoming energy can be tolerated. For space heating, water heating, and electricity generation, however, there must be some means of smoothing out the fluctuations of the solar radiation (insolation). Various means for achieving this smoothing through energy storage have been proposed and utilized. These means include biomass, hydrogen generation, electric batteries, phase-change media, and thermal mass. For small-scale space heating applications, thermal mass of some sort is used to store energy from times of incoming sunlight for the dark hours of night or even for sunless days.

The effectiveness of any material as a thermal mass depends on its density, thermal conductivity, dimensions, location, and specific heat.

The specific heat is simply the number of Btus needed to raise the temperature of 1 pound of the substance by 1 degree Fahrenheit. Of course, on cooling, 1 pound of the substance will give to its surroundings a number of Btus equal to the specific heat for each degree Fahrenheit of cooling. Water has a specific heat of 1.0 Btu/lb • °F, by the definition of the Btu. All other common substances have lower values. Several common building materials are listed in Table 6.3 along with some of their heat-storage properties. Within a

Table 6.3 HEAT STORAGE PROPERTIES OF SOME COMMON MATERIALS

Material	Specific Heat (Btu/lb • °F)	Density (lb/ft^3)	Volumetric Heat Capacity (Btu/ft^3 • °F)	Thermal Conductivity (Btu • in./hr • ft^2 • °F)
Water	1.0	62.4	62.4	4.2
Iron	0.11	490	54	320
Glass	0.20	170	34	4.0
Stone	0.21	160	34	3.0
Stone, loose	—	—	20	—
Wood, oak	0.57	51	29.1	1.4
Ice	0.46	57	26.2	15
Brick	0.22	112	24.6	4.6
Concrete	0.156	144	22.4	12
Wood, pine	0.67	31	20.8	0.7
Sand	0.195	100	19.5	2.3

Note: The volumetric heat capacity is the product of the specific heat times the density. The densities given are for solid material with no voids.

building, all materials contribute to heat storage. At times when the air is warmer than the objects, heat energy is transferred into them; at night when the room air cools below the temperature of the objects, heat energy is transferred back into the air. The walls, dishes in the kitchen cabinets, books on the shelves, and the furniture all contribute to smoothing out temperature fluctuations. The contribution of any item to the thermal mass can be calculated on a per pound or per cubic foot basis. In such estimates, it is also important to consider the thermal conductivity. For instance, a 6-in.-thick wall of solid wood will store about the same amount of heat energy as will a concrete wall of the same thickness if they are both raised in temperature by the same amount. However, the lower thermal conductivity of wood means that it will take much longer for the heat to flow either into or out of the wooden storage wall. This problem can be resolved, to some extent, by arranging the wooden storage medium so that more of the surface is exposed to airflow, as in 1-in.-thick layers with convective channels between them, or by stacking the wood loosely as in a pile of firewood.

For an active system with forced circulation of air or water, the thermal mass is usually in the form of a large tank of water or several tons of stones in an insulated container.

6.5 SOLAR SPACE AND WATER HEATING

The direct heating of residences and commercial buildings has become an important and increasingly popular use of the sun's energy during the past decade. The collectors used for this purpose, as well as for heating water, fall into two general categories: concentrating (or focusing) and flat-plate. The concentrating collectors, as typified by parabolic mirrors, can produce very high temperatures in a small region of space, as is needed for cooking or for the production of steam. They utilize primarily the direct, unscattered component of the solar radiation and are, therefore, usually mounted on a tracking mechanism such that during daylight hours they can always be pointed directly at the sun. Flat-plate collectors are much simpler in geometry and therefore easier and usually cheaper to build than concentrating devices. They utilize both the diffuse and direct components of the solar radiation and can be installed in a stationary position. These devices are well matched to the relatively low temperatures ($\leq$90°C) suitable for space and water heating; their efficiency drops rapidly as high temperatures are approached.

In addition to strictly concentrating and flat-plate collectors, there are a number of designs that fit somewhere midway between the two definitions. It is not uncommon for the performance of flat-plate devices to be enhanced by various arrangements of reflectors that concentrate sunlight on their surfaces; some concentrating devices, such as parabolic troughs, have broad acceptance angles for direct radiation and need not be constantly pointed directly at the sun.

Although solar heating systems have been under investigation and development by talented and dedicated individuals for generations, the widespread adoption of these systems has consistently been held back by economic factors. During recent decades, technological advances have seldom driven the cost to the consumer of capturing elusive sunlight below the cost of fuel. This situation now is changing; fuel costs are inflating, fossil fuel reserves are diminishing, and the size of the world's population to be supported by finite fuel resources is increasing. Space heating by active systems may now be approaching economic competitiveness with the use of fuels. The heating of water for domestic use is much more favorable economically because the system provides benefits 12 months of the year, including the hot months of summer when a space heating system is idle. Passive designs may now also be economical, because much of the construction is part of the building rather than an additional cost.

6.6 A FLAT-PLATE COLLECTOR SYSTEM

In designing a residential heating system, an important first step is to estimate the approximate sizes for the collector and heat-storage unit. As a guide, we can take the Federal Housing Administration (FHA) 1972 building and insulation standard (based on interior volume) of 1 Btu/ft^3 per degree day (see Chapter 9 for definition) as a measure of the heat loss of a house and, therefore, the heat that must be supplied to maintain a 65°F interior temperature. If the house in question has 1500 ft^2 of floor space and 8-ft ceilings, we shall then have a demand of 1500 ft^2 $\times$ 8 ft $\times$ 1 Btu/ft^3 per degree day = 12,000 Btu per degree day. On a winter day with an outside temperature of 15°F averaged over 24 hours, we shall then need 12,000 Btu per degree day $\times$ (65 $-$ 15) degree day = 600,000 Btu delivered from all sources of heat. It is probably reasonable to assume that about 100,000 Btu will be available daily from various internal sources (somewhat greater than the values shown in Table 6.4 for a conventional home); for simplicity, the solar heat gain through the windows can be ignored at this point. This leaves 500,000 Btu needed for this day. A typical solar system might be expected to supply 80% or 400,000 Btu of this amount. The remaining heat energy can be supplied by an auxiliary heating system. In

Table 6.4 INTERNAL HEAT SOURCES IN A CONVENTIONAL HOUSE, CORRECTED FOR HUMIDIFICATION

Source	Btu/hr	Btu/day
Lighting	440	10,560
Cooking	311	7,464
People	514	12,336
Television	195	4,680
Water heater	520	12,480
Dishwasher	140	3,360
Refrigerator–freezer	642	15,408
Clothes dryer	57	1,368
Other	195	4,680
Indoor plants	−92	−2,208
	2,922	70,128

Source: Adapted from Ross and Williams, p. 303.

addition to a collector, storage mass equivalent to about one day's collected energy, or 400,000 Btu, will be needed.

A collector such as the one shown in Figure 6.8 will be effective, incorporated into the system shown in Figure 6.9. The flat-plate collector shown includes in its design many features that stem directly from the principles of heat transfer previously discussed. The absorber is usually painted black so that the absorptivity of the solar radiation on the metal sheet is as high as

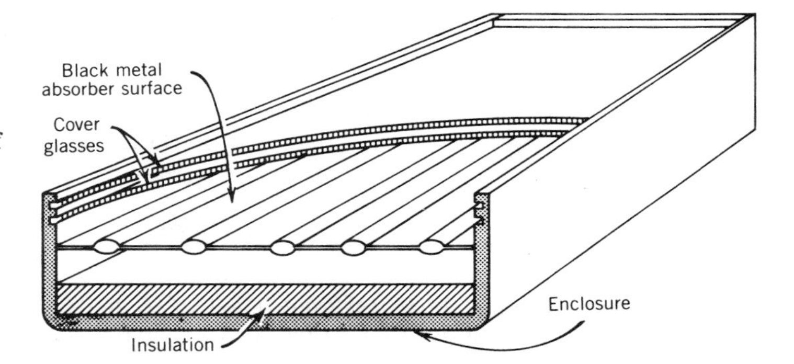

Figure 6.8 A cutaway view of a flat-plate solar collector with two cover glasses. A heat-transfer fluid is circulated through the tubular passages integrally formed into the metal absorber surface. (Not drawn to scale.)

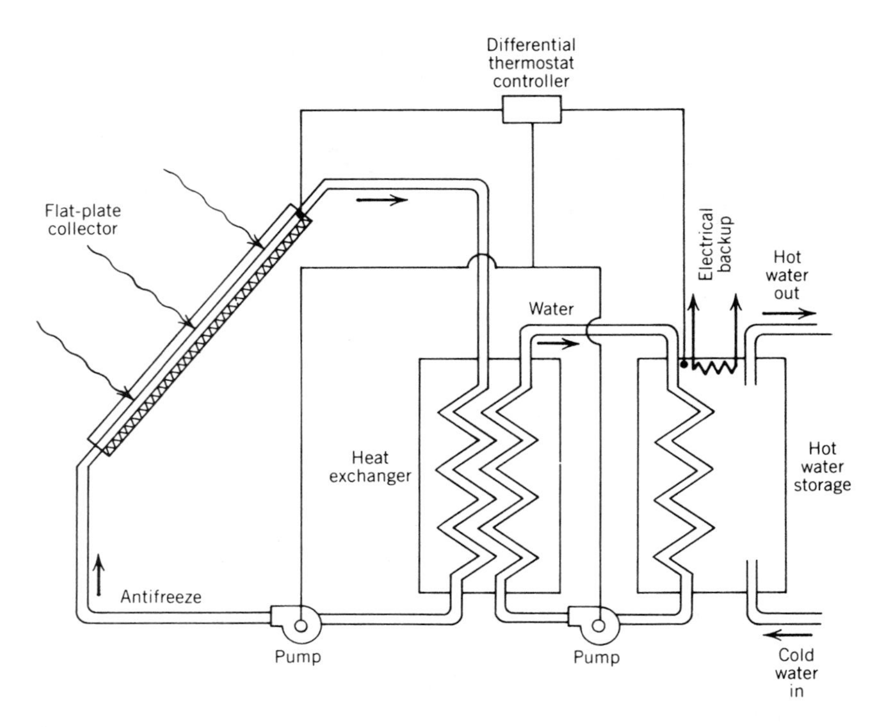

Figure 6.9 A circulating-liquid solar collector system that provides hot water for space heating and domestic use. In a typical installation the collector will be on the roof of the building, with the other components inside in a utility area.

possible. The tubes, or pipes, carrying the water (or antifreeze solution) are in good thermal contact with the absorbing sheet so that heat can be transferred by conduction to the fluid. Sometimes the pipes are directly soldered to the sheet, and sometimes they are part of the sheet construction (tube-in-sheet). The fluid carries the heat (by convection) to the thermal storage medium. The efficiency of the flat-plate collector is dependent on keeping the heat losses to a minimum. The double-glazed window on the flat-plate collector is designed to minimize heat losses. The effective *R*-value of a single sheet of glass is about 0.8 (hr • ft^2 • °F/Btu), whereas that for two sheets with an adequate space between them (0.5 in. or more) is about 1.7. The insulation on the back side of the collector is typically several inches of fiberglass or Styrofoam insulating material with a total *R*-value of 12 or so.

Earlier in this chapter, a value of 600 W/m^2 was given for the solar energy incident on a slanted collector at 40° latitude during an 8-hour winter day. This translates into

$$600 \text{ W/m}^2 \times 8 \text{ hr/day} \times 3.41 \text{ Btu/W} \cdot \text{hr} = 16{,}400 \text{ Btu/m}^2 \cdot \text{day}$$

incident on the collector surface. For a 50% collector efficiency, which is considered good in practice, the net yield would be 8200 Btu a day per square meter of collector surface. For the necessary 400,000 Btu, the required area is

$$\frac{400{,}000 \text{ Btu}}{8200 \text{ Btu/m}^2} = 49 \text{ m}^2 \text{ or } 530 \text{ ft}^2$$

Now for storage, either water or stones would be a common choice. Water stores heat at the rate of 1 Btu/lb • °F, or 62.4 Btu/ft^3 • °F. For stone, considering the voids in the storage unit, the heat capacity is about 20 Btu/ft^3 • °F. These numbers indicate that for water heated to 130°F and then used as a heat source until its temperature drops to 80°F, a volume of

$$\frac{400{,}000 \text{ Btu}}{62.4 \text{ (Btu/ft}^3 \bullet {}^\circ\text{F)} \times 50^\circ\text{F}} = 128 \text{ ft}^3$$

or about 1000 gallons would be needed. If stones are used for heat storage, the necessary volume, under the same ΔT conditions, would be

$$\frac{400{,}000 \text{ Btu}}{20 \text{ Btu/ft}^3 \bullet {}^\circ\text{F} \times 50^\circ\text{F}} = 400 \text{ ft}^3$$

or a cube about 2 m on an edge. In terms of weight, this amounts to about

$$400 \text{ ft}^3 \times \frac{1 \text{ yd}^3}{27 \text{ ft}^3} \times 1.35 \frac{\text{tons}}{\text{yd}^3} = 20 \text{ tons}$$

or two dumptruck loads.

Earlier in this section we indicated that the efficiency of a flat-plate collector decreases with increasing temperature, with values of 50% considered good for space heating applications. In discussions of solar heating systems, efficiency is usually defined as the ratio of useful heat energy delivered to a building or to the hot-water tank divided by the solar energy incident on the collector's front surface. There are several reasons why the efficiency actually achieved is less than 100%. Not all of the sun's energy penetrates the glass covers to the black collector surface, and once the collector surface attains a temperature higher than its surroundings, heat will be lost by conduction, convection, and radiation. The transmission of sunlight through glass is shown in Figure 6.10 as a function of the angle of incidence. The fraction not transmitted is partially reflected and partially absorbed. If the collector structure, plumbing or ducting, and storage unit are not well insulated, heat loss by conduction will be a cause of low efficiency. Some advanced designs use collector structures made of vacuum-insulated glass tubes to reduce heat loss by convection. The radiative heat loss becomes increasingly important at higher collector plate temperatures because of the steep temperature dependence expressed in Stefan's law, $P/A = \epsilon\sigma T^4$. Figure 6.11 shows the relative wavelength distributions for the solar spectrum and for the radiation from a collector surface at 200°F. The different wavelength distributions for sunlight and for infrared radiation are in accordance with Wien's displacement law, discussed earlier in this chapter. Because glass is relatively more transparent for the shorter wavelength radiation in the solar spectrum than for the infrared spectrum from the collector, radiant energy is transmitted through the cover glass into the collector box more easily than the infrared radiation can escape. This trapping of the sun's energy under a

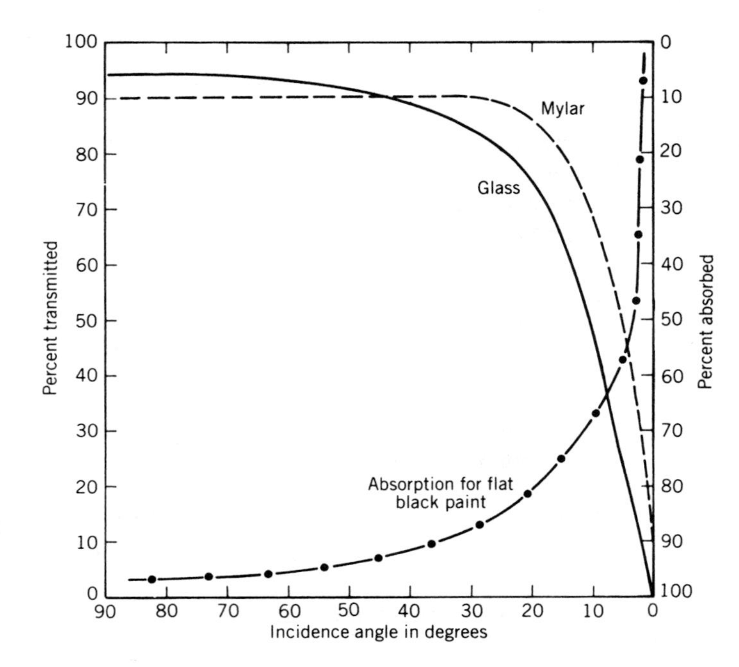

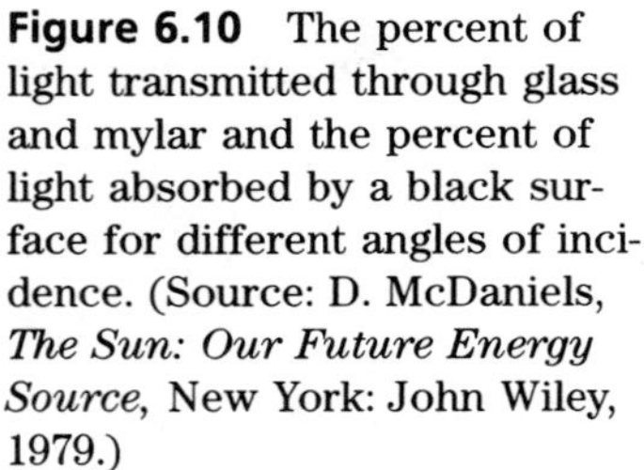

Figure 6.10 The percent of light transmitted through glass and mylar and the percent of light absorbed by a black surface for different angles of incidence. (Source: D. McDaniels, *The Sun: Our Future Energy Source*, New York: John Wiley, 1979.)

glass cover plate is often called the "greenhouse effect." It accounts, in part, for the high temperatures often experienced when an automobile is parked in the bright sunlight. The trapping of sunlight can be further enhanced through the use of "selective surfaces" or "selective coatings" on the collector plate surface. These specially prepared metal coatings have high absorptance for sunlight and low emissivity for infrared radiation.

Because the conductive, convective, and radiative heat losses all increase with temperature, the net system efficiency shows a temperature dependence,

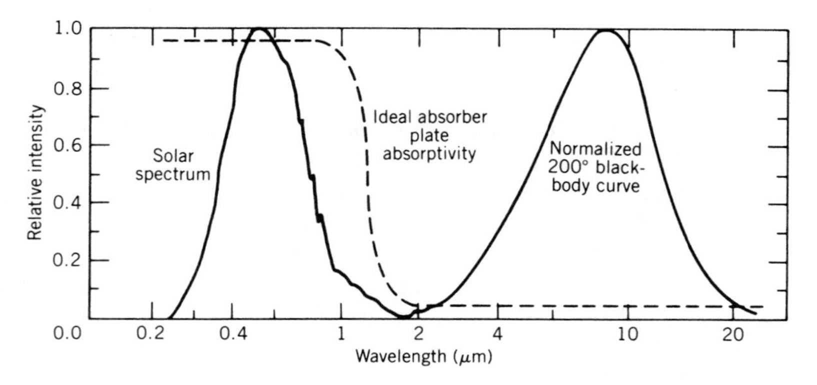

Figure 6.11 The wavelength distributions for radiated power in the solar spectrum and from a surface at 200°F. The absorption characteristics are also shown for an ideal selective surface that would have high absorption in the solar region and have very low emissivity in the long-wavelength region. (Source: D. McDaniels, *The Sun: Our Future Energy Source*, New York: John Wiley, 1979.)

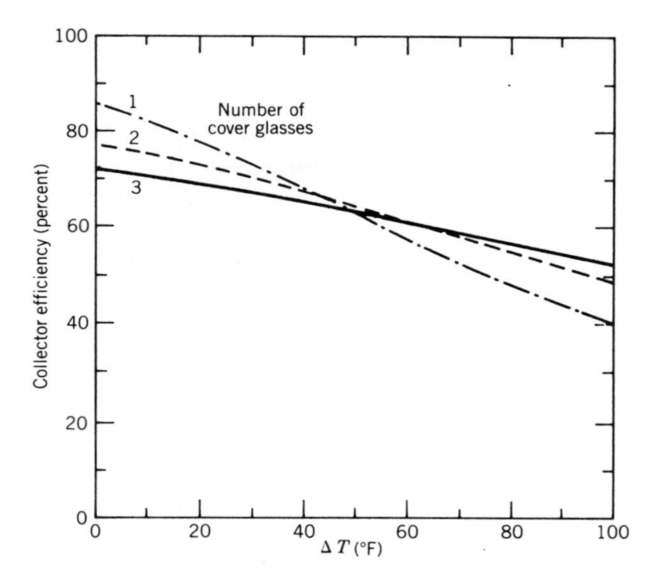

Figure 6.12 The efficiency of a flat-plate collector versus ΔT, the temperature difference between the inside and outside of the collector. At low ΔT the efficiency is highest with only one cover glass; at higher ΔT more cover plates produce an advantage by limiting heat loss.

as Figure 6.12 shows. From this figure it is apparent that flat-plate devices are best suited to the low temperatures suitable for space heating and are not generally useful for temperatures above about 200°F. At lower temperatures the collector performance is better with a single cover glass, because it is more important to admit the maximum solar energy than to inhibit conductive and convective heat losses. At higher temperatures the situation is reversed.

Example 6.2

Calculate the collector surface area necessary to heat 100 gallons of water a day (sufficient for a family of four) from 50°F to 120°F when the daily insolation on a slanted collector is 1000 Btu/ft^2. Assume 33% efficiency.

Solution

$$\text{heat needed} = 100 \text{ gal} \times 8 \text{ lb/gal} \times 70°\text{F} = 56{,}000 \text{ Btu}$$

Because $\eta = \frac{1}{3}$, there must be $3 \times 56{,}000 = 168{,}000$ Btu incident on the collector surface each day. This would require

$$\frac{168{,}000 \text{ Btu}}{1000 \text{ Btu/ft}^2} = 168 \text{ ft}^2 \text{ area}$$

The economic benefits of active solar systems for space and water heating for homes in the United States are very dependent on the location of the home and the assumptions made concerning the future costs of natural gas or electricity. The retrofitting of systems into older homes can also lead to plumbing complications. There may also be a lack of an adequate southern exposure. The *Energy in Transition 1985–2010* study (CONAES report) has taken the average cost of a collector as $\$20/ft^2$ (with \$8 of this being for the collector itself) and 0.12×10^6 Btu/ft^2 as the average useful solar heat provided annually. Based on an interest rate of 11%, the cost of this heat is about $\$20/10^6$ Btu, about three times the national average for natural gas and comparable to that for electrical resistance heat. There are, of course, local situations as well as anticipated increases in the cost of natural gas and electricity that make the case for active solar heating much more favorable.

6.7 PASSIVE SYSTEMS

The increasing efficiency of a solar collector with decreasing temperature, described above, suggests that if a collector system could be operated just above normal room temperature the efficiency would be at the maximum value consistent with useful heat collection. However, the storage of significant amounts of heat at these lower temperatures requires very large storage systems. These considerations have led to the design of passive solar buildings with south-facing windows functioning as excellent low-temperature collectors, with large thermal masses incorporated within the building. In this way the building itself can serve as both collector and heat storage unit.

By definition, passive systems operate only by natural heat-transfer mechanisms with no motor-driven blowers or pumps. No external sources of energy other than sunlight are required. The ancient cliff dwellings of Mesa Verde are an example of passive solar architecture that served a human population for centuries. It is interesting to consider the ways in which these dwellings might have been made even more effective if glass had been available to their builders. It is now obvious that the products of modern technology can play a role even in passive systems.

In many ways passive systems have an appeal not found in active systems. They lead to some independence from central power systems, the collector spaces can serve as pleasant living areas or as greenhouses, and they can be owner-designed and owner-built. The materials used in passive systems last for the life of the building; there is little to wear out or to demand maintenance as in other heating systems. The large thermal masses moderate the daytime heat of summer as well as the nighttime cold of winter. In many cases passive approaches apply as well to retrofitting of existing structures as they do to new construction. The methods are often simple: orientation of a building on its lot, building on a south-facing slope, location of windows, and attention to shadow lines from surrounding structures and from trees can all make a difference. Movable insulation on the windows will greatly enhance the performance of a passive system.

The practicality of using ordinary vertical south-facing windows as solar collectors can be seen from the data in Table 6.5, which indicate that during the heating season these windows receive nearly as much solar radiation per unit area as do more elaborately positioned collecting surfaces.

Table 6.5 INSOLATION IN Btu/ft^2 PER DAY ON SURFACES OF VARIOUS ORIENTATION FOR SELECTED DATES IN THE WINTER MONTHS UNDER CLEAR SKY CONDITIONS AT 40° N LATITUDE

	Perpendicular	Horizontal	Vertical South	60° South
October 21	2454	1348	1654	2074
November 21	2128	942	1686	1908
December 21	1978	782	1646	1796
January 21	2182	948	1726	1944
February 21	2640	1414	1730	2176
March 21	2916	1852	1484	2174

Note: The perpendicular surface is steered so that it is perpendicular to the sun's rays. The vertical surface is south-facing. The 60° south surface is south-facing and slanted back 30° from the vertical. This is nearly optimum for a fixed collector at this latitude for the coldest months. (Adapted from Kreider and Kreith.)

Example 6.3

Calculate the approximate net energy gain of a south-facing vertical, double-paned window as a passive solar collector. Assume clear sky weather conditions on December 21.

Solution

First, from Table 6.5, on a clear day there is 1646 Btu/ft^2 incident on the window surface. Of this perhaps 75% is transmitted. This leaves about 1230 Btu/ft^2 passing into the house through the window. The window will also transmit heat energy outward at a rate represented by the *R*-2 rating assigned to a double-paned window. If over a 24-hour day the temperature inside the house averages 40°F warmer than the outside temperature, the daily heat loss would be

$$\frac{Q}{A} = \frac{\Delta T \times \text{time}}{R} = \frac{40°\text{F} \times 24\ \text{hr}}{2°\text{F} \cdot \text{hr} \cdot \text{ft}^2/\text{Btu}} = 480\ \text{Btu/ft}^2$$

The net gain each day under these conditions would then be 1230 − 480 = 750 Btu/ft^2, indicating that a window can be a very effective collector indeed. In this case the efficiency is 750/1646 = 46%. The performance would be less favorable under average conditions of cloudiness, and more favorable with snow cover on the ground (which acts as a reflector) or if the windows are fitted with insulating covers at night. Of course, on days of low solar insolation or extreme cold, even south-facing windows can have a negative net energy gain.

In recent years a great deal of experience has been gained in passive solar design in both residences and commercial buildings. In general, the efficiency and cost-effectiveness of these installations seem to compare favorably with systems based on active solar collectors; however, it is difficult to evaluate these factors exactly because for passive systems the components and costs of the heating system cannot easily be separated from the building for purposes

(a)

(b)

Figure 6.13 (*a*) The south-facing wall of a passive solar home. (*b*) The interior of the passive solar home. The glass admits sunlight onto the heavy floor slab, which stores heat for night.

of accounting. Data on the performance of passive installations are now available in a number of publications. The performance of residential buildings has been studied in detail, and it is estimated that 70% or more of their heating requirement can be often obtained from the sun. Examples of passive solar architecture are shown in Figure 6.13.

6.8 SOLAR THERMAL ELECTRIC POWER GENERATION

One of the obvious applications of solar energy is the production of electric power, using the sun to heat the working medium of a conventional heat engine, which then drives an electric generator. The equations developed in Chapter 3 concerning the efficiencies of heat engines are directly applicable; one simply uses the sun to supply the heat energy to produce a certain temperature, T_h, of the working substance. It is clear that one of the basic problems in making use of a diffuse source of energy such as the sun is raising the temperature sufficiently, because the Carnot efficiency is $1 - T_c/T_h$, where T_c is the temperature at which the working substance is rejected from the engine. Except in special situations, this high temperature requirement means that solar collectors will have to be used that focus the energy falling on a relatively large area to a spot, or focus, relatively small in area. A variety of different schemes has been proposed, and some are currently being tested. We shall examine several of these approaches to understand the important ideas involved and to assess the cost of of electric power generated in this way.

The sun can also be used to generate electric power by utilizing the thermal gradients in the tropical oceans to drive immense heat engines and by making use of the energy in the wind and falling water. These somewhat specialized topics in solar energy are discussed in the following chapter, along with other alternative sources of energy.

There are two general categories of solar–thermal power devices. The first involves focusing the sun's energy by an array of reflectors or concentrating collectors onto a central receiver. Since a large amount of solar energy is concentrated onto a relatively small area, the working fluid of the heat engine can be raised to a rather high temperature with a resulting high efficiency for the heat engine. In the second type of power system, each collector is arranged to focus the solar light onto an individual receiver for that collector. The heated material, such as water, is then brought together from the various collectors to a central system to drive a heat engine. Both types of systems have their advantages as well as disadvantages.

Before describing some examples of these systems, it will be useful to learn how concentrating reflectors work. Figure 6.14 shows a parabolic surface and incident rays of light from the sun. Because the sun is so far away, the rays incident on the reflector can be considered parallel. On striking the surface, which is usually shiny metal or silvered glass or plastic, they are reflected with the angle of incidence (i) equal to the angle of reflection (r), as shown. Paraboloidal surfaces have the special geometrical property of having each of the rays so reflected meet at a common point or focus. It is at this focus that one would locate a receiver. The distance of the focus from the surface can be varied by changing the shape of the surface. If a relatively small aperture is needed for the mirror compared to the focal distance, a spherical mirror, which is generally easier to manufacture, will bring all the parallel rays to an approximate focus. For a spherical mirror the focal distance is just one half of the radius of curvature of the mirror.

It is also possible to use transparent lenses to bring the sun's rays to a focus. Unfortunately, lenses are generally expensive to fabricate; moreover, because of the reflections at the surface and absorption in the material, normally glass or plastic, they are seldom used. A Fresnel lens, however, has the

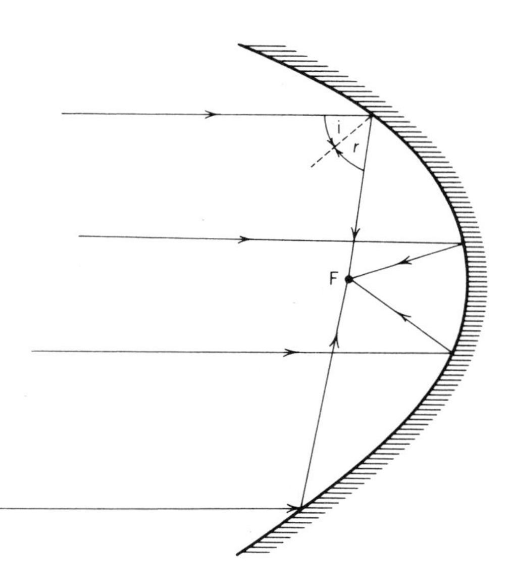

Figure 6.14 The parallel rays of light from the sun that are reflected from a parabolic surface will all be brought to a focus at a point.

focusing properties of a thick lens, but it can be very thin and inexpensively made from plastics.

Plane or flat mirrors are frequently used in solar applications because they are inexpensive. They, of course, do not individually focus the sun's rays; they present an image about as large as the mirror itself if the image is near the mirror, and a somewhat larger image far away because of the finite size of the sun.

Some general statements can be made about the temperatures that can be reached with various collectors. It is useful to discuss concentrating collectors in terms of the concentration ratio (CR), the ratio of the net collecting aperture area to the area of the receiver. Flat-plate collectors, for example, would have a CR equal to 1, and the maximum temperature that can be achieved is about 100°C (373°K). The corresponding Carnot efficiency, assuming $T = 20°C$, is about 20%.

Long, cylindrical parabolic collectors (troughs) have a CR that ranges up to about 50, which corresponds to maximum temperatures of about 400°C. Parabolic or spherical dishes have a CR between about 30 and 1000, which corresponds to a maximum temperature of about 1200°C. There is no fundamental reason, however, that larger dishes with higher CR values and higher temperatures cannot be built. The economics of such units are not favorable for driving a heat engine, but other uses such as a solar furnace might justify these devices. The record for high temperatures was set (over 3000°C) at Odeillo, France, where a 1-MW solar furnace was constructed by having one side of a building made into a huge paraboloidal reflector 40 m high and 54 m across.

The final type of system that can be useful for solar thermal electricity is the power tower, with a field of reflectors, or heliostats, on the ground all positioned to reflect the direct sunlight onto a central receiver. With several hundred heliostats, each having several reflecting surfaces, a CR of up to several thousand can be achieved with maximum temperatures between 1000 and 2000°C.

Flat-plate collectors are usually arranged to take advantage of both the direct and indirect components of the solar radiation, and they are normally fixed in some optimized orientation. Parabolic troughs will focus only the direct light of the sun, and it is necessary occasionally to change their orientation. A parabolic dish or field of heliostats also focuses only the direct light of the sun, but because both have a point rather than a line focus, their orientation must be changed by rotation in at least two directions to accommodate changes in time of day and season.

Heat engines driven by solar collectors have a long history. However, it is only within the last 15 years or so that the U.S. government, through the Department of Energy, has contracted with various industries and public utilities to construct systems that are appreciable in size and that will produce 10 MW or more of electric power. Since 60 to 70% of the capital cost of the system is in the collectors, it is hoped that innovative designs and mass production techniques for the various collectors will result from these pilot projects now being constructed. As with many applications of solar energy, the problem is not to show that a device can work; the problem is to show that an economically competitive system can be constructed that has solved the numerous technical and practical problems, such as heat storage and equipment maintenance. Several systems will be discussed to provide a perspective on what is involved.

POWER TOWER

The general idea of a solar, or power, tower is to collect the light from many reflectors at a central point to achieve high temperatures. The Sandia National Laboratory outside of Albuquerque, New Mexico, has been testing various heliostats and, in fact, has constructed a 5-MW thermal facility that consists of 222 heliostats, each of which has 25 flat mirrors. The heliostats must continuously move to track the sun across the sky, so that the reflected sunlight from all of the mirrors falls on the receiver. There are two drives on each heliostat: rotation about a central vertical axis and tilt about a horizontal axis. The tower is 61 m (200 ft) tall and the receiver at the top has a peak flux of about 2.5×10^6 W/m^2 for a concentration ratio of about 2630.

Based largely on the experience gained at the Sandia Laboratory, a 10-MW solar electrical power plant has been constructed at Barstow, California. A photograph of this facility is shown in Figure 6.15, and a schematic view is

Figure 6.15 A photograph of the solar-thermal power plant in Barstow, California, described in Figure 6.16.

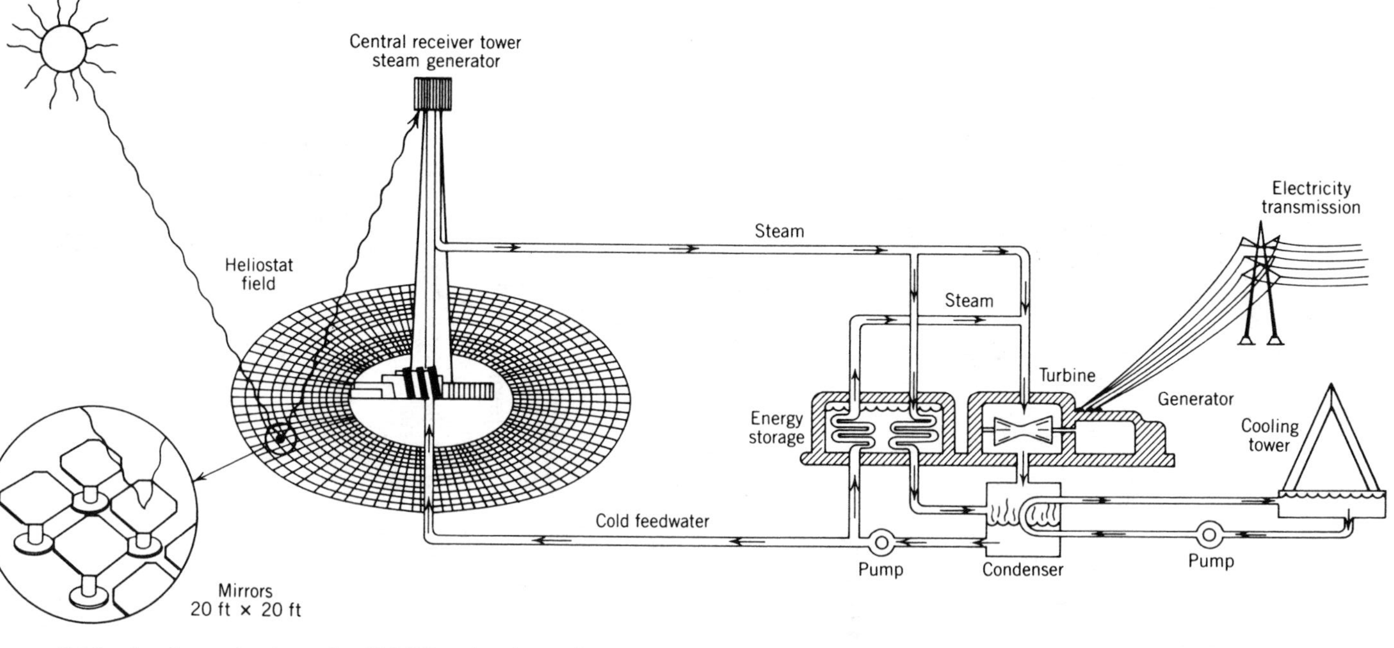

Figure 6.16 A schematic view of a 10-MW_e solar-thermal power plant near Barstow, California. The receiver and boiler that absorb the sunlight reflected from 1900 heliostats are at the top of a 90-m tower. The heliostats are each steered by computer control to reflect the solar radiation into the receiver. The steam from the boiler can be either delivered directly to the 10-MW_e turbine and generator or to storage. The storage system can provide steam for 4 hr of generation at a level of 7-MW_e without sunlight. (Figure supplied by the Solar Energy Research Institute.)

shown in Figure 6.16. There are 1900 heliostats, each about 20 ft by 20 ft, surrounding a tower that is 295 ft tall. As the figure shows, an energy storage system is provided that will permit the unit to continue generating about 7 MW for 4 hours without sunlight. The capital cost of such a power plant is high compared to a coal-fired or nuclear reactor power plant. The Barstow plant was originally to have cost $123 million, which corresponds to about $12,000/KW compared to about $1000/kW (1980) for coal or nuclear power. The cost of the electric energy produced, assuming an interest rate of 15% on the capital invested, would be about $1/kW • hr. This is about 10 to 20 times the cost of electricity for residential use at the present time. As more power towers are constructed, economies will be realized and technological improvements made that will bring the cost down, but it will be many years before they are competitive.

It may be that the units to be built in the future will combine electric power production with steam generated for industrial processes or space heating. There is a solar electric unit being studied for the Fort Hood Army Base in Texas that combines 1 MW of electric power production with space heating and air conditioning for the whole base. There also will be heat storage in two large tanks of various salt solutions.

PARABOLIC DISHES AND TROUGHS

In systems consisting of parabolic dishes and troughs, there is a receiver for each dish or trough, and the heated working substances must be collected from each receiver and generally taken to a central collection point to drive a steam turbine. A system of parabolic dishes is shown in Figure 6.17. Parabolic troughs are shown in Figure 6.18. This particular unit is designed to produce steam for gauze bleaching and will supply about 1.5 billion Btu each year. The long receiver pipes at the focus of the parabola can be seen as well as the collection points at the ends of the trough.

6.9 THE DIRECT CONVERSION OF SOLAR ENERGY TO ELECTRICAL ENERGY

The process of using the sun's energy to heat a medium such as water to provide steam to power a heat engine that in turn rotates an electric generator is a usable but rather expensive and cumbersome scheme. There are devices that can generate electricity much more directly from the sun's rays, thus avoiding many of the problems associated with large, complicated, mechanical equipment. The devices fall into two general categories: thermoelectric and photovoltaic.

THERMOELECTRICITY

Thermoelectricity has been known for a long time. It was originally observed by Seebeck in 1821. When two dissimilar metals in the form of wires are joined together at both ends, an electric current will be generated in the wires if the two junctions are kept at different temperatures. The induced voltage that drives this current increases as the temperature difference increases until a

Figure 6.17 A system of parabolic dish reflectors being used at a joint DOE–Georgia Power Company project in Shenandoah, Georgia. The solar energy collected will be used to provide electric power, process steam, and air conditioning for a knitwear factory. The 114 solar dishes, each 23 ft in diameter, furnish a total of 3 MW of thermal power.

difference of several hundred degrees Kelvin is reached. Such devices, called thermocouples, are widely used for measuring temperatures. One junction is kept at a fixed temperature so that a measurement of the induced voltage will provide a measurement of the temperature of the other junction.

In principle, the sun's energy could be used to heat one junction and electric energy removed from the circuit to perform useful work. It is difficult, however, to construct a useful device because the voltages are quite small, and many elements in series would be needed to provide useful voltages and power levels. Practical devices have an output of about 300 to 400 μV per degree Celsius, so that if a temperature difference of 400°C is achieved, 0.1 to 0.2 V is produced per element. Further complications arise because a high electrical conductivity for the thermocopule materials is desirable so that there will be a minimum internal resistance. At the same time, a minimum thermal conductivity is desirable so that heat is not transferred by conduction from the hot junction to the cold junction. Unfortunately, for almost all materials thermal conductivity is high when the electrical conductivity is high. A number of materials have been investigated for solar applications, including semiconductors, whose properties will be discussed shortly. It appears that an efficiency of about 5% is the

Figure 6.18 Two of the parabolic trough solar collectors used by a manufacturing company to provide 1.5×10^9 Btu annually of thermal energy. The rotation mechanism allows the collectors to track the sun.

maximum that can be expected with presently known materials and technology. In an actual system incorporating a flat-plate collector to heat one junction of the thermoelectric generators and cooling fins on the back side to keep the second junction cool, overall efficiencies of about 0.8% have been achieved. Although continued research on thermoelectric generators is warranted, they do not appear at the present time to be as promising as photovoltaic cells for the direct production of electric power from solar power. There are a few applications, however, primarily military, in which thermoelectric generators are powered by the burning of petroleum fuels to produce electric power quietly and reliably in small amounts as needed, for instance, by communications equipment.

PHOTOVOLTAIC CELLS

The principle by which photovoltaic cells produce electricity from sunlight is completely different from that of thermoelectricity. The energy of the sun available at the surface of the earth, as discussed earlier in this chapter, is in the form of light or electromagnetic radiation. Such waves can act, in some cases, like a collection of bundles or packets of electromagnetic energy, each similar in many ways to a particle. When light behaves in this fashion, these packets of energy, called photons, can strike an electron in some material and free it from the bonds that hold it to an atom. Under certain situations these freed electrons can be made to flow in an external circuit and, hence, produce electricity.

Photovoltaic cells, most commonly made of silicon, a material called a semiconductor, have been widely used in recent years to measure the intensity of light and are well-known for providing electric power on satellites and space vehicles that have been launched by the United States and other countries.

To understand how photovoltaic cells work and how they might become a very important method of obtaining electric energy directly from solar energy, we must understand what energies are associated with photons and what happens inside a semiconductor.

The energy associated with a photon, as described earlier, is the product of Planck's constant, h, and the frequency, f, of the electromagnetic energy, that is, $E = hf$. Planck's constant came from the work of Max Planck, in 1900, in understanding the characteristics of thermal radiation. The constant ($h = 6.63 \times 10^{-34}$ J • sec) has since played a fundamental role in understanding atomic physics.

Example 6.4

Calculate the energy associated with photons that have a wavelength, λ, of 550 nm (the middle of the visible portion of the light from the sun).

Solution

$$E = hf = h\frac{c}{\lambda}, \quad \text{since } f = \frac{c}{\lambda}, \text{ where } c \text{ is the velocity of light}$$

$$E = \frac{6.63 \times 10^{-34}\,\text{J} \cdot \text{sec} \times 3 \times 10^{8}\,\text{m/sec}}{550 \times 10^{-9}\,\text{m}}$$

$$E = 3.62 \times 10^{-19}\,\text{J}$$

or

$$E = \frac{3.62 \times 10^{-19}\,\text{J}}{1.60 \times 10^{-19}\,\text{J/eV}} = 2.26\ \text{eV}$$

One of the most frequently used materials for semiconductor devices and photovoltaic cells is silicon (Si). In the silicon atom there are four outer (valence) electrons that can participate in bonding one silicon atom to others in a crystal. Because all of the available electrons are participating in this covalent bonding, there are no free electrons available in pure silicon crystals for conducting electricity. Figure 6.19 illustrates the band structure of silicon. All of the electrons are normally in the valence band, and it is completely filled. The conduction band is normally completely empty and is separated by an energy gap (E_g) of 1.11 eV from the valence band. If a photon of light strikes one of the electrons in the valence band, the electron can receive enough energy to take it from the valence band across the band gap up to the conduction band. Once the electron is in the conduction band, it can move under the influence of an electric field and produce an electric current.

A photovoltaic cell is more than just a pure crystal of silicon; however, the requirement persists that to have conduction electrons, the photons must have an energy greater than E_g (1.11 eV). This corresponds to some minimum frequency (or maximum wavelength) that the solar radiation can have and still provide power from a silicon photovoltaic cell. This wavelength, λ, can be calculated as follows:

$$E_g = hf_{\text{min}} \quad \text{and} \quad \lambda_{\text{max}} = \frac{c}{f_{\text{min}}}$$

$$\lambda_{\text{max}} = \frac{ch}{E_g} = \frac{3 \times 10^8\,\text{m/sec} \times (6.63 \times 10^{-34}\,\text{J} \cdot \text{sec})}{1.11\,\text{eV} \times (1.60 \times 10^{-19}\,\text{J/eV})}$$

$$\lambda_{\text{max}} = 1.12 \times 10^{-6}\,\text{m} = 1.12\ \mu\text{m}$$

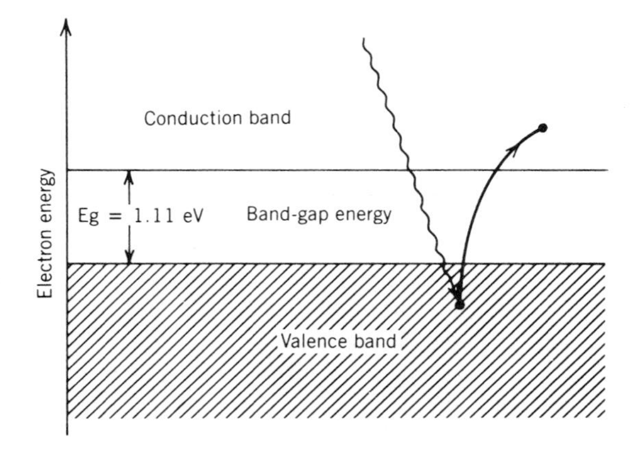

Figure 6.19 Energy band structure in silicon. Because there are normally no electrons in the conduction band, no electrical conduction can take place. An electron can be elevated to the conduction band by an incident photon, leaving behind a vacancy.

Only about 77% of the solar energy is contained in radiation with a wavelength less than 1.12 μm (see Figure 6.2). Unfortunately, the efficiency of silicon solar cells is far less than 77% because any electron energy in excess of the 1.11 eV does not go into electric power but only heats the crystal. In fact, about 43% of the average absorbed photon energy goes into heat. There are other losses in the crystal as well as light being reflected from the exposed surface of the crystal. Figure 6.20 shows the maximum possible efficiency that can be achieved with a variety of possible photovoltaic materials. We can see that at 0°C (273°K) a maximum theoretical efficiency of about 24% is possible with silicon; this would be appreciably reduced as the crystal is heated. There are other losses in practice, such as the internal resistance of the cell, that further reduce the overall efficiency of a silicon cell to about 10 to 14%.

The presence of electrons in the conduction band and holes, or vacancies, in the valence band will permit some electrical conductivity; however, without the presence of an electric field, the electrons would eventually drop back into the holes, and the cell would not produce electric power. Photoconductivity by itself can be used, though, for determining the intensity of light. To make a power-producing photovoltaic cell, a combination of two different types of semiconductors is needed.

If an impurity, such as arsenic, antimony, or phosphorus, is added to the silicon crystal, the intrinsic, or pure, silicon crystal becomes a doped semicon-

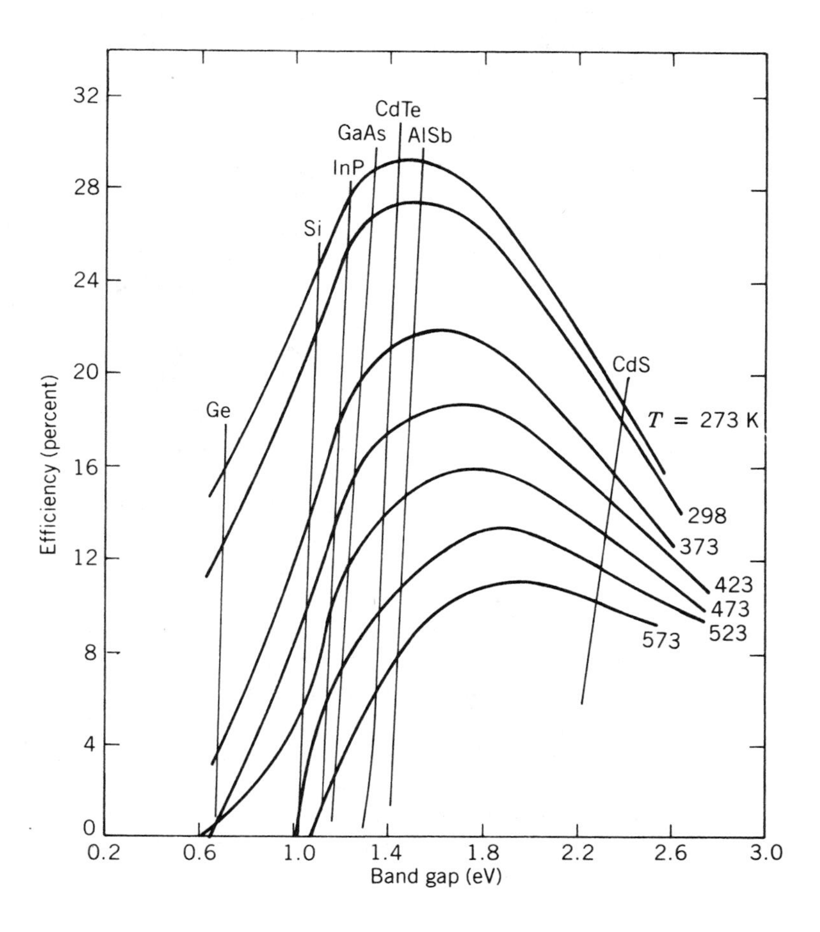

Figure 6.20 Photovoltaic conversion efficiency for several semiconductor materials having different band gaps. For all materials, the efficiencies drop as the temperature increases. Actual devices produce lower efficiencies than the theoretically maximum values.

ductor. Arsenic has five valence electrons, but only four of them are needed to provide the same covalent bonding that silicon has when arsenic replaces a silicon atom in the lattice. The fifth electron, not needed in bonding, is free to participate in the conduction of electric current. Because the electron is negatively charged, such semiconductors with an excess of electrons are called n-type. Normally, only very small amounts of arsenic are added (doped), perhaps 1 part per million.

There is also a possibility of doping the silicon crystal with an atom that has only three valence electrons, such as boron, aluminum, or indium. Since four electrons are needed for bonding the atom in the silicon lattice, the electron vacancy (or hole) is free to move about and contribute to electric conduction. Doping with such atoms produces positive, or p-type, carriers, and such material is called a p-type semiconductor.

If p-type silicon is put in contact with n-type silicon, a p–n junction is formed that is crucial to the function of a photovoltaic cell. In the region of the junction, a certain number of holes from the p-type material and electrons from the n-type material will combine by normal diffusion and neutralize each other. In this process the electron falls into the hole. Because the p-type and n-type materials were both neutral to begin with, the loss of holes from the p-type material and the electron from the n-type material, due to the recombination in the junction region, leaves behind negative and positive charges, respectively. The presence of these net positive and negative charges on either side of the junction creates an electric field across the junction. This electric field is now capable of taking any electron-hole pairs created in the region of the junction by incident solar radiation and forcing them to move in opposite directions through the rest of the silicon crystal and any external electric circuit. Thus, electric current is made to flow through an external load, and solar energy has been converted into electric energy.

To manufacture a silicon solar cell, one first has to obtain very pure silicon. Even though silicon is very abundant in the earth's surface, it is relatively expensive to purify it sufficiently to make a semiconductor. The cost of a kilogram of pure silicon is about $100. Crystals then have to be grown from the molten silicon. There are several processes, but the most common method is to rotate a seed crystal slowly and withdraw it from the molten silicon. A crystal grown in this way becomes a cylindrical ingot up to several inches in diameter and up to several feet long. A p-type crystal can be obtained by doping the molten silicon with just the correct amount of an element such as boron. The ingot is then sliced by sawing it in a very slow and precise way into wafers that are a fraction of a millimeter thick. The n-type part of the cell is made by diffusing a chemical such as phosphorus into the crystal. This diffusion should only go in a fraction of a micron, since light must penetrate into the depleted region (the junction region where the n- and p-type materials are adjacent). Electrical contacts must be made to the two surfaces, the crystal packaged into some assembly, and an antireflection coating applied to the front face. A schematic view of a solar cell is shown in Figure 6.21.

A photovoltaic cell installation is shown in Figure 6.22. Forty or 50 solar cells are normally grouped together so that when connected in series electrically, they produce about 20 to 25 V. The array must then be put in an enclosure and generally covered with some nonreflecting material. The assembly must be oriented in an optimum way to receive the sun's energy, and it must be protected

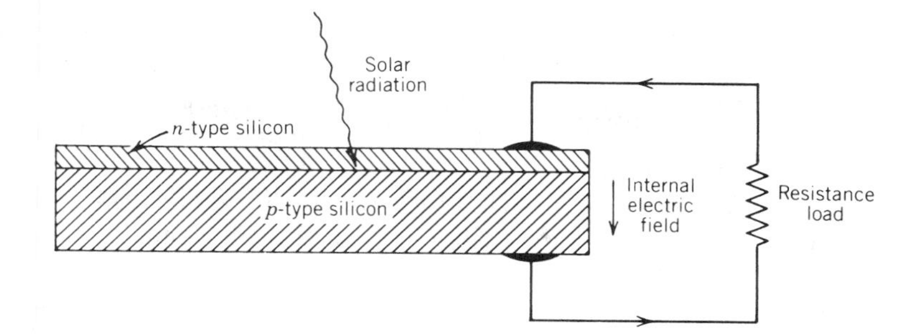

Figure 6.21 A schematic view of a solar cell made of n- and p-type silicon. If the incident radiation has sufficient energy ($\lambda < 1.12\ \mu m$), an electron can be taken from the valence to the conduction band. Because of the internal electric field, the electron will drift toward the contact on the n-type silicon to produce current in the external circuit and finally combine with a hole in the p-type silicon. The conventional current arrows shown are opposite in direction to the electron flow.

from the elements. To put alternating current into the power grid, a voltage regulator is usually used in conjunction with storage batteries, and finally an inverter converts the direct current to alternating current. The costs of these pieces of equipment must be added to the cost of the basic solar cells to arrive at a total capital cost per kilowatt.

Since solar cells were first developed in 1954 there has been a great deal of effort to increase the efficiency of the cells, develop new kinds of cells, and reduce the costs of manufacturing. The general aim of these efforts is to bring the costs of solar cells down from the present cost of about $10,000/kW of peak electric power to something comparable to the cost of a coal-fired power plant; i.e., about $1000/kW. This is equivalent to about $0.50/W for just the solar cell component of the facility.

One of the avenues of attack on the costs has been to try to develop less-expensive ways of growing the silicon crystals. The process called edge-defined film-fed growth (EFG) is one of the more promising developments of this kind.

Figure 6.22 These solar cells power an experimental irrigation project near Mead, Nebraska. A 10-hp pump is driven by 120,000 individual cells, which produce 25 kW at peak sunlight. The pump raises water out of a reservoir to irrigate 80 acres of corn and soybeans.

Here a ribbon of crystalline siicon is drawn up through a die by capillary action from the molten silicon. This produces a thin ribbon of silicon of about the correct thickness and of reasonable quality, so that solar cells with 10 to 12% efficiency can be made with far less expense than in the usual method of growing crystal ingots.

Amorphous silicon technology has made perhaps the greatest progress in the past 10 years or so. Here the solar cell is made in a continuous-film deposition process. The silicon atoms are randomly oriented (noncrystalline) on a glass substrate. The advantage of this process is that it is relatively inexpensive and uses negligible amounts of silicon because the thin film is only about a micron thick. This type of solar cell is widely used for watches, calculators and walk lights.

Several materials other than silicon can be made into solar cells, as indicated in Figure 6.20. Polycrystalline thin films of cadmium telluride, copper indium deselenide (CIS), and gallium arsenide are all being investigated, and significant progress has been reported in lowering costs and increasing efficiency compared to crystalline silicon. The fabrication process lends itself to low-cost automation. Still another direction that is being taken is to stack three cells on top of one another in order to capture a larger portion of the solar spectrum. The top cell is designed to absorb the high energy (blue part) of the spectrum and permit the remaining portion of the spectrum (mostly in the red) through to be absorbed in the lower cells. Figure 6.23 shows the progress that has been made in lowering the cost ($\$/W_p$) as the accumulated production increased through 1988, where W_p refers to peak electric power in watts. The

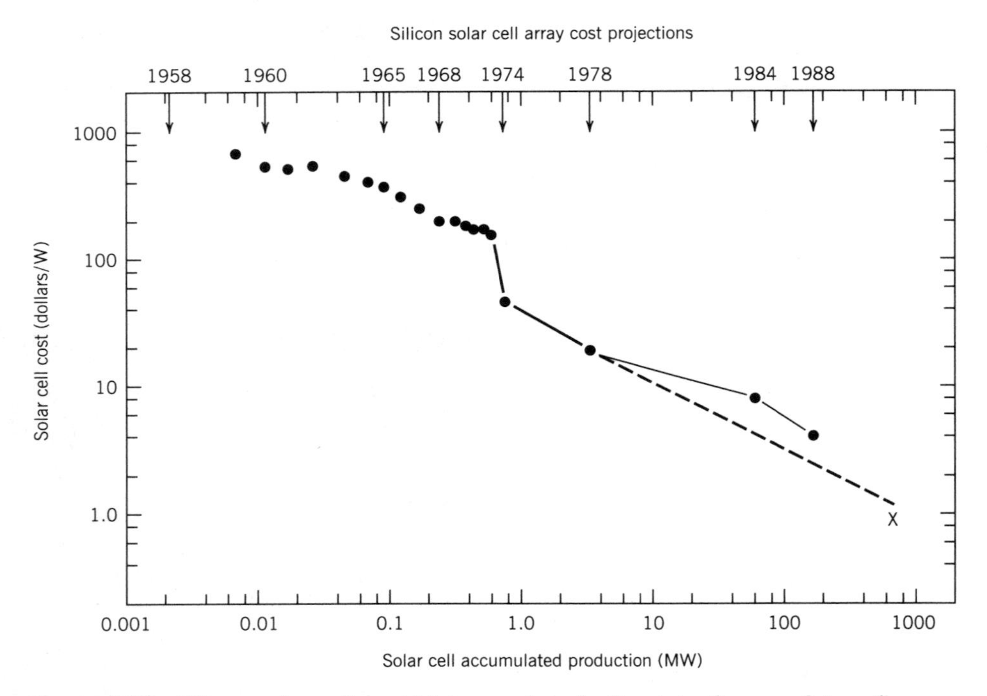

Figure 6.23 Silicon solar cell cost history and projection into the near future (in 1988 dollars). The dashed line is a projection based on the 1974–1978 experience. The Department of Energy goal for 1986 is shown by an X.

peak electric power is produced under a full noonday sun. It can also be seen that projecting costs into the future can be difficult. While the Department of Energy projection in 1978 (dashed line) follows the realized trend, the 1986 goal has not yet been achieved.

Another method of making photoelectric cells more competitive economically is to use them in conjunction with concentrating collectors. In principle, if a collector such as the parabolic reflector discussed earlier is used to focus the sunlight onto the solar cell, a gain of 100, for example, can be achieved and the area needed by the solar cell reduced by a factor of 100 for a given amount of solar energy. Although there are clear advantages to such a system, the cost of the collector and the tracking mechanism must be considered. Also of concern is the fact that the solar cell would experience much higher temperatures than when used without a collector, and as can be seen in Figure 6.20, the efficiency of all the photovoltaic cells shown goes down steeply with increasing temperature. To keep the temperature from rising, water cooling can be used. M. H. Ross and R. H. Williams have considered in some detail the design of a solar cell collector system for a community where the electric power from the photovoltaic cells would provide a large fraction of the electrical needs. Further, the heated water from the cooling of the cells would provide a large fraction of the space heating and hot water needs of the community. Diesel engine–powered generators would supplement the electric power from the solar cells, and the heat from the diesel engines would also be used for water and space heating.

As the price of solar cells has come down, the world market has increased from almost nothing in 1976 to 35.2 MW_p in 1988, and of this total about one third were manufactured in the United States. Most of the cells (40%) were amorphous silicon, with single crystal and polycrystalline cells accounting for most of the rest. These cells were used for everything from watches and calculators to outdoor lighting, microwave repeaters, remote pumps, satellites, and navigational aids. There are a number of grid-connected photovoltaic systems in the United States, but most of them are around a few kilowatts. There is one 6.2-MW facility in California that has been putting energy into the power grid since 1984. This particular facility has two-axis tracking of flat plate modules with an efficiency of 11%.

What is the future of photovoltaic electricity in the United States? While it is encouraging to see the price reductions and the increase in the market for solar cells, as is shown in Figure 6.23, it should be remembered that 35 MW or 6.2 MW is only a very small fraction of one 1000-MW_e coal-fired plant. In order to make major inroads into the 65,000 MW_p additional power capacity needs by the United States by the year 2000, some further improvements will be needed in the costs, efficiency and reliability of solar cells. If the price could be reduced to under \$1/W_p, the market should open up very quickly. Many think that polycrystalline thin films offer the greatest hope for such reductions.

Another way of stating the problem is that the cost of wholesale electricity from a new coal-fired plant in the United States is \$0.08 to \$0.20/kW • hr at the present time while photovoltaic power generation is between \$0.50 and \$1/kW • hr, depending on the size of the system. Arguments could be made that the total cost of coal-fired electricity should be increased to reflect the costs to society of the emissions of pollutants such as SO_2 and greenhouse gases such as CO_2. Even without corrections of this kind, solar cells seem destined to become a significant source of electric power by the year 2000 and beyond.

PROBLEMS

1. How large (in square meters and square feet) would a cast-iron (ϵ = 0.5) stove surface area have to be to radiate 10,000 Btu/hr at a surface temperature of 450°F? (Incidentally, at this temperature the radiant heat energy output is about two thirds of the total useful heat output of the stove.)

2. Assume that a solar collector has an area of 600 ft^2, that the incident solar energy is 1200 Btu/ft^2 • day, and that 85% of this incident solar energy is transmitted through the single glass cover plate. Also assume that the collector loses 50% of the energy coming in through the cover plate because of conductive and radiative heat losses from the collector assembly. Now, how large a house (floor area) could this collector heat if the house requires 4 Btu/ft^2 of floor area per degree day? Consider a 24-hour day with T_{inside} = 65°F and T_{outside} = 35°F.

3. Consider a tank of water for storing solar energy. It is to store 200,000 Btu between the water temperatures of 170°F and 80°F.
   (a) What is the tank volume in cubic feet?
   (b) If the tank is cubical in shape, what is the dimension (in feet) of one edge of the cube?
   (c) What is the tank capacity in gallons?

4. Starting from Example 6.3, calculate how many square feet of vertical south-facing windows are needed to provide the 500,000 Btu a day needed by a house under the given conditions. How would this answer change under conditions of partial cloudiness, so that only 50% as much energy is incident as on a clear day? Would total reliance on passive solar energy be practical under the latter conditions?

5. A typical American home requires electrical service capable of providing about 100 amperes at 240 volts for a peak power of 24,000 W.
   (a) About how large an area of photovoltaic cells would be needed to meet this peak demand if it occurred at a time of full noonday sun?
   (b) About how much would this array cost if purchased today?

6. A large (150-unit) apartment building is situated in a location of bright, sunny summers and cold, cloudy, completely sunless winters and has been designed to the most modern standards of energy conservation. The building has a total volume of 10^6 ft^3 and requires heat only in the amount of 50 Btu per degree day per 1000 ft^3.
   (a) Calculate the number of Btus needed for a 6000-degree-day heating season.
   (b) Calculate the volume in cubic feet of a perfectly insulated water tank needed to store enough heat for the entire heating season. Assume that at the beginning of this period the water is at 190°F and at the end the temperature has been lowered to 70°F. Also estimate the dimensions of the tank if it is cubical in shape. Given: 62.4 Btu/ft^3°F for water.
   (c) Calculate the area in square feet of the solar collector needed to reheat the water during the 200-day sunny season. Assume 2000 Btu of solar energy incident per square foot per day and an efficiency of 50% for the collector. Estimate the dimensions of the collector if it is square in shape.

7. Calculate the land area in square miles necessary for a 1000-MW$_e$ solar–thermal power plant to provide electricity continuously for a large city. Assume the collector efficiency, averaged over day and night and all seasons, is 10% and the thermodynamic efficiency of the boiler–turbine–generator system is 33%. Also assume that there is no problem with energy storage for the periods when the sun is not shining. Take the mean daily solar radiation incident to be 6 kW • hr/m^2 of horizontal area.

8. A wood-burning stove has an emissivity of 0.80, a temperature of 300°C, and an effective surface area of 1.50 m^2. How much power does it radiate?

9. The volumetric heat capacity of water (62.4 Btu/ft^3°F) is 2.8 times that of concrete (22.4 Btu/ft^3°F). If a 10 ft^3 volume of water is to be replaced by a concrete heat-storage facility, what is the necessary volume of concrete? How does its weight compare to that of the water?

10. A solar collector collects 80,000 Btus of heat in one day, and this is stored in a water tank containing 4000 pounds of water. Neglecting losses, how much does the water temperature increase in °F?

11. A new material for possible use in photoelectric cells has an energy gap (E_g) of 1.06 eV. What is the maximum wavelength in meters that would work for light on such a photo cell?

12. From the average solar energy incident on a horizontal square meter in the United States in 1 8-hour day at 40 degrees latitude, calculate the total amount of solar energy incident on the United States in 1 year in QBtu (the total area of the United States is 3.62×10^6 mi^2). What fraction of this total would be needed to supply the total yearly energy needs of the United States?

13. Every QBtu of electric energy generated by photoelectric cells will replace about 3 QBtu of coal burned at power plants. How many fewer tons of CO_2 will be added to the atmosphere for every QBtu of photoelectric energy used rather than coal? Assume coal is pure carbon.

14. A homeowner is considering putting a horizontal solar panel on her roof to heat water for domestic use. An average increase of water temperature each day from 60°F to 120°F is wanted for 150 gallons. How large a panel would be needed on a clear winter day at 40° latitude if the overall efficiency is 50%?

SUGGESTED READING AND REFERENCES

1. Brinkworth, B. J. *Solar Energy for Man.* New York: John Wiley, 1972.
2. Kreider, J. F., and Kreith, F. *Solar Heating and Cooling.* New York: McGraw-Hill, 1977.
3. Krenz, Jerrold H. *Energy—Conversion and Utilization,* Second Edition. Boston: Allyn and Bacon, 1984.
4. Mazria, Edward. *The Passive Solar Energy Book.* Emmaus, Pa.: Rodale Press, 1979.
5. McDaniels, David K. *The Sun.* New York: John Wiley, 1979.
6. Ross, Marc H., and Williams, Robert H. *Our Energy: Regaining Control.* New York: McGraw-Hill, 1981.
7. Wieder, Sol. *An Introduction to Solar Energy for Scientists and Engineers.* New York: John Wiley, 1982.
8. Final Report of the Committee on Nuclear and Alternative Energy Systems (CONAES). *Energy in Transition 1985–2010.* San Francisco: W. H. Freeman, 1979.
9. Carlson, D. E. *Photovoltaic Technologies for Commercial Power Generation.* Annual Review of Energy. **15** (1990), p. 85.
10. *Photovoltaics—Entering the 1990's.* Solar Energy Research Institute. SERI/SP-220-3461.
11. Weinberg, Carl J. and Williams, Robert H. *Energy from the Sun.* Scientific American, **263** 3 (September 1990), pp. 147–163.

CHAPTER SEVEN

ALTERNATIVE SOURCES OF ENERGY

7.1 INTRODUCTION

The burning of fossil fuels, the fissioning of uranium in reactors, and the direct use of the sun's energy all represent technologies proven, to some degree, to be capable of delivering useful power to populated areas of the globe. In addition to these methods, there are other energy technologies that have less well-defined promise either because of incompletely developed energy conversion apparatus, because the resources are of unknown or limited magnitude, or because applications have so far been local rather than widespread. This chapter discusses hydroelectric power, ocean thermal energy conversion, wind power, geothermal energy, and tidal energy. The capturing of the sun's energy in biomass is covered in Chapter 10. Of the alternative energy sources, only geothermal energy and tidal energy are of nonsolar origin; the others are indirect ways of harnessing the power in the sun's radiation. Solar radiation is a product of nuclear reactions in the sun, and geothermal energy is produced by the decay of radioactive nuclei beneath the earth's surface. This leaves only tidal energy as energy of nonnuclear origin.

7.2 HYDROELECTRIC POWER

About 2000 years ago, during the first century B.C., it was discovered that the force of falling water acting on a waterwheel could be used to ease human labor. Well before the discovery of electricity, various types of ingeniously contrived waterwheel mechanisms were connected by rotating shafts and cogwheels to mills for grinding grain and sawing wood. By the 13th century water power was used to operate hammers in the ironworks of western Europe, and by the 16th century the waterwheel was the primary source of industrial energy in that part of the world. The steam engine eventually replaced water power in many applications, but in selected locations and for certain purposes, water power continued to be the preferred energy source. As our nation developed, mills were established at sites with reliable water flow of sufficient volume and velocity. Communities then grew up around these mills. Dams were built at many such sites to impound the flowing water and to even out the flow over the seasons. After electricity came into practical use, waterwheels were used to drive generators; it then became reasonable to locate hydroelectric plants at a distance from population centers and to transport the electricity over power lines hundreds of miles to the point of demand. Now we use water power almost entirely for the generation of electricity, even if there is only a modest distance between the point of generation and the point of use.

Water power is, of course, a consequence of the natural cyclical transport of water between the earth's surface and atmosphere. This hydrologic cycle, shown in Figure 7.1, involves the evaporation of the earth's surface water when it is heated by sunlight, followed by precipitation and the downward course of the water in rivers and streams under the force of gravity. In a sense the water is the working fluid in an enormous heat engine powered by sunlight.

The available energy of water stored at a height above a generator is potential energy in the earth's gravitational field. The solar energy that went into lifting the water upward in this gravitational field can be made available again when, for example, the water falls downward onto a waterwheel. As each

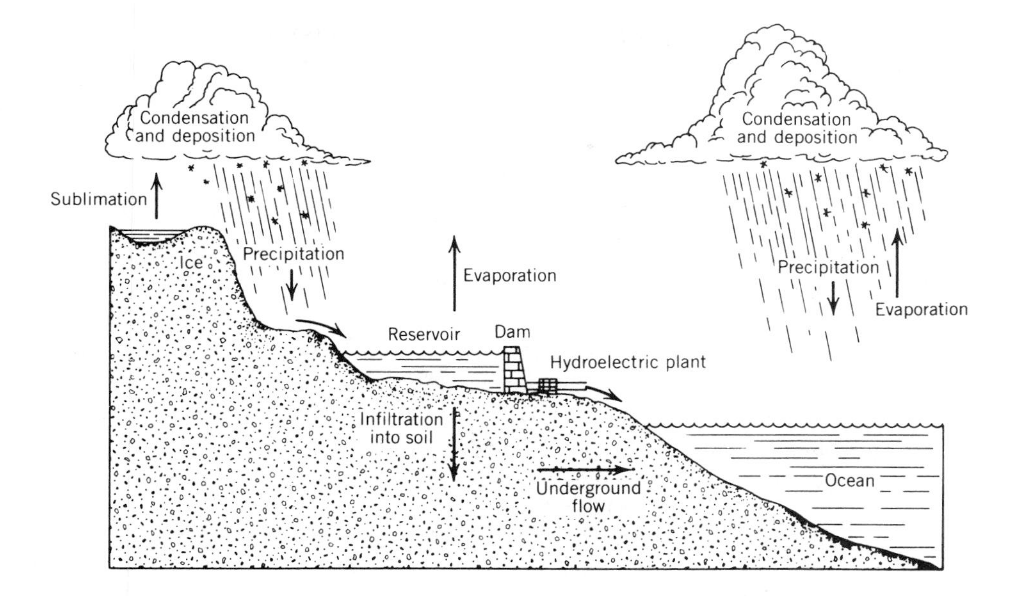

Figure 7.1 The hydrologic cycle. Electricity is produced in the hydroelectric plant by the action of water against a turbine connected to a generator. In this way the stored potential energy of the water in the reservoir becomes electrical energy.

element of water of mass, m, falls freely at a distance, h, it attains a velocity, v, and kinetic energy, $\frac{1}{2}mv^2$, equal to its original potential energy, which is given by mgh. The acceleration of gravity, g, is 9.8 m/sec^2. Thus, the potential energy available for conversion to kinetic energy is 9.8 joules per kilogram of water (or anything else) per meter of height above the point where the kinetic energy is to be ultilized. The distance, h, is often called the head. Low-head hydroelectricity can be generated with h as small as 10 feet, whereas high-head hydroelectricity is generated with heads of hundreds up to greater than 1000 feet. The detailed design of the hydraulic turbines is different in the two cases. Modern hydroelectric installations can convert the potential energy of water to electric energy at an efficiency of 80 to 90%.

Example 7.1

Calculate the flow rate of water required to provide 1 kW of electrical power if the water falls a vertical distance of 90 m. Assume 80% conversion efficiency.

Solution

The potential energy in joules of a mass, m, at a height, h, is mgh, where $g = 9.8$ m/sec^2, m is in kilograms, and h is in meters. If the flow rate is 1 kg/sec, the power in the stream of water after a fall of 90 m will be

$$\begin{aligned}\text{power} &= \frac{\text{energy}}{\text{time}} = \frac{mgh}{\text{sec}} = \frac{1\text{ kg}}{\text{sec}} \times \frac{9.8\text{ m}}{\text{sec}^2} \times 90\text{ m} \\ &= 881\text{ kg}\left(\frac{\text{m}^2}{\text{sec}^2}\right)\frac{1}{\text{sec}} = 881\text{ J/sec} \\ &= 881\text{ W}\end{aligned}$$

If this is converted to electricity at 80% efficiency, the electrical power produced will be $0.8 \times 881\ W = 705\ W$ at a flow of 1 kg/sec. To generate 1 kW (1000 W), a flow of

$$\frac{1000\ W}{705\ W\ \text{per kg/sec}} = 1.42\ \text{kg/sec}$$

will be necessary. This is equal to 1.42 liters/sec. To generate the same amount of electrical power at a low-head installation with $h = 3$ m would require a water flow 30 times greater, or 42.5 liters/sec.

There are many obvious advantages to hydroelectric power. There are no polluting emissions into the air or water, and no waste heat is rejected as thermal pollution. The operation of hydro plants relies only on renewable resources, and they have long lives and slow depreciation. They respond well to sudden changes in demand, making hydroelectricity well suited to matching peak loads. The dams can serve multiple purposes; water stored for irrigation, flood control, or a municipal drinking water supply can also power a hydroelectric plant as it flows to its other tasks. Figure 7.2 shows an example of a multipurpose installation.

On the negative side, there are several factors limiting the long-term future of hydroelectric power. In the United States, the principal hydroelectric sites

Figure 7.2 The Glen Canyon Dam on the Colorado River in Arizona. This installation produces 950 MW of electrical power as well as storing water for other purposes.

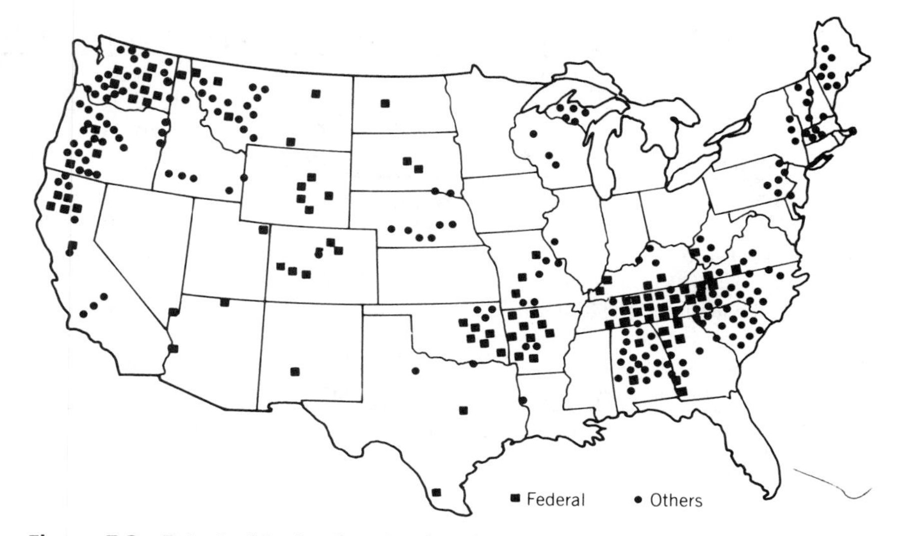

Figure 7.3 Principal hydroelectric plant locations in the United States. (Adapted from *River of Life,* Environmental Report 1970, U.S. Department of the Interior.)

are concentrated in the extreme Northwest and in the Southeast, with only limited possibilities in rapidly growing regions of the Southwest, where precipitation is scant, or near the population centers of the Northeast. This situation is illustrated in Figures 7.3 and 7.4 and in Tables 7.1 and 7.2.

Of the 180,000-MW hydroelectric potential in the United States, only about 30% has been developed. The remaining 70% includes sites that surely will be developed, but there is also enormous potential, as in Alaska, that is unlikely to be developed in the near term. The recent history of U.S. hydroelectricity is shown in Figure 7.5. On a percentage basis, it seems unlikely that the role of hydroelectricity in the United States will increase significantly in the near

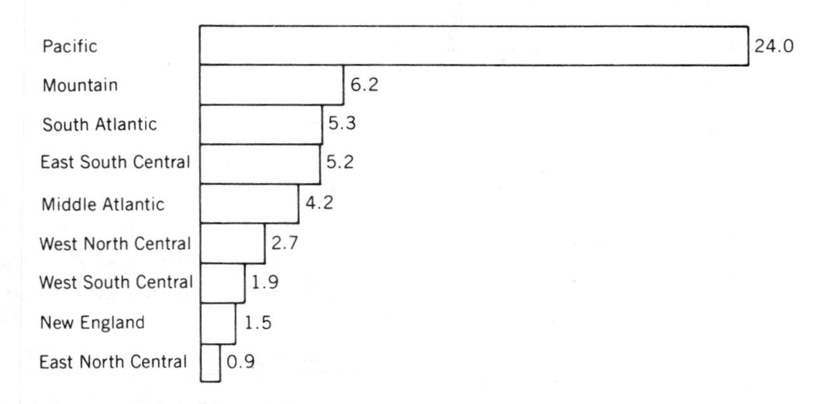

Figure 7.4 U.S. hydroelectric capacity in thousands of megawatts developed in various regions of the country as of 1970. (Source: H. S. Stoker; S. L. Seager; and R. L. Capener, *Energy: From Source to Use,* Scott, Foresman, 1975).

Table 7.1 SOME LARGE HYDROELECTRIC PROJECTS IN THE UNITED STATES

Project	First Year of Operation	Rated Capacity (MW)	Rated Capacity Planned (MW)
Grand Coulee, Washington	1942	6,480	10,230
John Day, Washington	1969	2,160	2,700
Niagara, New York	1961	1,950	—
The Dalles, Washington	1957	1,807	—
Chief Joseph, Washington	1956	1,500	3,699
NcNary, Oregon	1954	1,406	1,670
Hoover, Arizona–Nevada	1936	1,345	—
Glen Canyon, Arizona	1964	950	—

Source: Data from Bureau of Reclamation, United States Department of the Interior.

Table 7.2 HYDROELECTRIC POTENTIAL IN THE UNITED STATES

	Conventional Capacity (1000 MW)			
Region	Total Potential	Developed Capacity	Undeveloped Capacity	Percent Developed
New England	4.8	1.5	3.3	31.3
Middle Atlantic	8.7	4.2	4.5	48.3
East North Central	2.5	0.9	1.6	36.0
West North Central	7.1	2.7	4.4	38.0
South Atlantic	14.8	5.3	9.5	35.8
East South Central	9.0	5.2	3.8	57.8
West South Central	5.2	1.9	3.3	36.5
Mountain	32.9	6.2	26.7	18.8
Pacific	62.2	23.9	38.3	38.4
Alaska	32.6	0.1	32.5	0.3
Hawaii	0.1	—	—	—
Total	179.9	51.9	128.9	

Source: Data selected from "Final Environmental Statement for the Geothermal Leasing Program," Vol. 1, U.S. Department of the Interior, 1973, p. IV–170. Cited in Stoker, Seager, and Capener.

future. In fact, its role in supplying electricity in the United States has been declining for 40 years relative to other sources. On other continents, especially South America and Africa, there are very large potential hydroelectric resources. The South American and African resources are presently more than 99% undeveloped.

Some hydroelectric installations have lifetimes limited to perhaps 50 to 200 years because their storage volumes become steadily filled with silt that is washed downstream by the rivers that feed the reservoirs. This is a severe problem, with no solution in sight, for many of the world's largest hydroelectric facilities. It is less of a problem for other facilities fed by streams flowing over beds of rock rather than of soil. Once a reservoir has been filled with silt to

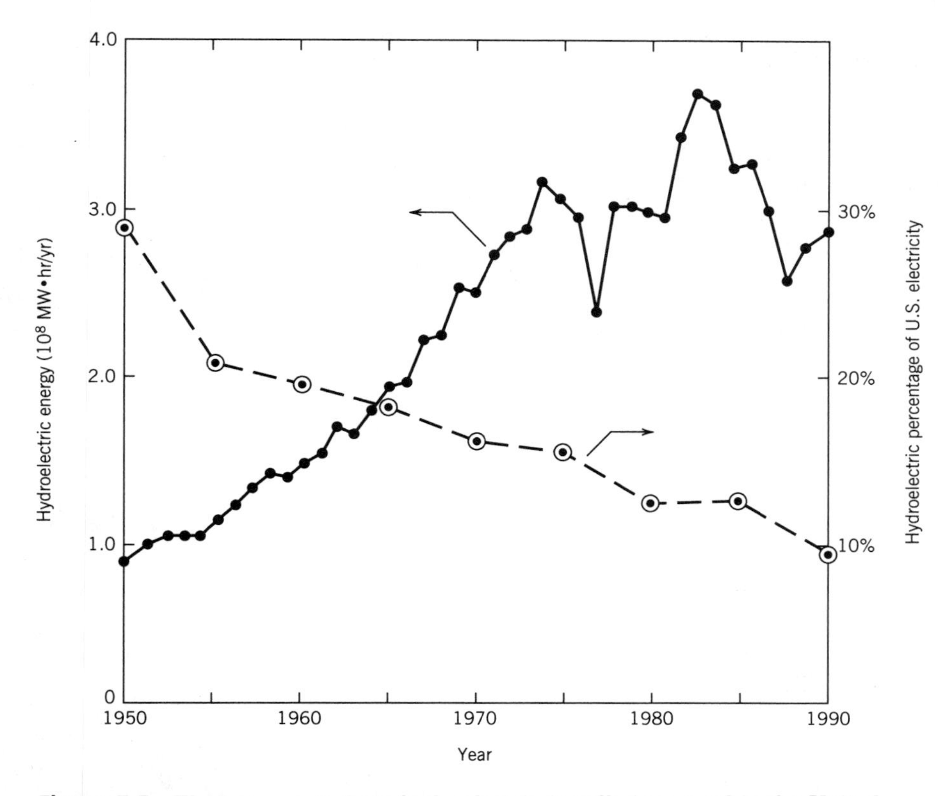

Figure 7.5 Electric energy from hydroelectric installations used in the United States, including net imports of hydroelectricity (solid line). The percentage of the United States electricity from hydroelectric sources is shown as the dashed line. (Data from *Annual Energy Review 1990,* U.S. Department of Energy)

the point that it is no longer useful for water storage, the maintenance of the dam will be a continuing burden. Downstream areas must be protected from the sudden release of enormous volumes of silt that could flow downstream in the event of dam failure.

Other objections to hydroelectric power include the loss of free-flowing streams and of the land submerged by the reservoirs. Native aquatic life of many forms is disturbed, particularly salmon in the Northwest; downstream from the dams the stream flow may be intermittent in response to power demand. The shorelines of the reservoirs vary greatly in elevation with the seasons, often presenting an unsightly view and limiting recreational use of the reservoir.

Storage reservoirs are often situated upstream from major population centers. This represents a considerable risk in the event of dam failure, such as might be expected to result from a seismic disturbance. Catastrophic dam failures have occurred frequently throughout the past century. Between 1918 and 1958, there were 33 major dam failures in the United States with 1680 resulting deaths, an average of 42 per year. From 1959 to 1965, nine major dams failed throughout the world. Major dam failures involving loss of human life occurred in the United States in 1976 and 1977. We now have a number of population groups with more than 100,000 persons at risk from dam failure.

7.3 WIND POWER

Since the distant time when our ancient ancestors first learned to rig sails on their rafts, wind power has been put to human use. After this early discovery, the use of the force of wind on fabric sails advanced technologically to the point where all the world's navies and transoceanic commerce were powered by the wind. The large sailing vessels of the 19th century could extract as much as 10,000 horsepower from the wind; no larger wind machines have ever been developed. Stationary wind machines now represent the major use of the wind's energy, and their early development also dates back many hundreds, perhaps

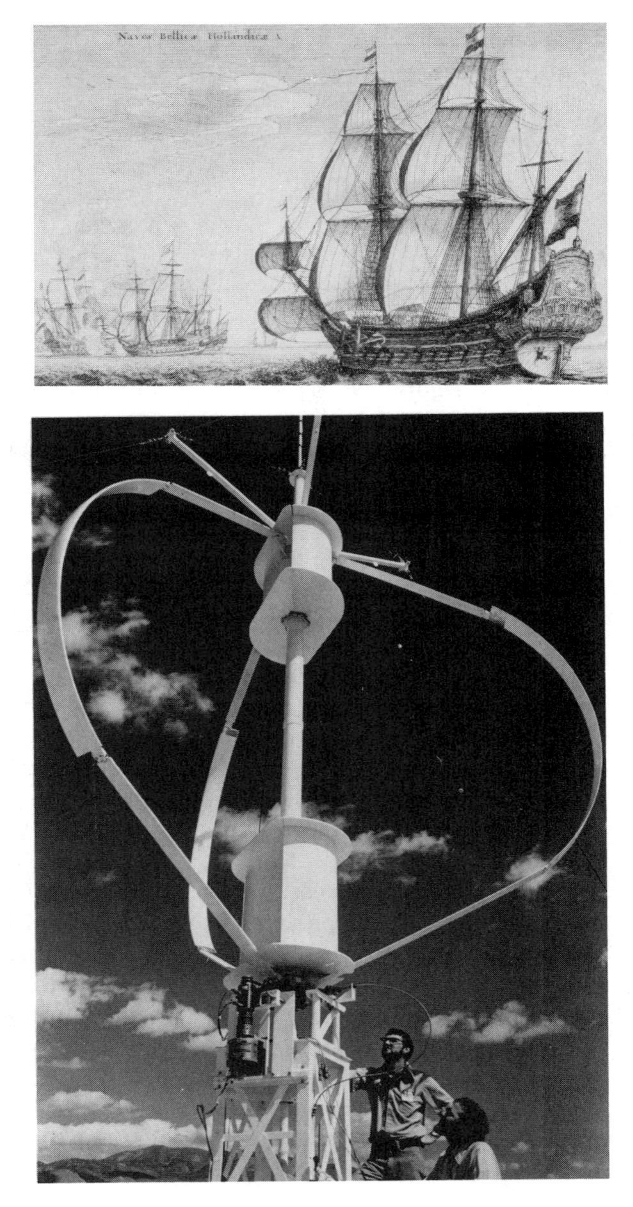

Figure 7.6 Two wind machines, one of a type developed centuries ago, the other modern. The Dutch warships of the 17th century could extract thousands of horsepower from the wind; the modern device shown produces 3 hp in a 20 mph wind.

thousands, of years. Windmills by the tens of thousands were operating in Europe during the 19th century, and several million were pumping water in the United States by the early 1900s. Starting about 1890, the use of windmills has been increasingly directed toward the generation of electricity and less toward their direct mechanical coupling with machinery such as mills. Today the development of wind machines is almost exclusively concentrated on electricity generation. Figure 7.6 shows two wind machines.

The earth's winds are a direct consequence of solar energy. On both local and global scales, these winds are generated because the sun heats certain areas of the earth's atmosphere and surface more than others. This differential heating induces both vertical and horizontal air currents, with the patterns of the currents modified by the earth's rotation and by the contours of the land. The familiar land–sea breeze cycle is an example of how winds are produced. During the daytime, the sun shines on the land causing it to become warmer than the nearby sea. As it is heated, the air over the land rises and is replaced by an onshore breeze that moves air in from over the cooler sea. After the sun sets, the land cools faster than the sea. Then the now warmer air over the sea rises and is replaced by an offshore breeze that moves the heavier air out from over the land.

The power in the wind increases rapidly with velocity; it is proportional to the third power of the velocity. This is explained as follows: Each unit of air mass has kinetic energy proportional to the square of its velocity, and the amount of air mass moving past a given point (for instance, the location of a windmill) per unit time is proportional to the velocity. Thus, the amount of wind power (energy/time) at any point in space varies as v^3. The exact relationship of power to velocity, of course, also involves the density and moisture content of the air; however, to a reasonable degree of approximation, for average conditions one may use the expression

$$P/\text{m}^2 = 6.1 \times 10^{-4} v^3$$

to find power in kilowatts per square meter of cross section oriented perpendicular to the wind's direction, with the velocity, v, given in meters per second. This is the total power in the wind, which cannot all be extracted by any practical device. Any windmill built so that it severely interrupts the airflow through its cross section will reduce the effective wind velocity at its location and divert much of the airflow around itself, thus not extracting the maximum power from the wind. At the other extreme, a windmill that intercepts an exceedingly small fraction of the wind passing through its cross section will reduce the wind's velocity by only a small amount, thus extracting only a small amount of power from the wind traversing the windmill disk. All practical windmills fall somewhere between the two extremes. An exact theoretical analysis of windmill performance shows that no more than 59% of the kinetic energy of the wind is recoverable; modern windmills can attain an efficiency of perhaps 50 to 80% of the theoretical maximum. The most efficient type, the high-speed propeller, can typically attain 70% of the theoretical maximum efficiency, thereby utilizing about 42% of the energy in the wind. The American multiblade windmill, used for pumping water in rural areas, utilizes at most about 30% of the wind's power, and the picturesque Dutch four-arm type extracts about 16%. The American multiblade type is well adapted to driving water pumps, as it has relatively high starting torque, whereas the high-speed propeller type has low starting torque but is most efficient at the high rotational speeds suitable for small-scale electricity generation.

Example 7.2

Calculate the power produced per square meter of windmill disk area for a windmill operating at 70% of the theoretical maximum efficiency when the wind velocity is
10 m/sec.

Solution

$$\begin{aligned}\text{power (kW/m}^2) &= (0.7) \times (0.59) \times 6.1 \times 10^{-4}\, \nu^3 \\ &= (0.7) \times (0.59) \times 6.1 \times 10^{-4}\,(10)^3 \\ &= 0.253\ \text{kW/m}^2 = 253\ \text{W/m}^2\end{aligned}$$

Example 7.2 shows that with a wind velocity of 10 m/sec, a power output of about 250 W/m^2 can be achieved. This is far above the average performance to be expected year-round, however, as in most of the United States a wind velocity of this magnitude is experienced near ground only a few percent of the time. Typical annual energy output for a windmill can be expected to range from about 100 kW • hr/m^2 for relatively calm areas to approximately 500 kW • hr/m^2 for relatively windy areas. These numbers correspond to average levels of electric power production of only 11.4 W/m^2 and 57 W/m^2, respectively, over the entire 8760-hour year. Figure 7.7 presents average wind velocities at 50 m above the ground level, and Table 7.3 indicates how wind velocity is classified.

Example 7.3

Calculate the diameter of a windmill needed to supply the 5000-kW • hr of electrical energy needed annually by an American household. Assume average wind conditions in which 250 kW • hr/m^2 is produced annually by the windmill.

Solution

The area of the windmill disk can be obtained from

$$\frac{5000\ \text{kW} \cdot \text{hr}}{250\ \text{kW} \cdot \text{hr/m}^2} = 20\ \text{m}^2$$

The diameter corresponding to this disk can be obtained from

$$\text{area (A)} = \pi \frac{d^2}{4}, \qquad \text{or} \qquad d = \sqrt{\frac{4A}{\pi}}$$

$$d = \sqrt{\frac{4 \times 20\text{m}^2}{\pi}} = \sqrt{25.5\ \text{m}^2} \approx 5\ \text{m} = 16.5\ \text{ft}$$

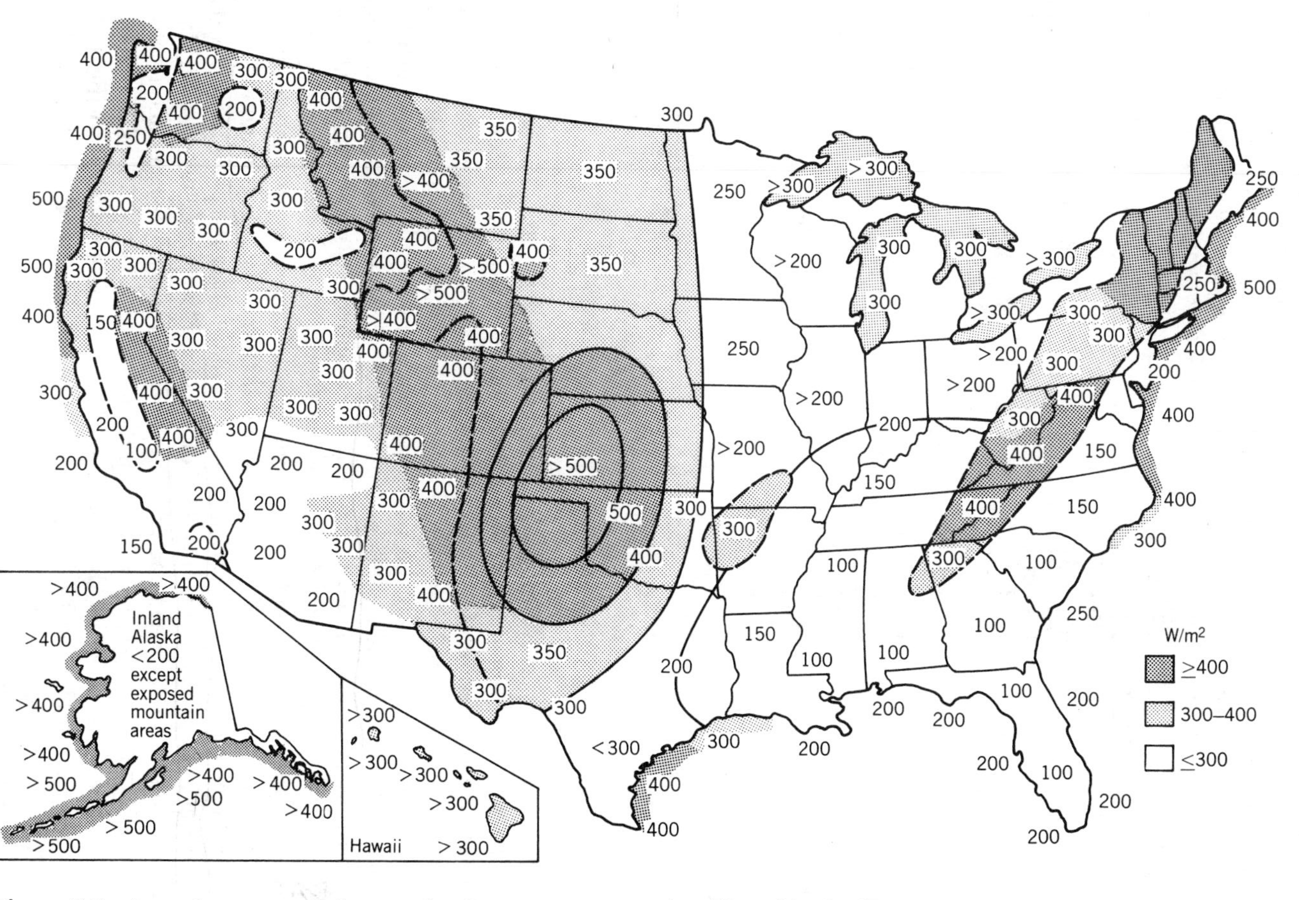

Figure 7.7 Annual average wind power (watts per square meter) at 50-m altitude. (Figure supplied by the Solar Energy Research Institute.)

Table 7.3 THE CLASSIFICATION OF WIND VELOCITIES: EVEN A GENTLE BREEZE IS ADEQUATE TO POWER A WINDMILL

Winds at Sea: The idea of more precise and concise terms than were commonly used in log books to describe winds at sea occurred to a scarred British Navy veteran, Francis Beaufort, in 1805. (At 31 years old he had already seen a lot of service: in one fight he was wounded by three sword slashes and 16 musket balls.) Later as hydrographer of the navy and a knight, he lived to see his scale of winds generally approved. It and the earlier but still widely favored table follow.

Beaufort Scale			**Logbook Term**	
Description	**Miles per Hour**	**Force**	**Description**	**Miles per Hour**
Calm	Less than 1	0	Calm	Less than 1
Light air	1–3	1	Very light	1–3
Light breeze	4–6	2	Light	4–7
Gentle breeze	7–10	3	Gentle	8–12
Moderate breeze	11–16	4	Moderate	13–18
Fresh breeze	17–21	5	Fresh	19–24
Strong breeze	22–27	6	Strong	25–38
Moderate gale	28–33	7	Gale	39–54
Fresh gale	34–40	8	Whole gale	55–72
Strong gale	41–47	9	Hurricane	More than 72
Whole gale	48–55	10		
Storm	56–65	11		
Hurricane	More than 65	12		

Source: From Peter Freuchen, *Book of the Seven Seas* (New York: Julian Messner, 1957). Copyright © 1957, renewed 1985 by Dagmar Freuchen Gale. Reprinted by permission of Don Congdon Associates, Inc.

A proposal for very-large-scale electric power generation has been put forth by Professor Heronemus of the Massachusetts Institute of Technology. His proposal to generate electricity economically by wind power would involve 300,000 towers, each 850-ft high (a 70-story building), and each with 20 generators powered by a two-blade, 50-ft diameter propeller. These towers would be distributed over the Great Plains from Texas to Canada at a density of nearly one tower per square mile. This would be an enormous undertaking with more than one tower per 1000 U.S. citizens, involving 6 million generators as well as energy storage apparatuses. (For comparison, there are now more than 100 million cars and trucks on the road in this country.) The large windmill system would have an average electrical output of more than 150,000 MW and would be capable of producing almost half of the electrical power used in the United States today. Another evaluation of the potential for large-scale electrical generation is given in Table 7.4, which indicates that 1.536×10^{12} kW • hr could be produced annually by the year 2000. This can be compared to the 1.97×10^{12} kW • hr of electrical energy now generated annually in the United States by fossil-fuel-burning plants.

In addition to using windmills to capture the wind's energy, there have been several attempts over the past centuries to extract energy from ocean waves. These waves receive their energy from the action of the wind on the ocean's surface. As early as 600 to 700 years ago, the Chinese operated iron ore crushing machines with wave power. One estimate of the power available

Table 7.4 MAXIMUM ANNUAL ELECTRICAL ENERGY PRODUCTION FROM WIND POWER (kW • hr) POSSIBLE BY 2000

Offshore, New England	318×10^9
Offshore, eastern seaboard Ambrose shipping channel south to Charleston, S.C.	283×10^9
Along the east–west axis, Lake Superior	35×10^9
Along the north–south axis, Lake Michigan	29×10^9
Along the north–south axis, Lake Huron	23×10^9
Along the west–east axis, Lake Erie	23×10^9
Along the west–east axis, Lake Ontario	23×10^9
Through the Great Plains from Dallas, Texas, north in a path 300 miles west–east, and 1300 miles long, south; wind stations to be clustered in groups of 165, at least 60 miles between groups (sparse coverage)	210×10^9
Offshore, Texas Gulf Coast, along a length of 400 miles from the Mexican border, eastward	190×10^9
Along the Aleutian Chain, 1260 miles, hydrogen is to be liquefied and transported to California by tanker	402×10^9
Total	1.536×10^{12}

Source: Data taken from "Solar Energy as a National Energy Resource," NSF/NASA Solar Energy Panel, December 1972. Cited in Gerard M. Crawley, *Energy* (New York: Macmillian, 1975).

from the ocean waves on our coastal areas predicts that wave-power apparatus on both ocean coasts from Canada to Mexico could generate 6×10^9 W, only about one tenth of our present hydroelectric capacity. Thus, it seems unlikely that wave power will ever contribute in an important way to the energy economy of the United States.

Any system of wind power must include some means of energy storage (see Chapter 8) if it is to deliver energy continuously and reliably in spite of variations in wind velocity. Methods commonly proposed are pumped hydro storage (a proven technology) and hydrogen production from water, with the hydrogen being burned as a fuel.

As with all energy technologies, various problems would be introduced by large-scale wind power utilization. The installations may be unsightly, they may interfere with the working of agricultural lands, they will interfere with television reception and microwave communication, and they may be hazardous to migrating birds. The economics of wind power are at best marginal at present, but some studies have reported that wind energy is now the least expensive form of solar energy. Figure 7.8 shows the cost of electricity predicted for a system of large 2.5 MW wind turbines if mass production is achieved. The indicated cost is similar to that now charged consumers in the United States.

7.4 OCEAN THERMAL ENERGY CONVERSION

The world's oceans constitute a vast natural reservoir for receiving and storing the energy of the sun incident on earth. These oceans take in solar energy in proportion to their surface area, nearly three times that of land. Water near the surface of tropical and subtropical seas is maintained by this solar radiation at

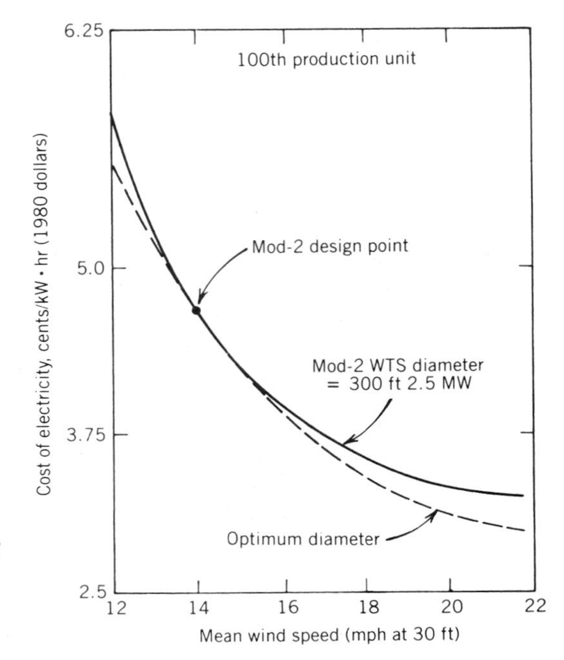

Figure 7.8 Effect of mean wind speed on economic performance of the Mod 2 wind turbine. (Figure supplied by the Solar Energy Research Institute.)

higher temperatures than the water at greater depth or higher latitudes. Some of this warm surface water is carried far from equatorial regions, either to the north or to the south, by ocean currents such as the great Gulf Stream or the Japanese Current. These movements of surface waters result from the action of prevailing wind patterns on the surface of the ocean; the courses of these currents are shaped further by the effects of the earth's rotation (Coriolis forces) and by the shorelines of continental land masses. It would be entirely possible to tap the energy in these currents for electrical power by using gigantic underwater turbines anchored to the ocean floor and connected to generators, much as windmills extract energy from moving air. The ocean currents are attractive for this purpose because they flow at a relatively steady velocity, in contrast to the intermittent nature of the wind. This eliminates the need for energy storage. The energy available is enormous; the Gulf Stream has a flow a thousand times as large as the Mississippi River and a maximum velocity greater than 4 miles per hour. Serious proposals have been put forth to construct ocean current turbines with the electricity produced then cabled to shore.

There is a larger effort, however, to extract energy from the oceans through the use of heat engines that exploit the thermal gradient between tropical surface waters and deeper layers. These temperature differences are very steady in time, persisting over day and night and from season to season, again eliminating the need for energy storage systems. Figure 7.9 shows the regions of ocean surface where these thermal gradients are of appreciable magnitude. The region of significant potential for ocean thermal energy conversion (OTEC) include the waters adjacent to many heavily populated coastal areas. For the United States, the OTEC potential is most promising for Florida, Puerto Rico, and Hawaii.

Example 7.4

Calculate the ideal thermodynamic efficiency for a heat engine operating between surface waters and water at 1000 m depth if the surface water temperature is 25°C and the deeper water is at 5°C.

Solution

$$T_c = 5°C = 278°K$$

$$T_h = 25°C = 298°K$$

$$\eta = \left(1 - \frac{T_c}{T_h}\right) = \left(1 - \frac{278}{298}\right) = (1 - 0.933) = 0.067$$
$$= 6.7\%$$

Of course, the efficiency of an *actual* heat engine will be even less than this.

Example 7.5

(a) Calculate how much power is made available to a heat engine by the cooling of 1000 gallons of water per second by 2°C. Express your answer in watts. Assume 1 gal = 3.8 kg.

(b) If this heat engine operates between heat source and sink temperatures differing by 20°C, as in Example 7.4, what is its theoretical maximum power output?

Solution

(a)

$$P_h\,(\text{cal/sec}) = (\text{g } H_2O/\text{sec}) \times (\Delta T, °C)$$
$$= 3.8 \times 10^3 \text{ g/gal} \times 10^3 \text{ gal/sec} \times 2°C$$
$$= 7.6 \times 10^6 \text{ g} \bullet °C/\text{sec}$$
$$= 7.6 \times 10^6 \text{ cal/sec}$$

$$P_h\,(W) = P_h\,(\text{cal/sec}) \times 4.184 \text{ J/cal} = 31.8 \times 10^6 \text{ W}$$
$$= 3.18 \times 10^7 \text{ W}$$

(b) From Example 7.4, $\eta = 0.067$

$$P_{out} = \eta \times P_h = 6.7 \times 10^{-2} \times 3.18 \times 10^7 = 2.13 \times 10^6 \text{ W}$$

or about 1/500 of the electric power outut of a large coal-burning or nuclear power plant.

The design of an apparatus for OTEC is a straightforward matter of devising a heat engine that operates at the modest temperature differential available. Such an engine can be achieved by using ammonia as the working fluid because it vaporizes and condenses at the available temperatures (see Figure 7.10). This

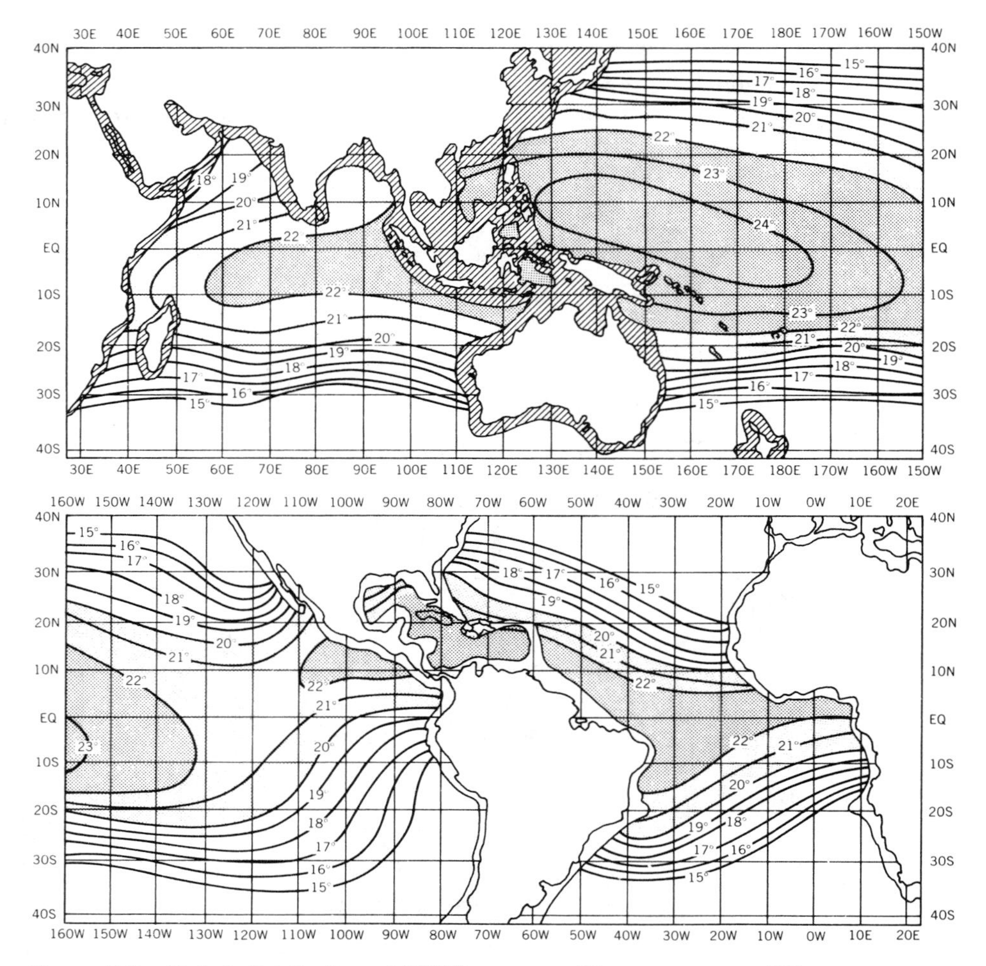

Figure 7.9 Global distribution of OTEC resource. The temperature difference (degrees Celsius) is shown between the surface and 1000-m depth. (Figure supplied by the Solar Energy Research Institute.)

is analogous to choosing water as the working fluid matched to the temperature differential between a fossil-fuel–fired boiler and a condenser cooled by ambient air or water. Other types of OTEC engines have also been proposed and are in the research and development stage. An OTEC plant is sketched in Figure 7.11. The energy extracted as electricity is, of course, derived from the cooling of the warmer surface water entering the heat engine. The energy extracted in a given time is proportional to the volume of warm water entering the engine in that time and the temperature drop that it experiences, reduced by the heat energy warming the colder water that cools the condenser.

The OTEC concept was first put forth theoretically in 1880 by d'Arsonval, and the first plant was constructed much later, in 1930, at Matanzas Bay, Cuba, by d'Arsonval's student, Georges Claude. This plant did succeed in converting the ocean's thermal energy into electrical energy, but as an engineering prototype it did not operate efficiently. However, on receiving an award from the

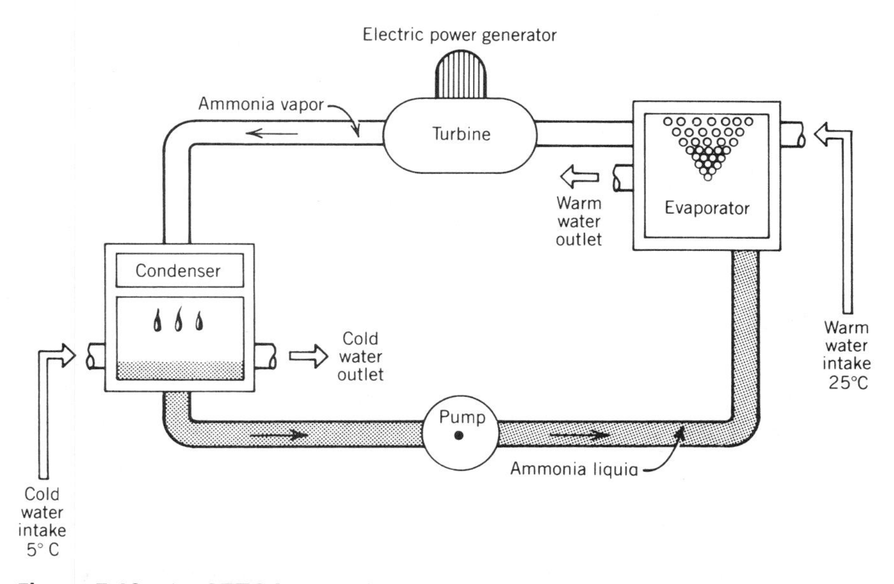

Figure 7.10 An OTEC heat engine using ammonia as a working fluid. The turbine is driven by the ammonia vapor and is connected to a generator to produce electricity. The warm water is drawn from the ocean surface; the cold water from a depth of 1000 m. (Figure supplied by the Solar Energy Research Institute.)

American Society of Mechanical Engineers in 1930, Claude said: "I affirm my faith in the realization of wonderful plants running ceaselessly throughout the year, unaffected by the seasonal scarcity of water in streams or variations in the cost of coal; and I hold all this is not the task of a remote future, but one of tomorrow."

Now, some sixty years later, as the end of abundant fossil fuels is in sight, OTEC is becoming increasingly attractive. Some idea of the expected scale of OTEC development in the United States can be gathered from Figure 7.12. The cost of OTEC-generated electricity is expected to be comparable to that of

Figure 7.11 An OTEC plant designed by TRW, Inc. This free-floating ocean platform can supply 100 MW of electricity. The platform is 100 m in diameter. A 15-m diameter pipe sucks up cold water from a depth of 1200 m.

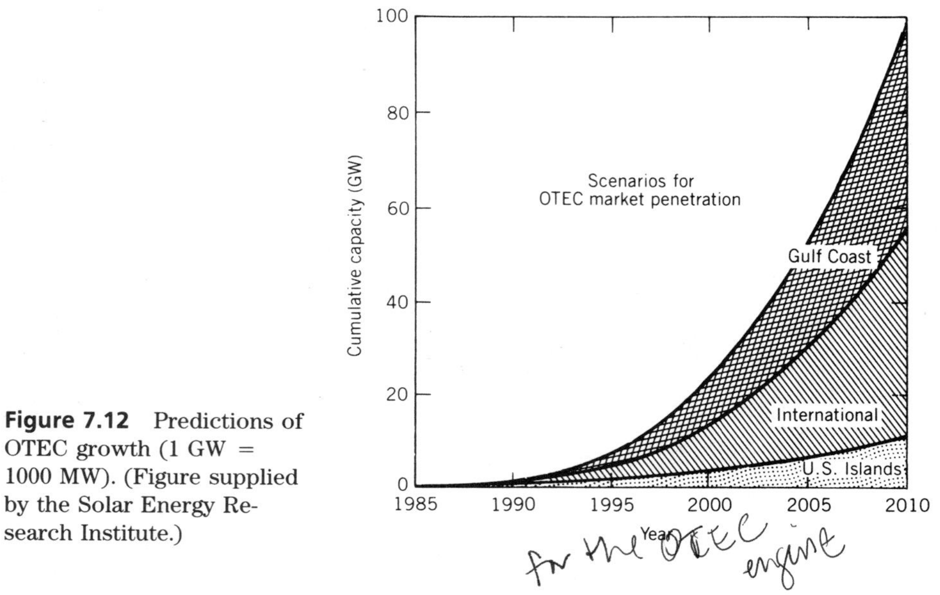

Figure 7.12 Predictions of OTEC growth (1 GW = 1000 MW). (Figure supplied by the Solar Energy Research Institute.)

electricity produced by coal-burning plants. The technology is especially valuable for islands, such as Hawaii or Puerto Rico, where electricity is now generated by oil-burning plants. It is estimated that in Puerto Rico more than 80,000 barrels of oil could be saved each day by generating 2000 MW of OTEC electricity.

7.5 GEOTHERMAL ENERGY

We are all familiar with geothermal energy in the form of the spectacular displays of volcanic eruptions and natural geysers, as well as in the less dramatic example of the numerous hot springs found throughout the world. The technology has emerged within this century for commercial production of electricity at many geothermal sites, such as in Italy since the early 1900s, in New Zealand since the 1950s, and in California starting in 1960. In addition to this large-scale use for electric power, geothermal energy has been used much longer for recreational and therapeutic purposes at natural hot springs, and for space heating at many locations where hot water can be brought to the surface. Figure 7.13 shows some natural geothermal sites.

There is a continual flow of heat energy outward from the molten interior of the earth to the cooler surface. Although there are local regions of intensely concentrated geothermal energy flow, the geothermal flux over the earth's surface is exceedingly small on average, less than one-thousandth that of incoming solar energy. The high temperatures in the earth's interior are thought to be due to radioactive decay of long-lived radioactive elements that were incorporated into the earth at the time of its formation, as well as, in part, to frictional effects resulting from the solar and lunar tides.

Geothermal energy is conducted outward from the earth's interior to the surface layers at a rate of only about $\frac{1}{16}$ W/m^2 averaged over the earth's surface area. This leads to the conclusion that in order for us to make large-scale beneficial use of the thermal energy stored in rock and water immediately under the earth's surface, we must withdraw energy more rapidly than it is

Figure 7.13 Some natural geothermal sites.

being replenished in any given area. Thus, as we use geothermal energy, we are effectively mining what is in the short term a nonrenewable energy resource, just as we mine fossil fuel resources. For example, at The Geysers in California (Figure 7.14), presently the largest geothermal energy production site in the world, heat energy is withdrawn at an estimated 80 times the rate at which it is being replenished. This is in an area with natural heat flow vastly greater than the average over the earth.

Although the normal geothermal gradient in the earth's crust is a temperature increase of about 30°C per kilometer of depth below the earth's surface, in areas such as Yellowstone National Park the molten rock penetrates upward into the earth's crust fairly close to the surface. In such areas, usually those of relatively recent geologic activity, high-temperature regions are found close to the surface.

The heat energy extracted from geothermal sites is used mostly to drive turbines to produce electricity. In the United States geothermal energy now contributes less than $\frac{1}{10}$ QBtu (compared to 82 total) of energy annually. The only large U.S. geothermal electric power plant is at The Geysers, north of San Francisco, where 1363 MW_e capacity was achieved in 1985. By 1990, the total U.S. geothermal industry was producing 2800 MW_e at 4 to 6 cents per kW • hr. Only about 15 MW_t is used in the United States for space heating and industrial process heat.

The total geothermal resource is diverse in nature and adaptable to various applications, although most studies project that in the coming decades electricity generation will predominate. The resource has been divided by one study (CONAES) into six categories. They are:

Hot Water Reservoirs These are geothermally heated reservoirs of underground water, very large in magnitude in the United States, but not generally

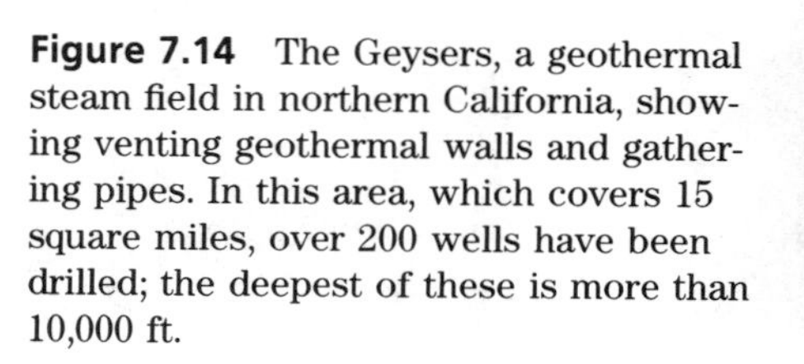

Figure 7.14 The Geysers, a geothermal steam field in northern California, showing venting geothermal walls and gathering pipes. In this area, which covers 15 square miles, over 200 wells have been drilled; the deepest of these is more than 10,000 ft.

appropriate for electricity generation. Hot water has been used to produce electricity in some locations outside the United States. The hot water may also be appropriate for space heating.

Natural Steam Reservoirs This is the type of geothermal resource found at The Geysers, where the naturally occurring steam is used to drive turbines. This is a highly desirable, but very rare, type of resource, unlikely for that reason to contribute in a major way to our national energy budget.

Geopressured Reservoirs This is brine completely saturated with natural gas under considerable pressure because of the weight of the overlying land. It may be an important energy resource because of both heat energy and the production of natural gas.

Normal Geothermal Gradient Even in dry rock the normal geothermal gradient (30°C/km) produces useful temperatures any place on the globe. Drill holes 20,000 feet deep are achievable, corresponding to temperatures of about 190°C above the surface temperature. This temperature is adequate for electricity generation. Although the normal geothermal gradient resource is truly enormous in magnitude, no technology has yet been developed to extract useful energy from this resource. Attempts have been made near Los Alamos, New Mexico, and elsewhere to circulate a working fluid through dry rock, but no commercially promising results have yet been reported.

Hot Dry Rock This is technically the same as the normal geothermal gradient resource, but at the more favorable locations where the geothermal gradient in the earth's crust is greater than 40°C/km. It is estimated that such conditions are found under 5% of the U.S. land area. As in the case of the previous category, no commercial successes have been reported in circulating a working fluid through a system of channels, natural or man-made, in hot dry rock.

Molten Magma No technology yet exists to exploit the high temperatures in molten lava found at volcanic locations. The high temperatures are attractive for electricity generation. Possible resources in the United States are in Hawaii and Alaska.

Example 7.6

The Geysers geothermal site near San Francisco, California, covers an area of 70 km^2, and the thickness of the subsurface zone from which heat is recoverable is 2.0 km. In this zone the temperature is 240°C and the volumetric specific heat is 2.51 $J/cm^3 \cdot °C$.

(a) Calculate the heat energy content in joules (at temperatures above the mean annual surface temperature of 15°C).

(b) For how many years can this site provide power for a 2000-MW_e plant if 1.9% of the thermal energy can be converted to electricity? These numbers are representative of plans for the future of The Geysers site.

Solution

(a) The volume of the zone is given by the area times the thickness:

$$V = 70 \text{ km}^2 \times 2.0 \text{ km} = 140 \text{ km}^3$$

The heat content is given by the volume times the volumetric specific heat times the temperature change:

$$Q = V \times (\text{v.s.h.}) \times \Delta T$$

$$V = 140\ \text{km}^3 \times 10^{15}\ \text{cm}^3/\text{km}^3 = 1.4 \times 10^{17}\ \text{cm}^3$$

$$\text{v.s.h.} = 2.51\ \text{J/cm}^3 \bullet {}^\circ\text{C}$$

$$\Delta T = 240^\circ\text{C} - 15^\circ\text{C} = 225^\circ\text{C}$$

$$Q = 1.4 \times 10^{17}\ \text{cm}^3 \times (2.51\ \text{J/cm}^3 \bullet {}^\circ\text{C}) \times 225^\circ\text{C} = 7.91 \times 10^{19}\ \text{J}$$

(b) For each year of operation, the electrical energy produced will be

$$P \times t = E$$

$$2000\ \text{MW}_e \times 1\ \text{yr} = 2000\ \text{MW} \bullet \text{yr}$$

To produce this much electrical energy at an overall efficiency of 1.9% requires that heat energy be extracted each year in the amount of

$$\frac{2000\ \text{MW} \bullet \text{yr}}{0.019} = 105{,}263\ \text{MW} \bullet \text{yr} = 1.053 \times 10^{11}\ \text{W} \bullet \text{yr}$$

In units of joules this is

$$1.053 \times 10^{11}\ \text{W} \bullet \text{yr} \times (3.15 \times 10^{7}\ \text{sec/yr}) \times 1\ \text{J/W} \bullet \text{sec} = 3.32 \times 10^{18}\ \text{J each year}$$

To obtain the number of years for which a 2000-MW_e plant can be operated before the heat reservoir is exhausted, divide the available energy by the amount consumed per year:

$$\frac{\text{available energy [from part (a)]}}{\text{amount consumed per year}} = \frac{7.91 \times 10^{19}\ \text{J}}{3.32 \times 10^{18}\ \text{J/yr}} = 23.8\ \text{yr}$$

Of the geothermal resources listed, only hot water and natural steam reservoirs have been put to use. For these two categories the environmental effects are, therefore, fairly well-known; for the other four categories, the probable effects are highly uncertain. Many of the hot water reservoirs, particularly those that are at higher temperatures and more saline, pose the potential for soil salination if the extracted water is not reinjected into the ground after its heat is extracted. There is also a risk of subsidence and aquifer disruption when large amounts of water are extracted from the ground. The gaseous air pollutant hydrogen sulfide is liberated into the atmosphere by some hot water reservoirs as well as by natural steam reservoirs. Other possible environmental effects include induced seismicity if water is injected into dry rock formations or if nuclear explosive fracturing techniques are used in normally impermeable rock formations.

Some idea of the size of the resource can be obtained from Table 7.5. The

Table 7.5 ESTIMATED U.S. GEOTHERMAL RESOURCES

Reservoir Type	Total Resource, Identified and Yet Undiscovered (QBtu[a])	Total Potentially Producible Thermal Energy–Accessible Reservoirs (QBtu[a])
Hot water	12,017	6,009
Natural steam	179	45
Geopressured[b]	73,363	2,421
Normal gradient	1,253,000	12,530
Hot dry rock	163,500	1,635
Molten magma[c]	3,500	35
Total	1,505,559	22,675

[a] To 6-km depth, $T \geqslant 80°C$, national parks excluded.

[b] Heat content of fluid only, $\geqslant 50°C$, onshore.

[c] Practical heat-extraction technology not yet developed.

Source: Adapted from CONAES, 1980.

first two categories, hot water and natural steam, are characterized by proven technology and are, thus, fairly accurately represented in this table; the other four are speculative and, thus, much more uncertain, especially with regard to the numbers in the second column. If these numbers are compared to the 1990 national energy budget of 82 QBtu, it can be seen that although natural steam and molten magma may have local importance, they do not appreciably affect the nation's energy future. The other four resources represent extremely large amounts of thermal energy.

Various assumptions can be made in projecting the future of geothermal electrical generation in the United States. In Table 7.6, projections to 1990 and 2010 are shown, which were made in 1980 under three different scenarios. The lowest projected number, 2930 ME_e for 1990, is very close to the actual electrical generation, 2800 MW_e, achieved that year; the largest number, 60,900 MW_e in 2010, is only about 10% of our present installed capacity powered by nuclear, fossil fuels, and hydroelectricity.

Predictions of the lifetime of any particular geothermal resource are highly uncertain, but as Example 7.6 shows, lifetimes of 20 to 50 years or perhaps as much as 100 years in some places are obtained when a geothermal site is fully developed. Geothermal energy is clearly not a renewable resource in the same sense that solar energy is renewable. Over a very long time span, however, such as hundreds or thousands of years, the heat energy withdrawn from a given reservoir may be replaced by natural processes.

Table 7.6 ESTIMATED INSTALLED GEOTHERMAL ELECTRICAL GENERATING CAPACITY IN THE UNITED STATES UNDER THREE DIFFERENT SCENARIOS (MW_e)

	1990	2010
Business as usual	2930	18,870
Enhanced supply	4090	32,170
National commitment	8270	60,900

Source: Adapted from CONAES, 1980.

It has been demonstrated at The Geysers and elsewhere that geothermally produced electricity is cost-competitive with electricity produced by other means. Some estimates for both proven and speculative technologies are given in Table 7.7, where the first two reservoir types, both proven, produce electricity at lower cost than do other energy sources, geothermal as well as conventional.

In making estimates of the future of geothermal energy, it is apparent that the limiting factor is not the size of the resource, because it is so enormous, but rather other factors, such as the economics relative to other energy sources, the technology of the future, and even population distribution. The largest resource, normal gradient heat stored in the earth's crust, has never been successfully tapped for energy, except for small demonstration projects. It is a diffuse distribution of thermal energy, and the various attempts so far to circulate water or steam through drill holes into deep beds of rock have not been encouraging. The more concentrated geothermal resources often lie far from population centers; in the United States the attractive sites are almost exclusively in the western states, including Hawaii and Alaska.

When considered in its entirety, it is apparent that although of local importance, the geothermal option is not likely to rank among the most important contributors to the national energy budget in the near future. The long-term prospects may be more encouraging, depending on the success of technologies that are now only speculative.

7.6 TIDAL ENERGY

Tidal energy differs from all other energy sources in that the energy is extracted from the potential and kinetic energies of the earth–moon–sun system. The well-known ocean tides result from this interaction, producing variations in ocean water levels along the shores of all continents. On the U.S. coasts, the vertical tidal range varies from about 2 ft in Florida, to 18 ft and more in Maine.

Table 7.7 COST ESTIMATES FOR ELECTRICITY GENERATED BY GEOTHERMAL AND OTHER ENERGY SOURCES

Reservoir Type	Temperature (°C)	Cost (1975 dollars/kW • hr)
Hot water	150–270	0.015–0.045
Natural steam	240	0.011–0.014
Geopressured	180–200	0.020–0.035[a]
Normal gradient	80–200	0.030–0.060
Hot dry rock	180–300	0.020–0.040
Nuclear		0.028[b,c]
Coal		0.033[b,d]
Oil		0.038[b,e]

[a] With credit for methane produced at $\$2/10^6$ Btu.

[b] For comparison, in 1976 dollars (fuel plus transportation charges).

[c] U_3O_8 at $35/lb plus charges for enrichment and fabrication.

[d] Coal at $\$1.00–\$1.60/10^6$ Btu.

[e] Oil at $\$2.30/10^6$ Btu.

Source: Adapted from CONAES, 1980.

As the water level fluctuates twice daily through this range, it alternately fills and empties natural basins along the shoreline, suggesting that the currents flowing in and out of these basins could be used to drive water turbines connected to generators. The technology employed is very similar to that of low-head hydropower.

To enhance the efficiency of this process, damlike structures can be built across the mouths of natural basins, with gates or channels to direct all of the natural flow through the turbine locations. The gates can be opened or closed in sequence with the tides permitting water flow only when there is sufficient head to power the turbines. The turbines are designed so that the flow of water both into and out of the basin produces electricity. Because of the intermittent nature of this flow, the effective duty factor of such an installation is less than 100%. A tidal power station produces only about one third as much electrical energy as would a hydroelectric power plant of the same peak capacity operating continuously.

The two large tidal power stations now operating are both outside of the United States. A 1-MW plant, on the White Sea in Russia, was completed in 1969, and a 240-MW commercial plant was completed on the estuary of the Rance River near St. Malo, France, in 1967 (see Figure 7.15). The larger plant has turbines that can also serve as pumps; thus, the installation can function as a pumped hydro storage facility to even out the loads on a large electricity generating and distribution system. In this way water pumped into the basin during times of low power demand increases the head on the turbines at other times.

For the United States, the most attractive sites are in Alaska and at the Bay of Fundy region in the northeastern United States and southeastern Canada. If this latter site, probably the most favorable in the world, were fully developed, it would provide about 15,000 MW_e to the United States and a similar amount to Canada. The U.S. share would certainly be locally important in the New England states, but on a national scale it would provide at most a few percent

Figure 7.15 The Rance River tidal power installation on the northern coast of Brittany. This 750 m long dike impounds tides that reach a height of more than 13 m.

Table 7.8 THE POTENTIAL OF SOME ALTERNATIVE ENERGY SOURCES IN THE UNITED STATES

Source	Time Scale	Estimated Projection (MW_e)
Hydroelectric	—	180,000
Wind	By 2000	46,000
OTEC	By 2000	25,000
Geothermal	By 2010	20,000–60,000
Tidal	—	15,000
Average electrical power production	1990 actual	320,000

of our need. Although a power station has been under consideration and planning for decades, to date only a 20-MW demonstration tidal plant has been built at this site. Worldwide, the most favorable tidal-power sites represent only about 63,000 MW_e, or about 50 times less than the world's potential hydroelectric power capacity.

7.7 SUMMARY OF THE POTENTIAL OF SOME ALTERNATIVE ENERGY SOURCES

For comparison, some estimated values are listed for various alternative technologies along with the present U.S. electrical power production (Table 7.8). The sum of the probably (extremely) optimistic projections for these alternative technologies is seen to be scarcely adequate for present electrical power needs.

PROBLEMS

1. Calculate the pressure in pounds per square inch at the bottom of a static column of water 300 meters high.

2. (a) How much electric *energy* can be generated by the water in a lake 2 km wide by 8 km long by 100 m deep if all the water falls through a vertical distance of 500 m? Assume that the generator is 90% efficient. Express your answer in joules.

(b) What would the *power* output be if the lake were drained over a period of one year? (One year $= 3.15 \times 10^7$ seconds.) Express your answer in megawatts.

(c) How large a community would this serve at the typical rate of 1 MW per 1000 people?

(d) At \$0.05/kW • hr, what is the value in dollars of this electrical energy?

3. **(a)** Calculate the energy in joules made available when 1 kg of water falls 30 m if 90% of the energy can be converted to a useful form.

(b) Calculate the energy in joules made available when 1 kg of water is cooled by 2°C if 3% of this energy can be converted to a useful form.

(c) Compare these two numbers. Does this say something about the relative amounts of water that must pass through a hydroelectric plant and an OTEC plant?

4. If a windmill produces 23 kW of electric power at a wind velocity of 10 miles per hour, how much power will it produce at 15 miles per hour?

5. A windmill has a diameter of 2 m. It operates at an efficiency of *60% of the theoretical maximum* when connected to an electrical generator.
   (a) What is the electric power output at a wind velocity of (1) 10 mph? (2) 20 mph? (3) 40 mph?
   (b) How many 60-W lightbulbs can be supplied with electricity under conditions (1), (2), and (3)?
   (c) What is the value in dollars each year of the electricity generated at a constant wind velocity of 10 mph? Would this installation be worth $4000 to the average consumer? Assume $0.05/kW • hr.

6. How much thermal energy in joules is made available by cooling 1 m^3 of rock from 240°C to 100°C? The specific heat is 2.4 J/cm^3 • °C.

7. A geothermal-powered steam turbine operates between a steam temperature of 210°C and an environmental temperature of 25°C. What is its maximum (ideal) efficiency? What percentage of the total steam energy must be discharged as waste heat?

8. Calculate the overall efficiency of an OTEC plant that operates with the ideal Carnot efficiency between the temperatures of 20°C and 5°C, but which uses two thirds of the energy extracted to run pumps and make up other losses.

9. (a) Starting from the results of Examples 7.4 and 7.5, estimate the number of cubic meters of water that would flow per second through an OTEC plant large enough (1000 MW_e) to provide electricity to Miami. Use the approximation of an ideal heat engine as in the examples.
   (b) If this water flows at a velocity of 4 m/s, what would be the necessary diameter of the circular pipes?

10. A tidal basin with an area of 14 km^2 and a depth of 12 m empties in 6 hours with the water passing through turbines.
    (a) How many m^3/s must flow on average during this 6-hour period?
    (b) How many square meters of area must the turbine pipes have if the flow velocity is 7 m/s?

SUGGESTED READING AND REFERENCES

1. Heyden, H. C. "Rosetta Stones for Energy Problems." *The Physics Teacher,* **19** 6 (September 1981), pp. 374–383.

2. Stoker, H. S.; Seager, S. L.; and Capener, R. L. *Energy from Source to Use.* Glenview, Ill.: Scott Foresman, 1975.

3. Hubbert, M. K. *Perspectives on Energy.* C. Ruedisili and M. W. Firebaugh, Eds. Oxford: Oxford University Press, 1975.

4. "Assessment of Geothermal Resources of the United States—1975." D. E. White and D. L. Williams, Eds. Geological Survey Circular 726.

5. Final Report of the Committee on Nuclear and Alternative Energy Systems (CONAES). *Energy in Transition 1985–2010.* San Francisco: W. H. Freeman, 1979.

6. Moretti, P. M., and Divone, L. V. "Modern Windmills." *Scientific American,* **254** 6 (June 1986), pp. 110–118.

7. Smith, D. R. "The Wind Farms of the Altamont Pass Area." *Annual Review of Energy,* **12** (1987), pp. 145–183.

8. Britton, P. "How Canada is Tapping the Tides of Power." *Popular Science*, **226** 1 (January 1985), pp. 56–58.

9. Greenberg, D. A. "Modeling Tidal Power." *Scientific American*, **257** 5 (November 1987), pp. 128–131.

10. Solar Energy Research Institute. "The Potential of Renewable Energy." March 1990. SERI/TP-260-3674.

CHAPTER EIGHT

ENERGY STORAGE

8.1 INTRODUCTION

We have seen that energy sources are vitally important to our established patterns of life. It is not only the sources, however, that matter, but also the technology of energy conversion and the various means for delivering energy of the proper form when and where it is needed. Energy storage systems are receiving increased attention as we confront the exhaustion of abundant fossil fuels and turn toward alternative energy sources that may be intermittent in delivery even though the total energy available is of enormous magnitude.

Energy is stored in a number of *natural* situations that we commonly regard as primary energy sources. These include fossil fuels, representing energy stored as biomass hundreds of millions of years ago, nuclear energy stored away billions of years ago and now used in our reactors, and even solar energy from nuclear processes in the sun. We are now dealing with ways to convert energy received from these primary sources into forms that can be stored and transported for later use.

These storage methods then may be regarded as secondary energy sources, capable of delivering energy when and where it is needed. There is usually a net loss of energy in the conversion from primary source to stored energy, but this inefficiency is often warranted because of the many advantages of stored energy.

There are a number of situations that call for energy storage. In solar space heating, for example, the energy received during daylight hours on clear days must be stored for later use. A second example is the need for load leveling, as shown in Figure 8.1. A large power plant can be run efficiently at a nearly

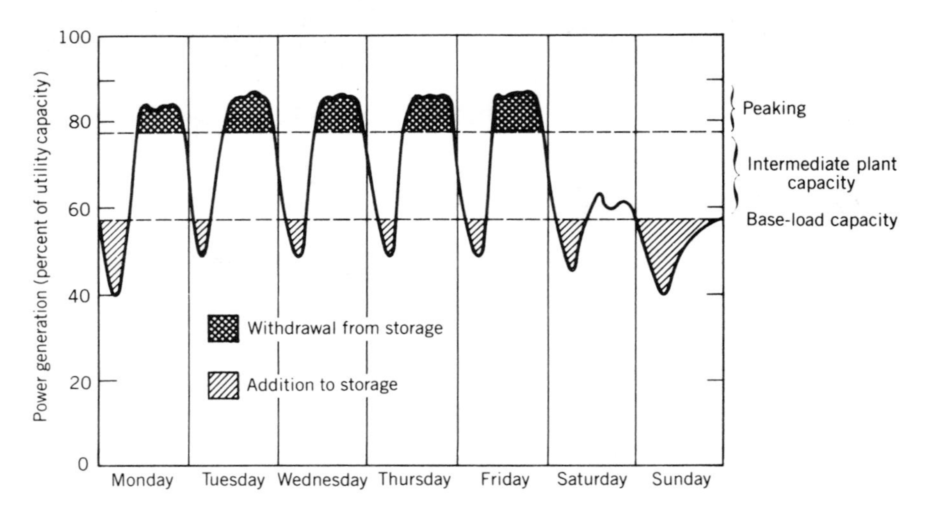

Figure 8.1 A typical week's power demand on an electric utility. For a utility with pumped hydro-storage, the storage capacity can be charged during periods of low demand and used to augment base-load plus intermediate-load capability during times of peak demand, thus reducing the need for expensive standby peaking power plant operation. The base-load plants operate efficiently at constant power output; the intermediate plants, usually older and less efficient, are operated only as needed. If there were no storage, the base-load capacity would be reduced and the intermediate capacity increased.

Table 8.1 ENERGY DENSITIES ACHIEVABLE WITH SOME ENERGY STORAGE TECHNOLOGIES CURRENTLY IN USE OR UNDER DEVELOPMENT

Storage Device or Medium	Energy Density (W • hr/kg)[a]
Gasoline	14,200
Lead–acid battery	25
Hydrostorage (water, 100 m head)	0.27 (270 W • hr/m^3)
Flywheel, steel	48
Flywheel, carbon fiber[b]	215
Flywheel, fused silica[b]	870
Hydrogen	38,000
Compressed air (70 atm)	21 (theoretical maximum, 2000 W • hr/m^3 actually deliverable with some augmentation by fuel)

[a] Not including weight of motors and generators or conversion efficiency.

[b] Under development.

constant power level, with the excess energy generated during periods of below-average demand being saved for periods of peak demand to augment power plant output. In addition to requiring that energy be available at the right time, we also need energy available at the right place. For example, if the energy is used to drive a vehicle, then a storage system must be devised that can be carried with the vehicle.

In this chapter we discuss a few representative forms of energy storage. In all the cases chosen, the concept has been proven, but the technology is not yet in widespread use.

Among the first considerations in evaluating the feasibility of any particular storage technology is some measure of the amount of energy that can be stored per unit mass or per unit volume. Some representative values that apply to various means of energy storage discussed here are given in Table 8.1, in terms of stored energy per kilogram. In using these numbers, the energy conversion efficiency between storage and task must also be taken into consideration. For instance, for gasoline the value of 14,200 W • hr/kg for the energy content of the fuel alone must be reduced to about 15% of this value, or 2130 W • hr/kg, to obtain the actual energy available for driving a motor vehicle.

8.2 Pumped Hydroelectric Energy Storage

Having already discussed hydroelectric generation of electricity using natural stream flow, we shall now see that the same technology can be used to recover energy from water previously pumped uphill by motor-driven pumps. The same mechanical devices used as motors and pumps when excess power is available can be used as turbines and generators during peak demand periods. The technology is proven; there are many such installations in the United States and elsewhere. A schematic diagram is shown in Figure 8.2, and an actual installation is pictured in Figure 8.3. As in the case of hydroelectric power, the energy recoverable from a raised body of water is proportional to both the volume of water and the height through which it falls onto the turbine. The problems of pumped hydro storage involve finding an appropriate site of suf-

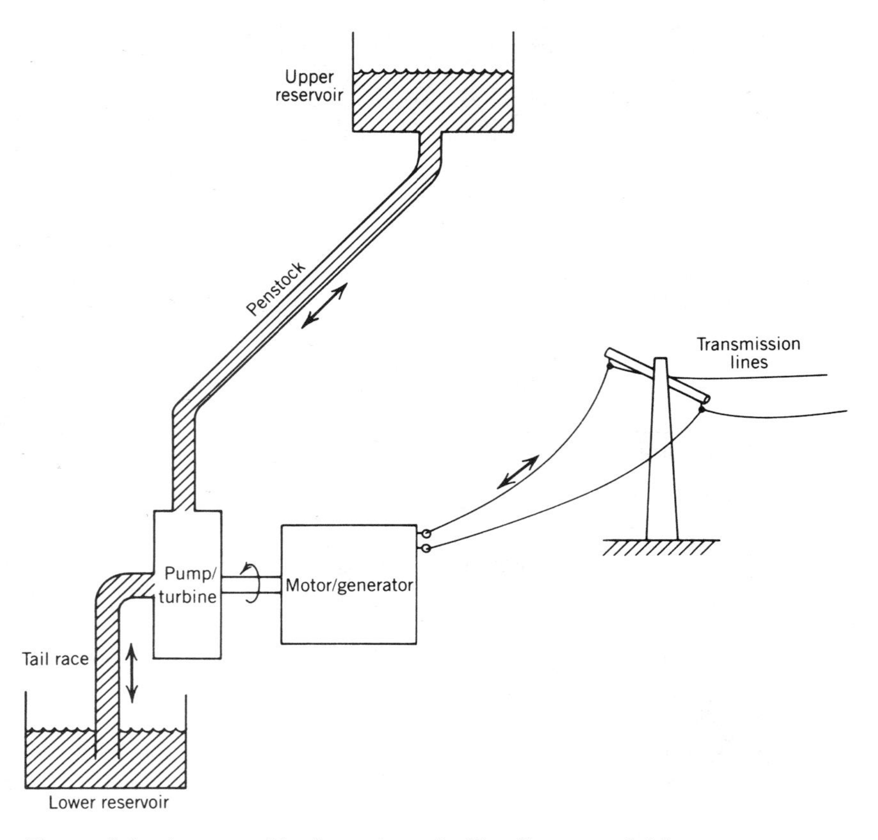

Figure 8.2 A pumped hydro-storage facility. Systems of this type are now in rather common use in the United States. During times of low power demand, the motor draws power from the transmission lines to drive a pump that raises water from the lower reservoir to the upper reservoir. This flow is reversed during times of peak power demand.

ficient area with a head of 100 m or so. Many of the problems of natural hydropower are not present with pumped systems. Typically, free-flowing streams are not lost, and the reservoirs are not susceptible to silting to the same degree as are reservoirs that impound natural watercourses.

The general advantages of pumped hydro storage are so great that underground pumped hydro-storage facilities are now being created by excavating lower reservoirs out of solid rock as much as 1000 ft underground. One such system, shown partially in Figure 8.4, is being constructed by Pacific Gas and Electric Company in the Sierra Nevada east of Fresno, California. This system will provide 1.12 million kW of pumped hydro storage to meet peak loads. Such a system, with the water cycled between artificially created upper and lower reservoirs, does not require any connection to a natural body of water. It achieves very high energy density per unit of water pumped because the vertical distance is so much greater than that typically available between reservoirs on the surface.

The rather high efficiency of a hydroelectric system, discussed in Chapter 7, also applies to both the pumping and generating functions in pumped hydro-

Figure 8.3 The world's largest hydro-storage facility at Ludington, Michigan, uses Lake Michigan as the lower reservoir and an artificial lake 250 ft higher as the upper reservoir. This plant can deliver 2000 MW at full power and can store 15 million kW • hr of energy.

storage systems. If each of these two functions is 80% efficient, their product, 64%, will be the recoverable fraction of the energy expended in charging the reservoir. This is about what is achieved in practice.

Example 8.1

A coal-burning power plant produces electricity at 36% efficiency. If part of this electrical energy is stored in a pumped hydro-storage facility for later use by consumers, what is the overall efficiency from combustion of coal to electrical energy at the consumers' location?

Solution

$$\begin{aligned}\text{overall efficiency} &= (\text{original production efficiency}) \times (\text{efficiency of hydro storage}) \\ &= 0.36 \times 0.64 = 0.23 = 23\%\end{aligned}$$

Because of the lower overall efficiency for producing and delivering electricity through a hydro-storage facility as compared to that coming directly from a power plant, each unit of once-stored electrical energy is more expensive to the power company. The added expense must be balanced against other investments and fuel costs that would be necessary to meet peak demands; and in many cases it is more attractive to provide the storage facility instead of increased primary generating capacity that would be idle much of the time.

Figure 8.4 Workers install one of three powerful pump turbines at the Helms Pumped Storage Project powerhouse, a facility of the Pacific Gas and Electric Company. The underground chamber is longer than a football field and is hewn out of solid granite 1000 ft below the surface of a Sierra Nevada mountain.

8.3 HYDROGEN AS A SECONDARY FUEL

On the scale of the universe, hydrogen is extremely abundant, accounting for perhaps 90% of the number of atoms and three fourths of the mass of the universe. But in the earth's atmosphere hydrogen is almost vanishingly rare as a free element, being present at a level of less than 1 part per million. This is because gaseous hydrogen is the least massive of all molecules; it is so light that at normal atmospheric temperatures hydrogen has a thermal velocity sufficient to escape the earth's gravitational field and drift off into space, a freedom not enjoyed by heavier molecular constituents of the atmosphere, such as oxygen and nitrogen. Whatever hydrogen is naturally present on earth is essentially all tied up in chemical compounds such as water (H_2O) or methane (CH_4), and any hydrogen made available for use must be separated from these molecules.

Hydrogen (H_2) is a colorless, odorless, diatomic gas of low density; it weighs only one eighth as much as natural gas, and as a fuel has nearly three times as much energy content per unit mass but only about one third as much per unit volume at normal temperatures. Hydrogen is well suited to the role of a secondary fuel. It can be produced from water by electrolysis wherever electricity is available and burned later at any location accessible by pipeline or provided with cryogenic storage. When hydrogen burns, it combines with oxygen from the air to form pure water; there are no other combustion products except possibly small amounts of nitrogen oxides formed from the air in the high-temperature combustion zone. Thus, the fuel cycle is a recycling of hydrogen (and oxygen) from water to water, with energy put in at one stage and taken out at another. The hydrogen is not destroyed in the process in the same sense that hydrocarbon molecules in fossil fuels are destroyed by burning. The recycling time for the complete water–hydrogen–water cycle is likely to be short, on the order of days or weeks, in contrast to the fossil fuel–carbon dioxide–fossil fuel cycle time, which is in excess of a million years.

As a secondary fuel, hydrogen has the potential to serve as a storage medium for absorbing excess electrical power during off-peak periods and delivering power during peak demand periods. For this type of application, it is proposed that hydrogen be stored as a liquid at a very low temperature. It is liquid below −253°C (20°K), and as a liquid has an energy density per unit volume about a thousand times greater than as a gas at standard temperature and pressure. For a given amount of stored energy, a cryogenic hydrogen facility is much less expensive and less demanding of a special site than pumped hydro storage. The necessary technology for cryogenic storage of hydrogen has been proved in the U.S. space program, where hydrogen is used as a rocket fuel. There is an obvious problem in applying hydrogen manufacture and storage as a means of leveling loads on an electrical system. The production of the hydrogen is perhaps 70% efficient in energy use, and the use of hydrogen as a boiler fuel would be perhaps 40% efficient in producing electrical energy, leading to an overall efficiency of no more than about 25% after the energy costs of maintaining cryogenic storage are considered. This efficiency is considerably lower than that achieved with pumped hydro storage. If the hydrogen is used as a fuel for producing direct heat energy rather than electricity, then the overall efficiency can be much more favorable. A possibility for relieving the problem of low thermodynamic efficiency in conventional power plants lies in using hydrogen and oxygen in fuel cells to produce electricity directly. In a hydrogen–oxygen fuel cell, these two gases are combined in the presence of a catalyst to form water and electrical power; the process is basically the reverse of that in an electrolytic cell. On a laboratory scale, such fuel cells have been operated to produce electrical energy with an efficiency of greater than 85%; however, they are not yet in common usage.

Hydrogen is currently produced in abundance in the United States and elsewhere for use in many industrial processes, not as a fuel but as a chemical raw material. In one important application, it is combined with nitrogen taken from the air to form ammonia (NH_3) for use as a fertilizer. On a worldwide basis, between 5 trillion and 10 trillion cubic feet of hydrogen are now produced annually. Almost all of this production is based on the catalytic reaction of fossil fuel hydrocarbons, such as methane, with steam to produce carbon dioxide and hydrogen. In this way our fossil fuel reserves form a feedstock for the fertilizer industry. Although it is now generally more expensive, the electrolytic decomposition of water is an excellent source of both hydrogen and oxygen wherever electrical energy is available in quantity at low cost. This method of production will most likely provide whatever hydrogen will be used as a secondary fuel. Hydrogen produced by electrolysis is very pure; purities of 99.9% are obtained with electrolytic cells containing alkaline water. The yield of hydrogen is about 7 $ft^3/kW \cdot hr$ of electrical energy put into the cell, resulting in an energy conversion efficiency of about 67%, based on a heat value of 325 Btu/ft^3 for hydrogen gas.

Example 8.2

The heat of combustion of hydrogen, H_2, can be expressed (in mixed units) as 129 Btu/g. Hydrogen burns according to the formula

$$2H_2 + O_2 \rightarrow 2H_2O + \text{heat energy}$$

How much water would be formed by this reaction in producing 1 million Btu to heat a home on a winter day?

Solution

Oxygen has an atomic weight of 16 and H_2 has a molecular weight of 2, so 18 g of water is formed for every 2 g of hydrogen gas burned. The amount of hydrogen consumed in releasing 10^6 Btu is

$$\frac{10^6 \text{ Btu}}{129 \text{ Btu/g}} = 7752 \text{ g}$$

This will form

$$\frac{18}{2} \times 7752 = 69{,}767 \text{ g } H_2O$$

which is equal to about 70 liters or 18 gallons.

If hydrogen as a fuel is ever to displace as much as 10% of our national energy budget of 80 QBtu, the production of the required amount of hydrogen by electrolysis would place enormous demands on our electrical generating capacity. At an efficiency of 67%, it would take the entire output of 400 1000-MW_e power plants operating continuously to produce only 8 QBtu of hydrogen fuel each year. This is almost twice the current total national demand for electricity.

The technology necessary to transport hydrogen wherever it is needed in the United States has been proven by the existing network of a quarter of a million miles of natural gas pipelines now operating in this country. Because of its lower heat content (about one third that of natural gas), the volume of hydrogen that must be transported is three times that needed for natural gas for the same amount of energy. Fortunately, the lower density and viscosity of hydrogen permits three times the flow rate of hydrogen through a given pipeline, meaning that existing natural gas pipelines would be adequate for hydrogen transport. It should be noted that pipeline transport of fuel gases is highly efficient; a single 36-in. pipeline, whether carrying natural gas or hydrogen, has an energy transport capacity equivalent to 10 high-voltage overhead electric transmission lines. The pipeline system, being underground, occupies less land area and is relatively inconspicuous.

It seems clear that the cost of hydrogen as fuel will never be as low as the current cost of natural gas. On an energy basis, because of inefficiency in its production, hydrogen will always be more expensive than the electricity that produced it, if the price comparison is done at the production site. However, when hydrogen is used as an energy transport medium over long distances, as from a nuclear-electric or OTEC power plant to customers 1000 miles away, hydrogen is projected to be less expensive in terms of its energy content (dollars/Btu) than electrical energy transported over the same distance. This is because pipeline transport and distribution of fuel gases is so much less expensive than the transport and distribution of electrical energy. If the energy is to be used not as a fuel for direct heat but to power machinery or electrical devices, the direct transmission of electricity may be more cost-effective because of the energy losses in converting the energy of hydrogen fuel back to electricity.

The energy density of gaseous hydrogen, or any fuel stored as a gas, is very low compared to gasoline, and, for automobiles and other vehicles, liquid hydrogen has problems of safety and a limited holding time in cryogenic containers. For airliners, liquid hydrogen could be an attractive fuel, especially in consideration of the very high energy content on a weight basis. Liquid hydrogen has about three times the energy of jet fuel (kerosene) by weight, but for a given stored energy, the liquid hydrogen tanks would have to be about three times as large as for kerosene. Developmental work is under way on storage of hydrogen as a metallic hydride at an energy density comparable to that of liquid hydrogen but at room temperature. This could open the way to the use of hydrogen as an automobile fuel; however, not all of the technical problems have yet been overcome.

There is widespread and serious concern about the safety of hydrogen as a fuel, with the burning of the dirigible *Hindenberg* in 1937 frequently cited as an example of the danger. In some respects hydrogen is a more dangerous fuel than natural gas, because it forms an explosive mixture with air over a very wide range of concentrations, from 4 to 75% hydrogen; natural gas is flammable only in the range of 5 to 15% concentration in air. In addition, the ignition energy for hydrogen–air mixtures is very small (about 2×10^{-5} J), only about one fifteenth as much as for natural gas or a gasoline–air mixture. This small ignition energy can easily be provided by a spark of static electricity. On the side of greater safety, any hydrogen from a leak will dissipate more rapidly into the atmosphere than will methane because of hydrogen's lower density. Early detection of leaks can be aided by adding an odorant to hydrogen, as is now done routinely with natural gas. Open hydrogen flames can be especially dangerous because they are nearly invisible; this may dictate adding an illuminant to the gas to make the flames more easily visible.

8.4 ELECTRIC STORAGE BATTERIES

Electric batteries of various descriptions have been used for energy storage for more than a century. They are now widely employed in applications ranging from hearing aids to submarine propulsion. The types of batteries that can be readily recharged are often designated as either secondary batteries or storage batteries to distinguish them from primary batteries, which cannot be recharged after use.

The suitability of a particular battery type for a given task depends on several factors. Batteries are usually characterized by their energy storage capacity per unit mass, given the name of *energy density* (W • hr/kg), and by their ability to deliver a given amount of power per unit mass, known as *power density* (W/kg). The first factor would be important in determining the range of a vehicle, whereas the second would be related to the maximum acceleration. Where storage volume is limited, as in a submarine, the volumetric energy density (W • hr/m^3) can also be an important factor. In addition, high permissible charging currents are desirable to reduce recharging time, and batteries that operate at normal ambient temperatures reduce the necessity of using energy to maintain elevated battery temperatures as well as the possible hazards in some applications if the batteries must be operated at several hundred degrees Celsius. In order for investment in storage batteries to be economically attractive, the batteries must endure large numbers of charge–discharge cycles without deterioration, and they must also be long-lived whether used frequently or infrequently. In any of the applications in which daily cycles are experienced, the batteries must survive many thousands of such cycles extending over periods of years to justify their expense.

Example 8.3

A practical motor vehicle needs about 30,000 W • hr of deliverable energy for a range of 100 miles, and for reasonable performance about 20 hp must be deliverable to the drive wheels. The efficiency of energy delivery from batteries to drive wheels is about 40%.

(a) How many kilograms of lead–acid batteries are needed to match the range requirement?

(b) How many kilograms of lead–acid batteries are needed to match the power requirement?

Solution

(a) Deliverable energy needed = 0.4 × stored energy, or

$$\text{stored energy} = \frac{30{,}000\ \text{W} \cdot \text{hr}}{0.4} = 75{,}000\ \text{W} \cdot \text{hr}$$

From Table 8.2, the energy density of lead–acid batteries is 40 W • hr/kg.

$$\text{Batteries needed} = \frac{75{,}000\ \text{W} \cdot \text{hr}}{40\ \text{W} \cdot \text{hr/kg}} = 1875\ \text{kg} = 4125\ \text{lb}$$

(b) Required power from batteries

$$= \frac{20\ \text{hp}}{0.4} \times 746\ \text{W/hp} = 37{,}300\ \text{W}$$

From Table 8.2, lead–acid batteries can deliver 70 W/kg. Therefore,

$$\text{required batteries} = \frac{37{,}300\ \text{W}}{70\ \text{W/kg}} = 533\ \text{kg} = 1172\ \text{lb}$$

A typical lead–acid car battery weighs about 60 lb; hence, 70 of them would be needed for the energy requirement or 20 of them for the power requirement.

Of the various storage batteries available or under development, the familiar lead–acid battery we use in the electrical systems of our automobiles is certainly the most widespread. Its use dates back at least a century, and its form has not changed a great deal since the turn of the century. This battery is

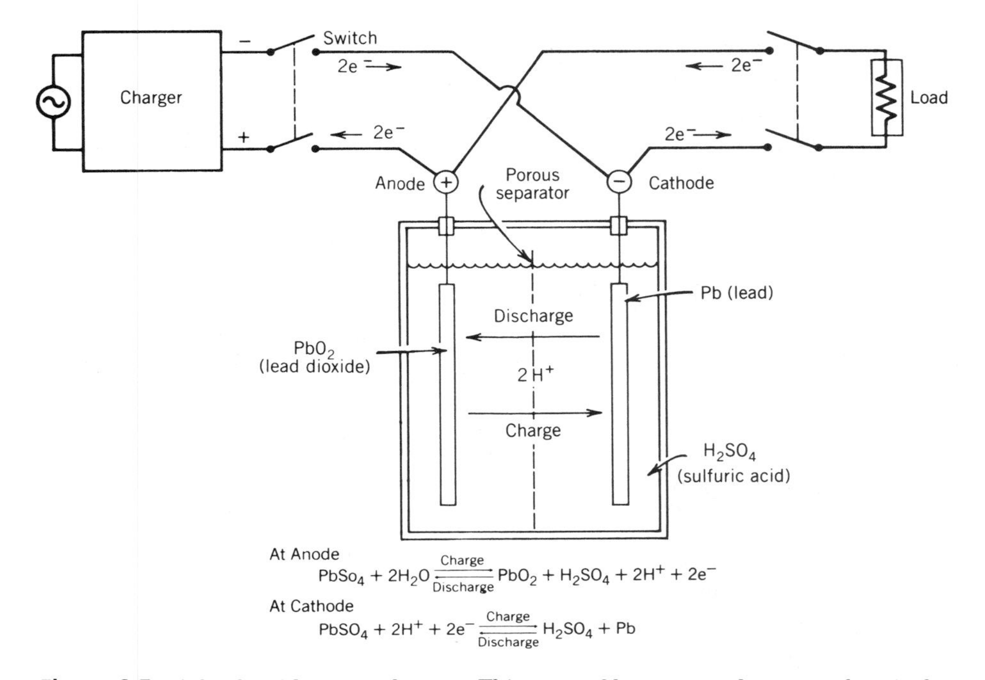

Figure 8.5 A lead–acid storage battery. This type of battery produces an electrical potential difference of 2 V between a pair of plates. Both electrodes become converted to lead sulfate during discharge; they are restored by charging. A practical battery is made of several pairs of plates rather than the single pair shown here.

shown schematically in Figure 8.5, and it is listed in Table 8.1 in its present state of development and in Table 8.2 in an advanced form. It is characterized by modest energy density and power density as compared to other battery types in the latter table; the cost per unit energy stored and the cycle life do not differ greatly from the other batteries. It operates at ambient temperature and is the only type of storage battery currently in use in large quantity for applications requiring substantial amounts of stored energy. The sodium–sulfur cell, shown in Figure 8.6, promises more than twice the energy density of the lead–acid battery and nearly half again the power density. It operates at temperatures hundreds of degrees Celsius above ambient, suggesting difficulties in adapting this type of cell to powering vehicles.

Table 8.2 SEVERAL TYPES OF STORAGE BATTERIES

Battery Type	Operating Temperature (°C)	Energy Density (W • hr/kg)	Power Density (W/kg)	Cycle Life	Cost ($/kW • hr)
Lead–acid	Ambient	40	70	>1000	70
Nickel–iron	Ambient	55	100	>2000	100
Nickel–zinc	Ambient	75	120	800	100
Zinc–chlorine	30–50	90	90	>1000	75
Sodium–sulfur	300–350	90	100	>1000	75
Lithium–iron sulfide	400–450	100	>100	1000	80

Note: These batteries are either available now or are expected to be available shortly.

Source: Adapted from *Scientific American,* December 1979.

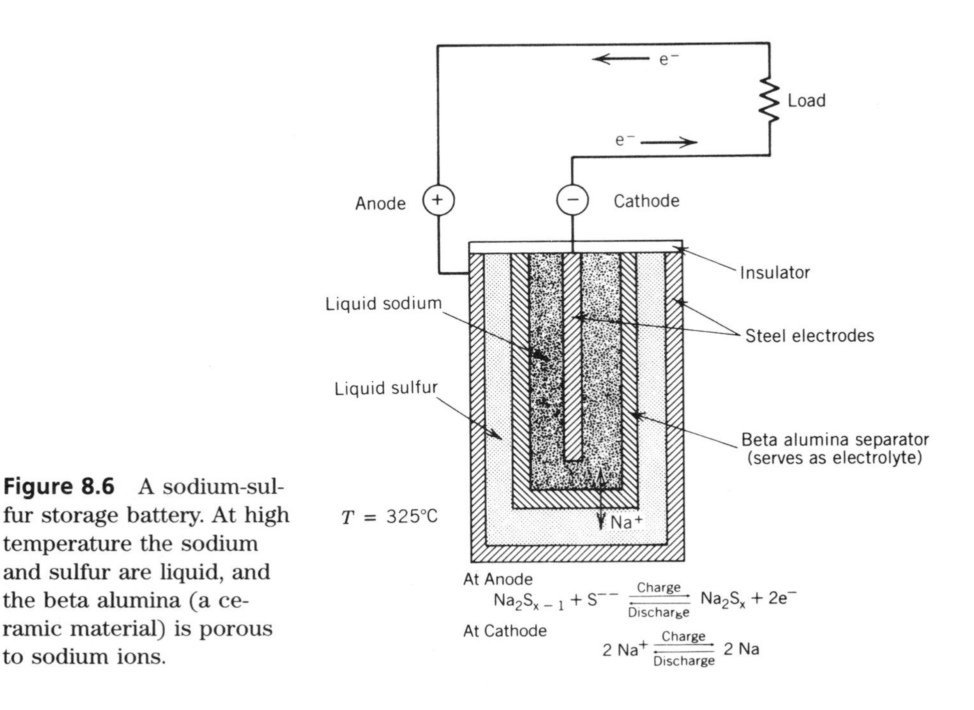

Figure 8.6 A sodium-sulfur storage battery. At high temperature the sodium and sulfur are liquid, and the beta alumina (a ceramic material) is porous to sodium ions.

Early in this century, electric cars powered by lead–acid batteries outnumbered combustion-powered vehicles, but the vastly superior energy density of gasoline soon led to its dominance in transportation, and electric cars all but disappeared. The higher efficiency of the electric vehicles (about 40%) compared to that of the combustion-powered vehicles (about 15%) somewhat compensates for the fact that the energy stored in a battery is no more than about 1% of that in an equivalent weight or an equivalent volume of gasoline. Nevertheless, gasoline-powered vehicles still have much greater range and greater acceleration. It seems unlikely that this situation will change even with the best of batteries under development. The advantages of the electric car lie not in range or power but in quietness, freedom from maintenance, the possibility of regenerative braking, and, most of all, the ability to derive energy for transportation from any energy source capable of driving an electric generating plant. Even in vehicles, batteries can form a part of a load-leveling storage capacity for large electric utilities if the vehicle batteries are customarily charged during off-peak hours.

Many of the factors inhibiting the use of batteries as energy storage for automotive transport are not severe problems for systems of batteries used for off-peak energy storage by public utilities. The low energy and power densities of batteries can be accommodated rather easily in permanently established sites set up at various locations within an extended power grid, such as that shown in Figure 8.7. When space and weight are not limitations, the high temperature requirements of some of the newer batteries are also lesser problems that can be met by insulated constant temperature enclosures. In off-peak storage applications, batteries respond flexibly to changes in load, and they are easy to situate in modular units wherever needed in a utility distribution system. They present almost no environmental threat and occupy rather little land area.

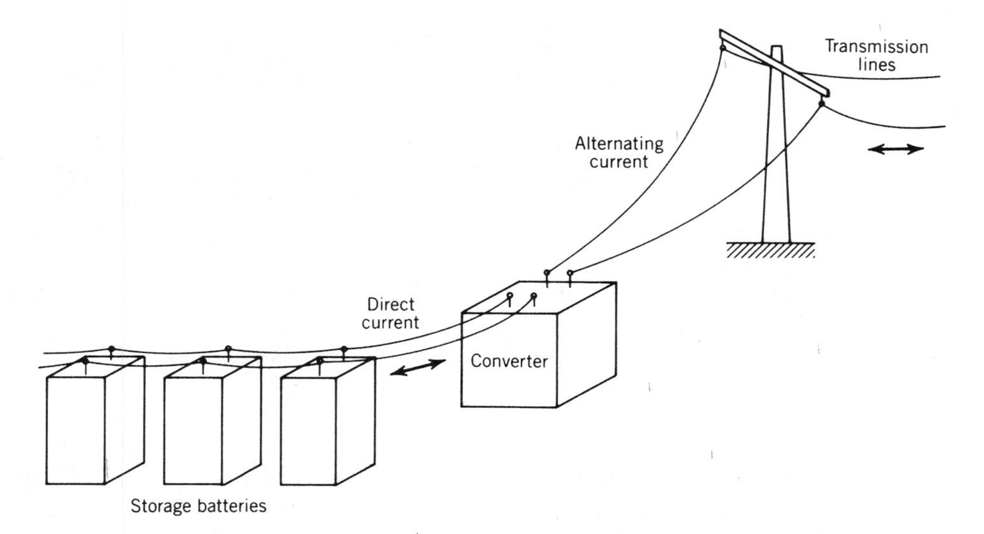

Figure 8.7 A storage battery installation for an electric utility. The batteries are charged by power from the transmission lines during times of low demand. During high demand periods, energy flows from the batteries back to the transmission lines. A 1-acre installation of this type could store up to 400 MW • hr of energy and deliver 40 MW of power.

8.5 FLYWHEELS

Most of us are familiar with the flywheel in one or more of its common forms: a potter's wheel, a child's top, a gyroscope, or a toy automobile. In these examples a wheel-shaped mass is set into rotation through human effort and the rotational energy continues to sustain this motion until it is dissipated by frictional forces. The flywheel is obviously an energy storage system; it takes energy from some source and delivers it later when needed, unless the rotational motion has first ceased because of friction either in the bearings or because of some surrounding medium such as air. On a larger scale, flywheels have been used to store energy for intermittently powering large magnets used in physics research and for powering urban transit buses. There is now increasing interest in the use of flywheels to store energy between periods of peak demand in large-scale electric distribution systems and to provide a portable source of stored energy for a new generation of automobiles.

The energy stored in a rotating flywheel is the sum of the kinetic energies of the individual mass elements that comprise the flywheel. When the wheel is in rotation, each of these mass elements has kinetic energy proportional to the square of its velocity, and the velocity of any point on the wheel is proportional to the product of the rotational speed (as in revolutions per second) and the distance from the axis of rotation. Thus, the energy stored by a flywheel varies as the square of the rotational speed multiplied by a factor that includes the wheel's mass and the distribution of the mass elements relative to the axis of rotation. This latter factor is known as the moment of inertia (I); for a solid disc of uniform thickness rotating about an axis through its center and perpendicular to the plane of the disc, it is given by $I = \frac{1}{2}mr^2$, where m is the total mass of the disc and r is the outer radius. For a wheel with essentially all of its mass located at the outer edge, as in a bicycle wheel, the moment of inertia is $I = mr^2$. The moment of inertia is a measure of a body's ability to resist changes in rotational velocity. A large moment of inertia indicates that considerable energy must be supplied to a flywheel to increase its rotational frequency, and a large amount of energy must be taken from the flywheel to reduce its rotational speed. In this way a heavy flywheel (one with large I) attached to the crankshaft of a gasoline engine smooths out the rotational motion between power strokes. Several values of I are shown in Figure 8.8 for flywheels of different shapes.

For any rotating mass, the kinetic energy is given by

$$\text{K.E.} = \tfrac{1}{2}I\omega^2 \quad \text{where } \omega = 2\pi \cdot (\text{rev/sec})$$

In this equation the kinetic energy is in joules if I is calculated by using mass in kilograms and radius in meters. The 2π corrects the angular speed in revolutions per second to the angular units of radians per second commonly used in scientific calculations.

We have seen that the energy stored by a given flywheel is proportional to the square of the rotational speed no matter what the configuration of the flywheel. Thus, for best energy-to-mass ratios, the flywheel must be spun to the maximum possible speed. The limit on this speed is the velocity at which the centrifugal forces within the flywheel exceed the strength of the material, resulting in the flywheel being torn apart. This maximum speed, then, depends

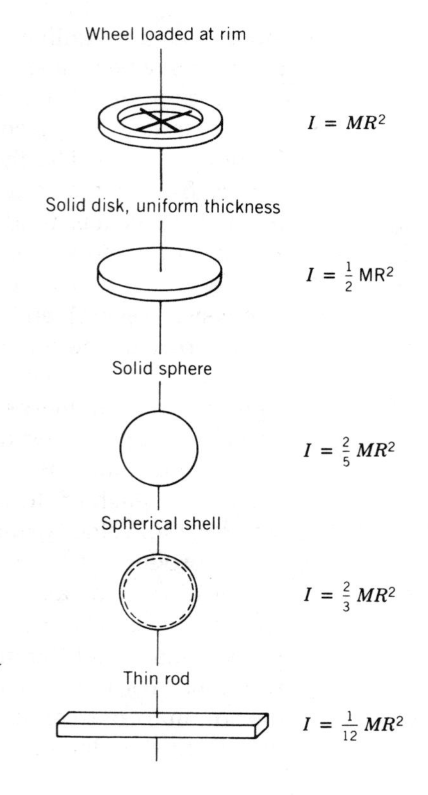

Figure 8.8 Moment of inertia, I, for bodies of various shapes. The mass of the body is given by M, the radius by R, and rotation is about an axis through the center of mass and perpendicular to the plane of rotation.

on the ability of the flywheel to resist tensile forces. Thus, the most dense materials are not necessarily the best for high-energy density storage at high rotational speed; tensile strength is more important than mass. To date, all flywheels used for powering buses and automobiles have been made of steel, but more exotic materials now emerging from the laboratory promise much higher energy density. One of these materials, fused silica fiber, has a ratio of strength to mass about 20 times greater than that of steel and, as a consequence, promises a specific energy (W • hr/kg) about 20 times that achievable with a steel flywheel and about 40 times greater than that of a lead–acid battery. Some of these values are shown in Table 8.1. With these newer materials, the useful energy that can be carried aboard a vehicle would, with its drive system, have mass comparable to that of the fuel plus engine and drive train of a petroleum-powered vehicle having similar acceleration, speed, and range. The flywheel-powered vehicle has the advantage that during downhill driving and braking, part of the kinetic energy of the vehicle can be put back into the flywheel storage by regenerative braking. A possible drive system of a flywheel-powered vehicle is shown in Figure 8.9.

For a flywheel to be a generally useful energy storage device, it not only must store the required energy, but it also must have a sufficiently long run-down time between periods of energy use so that the energy will be available when needed. The required long run-down times are achieved by using the best possible bearing systems and by operating the flywheel in a vacuum chamber to eliminate air drag on the rapidly rotating flywheel. Run-down times of 6 to 12 months are considered achievable. This is sufficiently long for automotive systems or for storage of energy from windmills during calm periods. For electric utility energy storage, as in the latter application, flywheel systems would use much less land area than pumped hydro storage, now the most widely used method.

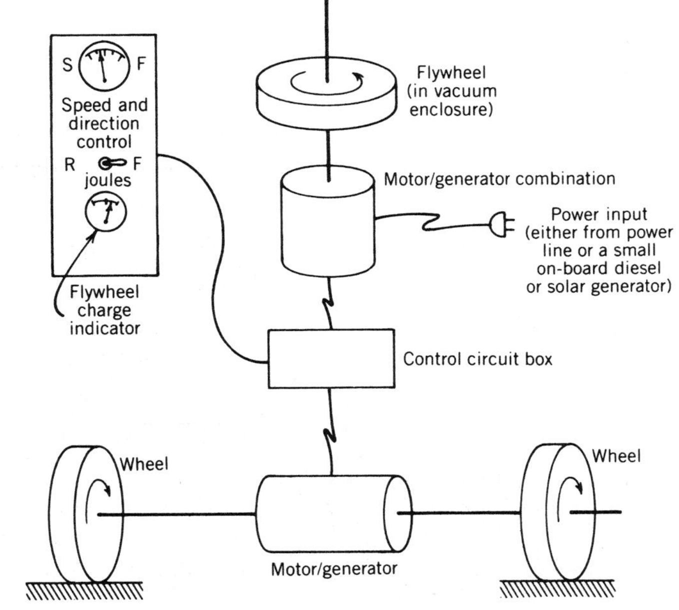

Figure 8.9 A schematic view of the drive system of a flywheel-powered motor vehicle. The motor/generator units can be operated in two modes, so that the flywheel can be charged either from an external source or whenever the vehicle is being braked electromagnetically. The rotation axis is vertical to reduce gyroscopic effects.

Example 8.4

For a solid disc flywheel with a radius of 50 cm and a mass of 140 kg, calculate the rotational speed in revolutions per second and in revolutions per minute corresponding to a useful stored energy equivalent to 10 kg of gasoline burned in an internal combustion engine. Assume that the flywheel system is 80% efficient in delivering energy and that the gasoline engine is 15% efficient. Could this be a steel flywheel?

$$1 \text{ kg gasoline} = 14{,}200 \text{ W} \bullet \text{hr}$$

$$1 \text{ W} \bullet \text{hr} = 3600 \text{ W} \bullet \text{sec} = 3600 \text{ (J/sec)sec} = 3600 \text{ J}$$

Solution

Gasoline

One kilogram gasoline delivers 2130 W • hr after 15% conversion efficiency is considered.
Delivered energy from 10 kg gasoline is

$$= 10 \text{ kg} \times 2130 \text{ W} \bullet \text{hr/kg} \times 3600 \text{ J/W} \bullet \text{hr} = 7.67 \times 10^7 \text{ J}$$

Flywheel

Because the flywheel has a conversion efficiency of 80% from flywheel to useful work, the stored energy must be

$$\frac{7.67 \times 10^7 \text{ J}}{0.8} = 9.59 \times 10^7 \text{ J}$$

and

$$\begin{aligned} \text{stored energy} &= \tfrac{1}{2} I\omega^2 = \tfrac{1}{2} \times (\tfrac{1}{2} MR^2) \times [2\pi(\text{rev/sec})]^2 \\ &= \pi^2 MR^2 \times (\text{rev/sec})^2 \end{aligned}$$

or

$$\begin{aligned} (\text{rev/sec}) &= \sqrt{\frac{1}{\pi^2 MR^2}(\text{stored energy})} \\ &= \frac{1}{\pi R}\sqrt{\frac{(\text{stored energy})}{M}} \\ &= \frac{1}{3.14 \times 0.5}\sqrt{\frac{9.59 \times 10^7 \text{ J}}{140 \text{ kg}}} \\ &= \frac{1}{1.57}\sqrt{6.85 \times 10^5} = \frac{827}{1.57} = 527 \text{ rev/sec} \\ &= 527 \text{ rev/sec} \times 60 \text{ sec/min} = 31{,}622 \text{ rpm} \end{aligned}$$

If 9.59×10^7 J are stored in a 140 kg flywheel, the energy density is

$$\frac{9.59 \times 10^7\,\text{J} \times 1/3600\,\text{W} \cdot \text{hr/J}}{140\,\text{kg}} = 190\,\text{W} \cdot \text{hr/kg}$$

a value much larger than possible with a steel flywheel according to the values given in Table 8.1. A flywheel made of carbon fibers or fused-silica fibers could match the energy density requirement.

8.6 COMPRESSED AIR

It has long been recognized that tremendous amounts of energy can be stored for future use by compressing various gases. The pressurized gases can be made to do useful work as they expand against a piston or through a turbine, or sometimes they do work in an undesirable way against surrounding objects. For example, water heater explosions caused by excessive steam pressure still cause fatalities at a significant rate in this country; in the past century exploding steam boilers have killed more than a hundred persons a year. Of course, these examples of accidents do not illustrate the deliberate storage and release of the energy in a pressurized gas; however, they do indicate the magnitude of the energy involved. In more recent times, we have the examples of CO_2 cartridge-powered pellet guns, pressure accumulators used for control power aboard military aircraft, and compressed-air motors used to start diesel engines on construction machinery. In the last example, the compressed air is delivered from a small tank pumped up by a compressor when the engine is operating, much in the way an automobile charges its own battery. Compressed-air-powered tools are becoming increasingly popular both in the automotive and construction trades.

With the recent demands for off-peak storage systems to enable public utilities to meet peak power requirements without expanding their primary generating capacity, compressed-air energy storage is being adapted to this purpose on a large scale. As indicated in Table 8.1, compressed-air storage has many times the capacity of a typical pumped hydro-storage system when calculated either on a weight or volume basis. Although there is at this time only one large commercial compressed-air energy storage installation for this purpose (in Germany), others are being planned at various locations. In the existing facility, the storage reservoir is a large underground cavity in a salt deposit. The storage volume is 300,000 m^3 (for comparison, a typical American house encloses about 300 m^3), and the air is compressed to 70 atm (about 1000 lb/in.^{2}) by electrically driven compressors. At times of peak power demand, this system can deliver up to nearly 300 MW for 2 hours by using the compressed air to drive a turbine coupled to a generator (see Figure 8.10). As with all energy storage systems, the efficiency is less than 100%. A major contribution to the system's inefficiency is the energy input required to cool the compressed air as it is put into storage and the fuel input needed to heat and expand the cool air taken from storage as it enters the turbine. If the air, which gets hot upon compression, were not cooled before being forced into the storage cavity, the high temperatures would fracture the rock and soften the salt walls of the cavity.

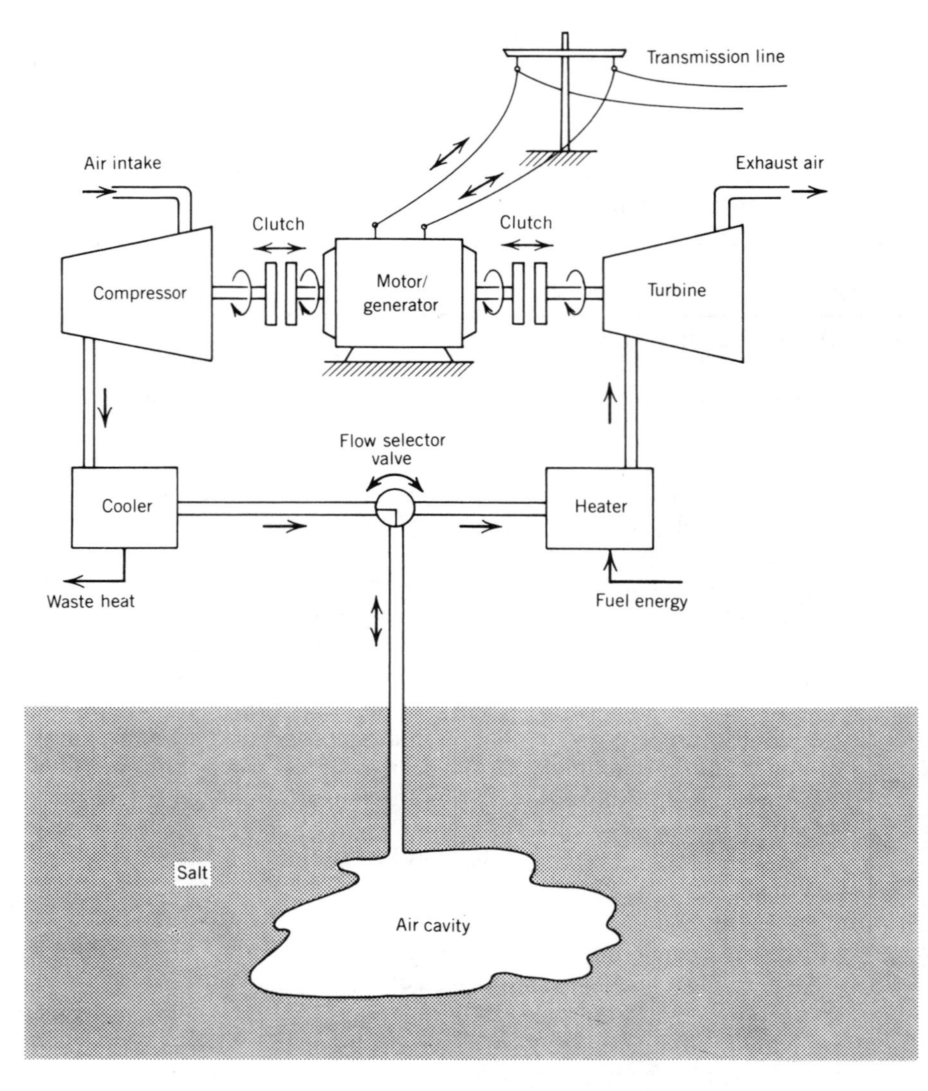

Figure 8.10 A simplified diagram of a compressed-air energy storage facility of the type now operating in Germany. The air cavity is a cavern deep underground in a natural salt deposit. The weight of the overburden is sufficient to withstand the high pressure in the cavern.

Certainly, advances in compressed-air storage technology will include provisions for storing and recycling the waste heat from the compression stage for use in reheating the air during expansion. Several schemes for accomplishing this are under study and, when put into practice, should significantly improve the efficiency of large-scale energy-storage systems. A second possibility for improving the system efficiency would be to use the exhaust heat from the power generating turbine to raise the temperature of the air being taken from storage to drive the turbine.

Example 8.5

When air is compressed, its temperature increases. An equation commonly used to relate the temperature, T_2, of a gas after compression to its previous temperature, T_1, is

$$T_2 = T_1 (P_2/P_1)^{(n-1)/n}$$

where P_1 and P_2 are the pressures before and after compression and where n = 1.4 for air. The temperatures are measured in degrees Kelvin.

(a) Calculate the temperature in degrees Celsius of the air if it is compressed from a pressure of 1 atm to a pressure of 100 atm and if the initial temperature was 20°C.

(b) Would the compressed air have a temperature capable of melting salt? (The melting point of NaCl = 801°C.)

Solution

(a) $T_1 = 20°C = 293°K$
$P_1 = 1$ atm
$P_2 = 100$ atm
$n = 1.4$

Then

$$T_2 = T_1 (P_2/P_1)^{(n-1)/n} = 293(100/1)^{(1.4-1)/1.4} = 293 \times 100^{0.286}$$

By using the y^x function on a calculator

$$100^{0.286} = 3.73$$

so

$$T_2 = 293 \times 3.73 = 1093°K = 820°C$$

(b) Because the air temperature is hotter than the melting point of salt, it would be capable of melting the salt.

PROBLEMS

1. Consider whether the following represent energy *sources* or energy *storage*: pumped hydro storage, hydrogen, and firewood. Use the concept of *energy conversion efficiency* in your discussion.
2. In a sense, all of what we consider as energy *sources* may also be considered as energy *storage*. Discuss solar energy, nuclear energy, coal, and electric batteries in this context.
3. Estimate that 1 gallon of gasoline in a chainsaw is enough for cutting a cord of wood. Estimate 20×10^6 Btu/cord. What is the ratio of fuel energy out to fuel

energy in to the process? What is the ratio of the costs of the two fuel amounts in your community?

4. How many kilograms of lead–acid batteries would you need in your basement to provide for backup lighting (10 100-W bulbs for 12 hours)?

5. Calculate the volume in cubic meters (and tonnes) of water at an elevation of 75 m that would have potential energy equal to the energy available from burning 1 tonne of H_2.

6. Compare the energy stored in a 100-g stick of dynamite (4.3×10^9 J/ton) to that stored in a jelly doughnut. (Estimate 100 g at the energy density of glucose: $C_6H_{12}O_6$, 674 kcal/mole; see Chapter 10.) How do the power outputs compare?

7. Consider the application of pumped hydro storage in a utility system. Would this be an appropriate technology in a system having as its primary source of energy coal? Nuclear? Hydro? Solar photovoltaic? Geothermal? Discuss.

8. Hydrogen gas at standard temperature and pressure occupies 22.4 liters (0.0224 m^3) per 2 g. If stored at 5 atm pressure, this 2 g will only occupy one fifth as much volume. How large a storage tank (in cubic meters at 5 atm) would you need to store the 10^8 Btu needed to heat your home over the winter? (See Table 8.1.)

9. A standard automobile battery is rated at 12 V and 70 ampere hours. Translate these figures into joules of stored energy using the relationships given in Chapter 1. For short periods the battery can deliver a current of 300 A. What is the delivered power in watts?

10. Consider the safety aspects of flywheel energy storage aboard automobiles. How might you design such a system so that the stored energy would not be released destructively in the event of a collision?

11. Calculate the energy in joules stored in the rotational motion of your bicycle wheels when they are rotating at 5 rev/sec. Assume $M = 3$ kg each and $R = 0.35$ m. See Figure 8.8. How much elevation gain could this stored energy provide if the bicycle and rider have a combined mass of 90 kg?

12. In an approximate sense, compressed air has the potential for doing work in the amount of one half of the volume stored multiplied by the pressure above atmospheric. Calculate the number of foot-pounds of energy stored in the compressed air tank of 4 ft^3 volume at your local service station when the gauge pressure (pressure above atmospheric) is 120 lb/in^2. Compare this to the energy liberated by exploding 10 g of TNT (1 kg TNT $= 4.3 \times 10^6$ J).

13. If homes were heated by burning hydrogen gas, would this also provide a suitable source of domestic water, thus reducing the need for water distribution systems?

SUGGESTED READING AND REFERENCES

1. Cohen, R. L., and Wernick, J. H. "Hydrogen Storage Materials: Properties and Possibilities." *Science* **214** 4525 (December 4, 1981), pp. 1081–1087.

2. Devins, Delbert W. *Energy: Its Physical Impact on the Environment.* New York: John Wiley, 1982.

3. Gregory, Derek P. "The Hydrogen Economy." *Scientific American,* **228** 1 (January 1973), pp. 13–21.

4. Kalhammer, F. R. "Energy Storage Systems." *Scientific American,* **241** 6 (December 1979), pp. 56–65.

5. Post, R. F., and Post, S. F. "Flywheels." *Scientific American,* **229** 6 (December 1973), pp. 17–23.

6. Stoker, H. S.; Seager, S. L.; and Capener, R. L. *Energy from Source to Use.* Glenview, Ill.: Scott, Foresman, 1975.

7. McLarnon, F. R., and Cairns, E. J. "Energy Storage." *Annual Review of Energy,* **14** (1989), pp. 241–271.

8. Sperling, D., and DeLuchi, M. A. "Transportation Energy Futures." *Annual Review of Energy,* **14** (1989), pp. 375–424.

CHAPTER NINE

ENERGY CONSERVATION

9.1 INTRODUCTION

The formal concept of the conservation of energy was introduced in Chapter 1, where it was stated that the total energy in a closed system can never change. The energy can change from one form to another, but it can neither be created nor destroyed. In this chapter, as well as in the mind of the general public, the word *conservation* has a quite different meaning. Here it is taken to mean the use of less energy to accomplish a given task. When energy is used, its form is changed in such a way that its availability to do work or perform a useful function is diminished. For example, if a piece of firewood is burned in a stove, its stored chemical energy is converted to heat energy that warms the room and then warms the atmosphere of the earth slightly before being radiated out into space. The question now being addressed is how can less firewood be burned and still maintain the room at the desired temperature?

There are two aspects of energy conservation. The first is related to the question of how the room can be heated to the same temperature with less wood being burned, for example, by the design of a more efficient stove, improvement of the house insulation, or some other change. In this aspect of energy conservation, less energy is used with no sacrifice on the part of the individual or society, as the same end has been accomplished. It is clear that a number of energy-conserving devices and processes can be introduced in homes and industry, and, as we shall see, the savings in energy costs can warrant rather extensive capital expenditures. Energy efficiency is another name given to this first aspect of energy conservation.

The second aspect of energy conservation involves some sacrifice on the part of the user. In the case of the wood stove, the homeowner may choose simply to have a lower room temperature and burn less wood. There are, of course, many such situations, for example, choosing to ride a bicycle rather than driving a car or hanging laundry out to dry in the sun rather than using a clothes dryer. Some situations may involve a significant compromise, such as choosing a light, fuel-efficient car over a heavier one while recognizing that there may well be some loss of safety in a collision.

The first approach to energy conservation involves primarily questions of economics and technology. The problem of insulating homes is discussed in the next section in some detail. The materials and technology needed for making sizable reductions in the fuel needed to heat a home are all available, and the question that homeowners face is one of how much money can be reasonably invested in view of the anticipated fuel savings. Frequently the number of years of fuel saving required to offset the capital expenditure is used as a gauge of whether or not an investment in energy conservation is economically sound. For example, if the cost of the fuel that is saved by a $100 investment is $10 a year, the "payback" time is, thus, naively estimated as 10 years. There are several complications, however, in deciding what is a reasonable payback time. The obvious comparison with existing rates of return on investments is sometimes misleading. In the above example, one might hesitate to invest in the conservation measure if 10% or more return could be earned on the capital invested elsewhere. The complications arise partially because one does not know what interest rates or fuel costs will be in future years, and income tax benefits, maintenance, and replacement cost must all be considered. Most indications are that the cost of energy is going to increase with time faster than the general inflation rate. The real cost of the energy that should be considered is not the present cost but the cost of discovering and producing a new barrel of oil or a new volume of natural gas, and these costs are larger at the present time than are the prevailing costs of the fuel. A calculation based on the present cost of energy is misleading for this reason. The fact is that an investment in some energy conservation measure will continue to save energy year after year as long as the measure is employed, and even apparent payback times of 20 or 30 years are frequently consistent with a good investment practices.

We saw in Chapter 1 that the total energy consumed in the United States increased approximately exponentially from about 1850 to 1975. The conventional projections made around 1972, based on the historic growth rate of about 4.3% a year, indicated that the nation would be using about 160 QBtu a year in the year 2000 compared to 78 QBtu in 1978. It is now clear that such projections are not warranted, and one important reason is energy conservation. After the oil embargo in 1973, the cost of energy rose very steeply. The increased fuel costs, coupled with federally sponsored conservation programs and help from public utilities and various community organizations, have provided the incentives and means for a massive drive toward reduction in the use of energy. In 1973 the total energy consumed in the United States was 74.3 QBtu, and the ratio of energy consumption to gross national product was 27.1×10^3 Btu/$ using 1982 dollars as a standard. Since 1973 this ratio has come down steadily, and in 1990 it was 19.6. While this reduction is a remarkable achievement, the ratio is still higher than in most industrialized countries. The total energy consumption in the United States has gone up and down since 1973, but it rose from 70.5 QBtu in 1983 to 81.5 QBtu in 1990. The National Academy of Sciences Committee on Nuclear and Alternative Energy Systems (CONAES) study indicated that with a very aggressive energy conservation policy and some sacrifice in life style, the nation could use as little as 58 QBtu in 2010.

Table 9.1, put together by Amory Lovins, shows predictions at various times since 1972 for the total U.S. energy demand in the year 2000 or 2010 by various people or groups. The point is that what was considered a hopelessly low prediction in 1972 is now much higher than conventional wisdom predicts.

Predicting future energy demand is a serious part of the business of the energy companies and public utilities, since tremendous investments must be made years in advance to meet demand. What has surprised many is the rapid response by the American people in the form of energy conservation to the increasing costs of petroleum, electricity, and natural gas. It is clear from Table 9.1 that some analysts feel we can continue this reduction process for many years. In the remaining sections of this chapter we explore the major ways in which such energy conservation can come about. A barrel of oil saved is entirely equivalent to a barrel of oil found, with the added benefit that saving oil generates no air pollution or other environmental degradation.

In discussing ways in which the national energy demand can be reduced, it is useful to remember how we use energy in the United States. At present, buildings and appliances (including commercial) use about 38% of the total energy consumed, industry 36%, and transportation 26%.

Table 9.1 PREDICTIONS SINCE 1972 BY VARIOUS GROUPS OF THE U.S. ENERGY DEMAND IN THE YEAR 2000 OR 2010 IN UNITS OF 10^{15} Btu[a]

Year of Forecast	Beyond the Pale	Heresy	Conventional Wisdom	Superstition
1972	125 (Lovins)	140 (Sierra)	160 (AEC)	190 (FPC)
1974	100 (Ford zeg)	124 (Ford tf)	140 (ERDA)	160 (EEI)
1976	75 (Lovins)	89–95 (Von Hippel)	124 (ERDA)	140 (EEI)
1977–78	33 (Steinhart)	67–77 (NAS I, II)	96–101 (NAS III, AW)	124 (Lapp)

Abbreviations: Sierra, Sierra Club; AEC, Atomic Energy Commission; FPC, Federal Power Commission; Ford zeg, Ford Foundation zero energy growth scenario; Ford tf, Ford Foundation technical fix scenario; Von Hippel, Frank Von Hippel and Robert Williams of the Princeton Center for Environmental Studies; ERDA, Energy Research and Development Administration; EEI, Edison Electric Institute; Steinhart, 2050 forecast by John Steinhart of the University of Wisconsin; NAS I, II, III, National Academy of Sciences Committee on Nuclear and Alternative Energy Systems (CONAES); AW, Alvin Weinberg study done at the Institute for Energy Analysis, Oak Ridge; Lapp, energy consultant Ralph Lapp.

[a] Amory Lovins put together this table showing the downward drift in forecasts.

Source: *Science*, **208,** 1353 June 20, 1980.

9.2 SPACE HEATING

Since most of the commercial and residential buildings in the United States were constructed when energy was relatively inexpensive, little effort was devoted to providing good thermal insulation, efficient furnaces, a proper orientation of windows relative to the sun, and other considerations to minimize heating bills. With monthly expenditures for heating in a winter month for a medium-sized residence now running as high as several hundred dollars in some parts of the country, it has become crucial for many families to increase the efficiency of their heating systems and to conserve the heat energy provided. About 20% of the energy consumed in the United States is used for space heating.

Heat is provided, of course, to keep the temperature inside buildings at a comfortable level. Figure 9.1 shows the basic problem for a simplified house. In winter, heat is conducted outward through the walls, windows, roof, and basement (or foundation), so that there must be a source of heat within the house to maintain the required temperature difference between the inside and outside. In addition to the conductive losses of heat, there is a certain amount of air exchange or infiltration from the outside, and it takes a surprising amount of energy to compensate for such convective heat losses. It is estimated that a typical U.S. house loses 30 to 40% of its total heat loss in this manner. An oil- or gas-fired furnace or electric resistance heating are the usual sources of heat. Figure 9.1 indicates that the sun can also supply heat, and we saw in Chapter 6 (Table 6.4) that appliances, lights, and people are important sources of heat. If a house is kept at a fixed temperature, the rate at which heat is supplied will

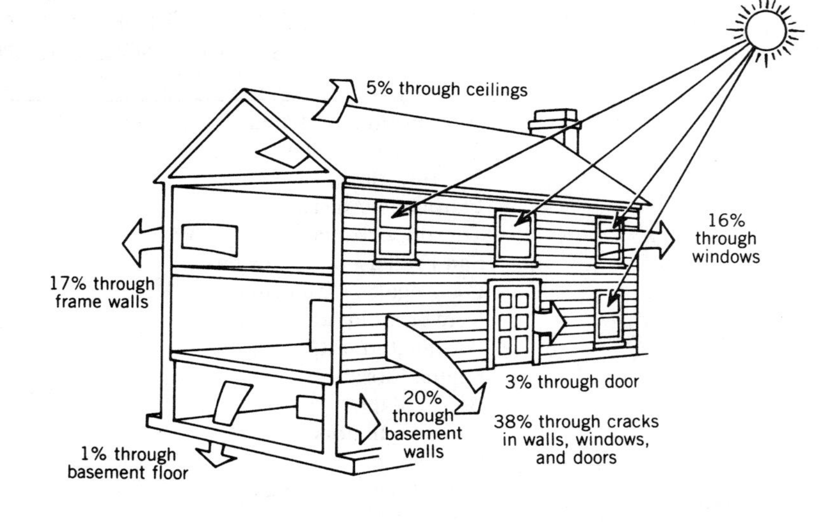

Figure 9.1 Heat losses from a typical conventionally insulated house. Solar energy entering through the windows can be a significant source of heat.

be just equal to the rate of heat loss. We shall treat the various components of the problem of building heating separately, beginning with the important question of how the thermal insulation can be improved.

THERMAL INSULATION

As we discussed in Chapter 6, the number of Btu/hr conducted through a wall is given by

$$\frac{Q}{t} = \frac{kA(T_2 - T_1)}{l}$$

where k is the conductivity of the wall in

$$\frac{\text{Btu} \cdot \text{in.}}{\text{hr} \cdot \text{ft}^2 \cdot {}^\circ\text{F}}$$

A is the area of the wall in square feet, l the thickness of the wall in inches, and $(T_2 - T_1)$ the temperature difference in degrees Fahrenheit between the inside and outside.

To facilitate calculations of the amount of heat lost per heating season, the number of degree days has been tabulated for various locations. For each day of the heating season the number of degree days accumulated is given by the average temperature difference between inside and outside. The inside temperature is conventionally taken as 65°F. For an entire heating season, the total number of degree days is obtained by summing these degree days; in other words, the total number of degree days per heating season is given by the total number of days that heat is required during each heating season times the average temperature difference between the outside and the inside, with the inside temperature taken as 65°F. In practice, of course, the temperature varies from day to day and throughout the day, but the degree days total cited takes all of these variations into account. Table 9.2 lists the annual degree days for a number of locations within the United States.

Table 9.2 DEGREE DAYS FOR THE HEATING SEASON IN VARIOUS LOCATIONS IN THE UNITED STATES

Birmingham, Alabama	2,780
Anchorage, Alaska	10,789
Barrow, Alaska	19,994
Fairbanks, Alaska	14,158
Tucson, Arizona	1,776
San Francisco, California	3,069
Los Angeles, California	2,015
Denver, Colorado	5,673
Washington, D.C.	4,333
Miami, Florida	173
Boise, Idaho	5,890
Chicago, Illinois	6,310
Des Moines, Iowa	6,274
New Orleans, Louisiana	1,175
Portland, Maine	7,681
Boston, Massachusetts	5,791
Detroit, Michigan	6,404
Minneapolis, Minnesota	7,853
International Falls, Minnesota	10,600
Kansas City, Missouri	4,888
Great Falls, Montana	7,555
Reno, Nevada	6,036
Newark, New Jersey	5,252
Rochester, New York	6,863
New York, New York	5,050
Cleveland, Ohio	5,717
Portland, Oregon	4,143
Philadelphia, Pennsylvania	4,523
Memphis, Tennessee	3,006
Houston, Texas	1,276
Dallas, Texas	2,272
Richmond, Virginia	3,720
Seattle, Washington	4,438
Madison, Wisconsin	7,300
Cheyenne, Wyoming	7,562

Example 9.1

(a) Calculate the number of degree days accumulated in one day in which the average outside temperature is 17°F.

(b) Calculate the number of degree days accumulated during a 150-day heating season in which the average outside temperature is 17°F.

Solution

(a) degree days $= 1 \text{ day } (65 - T_{out})$
$= 1\,(65 - 17) = 48$ degree days

(b) degree days per season $= 150 \text{ days } (65 - T_{out})$
$= 150\,(65 - 17) = 7200$ degree days per heating season

In calculating heat-loss rates, it is also useful to use the concept of R-value, which was briefly discussed in Chapter 6. For a slab of a particular material the R-value is defined as

$$R = \frac{l}{k}$$

where l is the thickness of the slab and k is the thermal conductivity as given in Table 6.1. For a layered series of materials each having thickness $l_1, l_2, l_3 \ldots$ and conductivity $k_1, k_2, k_3 \ldots$, the total R-value for the combination is given by the sum of the individual R-values, or

$$R_T = R_1 + R_2 + R_3 + \ldots$$
$$= \frac{l_1}{k_1} + \frac{l_2}{k_2} + \frac{l_3}{k_3} + \ldots$$

In practice, the R-value for insulating materials is frequently stamped right on the material, so that the conductivity and thickness are not needed to compute an R_1 or an R_2. The usefulness of R-values is that they can be simply added for a layered composite wall to form a total R-value (R_T) that may be used to find the heat loss for a given wall area (A), temperature difference (ΔT) and time (t)

$$Q = \frac{A}{R_T}(\Delta T)(t)$$

The total R-value may also be used in the following equation to calculate the loss over an entire heating season for one wall with area A (ft^2).

$$Q = \frac{A}{R_T} \times (\text{degree day}) \times 24 \text{ hr/day}$$

The degree day must be multiplied by 24 to obtain hours, since our original equation was in Btu/hr for Q/t.

Table 9.3 lists some typical R-values for materials used in building construction. In some cases the R-values per inch are listed. To obtain the total R-value in such a case, the R-value per inch must be multiplied, of course, by the actual thickness in inches. As discussed in Chapter 6, in addition to the R-value of the actual materials, there are certain insulation properties (R-values) for the layers of air just inside and outside the building walls. Typical values for these air layers are indicated in Table 9.3, but it must be realized that the outside value varies considerably with the wind velocity (less for strong winds). As discussed earlier, the insulating properties of a single sheet of glass are almost entirely due to these insulating layers of air.

Example 9.2

Calculate the heat loss in Btu for one heating season with 6000 degree days for an insulated 2 × 4 stud wall with a ¾-in. slab of insulating sheathing on the outside with an R-value of 2.06, fiberglass insulation, and a ½-in. gypsum board on the inside. The wall is 20 ft long and 8 ft high. For this problem, ignore conduction directly through the studs; consider the entire wall area to be of the insulated construction.

Solution

First we calculate the total *R*-value:

Material	*R*-Value
Outside air layer	0.17
Insulating sheathing	2.06
Fiberglass 3.50* × 3.70	12.95
Gypsum board, $^1/_2$ in.	0.45
Inside air layer	0.68
Total *R*-value	16.31

We can now use the equation

$$Q(\text{Btu}) = \frac{A(\text{ft}^2) \times (\text{deg} \cdot \text{day}) \times 24\,(\text{hr})}{R\,(\text{hr} \cdot \text{ft}^2 \cdot {}^\circ\text{F/Btu})}$$

$$= \frac{(20)\,(8)\,(6000)\,(24)}{16.31} = 1.41 \times 10^6\ \text{Btu}$$

*The depth of a 2 × 4 is presently 3.5 in. If the stud wall were completely filled with insulation at an *R*-value of 3.70/in., the resulting *R*-value for the wall would be 12.95. If standard batts of fiberglass were used, it would be an *R*-value of 11.

To view the result of the above exercise in some perspective, we can calculate how much heat would have been used if there were no fiberglass insulation. Until recently, homes were frequently built without such insulation in walls. It was a feature that the prospective home buyer could not see and

Table 9.3 SOME REPRESENTATIVE *R*-VALUES FOR BUILDING MATERIALS

	R (hr • °F • ft²/Btu)
Plywood, $^3/_4$ in.	0.94
Insulating sheathing, $^3/_4$ in.	2.06
Fiberglass, per inch (battens or loose)	3.70
Mineral wool, per inch (battens or loose)	3.70
Polystyrene board, per inch	5.00
Polyurethane board, per inch	6.25
Urea foam, per inch	5.25
Gypsum board, $^1/_2$ in.	0.45
Poured concrete, per inch	0.08
Brick, common, per inch	0.20
Stone, per inch	0.08
Concrete block, sand and gravel, 12 in.	1.28
Concrete block, cinder, 12 in.	1.89
Outside air layer (depends on temperature and wind velocity)	0.17
Inside air layer	0.68
Glass, $^1/_8$ in.	0.03

for the most part did not care about. Repeating the above calculation with the R-value of 3.36 (the 12.95 contribution of the fiberglass removed) yields a value for Q of 6.85×10^6 Btu, almost 5 times as much as with the insulation.

The benefit of the addition of such insulation can be calculated by determining the money saved per year.

Example 9.3

Calculate the saving in dollars for one heating season resulting from the addition of the fiberglass insulation in Example 9.2. Assume the cost of natural gas is $\$3.95/10^6$ Btu and the furnace is 75% efficient.

Solution

The value of the natural gas saved for one season will be

$$= \frac{\text{heat energy saving in Btu} \times \text{cost/Btu}}{\text{efficiency of furnace}}$$

$$= \frac{(6.85 - 1.41) \times 10^6 \text{ Btu} \times \$3.95/10^6 \text{ Btu}}{0.75}$$

$$= \$28.65$$

The cost of the insulation alone in these examples is about \$25 in 1991. Thus, the payback time is less than a year, and for new construction the insulation is obviously a good investment. Even if the installation charges doubled the cost, it would be an excellent investment. Not all such insulation additions are so favorable economically, especially in existing buildings. In some cases access to the place where the insulation is to be installed is extremely difficult. For many houses the most easily accessible area is the attic, where insulation can frequently be simply laid between the joists. Handbooks in the past have frequently recommended an R-value of 19 for attic insulation, with much higher values recommended recently; however, the best value for an individual house can only be determined when the costs, degree days, and other available insulation are known and a calculation carried out. The R-values currently recommended by the government are indicated in Table 9.4. As an example of a detail that might affect a particular situation, there are sometimes recessed ceiling lights that penetrate the attic insulation, and they are commonly areas of conductive heat loss as well as sources of cold air infiltration.

As can be seen in Table 9.3, masonry walls have particularly poor insulation properties. For an exterior stone wall, it is desirable to put an inch or more of polystyrene board on either side of the wall or even in the middle of the wall during construction. Frequently brick facing is used on the exterior of a frame house. In such a case insulation can be added to the stud wall in back of the brick facing. The insulation properties of foundations and cellars can be dramatically improved by placing insulating boards around the exterior of the concrete walls both above and below ground level.

Double glazing of windows is an effective way to improve the insulating properties by a factor of about 2 over a single sheet of glass (a R-value of about 0.9 to about 2.3) if the spacing between the glass is about ⅜ in. or more.

Table 9.4 RECOMMENDED *R*-VALUES FOR NEW RESIDENTIAL CONSTRUCTION BY DEGREE DAYS AND TYPE OF HEATING SYSTEM (hr • ft^2 • °F/Btu)

Degree Days	Ceilings	Walls	Floors
Gas or Oil			
Above 7000	38	17	19
6001–7000	30	12	11
4501–6000	30	12	11
3501–4500	30	12	11
2501–3500	22	11	0
1001–2500	19	11	0
1000 and under	19	11	0
Electric Resistance			
Above 7000	38	17	19
6001–7000	38	17	19
4501–6000	30	17	19
3501–4500	30	17	19
2501–3500	30	17	11
1001–2500	22	12	0
1000 and under	19	11	0
Electric Heat Pumps			
Above 7000	38	17	19
6001–7000	38	17	19
5001–6000	30	17	19
4501–5000	30	12	11
3501–4500	30	12	11
2501–3500	22	11	0
1001–2500	19	11	0
1000 and under	19	11	0

Note: Ceiling insulation values in the chart are based on vented attic construction using ½-in. gypsum board interior finish. Floors are floors of heated spaces over unheated basements, garages, or crawl spaces. The insulation values are based on wood frame construction, ¾-in. plywood, and carpeting with pad. Values listed for walls refer to the total *R*-value of cavity insulation plus sheathing material. They are based on wood frame construction using wood siding and ½-in. gypsum board interior finish.

Since the added insulation comes from the relatively immobile layers of air clinging to both surfaces of the glass, each added layer of glass contributes an *R*-value approximately equal to that provided by the first single layer. Because the boundary air layers are typically several millimeters thick, the layers of glass must be separated by a centimeter or more for optimum thermal resistance. There are a number of ways of achieving two or more layers of glass. The use of storm windows that replace the screen windows used in the summer is one conventional method. A variety of commercially made double-glazed window assemblies are available that offer a permanently sealed package with a desiccant between the panes to absorb entrapped moisture. The payback time for double glazing is generally longer than for the simple addition of insulation to an attic, but many homeowners, particularly those with heating seasons of 5000 or more degree days, find double glazing economically advantageous. It is difficult to make generalizations about the addition of insulation to foundations, roofs, and walls of older homes, as each house faces somewhat different problems. Certainly for new construction, adequate insulation should be demanded of the builder, as later addition of the insulation is generally far more expensive. Putting some fraction of the house beneath the surface of the ground, particularly on the north-facing side, is an effective way to provide insulation for some homes.

In addition to building alterations, there are a number of interior decorating or style-of-living changes that can make a significant difference in the amount of energy consumed to heat a building. Draperies in front of the windows can be quite effective if they are heavy enough to stop the movement of air, and if they are closed at the top, bottom, and sides so that convection currents cannot transport warm air around to the window side of the drape. Obviously, the draperies should be opened during the day to let in direct sunlight, and they should be closed at night. Although they can be unaesthetic and troublesome, sheets of foam insulation placed in the windows at night can reduce heat losses of even double-glazed windows by a factor of 3 or so. There are also various mechanical schemes for inserting insulation at night between the sheets of glass on a window, but these are not generally suited to the average house.

AIR INFILTRATION

Air leaks around windows and doors represent a major component of heat loss for most homes. A full exchange of the air in the house once every hour by various ill-fitting doors, windows, furnace vents, chimneys, and so forth can account for a third of the total fuel bill for heat. The caulking and weather-stripping of doors and windows is usually the easiest and least expensive way to conserve energy in a house. Fireplace chimneys should have a damper that closes tightly when the fireplace is not in use. Glass fireplace doors also help considerably, since they can be closed both during and just after a fire, when the natural draft in the heated chimney tends to remove the heated air of the house up and out the chimney. An open fireplace often increases the fuel bill when it is used.

Dampers that automatically close off the flue from the furnace when it is not running can be purchased and are usually cost effective. Obviously, this must be a fail-safe system so that there is no chance of carbon monoxide or other gases from the combustion process filling the living space with noxious fumes.

By such measures the air exchange in a typical house can be reduced from once an hour to about once every five hours, with a reduction of greater than 10% in the total fuel consumed, but this is not without problems. Concern has been expressed about the danger of sealing a house too tightly. There are several gases that are present in some houses that can lead to health hazards: radon, carbon monoxide, and nitrogen dioxide, as well as vapors from insulating material involving formaldehyde.

Radon is a radioactive gas that is a daughter activity of uranium and thorium. Because all stones and concrete contain uranium and thorium to some degree, radon gas will be present and will diffuse out of masonry surfaces into the living space. The inhalation of radon gas with the formation of some of its alpha-emitting daughters is known to cause lung cancer. A more detailed discussion of the effects of radon appears in Chapter 11. Carbon monoxide can be formed whenever any fuel is burned. If combustion takes place in a heating stove, fireplace, furnaces, or water heater, the carbon monoxide fumes will be vented to the outside. On the other hand, the use of natural gas in cooking stoves or ovens that are not vented can lead to the accumulation of carbon monoxide in the house. As we discuss in Chapter 12, the hemoglobin in the blood has a great affinity for CO that tends to block the normal uptake of O_2 by the lungs. In sufficiently great concentrations, carbon monoxide can be lethal. The fumes from urea–formaldehyde foam insulation as well as radon and carbon monoxide gas have generally not been a major problem, since air infiltration has been sufficient to keep their concentrations down. However, if a house is sealed too tightly without some thought given to the presence of these gases, significant health problems could be encountered.

It is possible through the use of a heat exchanger to have an exchange of air to prevent unhealthful concentrations of these and other noxious gases and essentially not waste any energy. If a proper heat exchanger is provided, the warm air in the house can be used to bring the cold air from outside up to the interior air temperature with a very high efficiency. The basic device is shown schematically in Figure 9.2. A long set of tubes with a thin paper barrier between the two air streams is provided so that as long as the interior air is warmer than the exterior air, heat will be transferred from the outflowing interior air to the inflowing exterior air. The cold exterior air will finally arrive inside at the interior temperature. A small electric blower is usually employed to provide a forced draft.

There is another type of heat loss, related to humidification, when outside air is brought into the house. If outside air at 40°F and a relative humidity of 60% is brought into a house where the temperature is 70°F, the relative humidity will then be only 20% in the interior. Because this is usually considered too dry an environment for comfort and health, water is often evaporated in a humidifier to increase the humidity. The humidified air then exfiltrates from the house, requiring that the replacement air be continuously humidified. Of course, the evaporation of water involves energy (970 Btu/lb of water or 540 cal/g of

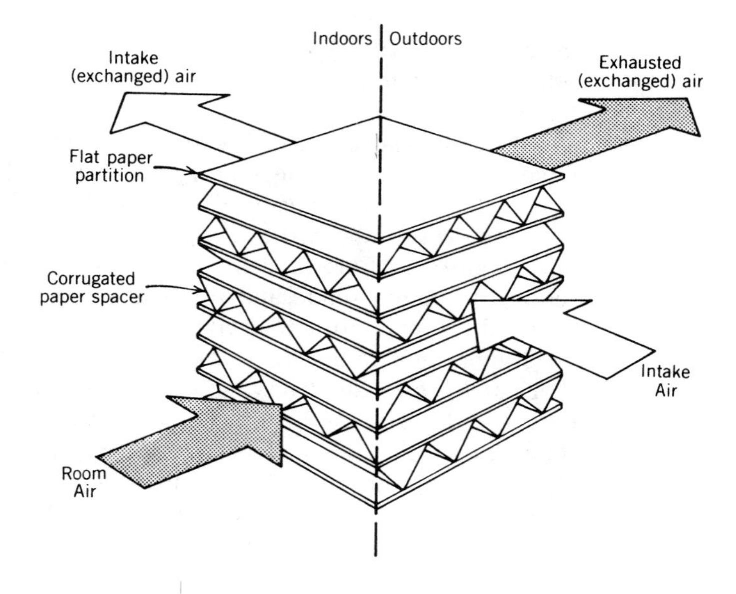

Figure 9.2 A commercially available heat exchanger for residential applications. The air flows shown are driven by an electric fan. Both heat and moisture are interchanged through the paper partitions in this device to effect a maximum energy savings.

water), and the energy must be supplied by the furnace. For example, if a house has a volume of 15,000 ft^3 and exchanges air once an hour with the outside, it requires 3800 Btu/hr to increase the relative humidity from 20% to 40%. If there is only one air exchange every five hours, then the heat needed for humidification is reduced by a factor of 5.

FURNACES, STOVES, AND FIREPLACES

A building is usually maintained at a comfortable temperature by a furnace that burns natural gas or fuel oil. The furnace heat is used to warm air or water that is circulated around the building to provide a reasonably uniform temperature throughout. The fraction of the chemical energy in the fuel that is transferred to the circulating medium is less than 100%. The rated (nameplate) efficiency of furnaces is usually in the range of 75 to 85%, but frequently the actual efficiency is no more than 50%. In order to have 1 Btu of useful heat with 50% efficiency, 2 Btu has to be released from the fuel. A schematic view of a hot air furnace is shown in Figure 9.3. Clearly, any heat from the combustion process that goes up the flue is wasted as far as space heating of the building is concerned. The electric energy needed to run the blowers or pumps to circulate the air or water must be taken into account in calculating the efficiency. Methods of improving the efficiency are discussed later.

Electric resistance heating has an efficiency of essentially 100% because all the heat generated can go directly into space heating, with venting being unnecessary. The major disadvantage of electric heating is the cost, since 1 Btu of electric energy costs about three to four times as much at the present time as 1 Btu of natural gas fuel energy. After the efficiency of a gas furnace is taken into account, electric heat is two to three times as expensive as gas. Overall, the production of electricity from fossil fuels has an efficiency of 35% or so and involves a complex distribution system with a large capital investment. It is, thus, not surprising that electrical resistance heating is expensive.

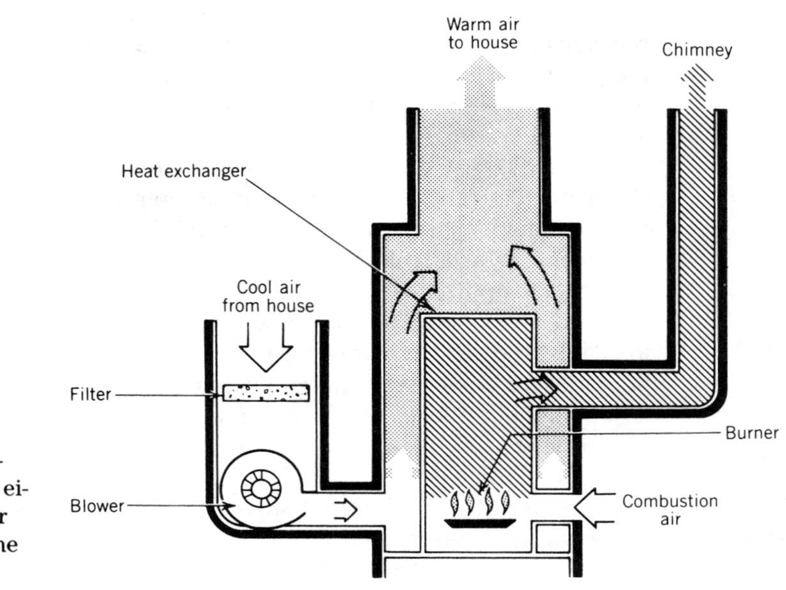

Figure 9.3 A hot-air furnace that can be fired by either natural gas or oil. Air circulation is forced by the electric blower.

Electrically powered heat pumps, discussed in Chapter 3, are an attractive way to heat buildings in certain sections of the country. Where the temperatures are not too low in the winter, the coefficient of performance of heat pumps is quite good, and they are considerably less expensive to run than electric resistance heating. If air conditioning is needed in the summer, the capital costs may be fully warranted.

One of the most important variable factors in the energy used in building heating is the setting of the thermostat. For many years Americans regarded 72°F as the normal indoor temperature for a heated house in the wintertime. Because of rising fuel bills, 65 to 68°F is now more typical. Temperatures around 60°F are not uncomfortable for many people if adequate clothing is worn. There is little excuse for office buildings to be maintained at more than 68°F in winter. A lot of misinformation has been circulated about the optimum setting for thermostats at night. Homeowners have been told for years that they should not set the temperature back too much at night, as it will require more energy to heat the house in the morning than was saved at night. There is no thermodynamic justification for such a statement. The fact is, of course, the lower the temperature inside, the less fuel is used. The main considerations are the discomfort of dressing and eating in the morning at a low temperature while the house is warming up, and not having the water lines freeze. A setting of 50°F at night is quite tolerable for many households. A clock-controlled automatic thermostat can be installed that takes care of these adjustments without dependence on the presence or memory of the homeowner. A nighttime setback from 68°F to 55°F can typically save about 20% of the fuel bill.

Certain adjustments can improve the performance of a hot air furnace, such as optimizing the time delay of starting the blowers after the combustion starts and the time the blowers run after the combustion stops. Replacing the pilot light in a furnace with an electronic ignition device helps to reduce not only gas consumption but also the heat in the furnace flue, which increases the draft up the flue when the furnace is not running. This latter point is less valid if the combustion air is taken directly from the outside.

The major improvement in furnace efficiency can be achieved by somehow recovering heat that goes up the flue. It is generally believed that the stack temperature should be about 300°F to have a sufficient draft and avoid condensation problems in the flue. In some furnaces, particularly older ones, the stack temperatures may be 600°F or over. A variety of flue heat recovery devices are available that basically act as heat exchangers to warm room air directly. Other devices take the combustion air from the outside and preheat it using a heat exchanger on the flue before it is injected into the combustion chamber. Commercially available furnaces that incorporate many of these features have efficiencies up to 95%.

Other less complicated ways of increasing the furnace efficiency involve insulating the air ducts or water pipes located in an unheated space. The overall potential for conserving energy by increasing the efficiency of furnaces is very large. Without the expenditure of large amounts of money, a do-it-yourself homeowner can, in some cases, raise the efficiency of a typical furnace from about 50 to 75%.

The use of wood-burning stoves and fireplaces is not basically an energy conservation measure unless they are used to heat small areas of a house while the main furnace thermostat is turned down. The major effect of wood burning is to transfer the energy source from oil or natural gas to wood. If one has access to a supply of wood at little cost, then money will be saved over the use of expensive fossil fuels, but more total fuel energy will probably be used. At the present time, for an average energy content per cord of wood and an average stove efficiency, if wood costs more than $90 per cord, oil is more economical; if wood costs more than $60 per cord, natural gas is more economical; and if wood costs more than $130 per cord, electricity is more economical. The efficiency of good wood-burning stoves ranges from 40 to 65%. The better-built airtight stoves will, with intelligent use, yield about 65%.

Fireplaces frequently take more heat from a home than they contribute. As mentioned above, glass doors will prevent the furnace-heated room air from going up the chimney when the fireplace is used, but the glass cuts off most of the infrared radiation that could contribute to the heating of the house. The infrared radiation absorbed by the glass raises its temperature, however, thus helping to heat the room air against its outer surface. Fireplaces constructed so that air can circulate by convection in back of the fireplace and then out into the room are quite efficient. In summary, fireplaces at their best are not nearly as efficient as wood stoves, but their aesthetic appeal cannot be denied.

The number of Btus that can be obtained from a cord of wood ranges from about 12×10^6 to about 30×10^6. The major variable is the density of the wood; higher density woods such as oak have a higher heat content per cord than lower density woods such as cedar or fir. When the heat content for dry wood is considered on a weight basis, almost all wood is the same, about 8600 Btu/lb.

SOLAR AND OTHER SOURCES OF HEAT

It is useful to perform an energy audit to understand the areas where energy conservation will be a good investment for a building. This is a process of adding all of the losses for space heating over a period of a year and trying to account for these by adding all known sources of heat. The major component, fuel for the furnace, can be obtained from statements from the gas utility or heating oil company. The other heat sources are the appliances, lighting, and people (listed in Table 6.4), solar energy, and wood burning. The miscellaneous items in Table 6.4 add up to approximately 7% of the total heat needed for an average home on a winter's day of about 1×10^6 Btu. The wood burning component can be calculated from the information in the preceding section. The solar energy input is important whether or not explicit passive solar plans have been incorporated into the building.

Information presented in Chapter 6 permits a reasonable estimate to be made of the solar energy input for a heating season. Table 6.5 contains the basic information needed. Although all windows receive some solar energy from diffuse light, the direct sunlight through south-facing windows is usually the major component. The insolation cited in Table 6.5 for vertical south windows must be multiplied by the fraction of the time the sky is clear during the heating season, which ranges for many areas from about 60 to 75% of the time. In addition, the transmission of the sunlight through the glass window panes must be taken into account (about 86% transmission per pane), as well as the fraction of the transmitted energy that is converted into heat. The conversion efficiency depends on the nature of the material the sunlight encounters in the room and, hence, varies widely but generally ranges from about 60 to 90%. The energy not converted into heat is reflected back through the windows.

Example 9.4

Calculate the solar energy contribution for one heating season to the heating of a home at a latitude of 40°N that has 300 ft^2 of double-paned south-facing windows. Assume clear skies 75% of the time and an absorption efficiency of 90%.

Solution

From Table 6.5, the total incident solar energy on a vertical south-facing surface from October 21 to April 21 can be estimated to be about 0.30×10^6 Btu/ft^2. Multiply this by a factor of 0.75 for clear skies, $(0.86)^2$ for transmission, and 0.90 for absorption.

$$0.30 \times 10^6 \frac{\text{Btu}}{\text{ft}^2} \times (0.75) \times (0.86)^2 \times (0.90) = 0.15 \times 10^6 \frac{\text{Btu}}{\text{ft}^2}$$

$$0.15 \frac{\text{Btu}}{\text{ft}^2} \times 300 \text{ ft}^2 = 45 \times 10^6 \text{ Btu per season}$$

The total heat needed for a season for a reasonably well-insulated single-family house in a region with 6000 degree days is around 90×10^6 Btu. In the above example solar energy provides about one half of this. The heat losses through the same windows without benefit of nighttime draperies or other special insulating devices would be about 25×10^6 Btu. In general, south-facing windows are an important component of a well-designed, energy-efficient house. Figure 9.4 shows an example of south-facing windows.

STANDARDS FOR HOME HEATING

How many Btus should be needed to heat a home? The answer obviously depends on the climate and the size of the house. One method of treating the

Figure 9.4 Proper window orientation is important for maximum use of solar energy for space heating. A wood-burning stove can shift the fuel problem from fossil fuels to wood and frequently offers a savings in heating expense if a convenient supply of firewood is available. (For safety, the stove should be positioned on an approved fireproof surface.)

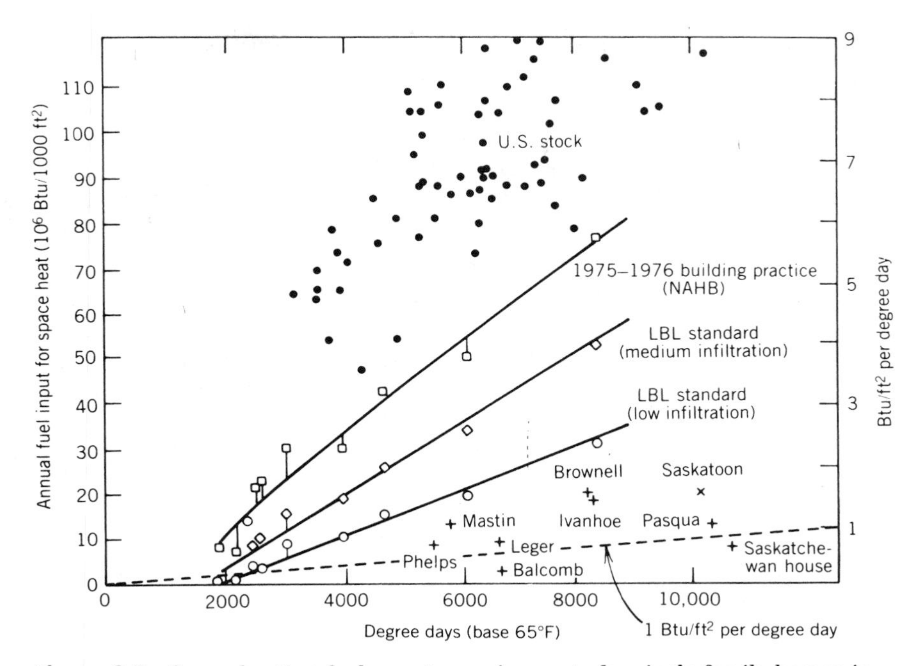

Figure 9.5 Space heating fuel energy requirements for single-family homes in the United States for 1 year in terms of 10^6 Btu/1000 ft^2 as a function of degree days. The black dots show typical existing housing requirements in 1978 in various parts of the country. Also shown are standards recommended in 1975–1976 by the National Association of Home Builders. The LBL standards refer to building performance standards put forth by Lawrence Berkeley Laboratory with 0.6 air changes per hour (medium infiltration) and the lower line (low infiltration) with 0.2 air changes per hour. The individual homes at the bottom of the plot illustrate what can be accomplished in specially designed, well-insulated homes in various locations. It appears that 1 Btu/ft^2·degree day is an achievable goal. (Source: Lawrence Berkeley Laboratory Report LBL-11990.)

size is to compute the heat needed relative to the number of square feet of floor area. Figure 9.5 contains data for various homes on the basis of the number of Btus per square feet (in units of 10^6 Btu/1000 ft^2) as a function of the degree days for the heating season. The data, compiled by a group at Lawrence Berkeley Laboratory, show the results for a group of typical houses for calendar year 1978 and some standards in effect today. For a particular house, a calculation shows how the space heating requirements compare with those shown in the plot.

9.3 WATER HEATERS, HOME APPLIANCES, AND LIGHTING

WATER HEATERS

Water heaters use about 10 to 20% of the total energy consumed in a typical home. Natural gas or electricity is used to heat water to a temperature in the range of 120 to 150°F. One of the major problems is that a tank of 30 to 80 or so gallons of water is maintained at this temperature. Even if the water is not used, heat losses through the thermal insulation of the tank require a continual expenditure of energy to keep the water up to the required temperature. The energy conservation steps that can be taken to reduce the amount of natural gas or electricity used are similar to those discussed for space heating.

(a) Lowering the temperature to about 100° or 120°F saves considerable energy, and this seems to be a high enough temperature for satisfactory clothes washing and bathing. Dish washing may require a higher temperature.

(b) Increasing the insulation on the tank is useful for the summer, but any heat that is lost from the tank to the house interior in the winter may help toward space heating. Insulating the hot water pipes can also provide useful savings. Commercially available kits for insulting the tank are quite inexpensive.

(c) Reduction of the use of hot water obviously saves energy. Flow-restricting heads for showers and sinks are useful.

(d) Electronic igniters can replace the gas pilot lights for some useful savings.

(e) A flue damper can be installed.

APPLIANCES

The typical amount of electric energy used in 1 year by various appliances is listed in Table 9.5. Aside from water heaters, the major consumers of electric energy are refrigerators, clothes dryers, and air conditioners. Some of our widely scorned symbols of energy waste, such as electric carving knives and toothbrushes, use a negligible amount of energy.

For a number of years refrigerators had the thickness of their insulation reduced to provide a larger usable interior volume for given exterior dimensions. This trend finally led to so little insulation that the outside walls would frost up. To prevent this problem, electric heaters were put in the exterior side

Table 9.5 ELECTRICAL ENERGY CONSUMPTION IN THE HOME

Appliance	Approximate Average Energy Consumed per Year (kW • hr)
Water heater	4200
"Quick recovery" water heater	4800
Standard 14-ft^3 refrigerator	1140
"Frost-free" 17-ft^3 refrigerator	2000
Washer	103
Dryer	1000
Black-and-white TV (3.6 hr/day)	120
Color TV (3.6 hr/day)	440
Air conditioner (on "high" for 24 hr/day for 1 month)	1400 (per month)
Electric blanket	80
Garbage disposal	30
Carving knife	8
Electric toothbrush	0.5
Total average electrical energy used per household in 1972	8000

walls, thus increasing the thermal gradient across the insulation, causing the refrigerator to run more frequently. This total disregard for energy consumption on the part of the manufacturers is changing in response to the greater energy awareness of the consumer and government-directed energy efficiency labeling requirements. As can be seen in Table 9.5, automatic defrosting is very costly in energy. At $0.06/kW • hr, the "frost-free" refrigerator will use $2400 worth of electricity over its 20-year lifetime, far more than the initial cost of the appliance.

Clothes dryers typically consume about 15% of the electric energy used in a household. Gas-fired dryers cost less to run than electric, but the energy consumption is about the same. Hanging clothes out to dry is a very real energy conservation step. Venting the hot air from an electric dryer into the interior of a house helps with space heating and humidification, but may present other problems.

Air conditioners are major consumers of electric energy. As with refrigerators, care should be exercised to select one with a good efficiency (energy efficiency rating or coefficient of performance) and to set the temperature as high as consistent with comfort. The insulation of homes for space heating also helps in reducing the energy consumption of air conditioners in the summer.

Although it is not listed in Table 9.5, the use of an electric or gas oven consumes a fair amount of energy per year. An average family might use about 1000 kW • hr for an oven in 1 year. The problem with ovens is that a large volume is being heated, but only a small fraction of the energy is going into the food. Small, well-insulated ovens are clearly more energy efficient. Microwave ovens have distinct advantages in that a much greater fraction of the electric energy consumed goes directly into heating the food. For example, to bake a few potatoes with a conventional oven might take 1 hour in a 4000-W oven with a 50% duty cycle. This is equivalent to 2000 W • hr of energy. A microwave oven of 1000 W, on the other hand, can bake the same potatoes in 15 minutes, which is equivalent to 250 W • hr of energy, a factor of 8 less than the conventional oven. About 50% of the electric energy in a microwave oven goes into the food.

LIGHTING

At present about 20% of all the electric energy (or 5% of the total energy) in the United States goes to provide lighting. Thus sizable savings can be accomplished with prudent energy conservation measures. Standards for the light levels in public schools have increased from 20 lm/ft^2 in 1952 to 60 lm/ft^2 [a lumen (lm) is a measure of the amount of useful light present]. Although 60 lm/ft^2 is useful where concentrated reading is taking place, many recently designed office buildings have lighting levels in the range of 80 to 100 lm/ft^2, including corridors and stairways. To make matters worse, many buildings have master light switches that turn on whole floors at a time. Late at night one can frequently see an entire multifloor office building glowing with light while a few janitors sweep the floors.

Table 9.6 lists the efficiencies of various sources of light in lumens per watt. Of particular interest for the home is the comparison of incandescent and fluorescent lamps. About four times as much useful light per watt is obtained with a fluorescent fixture as with an incandescent fixture. Although the energy saving is attractive and well worth the investment in the more expensive

Table 9.6 EFFICIENCIES OF SELECTED FLAMES AND LAMPS (ALL ENTRIES EXCEPT FOR FLAMES AND MANTLES ARE IN lm/W_e)

	Efficiency	
Light Source	**Lamp Alone**[a]	**Lamp Plus Ballast**
Sunlight	92 lm/W_t	
Open gas flame[b]	0.2 lm/W_t	
Gas mantle[b]	1 to 2 lm/W_t	
Incandescent lamps[c]		
40 W	12	
100 W	18	
Fluorescent lamps,[c] plus ballast[d]		
20 W, 24 in. plus 13 W	(65)	39
40 W, 48 in. plus 13.5 W	(79)	59
75 W, 96 in. plus 11 W	(84)	73
Metal halide, 400 W, plus 26 W	(80)	75
Sodium		
400 W high pressure, plus 39 W	(120)	109
180 W low pressure, plus 39 W	(180)	154
Mercury, 400 W, plus 26 W	(57)	53.5

[a] Numbers in parentheses are for the bare lamp, without ballast losses. W_t refers to thermal watts and W_e to electric watts.

[b] ***Source:*** *Handbook of Chemistry and Physics*, The Chemical Rubber Co., Cleveland, Ohio (1961), p. 2849.

[c] ***Source:*** *Handbook of Chemistry and Physics*, The Chemical Rubber Co., Cleveland, Ohio (1971–72), p. E-185.

[d] These are typical ballasts; however, 5-W ballasts are available for an extra 10 to 20% in price.

fluorescent fixture, there are grounds for objecting to the somewhat cold white light provided by such light sources. For bathrooms, kitchens, and work areas, however, fluorescent lights seem to be a wise choice, and new improved fixtures that offer a softer light are being developed. There has recently come on the market a wide variety of compact fluorescent tubes of various wattages to replace incandescent light bulbs. For example, a 15-W compact fluorescent tube is now available that provides the same light as a 75-W incandescent light bulb. The fluorescent light is not much larger than a conventional bulb, but it is rated to last 10,000 hours instead of the 1000 hours for the conventional bulb. While the compact tube cost $14 in 1991 compared to $0.50 for the incandescent light, its greater efficiency more than makes up for the price difference over its 10,000-hour life. (See Problem 17 at the end of the chapter.) In addition to this saving in cost, there is also a saving in the fossil fuels needed to provide the electricity, and hence a reduction in the emissions that result from burning the fossil fuels.

Obviously, shutting lights off that are not being used is a direct way to save energy. Institutions and businessses often seem to gang too many fixtures on one switch. Installing individual pull chains on light fixtures so that unneeded lights can be separately turned off has proved to be cost effective. In new construction, reducing the number of fixtures in the corridors and stairways, so that there is about 10 to 20 lm/ft^2, not only saves on the cost of the fixtures and electric energy but also reduces the need for air conditioning in the summer.

In any air-conditioned building the load on the air conditioner will be reduced by using more efficient or fewer lights.

THE ENERGY-CONSERVING HOUSE

It is difficult to assess accurately the effects on the nation's energy consumption of accomplishing all or some of the energy conservation measures for the home that have been discussed so far. Houses differ, climates differ, personal habits differ, and, of course, there is the question of how much capital investment can be made or should be made to effect the changes.

A study of a 1200-ft^2 northern California home was carried out to show explicitly, for one case, what the energy savings would be for various levels of investment in the types of conservation measures discussed in this chapter. Figure 9.6 summarizes the various steps taken and their cost. The major saving was in space heating, which went from 120×10^6 Btu/yr to 35×10^6 Btu/yr with an expenditure of about $1600, taking the house with *R*-11 in the ceiling as the starting point. From the experience of others, it appears reasonably easy to reduce space heating requirements in an average home by about two thirds using straightforward and not unduly expensive measures. The other conservation measures shown are also important ingredients in reducing the total energy (gas plus electricity) from about 235×10^6 Btu/yr to 100×10^6 Btu/yr with an investment of $2700.

At the present time in the United States, buildings and appliances consume roughly 38% of a total of about 80 QBtu annually, which amounts to 30.4 QBtu. If the 1/2.35 reduction ratio for our typical house were applied nationally, the 30.4 QBtu would be reduced to 12.9 QBtu at a cost of around $150 billion.

Although this simplistic calculation cannot be taken as a basis for a national energy plan, it is obvious that the potential for energy conservation in the home and in commercial buildings is very real and very important.

The question of how the capital can be raised to implement energy conservation measures has recently received at least a partial answer in the state of Oregon. There, public utilities have received approval to invest their money in home conservation projects and have these costs form part of the rate base. Under a plan of this type, an energy audit would be carried out by the public utility, and those projects that are cost effective would be implemented. The homeowner's approval is required, and the public utility provides an interest-free loan. The homeowner then enjoys the benefits of reduced energy consumption, and the public utilities do not need to invest as much capital in new power stations or natural gas distribution systems. When the homeowner sells the property, the loan is repaid from the money derived from the sale. The cost of the equivalent energy resource achieved by such a conservation scheme is less than half the cost of new electric energy ($0.017 versus $0.045/kW • hr). The utility can then sell this released energy to new customers at the prevailing retail price.

9.4 ENERGY CONSERVATION IN INDUSTRY AND AGRICULTURE

The industrial and agricultural sector of the energy economy is so diverse that it is difficult to discuss conservation measures in very explicit terms. There are construction, mining, manufacturing, and agriculture, each of which consumes

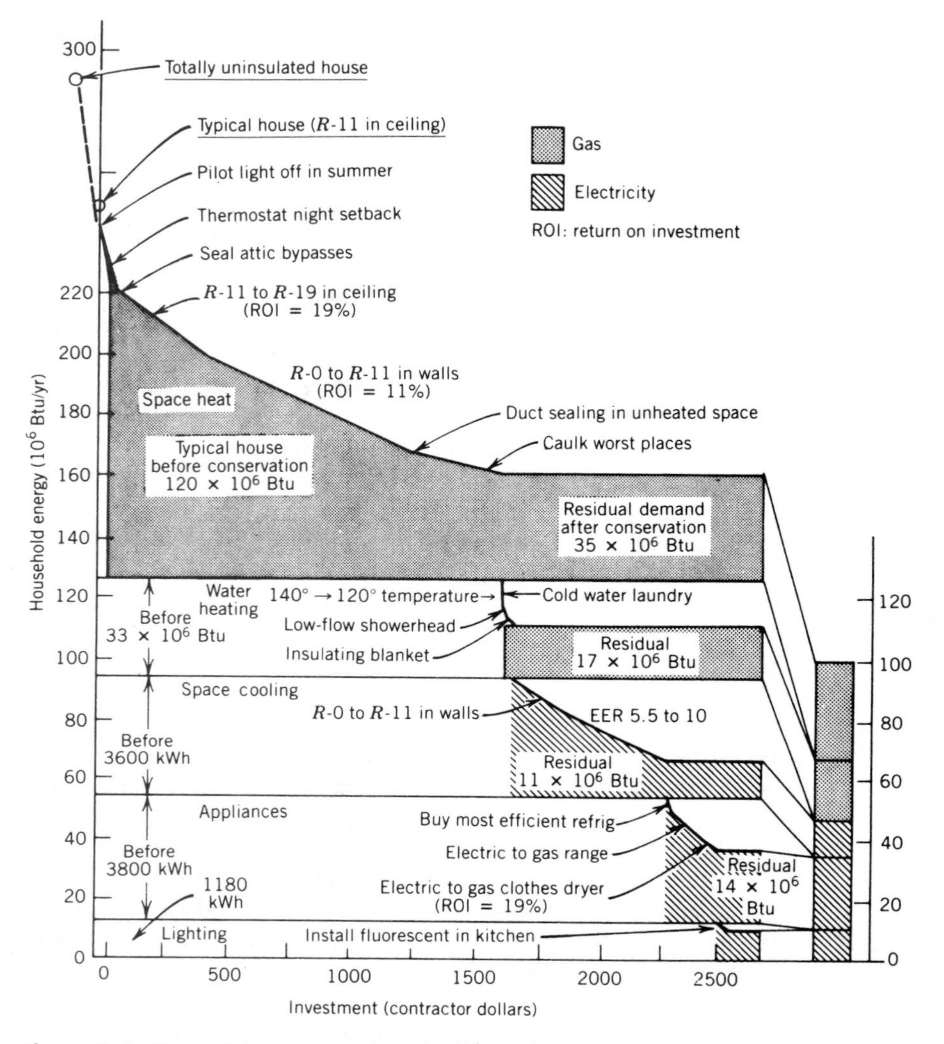

Figure 9.6 Potential energy savings in 10^6 Btu/yr for various energy conservation measures for a northern California house (1200 ft^2, 3000 heating degree days) in terms of the investment required. The shaded areas correspond to energy demand in the form of natural gas and the lined areas electrical energy demand. If all the indicated measures were adopted, the total energy demand per year would be reduced (compared to a totally uninsulated house) by a factor of 3 for an investment of $2700. (Source: Lawrence Berkeley Laboratory Report LBL-11650.)

energy in a variety of ways. However, conservation measures in some general areas can lead to increases in energy efficiency.

HOUSEKEEPING

This area involves the same kind of considerations and changes that were discussed for homeowners, since most of the same problems of space and water heating, appliances, and lighting are present in industry. In one sense the problems are more amenable to solution because larger units are involved with greater technical help available. On the other hand, the average worker in a factory may not be as enthusiastic about shutting off lights and so forth as the individual homeowner who is directly paying the bills.

WASTE HEAT RECOVERY AND COGENERATION

As we discussed in Chapter 3, it is possible, with various heat engines and processes using heat, to recover part of the heat that is rejected to the atmosphere or to some body of water. This heat can be used in a variety of ways, including for space and water heating. The efficiency of heat exchangers can be increased and better insulation provided on lines carrying heated water or steam. It has been estimated that a 5 to 10% increase in overall thermal efficiency in industry could be brought about in 10 years.

PROCESS CHANGES

In the smelting of metals (particularly aluminum), in uranium enrichment, and in the refining of chemicals and fuels, fundamental changes can be made in the process that may lead to a sizable reduction in the amount of energy used per unit of output. These changes generally involve some new technology and a capital investment. The potential for energy savings is very large, and as the cost of fuel increases, more of these investments will be cost effective.

RECYCLING

Certain materials, such as aluminum, require a large amount of energy for smelting. However, these energy costs can be, to a large degree, bypassed if recycled material is used. In addition to the ubiquitous aluminum can, it is possible to use aluminum scrap and so forth from manufacturing for an energy saving. Recycling is also effective for paper, tires, steel, glass, storage batteries, and other materials. In addition to this type of recycling, some manufacturing facilities can derive energy from various by-products, such as wood chips and sawdust in lumber mills and paper manufacturing, and methane from agricultural byproducts.

According to the CONAES study, industrial energy consumption per unit of output fell by 1.6% from 1950 to 1970 and by 20% from 1973 to 1979. This stepped-up efficiency can be attributed to increased energy costs. On this basis it is estimated that 40% less energy per unit output could be in effect for new plants by 2010. An obvious problem is that the cost of retrofitting is expensive, and an estimated one third of the existing plants will still be in use in 2010.

9.5 TRANSPORTATION

The ways in which energy is used in the transportation sector are shown in Table 9.7. As could have been assumed, trucks and cars dominate the use of energy, and these vehicles largely employ internal combustion engines fueled by gasoline or diesel fuel. The dynamics of automobile motion and fuel consumption are discussed in some detail in Chapter 12, so only the general aspects of the problem are brought up here.

Americans have been infatuated with the automobile for 70 or 80 years.

Table 9.7 DISTRIBUTION OF ENERGY USE IN THE TRANSPORTATION SECTOR

Mode	Percentage
Automobiles	50
Trucks	24
Air, passenger and freight	7
Military	5
Rail, freight	3
Water	4
Pipeline	4
Miscellaneous passenger, including bus and rail	3

We have more cars per capita, and we drive more miles per capita, than the citizens of any other country. In 1989 there were 1.48×10^8 passenger vehicles registered in the United States that went a total of 1.50×10^{12} miles with an average fuel consumption of 496 gallons per vehicle. There were 1.8 persons per vehicle in the United States compared to the next lowest of 2.1 in Canada and West Germany. The potential for energy conservation in automobile transportation in the United States is clearly tremendous.

The solution to the problem lies in two different areas: the greater use of more efficient modes of transportation and increased efficiency of the automobile. Because of the convenience of the auto, the U.S. public transportation system has deteriorated very badly in the last 50 years or so. The efficiency of various modes of transportation is shown in Table 12.2; it is clear that we have opted for an inefficient mode, because most travel is by little more than one occupant in an auto. Since 1973, there has been effort by the federal government as well as local governments to improve public transportation in the cities, but this has not as yet made a major difference in the use of the automobile. Car or van pooling, sharing rides, and so forth are widely used for commuting to work in areas where an inadequate bus or train system exists. We are also finally giving thought to urban design, so that the distances the average person must travel for employment or for shopping are reduced from those with the present unplanned sprawl.

It appears, however, that the major energy conservation gains that will be made in the near future in U.S. transportation will be made through increasing the fuel efficiency of the automobile. Because of the low cost of gasoline for many years both before and after World War II, the heavy, oversized, gas guzzler became the standard for automobiles. In 1973, the American car averaged about 13 miles to the gallon. This is far less than the smaller cars of the 1920s and early 1930s.

With the oil embargo of 1973, there was a renewed interest in fuel-efficient automobiles, and the car manufacturers, particularly those in Japan, started turning out smaller, more efficient cars. In 1975 the new cars produced or sold in the United States obtained an average of 15.8 miles per gallon (MPG), and the Energy Policy and Conservation Act of 1975 specified that the average of every manufacturer's new car fleet should increase to 27.5 MPG in 1985 and to 40 MPG by 1995. The detailed fuel economy of new cars and light trucks is shown in Figure 9.7. These performance figures are based on the Environmental

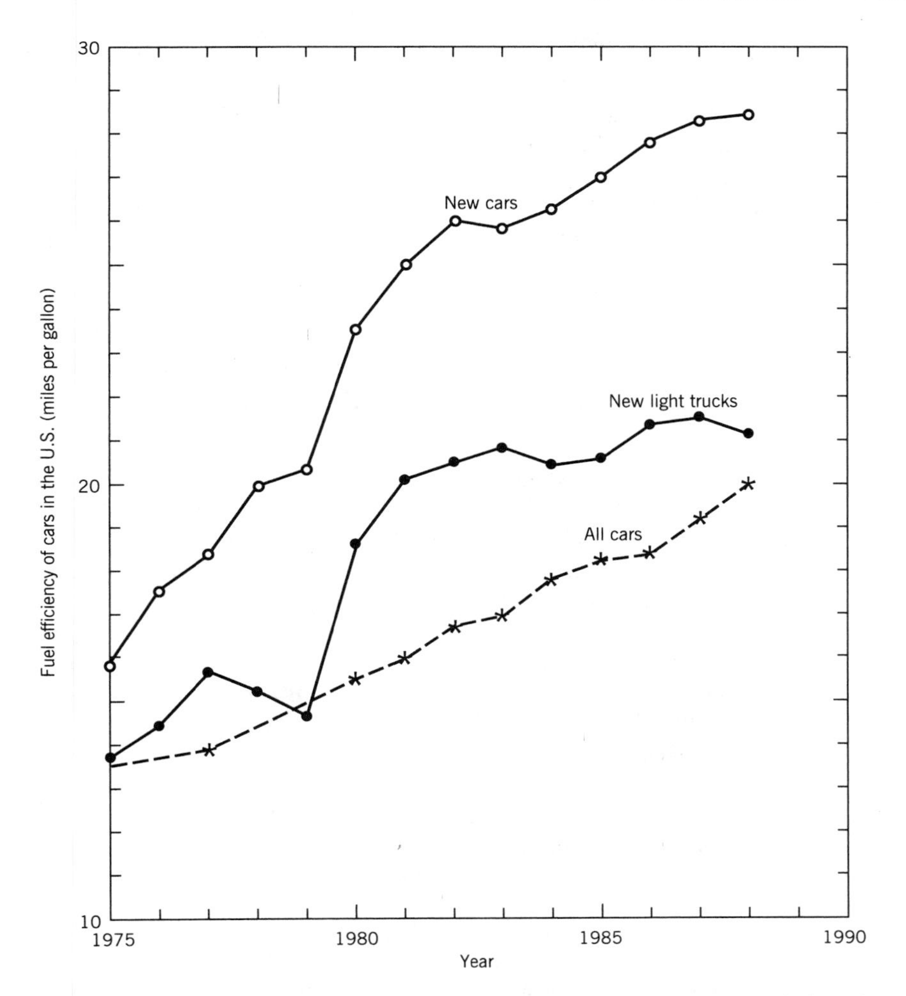

Figure 9.7 The sales-weighted fuel efficiency of new cars (open circles) and light trucks (closed circles) produced or sold in the United States since 1975. Also shown are the average fuel efficiencies of the entire fleet of cars in the United States (crosses). (Source: Energy Information Administration, *Annual Energy Review 1989,* and Ross, Marc "Energy and Transportation in the United States," *Annual Review of Energy,* **14** (1989), p. 131.)

Protection Agency standard of 55% city driving, 45% highway driving composite. There has been, however, a strong resistance on the part of American manufacturers to continue to the goal of 40 MPG by 1995. Arguments are made about the reduced safety of smaller cars, but the larger profits to be earned on the bigger cars may be another reason for manufacturers to want to forgo further increases in fuel economy.

Also shown in Figure 9.7 is the average fuel efficiency for the entire fleet of U.S. cars. These fuel efficiency rates, of course, are far lower than for the new cars since a large stock of older cars with poorer efficiency is still being driven. The upward trend of the MPG is, however, very significant.

There are cars now under development, such as the Volvo LCP2000, that can reliably achieve 60 to 80 MPG, and there are experimental cars that achieve even higher gas mileage. During the period from 1972 to 1985 the energy consumed annually by automobiles in the United States stayed about the same, 9.1 QBtu, as did the number of miles driven per car, about 9.6×10^3 miles per car. The fact that there were more cars on the road just offset their greater fuel efficiency. The technical aspects of fuel requirements and efficiencies are discussed in Chapter 12, and the important and related questions of auto emissions are detailed in Chapter 13.

PROBLEMS

1. Calculate the number of Btus conducted through a wall of a house in 8 hours. The wall has dimensions of 8 ft by 24 ft and has a total R-value of 16.2 hr • °F • ft²/Btu, including the effects of the interior and exterior air layers. The inside temperature for the period is 68°F and the outside temperature is 16°F.

2. Calculate the number of degree days in the heating season at a location where the average outside temperatures are distributed as follows: 60 days at 55°F, 50 days at 45°F, 30 days at 35°F, and all the other days at or above 65°F.

3. A 1500 ft² wall has an R-value of 11.0 hr • °F • ft²/Btu, including the effects of the interior and exterior air layers. How many Btus are lost through this wall in a 5600 degree day heating season?

4. Calculate the total R-value of a brick-faced stud wall that has the following layers: 2.5 in. of brick, 0.75 in. of insulating sheathing, 3.5 in. of urea foam, 0.25 in. of plywood, inside air layer R-value of 0.68, and outside 0.17.

5. A house in Reno, Nevada, consists of the following:
(a) 544 ft² of single pane windows
(b) 544 ft² of stud walls with a total R-value of 16.31
(c) 480 ft² of stone walls, 13-in. thick
(d) 1632 ft² of roof with a total R-value of 8.0
Neglecting the heat lost through the concrete slab floor, air infiltration, and solar energy input, calculate the total number of Btus needed to heat the home for one heating season.

6. Calculate the cost of heating the house in Problem 5 for one season with a natural gas furnace, assuming the cost of natural gas is $\$3.95/10^6$ Btu and the furnace is 65% efficient.

7. Calculate the savings in the cost of natural gas for one season if the house in Problems 5 and 6 has the roof insulation increased to an R-value of 19. Would the purchase of such insulation appear to be a wise choice if it costs $0.25/ft²?

8. What would be the payback period for double-paning windows in Anchorage, Alaska, if natural gas costs $\$4.50/10^6$ Btu, the furnace is 75% efficient, and the cost of double-paning is $2.50/ft²? Assume an R-value of 1 for single-pane windows and 2 for double-pane windows.

9. Estimate the percentage saving in the fuel bill if the thermostat is set back from 68°F to 50°F for 12 hours of the day and left at 68°F for the other 12 hours, assuming an outside temperature of 15°F all day.

10. How many Btus would be contributed to the heating of a house by the metabolism of four occupants over the 6-month heating period? Assume each person generates 100 W and they are in the house 50% of the time.

11. Enumerate effective ways in which your home could be made more energy efficient.

12. Enumerate changes you could make in your personal life that would lead to appreciable savings in energy.

13. For a few of the appliances listed in Table 9.5, estimate the life-cycle energy costs and compare them to the initial appliance cost. If you do not know the local cost of your electricity, you may use $0.06/kW • hr.

14. Calculate the annual energy cost in dollars for an electric blanket. Assume that it has a 135-W heating element, which is on 50% of the time during each 8-hour night, 160 nights each year. If use of the electric blanket permits you to lower your thermostat by 10°F during the night, estimate whether the electric blanket is an energy gainer or loser. Does it make sense to cover an electric blanket with ordinary blankets?

15. How could you improve the design of the typical gas-fired residential water heater? Consider electric ignition, flue dampers, insulating, tempering tank, insulated pipe connections, separate "instant" heater, and so forth.

16. Consider the relative ease and cost effectiveness of measures you might take to conserve energy in your present means of transportation as compared to your home.

17. You can purchase an energy-efficient 15-W compact fluorescent tube for $14 that is rated to last 10,000 hours. This provides the same light output as a 75-W incandescent light bulb that costs $0.50 and lasts 1000 hours. Based on an electric energy cost of $0.08/kW • hr, calculate the money saved over the 10,000-hour life by using the new fluorescent tube instead of the conventional light bulb. Include the cost of the light bulbs.

SUGGESTED READING AND REFERENCES

1. Hickok, Floyd. *Your Energy Efficient Home.* Englewood Cliffs, N.J.: Prentice-Hall, 1979.

2. Lovins, Amory B. *Soft Energy Paths.* New York: Harper & Row, 1977.

3. Ross, M. H., and Williams, R. H. *Our Energy, Regaining Control.* New York: McGraw-Hill, 1981.

4. Sawhill, John C. *Energy, Conservation and Public Policy.* The American Assembly. Englewood Cliffs, N.J.: Prentice-Hall, 1978.

5. Final Report of the Committee on Nuclear and Alternative Energy Systems (CONAES). *Energy in Transition 1985–2010.* San Francisco: W. H. Freeman, 1979.

6. Report of the Ford Foundation Study Group. *Energy: The Next Twenty Years.* Cambridge, Mass.: Ballinger, 1979.

7. "Energy for the Planet Earth." *Scientific American,* **263** 3 (September, 1990), pp. 54–163.

8. Rosenfeld, Arthur H., and Hafemeister, David. "Energy-efficient Buildings." *Scientific American.* **258** 4 (April 1988), pp. 78–85.

9. Ross, M. "Energy and Transportation in the United States." *Annual Review of Energy* **14** (1989), p. 131.

10. *Energy, Production, Consumption and Consequences.* Washington, D.C.: National Academy Press, 1990.

CHAPTER TEN

PLANT AND FOOD PRODUCTION

10.1 PHOTOSYNTHESIS

photo—(Gr. *phos, photos*, light),
synthesis—(L., a mixture, prop., a putting together,
fr. Gr. *synthesis*, deriv. of *syn-* + *tiuthenai*, to place)

An aspect of solar energy use not discussed in detail so far but essential to all plants and animals is the production of biomass by photosynthesis. Not only are we dependent on this process for our food, cotton and many other fibers for clothing, wood for lumber and paper, and innumerable other products from vegetable matter, but it is this same photosynthesis that converted the sun's energy into living organisms millions of years ago to provide the fossil fuels we use today.

As we saw in Chapter 6, the energy of the sun that reaches the earth is electromagnetic radiation with a spectrum that ranges from about 0.3 to 3 microns (μ) in wavelength. This corresponds to radiation from the near-ultraviolet through the visible to the infrared (or, in terms of angstroms, from 3000 to 30,000 Å). Figure 1.9 shows how this rather narrow portion of the electromagnetic spectrum is related to other types of electromagnetic radiation.

How does this process work by which plants use light energy to form vegetable matter? To understand the photosynthetic process, we must use the quantum, or corpuscular, description of electromagnetic radiation. As we saw in Chapter 6, the energy associated with a quantum of light is hf, where h is Planck's constant and f is the frequency of the light. The energy of a quantum, or photon, in the ultraviolet is sufficient to break a chemical bond; this is not generally so, however, in the middle portion of the spectrum, where solar intensity is at its peak and where most of the photosynthesis takes place. A photon in the visible portion of the spectrum, although not capable of breaking a chemical bond, has sufficient energy to raise an atom to an excited state. The excited atomic state may make it possible for bonding to take place with a neighboring atom. This is generally the way that photochemical reactions proceed. The photosynthetic process is a very special photochemical reaction sequence whereby light interacts with the molecules of water and carbon dioxide to form carbohydrates. The following generalized chemical equation expresses these ideas for the simplest carbohydrate:

$$CO_2 + 2H_2O \xrightarrow{\text{LIGHT}} CH_2O + H_2O + O_2$$

where 112 kcal of light energy is needed per mole of CH_2O formed. As discussed earlier, a mole (1 gram molecular weight) is the weight in grams equal to the molecular weight. For example, the mass number of oxygen is 16, so the mass number of the oxygen molecule, O_2, is 16 times 2, or 32. Hence, 1 mole of O_2 is 32 grams, and it contains Avogadro's number (6.023×10^{23}) of molecules.

The common carbohydrates are represented by the formula $C_x(H_2O)_y$. Some of the simple carbohydrates involved in photosynthesis are glucose, $C_6H_{12}O_6$, and sucrose, $C_{12}H_{22}O_{11}$. These are not hydrates, but consist of complex ring structures. The photosynthetic reaction leading to glucose is

$$6CO_2 + 6H_2O \xrightarrow{\text{LIGHT}} C_6H_{12}O_6 + 6O_2$$

and 674 kcal of light energy is needed per mole. This reaction does not occur in a single step but requires several steps in which the various components of the sugar molecule enter into the reaction At least two pigments in the plant are involved. One is chlorophyll, which absorbs light in the red part of the visible spectrum (a wavelength of about 0.7 μm). Another pigment generally absorbs light in the blue. With these two parts of the spectrum absorbed, the reflected light appears green; hence, this is the common color of plant leaves such as the corn shown in Figure 10.1. The light absorbed by chlorophyll leads to the formation of oxidants and reductants. These help in the production of energy-rich adenosine triphosphate (ATP), which is important for the conversion of carbon dioxide to carbohydrates. The second pigment is used in the release of oxygen that accompanies the carbon dioxide fixation.

The exact photosynthetic process is complex; it differs for different plants and involves a variety of pigments and chemical steps. It is the pigments in a plant that allow it to adapt to the light it happens to receive. Below the surface of the sea, for example, the light is mostly green, and red algae thrive because they can absorb the green light, allowing photosynthesis to take place; green algae, on the other hand, which would reflect rather than absorb green light, cannot grow.

Respiration, essentially the reverse of photosynthesis, also takes place and provides energy for the plant. The process of "burning" or oxidizing the carbohydrate molecule, releasing carbon dioxide, water, and energy, occurs continually. It is basic not only for the survival of the plant but also for the nutrition of the animals that eat the plant and derive energy from the carbohydrates.

At what rate can vegetable matter be produced by photosynthesis on the

Figure 10.1 The energy on which life depends enters the biosphere in the form of light. Photosynthesis is the process by which this light energy is converted to stored chemical energy.

surface of the earth? The answer to this question is important in assessing both how many people can be fed and whether growing wood for a fuel or grain for alcohol offers a significant source of energy for space heating and transportation. The amount of vegetable matter than can be grown certainly depends on the availability of sunlight to drive the conversion of CO_2 and H_2O into carbohydrates. In addition to limitations of sunlight, there are also limits placed by the availability of appropriate land, temperature, climate, and nutrients in the soil (nitrogen, phosphorus, and trace minerals), as well as by plant disease and insects. Overlooking these complications for the moment, let us examine the overall efficiency with which biomass may be produced from sunlight. In Chapter 6 the amount of solar radiation that finds its way to the surface of the earth was examined. The average solar energy per unit horizontal area and time at the top of the atmosphere was found to be 0.5 cal/min • cm^2. This value is averaged over day and night and over all latitudes. For the purposes of calculating plant production, it is convenient to know the number of calories in a day per square centimeter. Assuming that 47% of the solar energy incident on the atmosphere reaches the ground, the energy available for food production averages

$$0.5 \frac{\text{cal}}{\text{min} \cdot \text{cm}^2} \times (0.47) \times \frac{60 \text{ min}}{1 \text{ hr}} \times \frac{24 \text{ hr}}{1 \text{ day}} = 338 \frac{\text{cal}}{\text{cm}^2 \cdot \text{day}}$$

On a typical summer day a forest or field will have a somewhat higher value, about 500 to 700 cal/cm^2 • day. It is difficult to calculate theoretically what fraction of this light energy will end up as biomass. Only about 25% of light energy incident on the earth has the right wavelength to produce photosynthesis. Of this, 60 to 70% will be absorbed by the plant leaves if there is dense foliage. The amount of energy stored per carbohydrate unit synthesized is about 5 eV. The photons driving the process are mostly from the red end of the visible spectrum where the energy per photon is 1.7 eV. Most measurements suggest that about eight photons of incident red light are required per carbohydrate molecule formed. On this basis the energy input would be about 14 eV (8×1.7 eV) to have a stored energy of 5 eV. This corresponds to an efficiency of roughly 35% for the absorption process. Taking all of these factors together ($0.25 \times 0.70 \times 0.35 = 0.06$) gives an overall efficiency of about 6%.

A less theoretical but perhaps more practical approach to determining the efficiency of biomass production is to make direct measurements. Such measurements indicate that if there are 500 cal/cm^2 • day incident, the net potential plant production is about 70 g/m^2 • day. The gross production is 107 g, but respiration of 35 g reduces this to a net yield of 71 g.

Example 10.1

From the above data calculate the maximum efficiency with which solar energy is converted into biomass. Assume glucose ($C_6H_{12}O_6$) with 674 kcal per mole is the material produced.

Solution

First calculate the amount of energy stored per gram of glucose. Since carbon, hydrogen, and oxygen have atomic masses of 12, 1, and 16, respectively, a mole of glucose will have a mass of

$$12(6) + 1(12) + 16(6) = 180 \text{ grams/mole}$$

Since 674 kcal per mole of energy are stored, we have

$$674 \frac{\text{kcal}}{\text{mole}} \times \frac{1 \text{ mole}}{180 \text{ g}} \times \frac{10^3 \text{ cal}}{1 \text{ kcal}} = 3744 \frac{\text{cal}}{\text{g}}$$

With a net production of 71 $\text{g/m}^2 \cdot \text{day}$, this is equivalent to

$$71 \frac{\text{g}}{\text{m}^2 \cdot \text{day}} \times 3744 \frac{\text{cal}}{\text{g}} \times \frac{1 \text{ m}^2}{10^4 \text{ cm}^2} = 26.6 \frac{\text{cal}}{\text{cm}^2 \cdot \text{day}}$$

When this output of 26.6 $\text{cal/cm}^2 \cdot$ day is compared to the input of 500 $\text{cal/cm}^2 \cdot$ day, we have a photosynthetic efficiency of

$$26.6 \frac{\text{cal}}{\text{cm}^2 \cdot \text{day}} \times \frac{\text{cm}^2 \cdot \text{day}}{500 \text{ cal}} = 0.053 = 5.3\%$$

Since 25% of light has the correct wavelength and 60% of the incident light is absorbed by the plant's leaves, to have an overall efficiency of 5% means that 33% of the light energy actually absorbed is finally converted into biomass. This is quite consistent with our theoretical estimate. The actual net production of biomass in various locations in the United States is listed in Table 10.1.

Figure 10.2 shows the distribution of solar energy incident on the earth into various modes of absorption and scattering that take place in the atmosphere and on the surface of the earth. Of particular interest for present purposes is the 40×10^{12} W of solar power that goes into photosynthesis. This is only about 1/4000 of the total solar power incident. About half of the photosynthesis is thought to take place in the oceans, which occupy 70% of the surface of the earth. If we assume that the dry plant production resulting from photosynthesis is basically glucose with an energy content of 3744 cal/g, we have for the total plant matter produced each year

$$40 \times 10^{12} \frac{\text{J}}{\text{sec}} \times \frac{1 \text{ cal}}{4.184 \text{ J}} \times \frac{3.15 \times 10^7 \text{ sec}}{\text{year}} \times \frac{1 \text{ g}}{3744 \text{ cal}} = 8 \times 10^{16} \text{ g/yr}$$

Table 10.1 NET PRODUCTION OF BIOMASS FOR A SUMMER DAY AND THE APPROXIMATE CONVERSION EFFICIENCY OF SOLAR ENERGY

Location	Plant Production ($\text{g/m}^2 \cdot$ day)	Conversion Efficiency (%)
Potential maximum	71	5
Polluted stream	55	4
Iowa cornfield	20	1.5
Pine forest	6	0.5
Wyoming prairie	0.3	0.02
Nevada desert	0.2	0.015

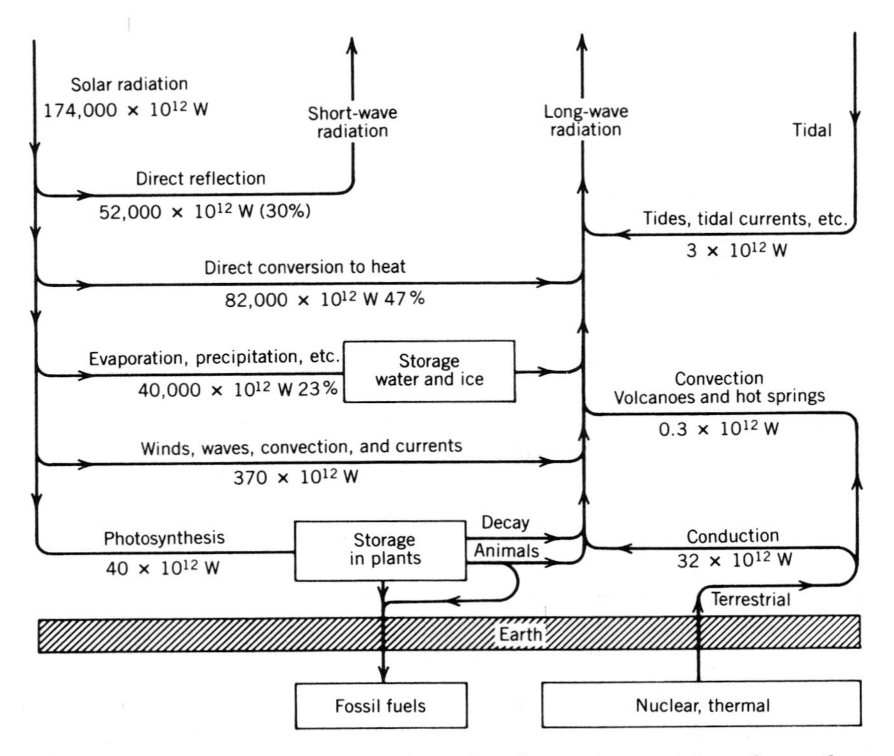

Figure 10.2 Natural energy flow (in units of power) to and from the earth.

Another estimate of the earth's yearly plant production is 320 $g/m^2 \cdot yr$ averaged over the total earth's surface. Since the earth has a radius of 6.37×10^6 m, the total production is

$$4\pi(6.37 \times 10^6)^2 \text{ m}^2 \times 320 \text{ g/m}^2 \cdot \text{yr} = 16 \times 10^{16} \text{ g/yr}$$

The factor of 2 difference in these two numbers is characteristic of the uncertainties of this type of information.

A more extensive list of the net biomass production levels in terms of $g/m^2 \cdot yr$ is presented in Table 10.2 for various types of plants in a number of locations around the world. The reason the world average is so low is that 70% of the earth consists of oceans, which produce an average of 100 $g/m^2 \cdot yr$.

Before addressing the question of how many people the earth can feed, we need to know the amount of food each person requires and how much yearly gross plant production is actually available to meet that need.

10.2 HUMAN FOOD REQUIREMENTS

We must eat to survive. Our food has stored chemical energy that is converted, for the most part, into thermal energy to keep our body temperature at the requisite level. A smaller amount of the chemical energy is given over to the performance of work by the use of muscles. The chemical reactions that release

Table 10.2 NET BIOMASS PRODUCTION LEVELS FOR NATURAL AND AGRICULTURAL ECOSYSTEMS

	Net Production ($g/m^2 \cdot yr$)
Natural Ecosystems	
Temperate terrestrial zone	1195
Oak–pine forest (New York)	1350
Beech forest (Denmark)	1450
Scotch pine (England)	1600
Grassland (New Zealand)	3200
Tropical terrestrial zone	
Forest (West Indies)	6000
Oil-palm plantation (Congo)	3700
Forest (Ivory Coast)	1340
Freshwater	
Freshwater pond (Denmark)	950–1500
Sewage ponds (California)	5600
Cattail swamp (Minnesota)	2500
Marine	
Algae (Denmark)	260–340
Seaweed (Nova Scotia)	2000–2600
Algae on coral reef (Marshall Islands)	4900
Open ocean (average)	100
Coastal zone (average)	200
Upwelling areas (average)	600
Agricultural Ecosystems	
Temperate Zone	
Corn (Minnesota)	1390
Corn (Israel)	3600
Corn (U.S. average)	2500–4000
Rice (Japan average)	1000–1200
Tropical Zone	
Sugarcane (Hawaii)	7200–7800
Sugarcane (Java)	9400
Rice (Ceylon average)	340–550
Rice (West Pakistan average)	560–700

Source: Data from George M. Woodwell, "The Energy Cycle of the Biosphere," *Scientific American*, September 1970.

this stored energy are basically the same as those of photosynthesis in reverse. For example, for glucose the reaction is

$$C_6H_{12}O_6 + 6O_2 \rightarrow 6CO_2 + 6H_2O + \text{energy}$$

These reactions take place in the cells of our body and are called metabolic reactions. For someone at complete rest, the rate of heat production is called the basal metabolic rate, and it is dependent on the characteristics of the individual as well as the surrounding temperature. For most of us the basal metabolic rate is generally between 65 and 85 W. This corresponds to about 1340 to 1750 kcal a day. Unfortunately, people concerned with nutrition use the word *Calorie* (or food calorie) for what in physics is a kilocalorie (10^3 calories), so one frequently sees these written as 1340 and 1750 Cal a day.

Example 10.2

If a person has a basal metabolism rate of 80 W, to how many kcal a day does this correspond?

Solution

$$80 \text{ W} = 80 \frac{\text{J}}{\text{sec}} \times \frac{1 \text{ kcal}}{4184 \text{ J}} \times \frac{3600 \text{ sec}}{1 \text{ hr}} \times \frac{24 \text{ hr}}{1 \text{ day}} = 1652 \text{ kcal/day}$$

It is sometimes forgotten in the design of heating or air conditioning systems for buildings that when many people congregate the combined effect of their body heat (metabolic rate) can represent significant thermal power. For example, in a lecture hall holding 300 students with a metabolic rate of 150 W each, there is 45 kW of thermal power. The problem in lecture halls is frequently one of removing heat rather than providing it.

The question of how many kilocalories must be eaten every day to carry on normal activities varies greatly with the circumstances. Table 10.3 indicates the energy expenditure per hour for various kinds of activities. Someone sleeping 12 hours and doing office work for 12 hours would need a total of 2580 kcal a day. A person chopping wood for 8 h would require 3600 kcal and sleeping 16 hours would require 1120 more for a total of 4720 kcal. The numbers displayed in Table 10.3 are only estimates, and are affected by the temperature of the environment, the weight of the individual, and, of course, the vigor with which one chops wood, for example. What is perhaps as relevant as the calories we need are the calories we actually consume.

Table 10.3 THE RATE AT WHICH A PERSON USES ENERGY FOR VARIOUS ACTIVITIES[a]

	Rate of Energy Use	
	kcal/hr	W
Sleeping	70	81
Office work	145	168
Walking slowly (3 km/hr)	170	198
Bicycling on the level (10 km/hr)	210	244
Walking briskly (6 km/hr)	300	349
Chopping wood	450	523
Skiing on the level (5 km/hr)	550	639
Running on the level (9 km/hr)	720	837
Running on the level (18 km/hr)	1300	1511

[a] Note that the rate of energy use is much greater than the rate at which a person can do work because of the limited efficiency with which a person can convert food energy into mechanical work. Most of the energy used is lost as heat.

Table 10.4 PER CAPITA FOOD SUPPLY BY ECONOMIC CLASS AND MAJOR REGION AND COUNTRY (1977)

	Per Capita Calorie Supply (kcal/day)			Per Capita Protein Supply (g/day)		
Economic Class and Region	**Vegetable Products**	**Animal Products**	**Total**	**Vegetable Products**	**Animal Products**	**Total**
Developed Countries						
United States	2266	1312	3578	33.1	73.3	106.4
Canada	1938	1429	3368	35.1	66.2	101.3
Western Europe	2267	1109	3376	40.7	54.1	94.8
South Africa	2507	414	2921	49.8	27.3	77.1
Japan	2399	547	2946	45.6	42.4	88.0
Oceania	2034	1364	3398	34.0	73.5	107.5
Developing Countries						
Latin America	2111	446	2557	38.6	26.8	65.5
Far East (excluding Japan and the People's Republic of China)	1914	114	2029	41.2	7.4	48.7
Near East	2372	249	2620	57.9	15.5	73.5
Africa (excluding South Africa)	2060	146	2205	44.1	10.9	55.0
Centrally Planned						
Eastern Europe and Soviet Union	2492	989	3481	51.7	51.6	103.3
China	2141	246	2386	51.0	11.4	62.5
World	2136	435	2571	44.8	23.9	68.8

Table 10.4 lists the kilocalories obtained per day from vegetable and animal products in various countries and groups of countries. The United States, with 3578 kcal a day, has the highest consumption rate of any group. When more kilocalories are taken in each day than are required for daily activity, the excess tends to accumulate as body fat. Many Americans and others are mindful of this fact and are restricting their calorie intake or are burning up more calories by jogging, chopping wood, or doing other forms of exercise.

On the other hand, there are other countries where the calorie intake is not sufficient to enable some to carry on normal productive daily activities. In Africa, for example, where on average 2205 kcal a day are consumed, many individuals are receiving not much more than the 1500 kcal a day or so needed for basal metabolism and, hence, are unable to work effectively.

There is more to human nutrition than just counting calories. The body is a complicated chemical processing plant and needs, in addition to calories and water, amino acids, fatty acids, various vitamins, and minerals. Protein is important to the human diet because it can provide needed amino acids. We each need about 70 g of protein a day for a normal balanced diet. This protein can come from the consumption of animal products and certain vegetable products, such as soybeans, that have a high protein content. As Table 10.4 shows, people in the United States and Oceania (Australia and New Zealand) have the highest protein intake mainly because of the high proportion of meat in their diet. It can also be seen that many regions (Latin America, the Far East, excluding Japan, and the Near East) have a level of protein intake that is marginal.

Hunger can be averted by sufficient caloric intake, but conquering malnutrition requires a balanced diet that includes sufficient protein and other necessary nutrients. Malnutrition can lead to scurvy, pellagra, beri-beri, and rickets, all of which are caused by certain nutritional deficiencies. Protein is more important to humans than it is to most animals. Herbivorous animals have the ability to synthesize their own amino acids from the plants they eat.

In attempting to assess the number of people the world can feed, it is important to recognize that protein is a necessary part of the diet and that, to a large extent, its source is animal products. The use of agricultural land to furnish animal products for human food is a very inefficient process. If beef cattle are fattened on corn, only a small percentage (≈10%) of the energy content of the corn will end up as body tissue on the animal. Only a small percentage (again ≈10%) of that will be utilized by a person eating the meat. The growing of grains to fatten cattle is a very luxurious use of land. There are, of course, land areas that are not suitable for cultivation of crops but that can serve for grazing cattle, sheep, or goats. In some studies of land devoted to world food production, equivalent vegetable calories are used that consist of the calories in the vegetables, grains, and cereals plus seven times the meat calories consumed. With such a combined figure, cultivated and grazing lands can be lumped together.

10.3 FEEDING THE WORLD'S POPULATION

We have examined the conversion of solar energy via photosynthesis into carbohydrates that serve as food. We have also assessed human food requirements. It would appear then to be a trivial exercise to take the known number of arable acres of the world with a certain annual yield of so many calories per year to arrive at an estimate of the number of people the earth can support. Complications arise, however, because the productivity of the land is not constant, but may increase as the use of irrigation water, fertilizers, herbicides, insecticides, and hybrid seeds increases. Nor is the number of arable acres constant. Some new land is being brought into production by clearing, and some presently used land is being taken out of production because of development (housing, roads, dams, mines, etc.), soil erosion, depletion, and salination. Because of these and other complications, the calculations must be at best only estimates.

One can attempt to calculate the food production per acre per year from the information presented above. We shall assume that the solar energy per unit area per day is 500 cal/cm^2 • day and that the efficiency for the conversion of solar energy into vegetable matter is 1.5% (this corresponds to the efficiency of an Iowa cornfield). The growing season varies widely, but we shall take a 120-day season that is something of an average between the colder and more tropical zones. The yield of vegetable matter for a year is then

$$500 \frac{\text{cal}}{\text{cm}^2 \bullet \text{day}} \times (0.015) \times 120 \frac{\text{day}}{\text{yr}} \times \frac{10^4 \text{ cm}^2}{1 \text{ m}^2} \times \frac{1 \text{ kcal}}{10^3 \text{ cal}} = 9000 \text{ kcal/m}^2 \bullet \text{yr}$$

In the metric system, 10^4 m^2 is called a hectare (ha), and most studies of world food production use this as the measure of land area. For our purposes a hectare is equal to 2.47 acres. With this conversion the above number becomes

$$9000 \frac{\text{kcal}}{\text{m}^2 \cdot \text{yr}} \times \frac{10^4\ \text{m}^2}{1\ \text{ha}} \times \frac{1\ \text{ha}}{2.47\ \text{acres}} = 3.6 \times 10^7\ \text{kcal/acre} \cdot \text{yr}$$

It is interesting to compare this calculation to the net production figures shown in Table 10.2. If we have 3700 cal/g, 9000 kcal/m^2 • yr is 2432 g/m^2 • yr, which is near the lower side of the 2500 to 4000 g/m^2 • yr indicated in Table 10.2 for average corn yields in the United States.

In computing how many people this acre will feed, let us assume that we are discussing cultivated cropland, and the protein component of the diet will come separately from grazing land. Let us also assume that one person requires 3000 kcal a day. Unfortunately, for almost all crops only a rather small fraction of the total biomass produced provides human food. Although this fraction depends largely on the crop involved, 5% is a typical value. For the *food* yield per acre per year

$$3.6 \times 10^7 \frac{\text{kcal}}{\text{acre} \cdot \text{yr}} \times (0.05) = 1.8 \times 10^6\ \text{kcal/acre} \cdot \text{yr}$$

For the food requirement for one person per year

$$3000 \frac{\text{kcal}}{\text{day}} \times 365\ \text{day/yr} = 1.1 \times 10^6\ \text{kcal/yr}$$

Taking the ratio of these two numbers indicates that 1 acre of cropland can support about 1.6 people.

The total land area of the world is about 32×10^9 acres. About one fourth (8×10^9 acres) is potentially arable, and about one fourth (another 8×10^9 acres) can be used for grazing. Roughly one half of the total area cannot be used for agriculture because it is wasteland, tundra, desert, and mountainous terrain. There are now about 3.5×10^9 acres under cultivation and approximately 5×10^9 acres used for grazing. The population of the world in 1990 was 5.3×10^9 people, which means that 0.7 acre per person was used for crops and 0.9 acres per person for grazing. The 0.7 acre per person corresponds to about 1.4 persons per acre, and this is reasonably consistent with the 1.6 persons per acre just calculated.

Assuming no change in the level of production per acre and that the amount of food available per person stays the same, it would appear that there is enough land available to support a population about twice the present size, or about 10.6×10^9 people. Unfortunately, not everyone is now being fed adequately, and it is not clear that there is enough water and fertilizer to sustain a yield for the new area put into production as high as that of the present land area. On the other hand, there has been a trend toward higher yields as more sophisticated agricultural techniques are introduced, especially in the developing countries.

Great advances have been made in recent years in food production. The compounded annual rate of growth of world food production was 2.6% from 1952 to 1979, considerably faster than the rate of population growth. In the earlier part of this period, the growth was to a large extent due to an expansion of the cultivated areas, primarily in the developing countries. In the last 20 years or so, the increases have come about mainly through increased yields brought about by increased use of fertilizer, use of hybrids and other high-yielding seeds, increased irrigation, pest control, and improved cultivation and harvesting practices.

The average corn yield in 1933 in the United States was 22.6 bushels per acre. This has increased rather uniformly with time, and in 1979 it was 100 bushels per acre. This kind of dramatic increase in yield has not been the general rule; nevertheless, food production in the developed countries has more than doubled since 1950, mostly due to an increase in yield. In recent years it has not been possible to sustain the historic increase rate for yields in the developed countries, and, in fact, in the last few years there has been a slight decrease in production. The developing countries have been slower in adopting the agricultural practices that lead to increased yields, but under programs such as the cooperative program between the Rockefeller Foundation and the Mexican government, significant progress has been made. Under this program wheat production in Mexico increased by a factor of 6.7 from 1945 to 1964. Overall yields in the developing countries are now 70% of the yields in the developed countries.

The question of how many people can exist on the earth has been the subject of much discussion over the years. In 1798, Thomas Malthus, an English clergyman and economist, published an essay stating that since populations tend to increase more rapidly than their food production, there would eventually have to be some disaster to reduce the population unless some form of planned parenthood were widely practiced. Malthus envisaged famine, war, and disease as the agents that would limit population. Under normal conditions, populations have tended to grow exponentially, that is, by a certain percentage increase every year. This type of growth, as we have seen, means that the absolute number of people added to the population annually always increases. The world population now has a 1.8% growth rate, which corresponds to a doubling time of 38.5 years. Food production, on the other hand, might increase for some limited time as long as land and other resources are available, but there is no reason to expect that it will increase exponentially over a long period.

For all of mankind's history on this planet, the human birth rate has exceeded the death rate, except for temporary situations when the population may have actually declined. We can now observe with virtual certainty that we are rapidly approaching the situation where the long-term death rates must at least equal birth rates, meaning no more continued population increase. This situation could be reached within one or two generations. The limiting factors on human population are still basically those considered by Malthus.

There have been many estimates of the ultimate possible population of the earth by people in various fields. From the information available, it is clear that the question has no exact answer, as there are too many variables that are poorly known. In addition to the complications discussed so far, there are others that relate to food from the sea, climatic changes, and a multitude of political and economic factors. A number of forecasts have projected the world's eventual population at about two or three times the present level. This would occur in some 40 to 80 years at the present growth rate. It is far from clear that such an expansion of the population can be considered desirable, since we are not adequately feeding the population we have now. During the length of time it takes you to read this sentence, four people will die from starvation or malnutrition. It is also difficult to imagine adding 100 million more people per year when we now have just a 40-day reserve of grain worldwide. In 1962 there was a 105-day reserve.

It is, of course, not certain that the limitation of the world's population will be brought about by a limit in the land available for food production. A series of critical resources are in short supply, including the energy needed for high-yield farming. Water is also limited. The consequences of a tripling of the population on air and water pollution, soil erosion, and political stability are all of great concern.

10.4 FUEL FROM BIOMASS

Until about 1880, the main source of energy for heating, transportation, and industrial processes in the United States was wood (see Figure 1.13). Although the industrialized countries now primarily rely on fossil fuels as an energy source, wood as well as agricultural and animal wastes are still major sources of energy in the developing countries. Is it possible for developed countries to return to this renewable source of energy for their basic energy needs? The process of using biomass as fuel is basically an alternative form of solar energy use; the energy storage problem is nicely solved through the production of vegetable matter by photosynthesis.

Wood and other dry vegetable matter can, of course, be directly burned as a source of heat energy (Figure 10.3). The vegetable matter could be purposely cultivated as a fuel, or it could be agricultural waste, municipal waste, sewage, algae, or seaweed from the oceans or freshwater ponds or lakes. It is also possible to convert vegetable matter into gaseous or liquid fuel as a substitute for natural gas, oil, and gasoline. The most publicized example of this general conversion process is gasohol, a mixture of 10% grain-based ethanol and 90% gasoline.

Before examining the details of the conversion processes, it is informative to obtain an overview of how much fuel energy could be provided by biomass each year in the United States and to compare this with the country's energy needs. One pound of dry plant material will yield about 7500 Btu when burned directly. This is equivalent to 4332 cal/g, which is quite close to the value for sucrose calculated earlier. The yield in terms of $g/m^2 \cdot yr$ is, of course, highly variable depending on the location, rainfall, type of crop, and agricultural practices. If we assume a value of 3300 $g/m^2 \cdot yr$ (15 tons/acre • yr), which is in the middle of the range for corn in the United States (see Table 10.2), the annual yield for all of the U.S. land harvested (350×10^6 acres) is

$$\frac{15 \text{ tons}}{\text{acre} \cdot \text{yr}} \times 350 \times 10^6 \text{ acres} \times \frac{7500 \text{ Btu}}{1 \text{ lb}} \times \frac{2000 \text{ lb}}{1 \text{ ton}} = 79 \times 10^{15} \text{ Btu}$$

We saw in Chapter 1 that the total energy used in the United States in 1990 was 81×10^{15} Btu. Although the comparison is far from exact, on the basis of the above calculation we could expect to furnish essentially all our energy from cultivated crops. Grazing and forest lands have been neglected in the calculation; however, because yields for these areas are low, they are not going to make major contributions. There are, of course, many problems with de-

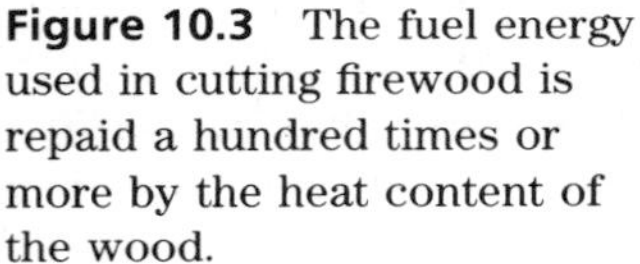

Figure 10.3 The fuel energy used in cutting firewood is repaid a hundred times or more by the heat content of the wood.

pending totally on biomass for energy. Our energy needs are for either electricity or a combustible gas that can be easily transported to homes for heating and cooking and for a liquid fuel that can be used for heating as well as for transportation. To convert the dry vegetable matter to liquid and gaseous fuels is somewhat inefficient, and it is certainly a complex and costly process at present. It may be better to burn the biomass in electric power plants. Using all of our cropland for fuel obviously presents a problem for food production, but recall that only 5% of a plant's biomass is suitable as human food. The magnitude of the biomass potential is significant compared to our energy needs, and it is worthy of closer study because it is a renewable source of energy.

The process of gasohol production will be discussed as an explicit example of converting biomass into a liquid fuel. Although other grains can be used, the principal crop utilized in the United States has been corn. After harvesting, the entire stalk and cobs are chopped up, ground, and mixed with water. The resulting material is then cooked to help convert the starches into sugars by enzymatic action. The sugars are then converted in a fermentation process to ethanol. Distillation removes the ethanol from the rest of the material; it then is blended, usually with unleaded gasoline, to make a product directly usable in unmodified auto engines. Figure 10.4 shows diagrammatically the process of conversion as well as the ways in which fossil fuel energy enters into the

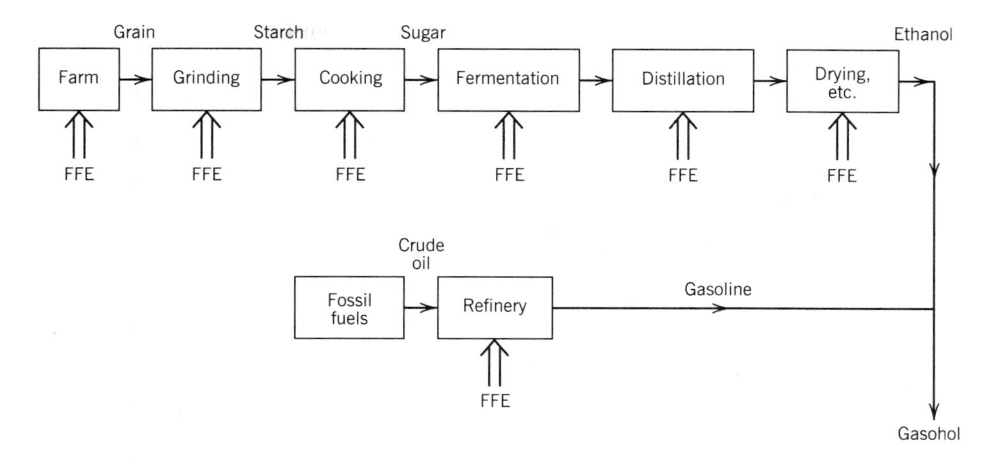

Figure 10.4 Schematic diagram of the production of gasohol from grain such as corn. In addition to the ethanol product, the stalks and cobs as well as by-products of the fermentation and distillation process have some economic value. Fossil fuel energy (FFE) is required for almost every stage of the process, including steps not indicated such as transportation.

various stages. Because there is usually a sizable surplus of corn in the United States, fuel tax exemptions have been provided by a number of states to encourage the use of corn to produce gasohol to augment the domestic supply of gasoline. Because of these subsidies, it has been unclear whether gasohol would be economically viable if left to a free market. It has also been unclear whether more Btus of liquid fuel are consumed in the process than are made because of the large use of fossil fuels in agriculture. The answer to these questions is apparently that gasohol is close to an energy break-even point, and much depends on whether just agricultural wastes are used and the details of the conversion techniques.

Brazil, which has very limited oil reserves, has made a major effort to shift to ethanol as a transportation fuel. Sugarcane, which is widely grown there, can be converted to ethanol in the fashion shown in Figure 10.4. Both methanol and ethanol have high octane ratings. To take advantage of this, engines must be modified to run with a higher compression ratio than with normal gasoline. Automobiles manufactured in Brazil are predominantly made to run efficiently on methanol or ethanol, and nearly one third of the 10 million cars there run on hydrated ethanol (192 proof). It has been estimated that if 2% of the land area in Brazil were devoted to sugarcane for ethanol, all the imported petroleum could be replaced.

In addition to producing a liquid fuel, biomass can also be converted into a very usable gaseous fuel, methane. Several processes are used, but the one that is most direct is simply the fermentation of organic matter by the action of bacteria in the absence of oxygen (anaerobic fermentation). The organic material used can be crops, agricultural waste (either vegetable or animal), waste from lumber mills, breweries, or other industries, algae, sludge from sewage treatment plants, or waste from municipal disposal sites. In the presence of water and absence of oxygen, such organic matter will ferment naturally, and 60 to 80% of the carbon in the organic material is converted into

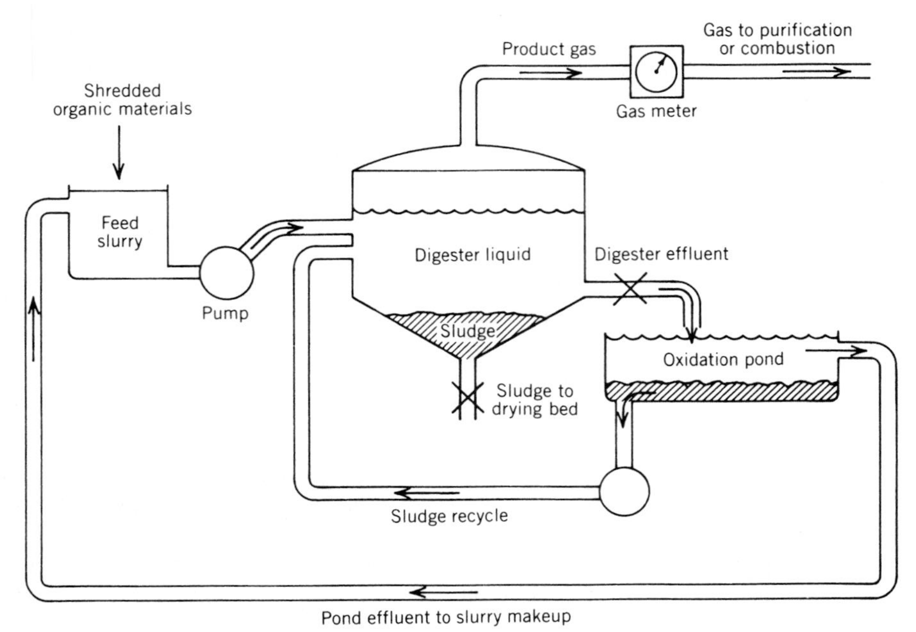

Figure 10.5 A unit for the continuous conversion of biomass by anaerobic fermentation into methane and other combustible gases. (Source: *Solar Energy as a National Resource,* NSF/NASA Solar Energy Panel, December 1972.)

carbon dioxide, methane, and some hydrogen sulfide and nitrogen. Figure 10.5 shows schematically the process involved. After purification the volatile gases are almost pure methane, with a heating value of about 1000 Btu/ft^3. One pound of dry organic material will produce 4.5 to 6.5 ft^3 of methane. This is 9 to 13 $\times$ 10^6 Btu per ton.

Such gaseous conversion processes have been in operation for some time at places such as sewage treatment plants and breweries, and they have provided a local but very useful source of methane. Although the conversion process can be costly, the efficiency (the ratio of energy in the product fuel to the energy in the incoming biomass) is usually reasonably high (50 to 70%); and if the organic material is waste that otherwise would be disposed of, there is a good chance that the economics will be favorable. It does not appear that growing crops especially for methane production will be competitive with natural gas, however, as long as the domestic supply of natural gas holds up.

An important but frequently overlooked aspect of agriculture in developed countries is its growing dependence on fossil fuel energy. In many ways a modern farm uses land to convert oil, natural gas, and coal into food, and the energy input from the sun is but a part of the picture. In the United States in 1900, one farmer was needed to feed five people. In 1974, 50 people were fed by one farmer. The 10-fold increase came about through the increased use of machinery, irrigation, pesticides, fertilizers, and better seeds, and all of these ingredients use large amounts of energy. The production and delivery of these ingredients also employ large numbers of people who should properly be included within the agricultural enterprise along with the farmers. The amount of food consumed in the United States increased from 150 $\times$ 10^{12} to about 300

$\times 10^{12}$ kcal/yr from 1940 to 1970. During this same period, the energy input into the food system went from about 670 $\times 10^{12}$ to 2250 $\times 10^{12}$ kcal/yr. The total energy subsidy to the American food system, including farming, packaging, distribution, and machinery, is indicated in Table 10.5. The subsidy is represented as a ratio of equivalent fuel energy to food energy produced. This heavy reliance on energy for high-yield farming has important consequences for considerations of ultimate food production as well as the net energy gain in the use of biomass to produce gaseous and liquid fuels.

10.5 BURNING OF MUNICIPAL WASTES

Solid municipal waste is a growing problem in the United States and the rest of the world. Current landfill sites are rapidly being filled and the availability of new sites is very limited, especially near cities. In addition to site availability problems, there are problems of groundwater contamination, odor, disease, toxic chemicals, and escaping methane. The waste of valuable mineral resources due to lack of recycling is also a significant problem. About 1000–1500 lb/yr of solid refuse is generated every year per person in the United States. Every day about 7.5 $\times 10^5$ m^3 of waste is added to the landfills in the United States that are still available.

A major fraction of the municipal wastes is the result of photosynthesis: paper (56%), food wastes (9.2%), plant waste (7.6%), and wood (2.5%). Glass (8.5%), plastic (3.5%), and metals (7.5%) make up the bulk of the rest. It is evident that sooner or later a system of separation and recycling of municipal waste will be needed with an environmentally accepted burning of the combustible fraction. The average heat content of waste is about 4300 Btu/lb (10,000 kJ/kg), which is about 32% that of good coal. While disposing of the waste is the primary goal, sufficient heat will be released by burning waste to make a worthwhile contribution to steam-generated electric power (see Problem 17). The incinerators must be carefully designed to dry the waste in addition to burning it. This is usually accomplished by injecting air into the firebox from beneath the grate to provide oxygen for complete burnout. The hot exhaust

Table 10.5 FUEL ENERGY SUBSIDY TO THE AMERICAN FOOD SYSTEM

Year	Fuel Energy Input / Food Energy Produced
1910	0.8
1920	1.5
1930	2.5
1940	4.4
1950	6.2
1955	6.4
1960	6.6
1965	8.0
1970	8.7

Source: Steinhart and Steinhart, 1974.

gases must be cleaned by electrostatic precipitators and/or bag houses to remove particulates, combined with wet or dry scrubbers for acid gases before they are released to the atmosphere. These processes are discussed in Chapter 13.

PROBLEMS

1. How many food calories (kcal) need to be eaten every day in order for a person with a basal metabolic rate of 75 W just to stay alive?
2. If the world population continues to increase at 1.8% per year, how many years will it take it to double?
3. Calculate the number of food calories (kcal) needed by a person in one 24-hour period if the person sleeps (70 kcal/hr) for 8 hours, hikes strenuously (400 kcal/hr) for 6 hours, and operates at 140 kcal/hr for the rest of the day.
4. A backpacker plans to take 4000 food calories for each day. She wants to get half her calories from fat (4100 kcal/lb) and half from carbohydrates and protein (both at 1800 kcal/lb). How many pounds of food must she carry for a 10-day trip?
5. The total energy consumption in the United States in 1990 was about 81×10^{15} Btu (81 QBtu) per year. From the following information, calculate the fraction of this energy that is used for food production. Population $= 250 \times 10^6$. Average food energy intake/person = 3578 kcal/day. The fuel energy input in the United States is about 8.7 times the food energy produced. Neglect imports, exports, spoilage, etc.
6. Glucose has the chemical formula $C_6H_{12}O_6$ and stores 674 kcal/mole. How many kilocalories are there in 50 g of glucose? The mass number of H is 1, C is 12, and O is 16.
7. A large, vigorous, university athlete consumes 5000 kcal a day, 365 days a year. Meat from corn-fed cattle provides 40% of these calories and vegetables the remaining 60%. Assume that 0.01% of incident solar radiation is available to an individual via the food chain through eating beef, and 0.1% is available through eating vegetables. How many hectares of cropland does it take to support this one individual for 1 year if the growing season is 110 days and if the incident solar energy equals 500 cal/cm^2 a day during the growing season? Note: 1 hectare = 10^4 m^2, or 2.5 football fields.
8. Discuss quantitatively the history of the U.S. food system in terms of the number of calories of fuel energy required to produce and deliver 1 calorie of food for human consumption over the past 75 years. Give reasonable explanations for the trends observed on a decade-by-decade basis. What would be an enlightened projection of this trend for the next 50 years?
9. Approximately how many sedentary persons would be required to maintain the temperature of a building by their body heat during a period when no electricity is available if the building is normally heated by a 12-kW electric heater operating continuously?
10. It is known that the temperate latitudes have more hours of sunlight per 4-month growing season than do the equatorial latitudes. Could this explain the higher potential photosynthetic yields in temperate latitudes as compared to equatorial regions during the 4-month summer?
11. List several factors other than the lack of available solar energy that often limit the productivity of land for growing food crops.
12. Discuss the probable wisdom of relieving the petroleum fuel shortage in the United

States by raising large amounts of corn by currently accepted U.S. agricultural methods and then burning the corn plants to generate electricity rather than burning oil in electricity-generating plants.

13. It has been frequently proposed that the United States could relieve the food shortage in countries such as India by exporting our advanced food production technology and our high-yield strains of food grains. Discuss the feasibility of this approach.

14. Discuss the factors that may cause the world's human population growth patterns eventually to depart from the past trends of increasingly rapid growth. Include in your discussion factors imposed by nature and by man's possible deliberate action.

15. Calculate the maximum number of people that could be provided with food under the following conditions: 1.5×10^9 ha cropland presently available worldwide; 400 cal/cm^2 incident solar radiation per day for a 120-day growing season each year as an average value on this cropland; 0.1% conversion efficiency of solar energy to human food; and 2500 kcal a day average human requirements.

16. Estimate the forest area needed to supply fuel continuously for a 1000-MW_e power plant. Note that this will require about three times as much thermal input as electrical output.

17. Estimate the number of Btus that would be generated annually in the United States if all municipal waste were incinerated. Assume 1000 lb per year per person of the waste is burnable at 4300 Btu/lb.

SUGGESTED READING AND REFERENCES

1. Barr, Terry N. "The World Food Situation and Global Grain Prospects." *Science* **214** 4524 (December 4, 1981), pp. 1087–1095.
2. Brinkworth, B. J. *Solar Energy for Man.* New York: John Wiley, 1972.
3. Brown, Lester R. "World Population Growth, Soil Erosion and Food Security." *Science* **214** 5424 (November 27, 1981), pp. 995–1002.
4. Chambers, R. S.; Herendeen, R. A.; Joyce, J. J.; and Penner, P. S. "Gasohol: Does It or Doesn't It Produce Positive Net Energy?" *Science* **206** 4420 (Nov. 16, 1979), pp. 789–795.
5. Gates, David M. "The Flow of Energy in the Biosphere." *Scientific American* **224** 3 (September 1971), pp. 88–100.
6. Harte, John, and Socolow, Robert H. *Patient Earth.* New York: Holt, Rinehart & Winston, 1971.
7. Loftness, Robert L. *Energy Handbook.* New York: Van Nostrand Reinhold, 1978.
8. *Man and the Ecosphere.* Readings from the *Scientific American.* San Francisco: W. H. Freeman, 1971.
9. Committee on Resources and Man, National Academy of Sciences, National Research Council. *Resources and Man.* San Francisco: W. H. Freeman, 1969.
10. Steinhart, Carol E., and Steinhart, John S. *Energy—Sources, Use and Role in Human Affairs.* North Scituate, Mass.: Duxbury Press, 1974.
11. Turiel, Isaac. *Physics—The Environment and Man.* Englewood Cliffs, N.J.: Prentice-Hall, 1975.
12. Woodwell, George M. "The Energy Cycle of the Biosphere." *Scientific American* **223** 3 (September 1970), pp. 64–74.
13. Mills, G. A., and Ecklund, E. E. "Alcohols as Components of Transportation Fuels." *Annual Review of Energy* **12** (1987), pp. 47–80.

14. Penner, S. S., Wiesenhahn, D. F., and Li, C. P. "Mass Burning of Municipal Wastes." *Annual Review of Energy* **12** (1987), pp. 415–444.

15. Aubrecht, G. *Energy.* Columbus, Ohio: Merrill Publishing Co., 1989.

16. Scrimshaw, Nevin S., and Taylor, Lance. "Food." *Scientific American* **243** 3 (September 1980), pp. 78–88.

17. Crosson, Pierre R., and Rosenberg, Norman J. "Strategies for Agriculture." *Scientific American* **261** 3 (September 1989), pp. 128–135.

CHAPTER ELEVEN

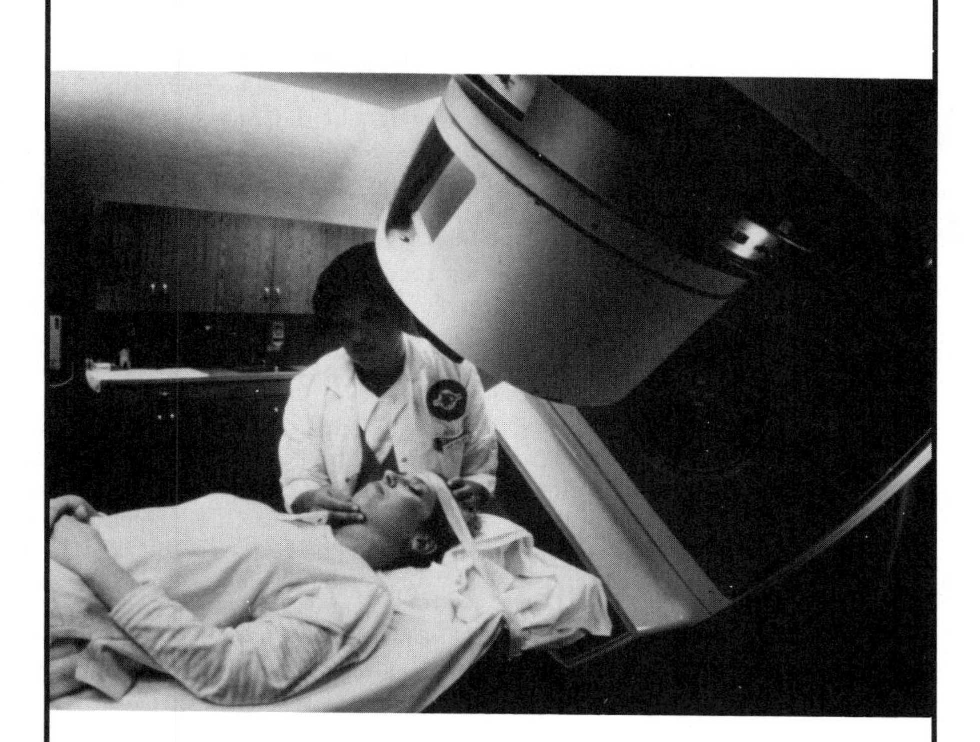

RADIATION AND MAN

11.1 INTRODUCTION

Ionizing radiation is, by definition, radiation capable of ejecting one of the bound electrons from an atom or molecule, thereby forming a positive ion—an atom or molecule that has one less electron than it does in its neutral state. If this atom or molecule is part of a biological system and the ionizing event leads to the breaking of molecular bonds, severe consequences can result, such as the scrambling of the genetic information carried by genes or the changing of normal cells into cells that can eventually become cancerous. There are many types of ionizing radiation: electromagnetic radiation of short wavelength; energetic electrons or positrons; energetic heavy charged particles such as protons, deuterons, and alpha particles; and many unstable particles such as muons and pions. Even uncharged particles such as neutrons can cause ionizing events by two-step processes in which some of the neutron kinetic energy is transferred to atomic nuclei that then recoil and create ionizing events.

Non-ionizing radiation (electromagnetic radiation with wavelength longer than that of ultraviolet light, sound waves, laser beams, etc.) is of lesser concern because it cannot disturb biological cells in such a destructive way as does ionizing radiation.

Man* has evolved in the presence of ionizing radiation that comes continually from cosmic rays and from the natural radioactivity in the ground and atmosphere. Although the effects of such radiation have been generally harmful, they have not been of sufficient magnitude to thwart the development of the human race or innumerable animal and plant species. In fact, radiation has caused some of the mutations needed for the evolutionary process to proceed as it has.

In the last 80 years, however, a number of different sources of ionizing radiation have been introduced by man. The exposure of the average person in the United States to man-made radiation at present is at roughly the same level as the exposure to radiation from natural sources. Some people receive exposures many times that of the natural background levels, and the average exposure is increasing with time.

It is important to understand the effects of these increased radiation levels, both with regard to long-term genetic consequences and the more immediate influence on cancer incidence.

The beginnings of man-made radiation occurred in 1895 when Professor Wilhelm Roentgen of the University of Würzburg in Germany discovered that radiation from a cathode-ray tube could penetrate a thousand-page book and expose a photographic plate on the other side. Within a few months of this discovery, x-rays were used to locate lead pellets from a gunshot wound in a man's hand. Rapid and widespread use of x-rays in various medical applications followed with little understanding of the harmful side effects that could be brought about by such radiation. At first, the known harmful effects were related to erythema (a reddening of the skin where the x-rays were incident), but later, in 1911, tumors were noted in people using x-rays, particularly radiologists. In 1949, it was noted that the death rate from leukemia (cancer of

* Here and throughout this book, the word *man*, and derivatives such as *man-made*, are intended to refer to the human race as a whole.

the blood) was nine times greater among radiologists in the United States than among other physicians.

In 1896, Henri Becquerel first reported penetrating radiation being emitted spontaneously from pitchblende, a uranium ore, and thus the discovery of natural radioactivity was announced. Shortly thereafter Marie and Pierre Curie achieved the concentration of radium and went on to do a variety of fundamental experiments delineating the nature of radium and its radioactivity. Both the Curies and Becquerel developed tumors from the radiation from natural radioactivity. Use and misuse of naturally occurring radioactivity was made in a number of areas in the years that followed.

The production of man-made radioisotopes became possible with the development of the cyclotron in 1931 by Lawrence and Livingston. The availability of energetic charged particles (mainly protons that had an energy of several MeV or more) allowed the production from stable isotopes of a wide variety of radioisotopes with half-lives from a fraction of a second to years. Radioisotopes could be produced with such accelerators, but at the same time prompt radiation in the form of neutrons and gamma rays presented health hazards to the people working around the accelerators.

As was discussed in Chapter 4, a milestone was passed in 1939 with the discovery of the fissioning of uranium and its application to nuclear reactors and, shortly afterward, to nuclear weapons. With the operation of the first reactor in 1942 by Enrico Fermi and his coworkers, it not only became possible to have a prolific source of neutrons, but also the reactor permitted a whole new set of artificial radioisotopes to be made that could not be made in practical quantities with charged particles from cyclotrons and other accelerators. The success of the Manhattan Project during World War II in developing reactors and in producing the nuclear weapons that were exploded in New Mexico and in Nagasaki and Hiroshima, Japan, in 1945 ensured that mankind and ionizing radiation were going to be intimately involved for many years. The genie was out of the bottle and in all likelihood would never be put back.

The subsequent years have seen the exposure of people to fallout from nuclear weapons testing, a sharp increase in the use of x-rays and radioisotopes in medicine and industry, extensive mining and milling of uranium, the involvement of many people with the handling of radioisotopes in connection with nuclear reactors and nuclear weapons fabrication, and the exposure of the general public to the radioactive emissions from nuclear reactors. These radiation exposures are in addition to those from natural radioactivities contained in building materials and released by coal burned in power plants. A vastly greater potential exists for the large-scale exposure of millions of people from the large inventory of nuclear weapons and the political instabilities of the world.

The ionizing radiation that we are concerned with in this chapter is quite diverse. Energetic quanta of electromagnetic radiation (x-rays or gamma rays) are of concern because of the common use of x-ray machines in medicine and industry, and because most of the radioisotopes emit gamma rays in their decay process. X-rays are generally produced by electrons accelerated by a high voltage in a vacuum striking a target or anode and being slowed down very abruptly. Figure 11.1 shows the arrangement schematically. When heated, the filament on the left will emit electrons that are accelerated by the potential difference of V volts from the filament (cathode) to the anode. The x-rays

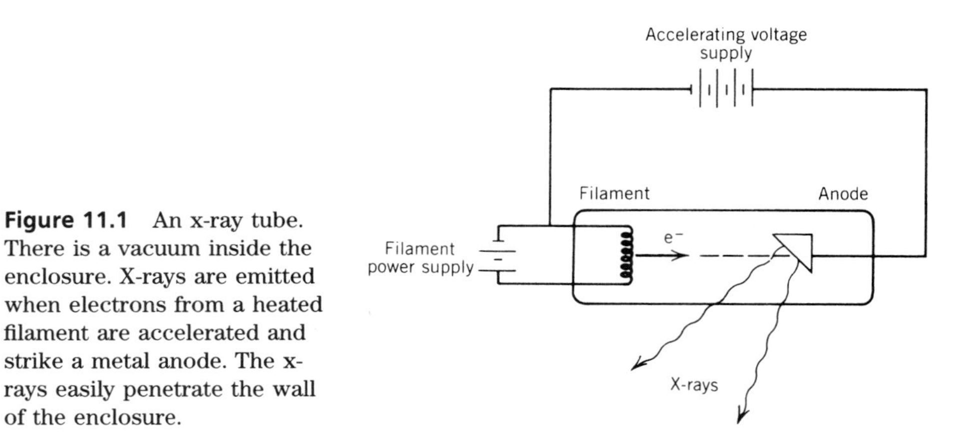

Figure 11.1 An x-ray tube. There is a vacuum inside the enclosure. X-rays are emitted when electrons from a heated filament are accelerated and strike a metal anode. The x-rays easily penetrate the wall of the enclosure.

emitted from the anode will have an energy ranging up to the full kinetic energy of the electrons. An x-ray tube using a potential difference of 100 kV will emit x-rays with energies up to 100 keV, which will be capable of causing many ionizing events. X-rays and gamma rays are generally very penetrating. X-rays of 100 keV will penetrate 6 cm of water, and x-rays of 1 MeV, 30 cm of water. Penetration depths in human tissue are similar.

Radioactive nuclei, as discussed in Chapter 4, can also emit electrons that have either a positive or negative charge and are called beta rays. Beta rays can have energies up to about 10 MeV, but most have energies more in the range of 1 MeV. A 1-MeV beta particle will penetrate only 0.4 cm of water or human tissue.

The other particles of concern include the heavy charged particles such as protons, deuterons, and alpha particles. Alpha particles are of particular interest because they are frequently emitted in the decay of radioisotopes heavier than lead. For example, ^{239}Pu, an important radioisotope for nuclear reactors and nuclear weapons, emits 5.16 MeV alpha particles, which only penetrate about 0.005 cm into water. Their range is short because they interact very strongly in matter, depositing their considerable energy in a small volume, creating a densely ionized region. There are heavier charged particles such as the nuclei of carbon and oxygen and so forth, and a whole series of unstable particles such as muons and pions that are associated with cosmic rays and high-energy accelerators. These latter two categories of particles can cause biological damage, but they are not common enough to be of major concern except around accelerators or in cosmic radiation.

There are also two common neutral particles, the neutrino and the neutron. Having either no mass or a very small mass, the neutrino does not cause appreciable ionization; in fact, it can pass through the entire earth without interacting. The neutron, however, with essentially the same mass as a proton, can cause significant biological damage because it scatters from the nuclei of atoms such as hydrogen, and the recoiling nuclei, being charged, can cause many ionizing events. Neutrons are found in nature only as a component of the cosmic radiation, but they are common around accelerators and nuclear reactors and are emitted copiously in nuclear weapon explosions.

There appears to be a general lack of understanding on the part of the public concerning the effects of ionizing radiation. On one hand, there seems to be an unnatural fear on the part of some whenever the words radiation or radioactivity are mentioned. On the other hand, many people are quite unconcerned about the effects of x-rays and even the effects of nuclear weapons. The essential facts about the effects of ionizing radiation on people are far better established than are the effects of many chemicals and other carcinogens to which the public is routinely exposed. To discuss the effects of ionizing radiation in a quantitative way, we must first define several quantities that relate to radiation intensity and the nature of its reactions in biological material.

11.2 UNITS OF EXPOSURE AND DOSE FOR IONIZING RADIATION

The roentgen is the unit that was originally defined for exposure to x-rays. It was defined to facilitate radiation measurements with a device called an ionization chamber, which is described later. The roentgen (R) is defined* as that quantity of x-radiation that produces 1 electrostatic unit of charge (1 esu, also called 1 statcoulomb) in 1 cm^3 of air at standard temperature and pressure (STP). In other words, the x-rays interact in 1 cm^3 of air and cause a certain number of ionizing events. When the charge of either sign released from these events equals 1 esu (1 coulomb (C) $= 3 \times 10^9$ esu), then the exposure has been 1 roentgen. Each ionizing event is said to produce one ion pair—the ion plus the electron.

Example 11.1

How many ion pairs are formed per cubic centimeter of air at STP by 1 R of x-radiation?

Solution

There will be 1.6×10^{-19} C (the charge on the electron or the ionized atom) released for every ion pair made, therefore

$$\begin{aligned} 1\text{ R} &= \frac{1\text{ esu}}{\text{cm}^3} \times \frac{1\text{ C}}{3 \times 10^9\text{ esu}} \times \frac{1\text{ ion pair}}{1.6 \times 10^{-19}\text{ C}} \\ &= 2.08 \times 10^9 \text{ ion pairs per cm}^3 \text{ of air} \end{aligned}$$

From Example 11.1, the roentgen is seen to be equivalent to having 2×10^9 ion pairs formed per cubic centimeter of air by x-rays; the equivalence does not directly specify anything about the energy deposited or the biological damage to any tissue that might have been irradiated. The roentgen unit can also be used to measure the exposure to energetic charged particles such as electrons, if appropriate ionization chambers are used. An ionization chamber that could be used for a precise measurement of the exposure from x-rays of a few hundred keV is shown in Figure 11.2. The charge liberated by the formation of ion pairs is measured by the electrometer for a particular x-ray

* An alternative definition consistent with the International System (SI) of units is that 1 roentgen is the amount of x-radiation that produces 2.58×10^{-4} coulomb of charge per kilogram of air. For air of density 0.001293 g/cm^3, the two definitions are equivalent.

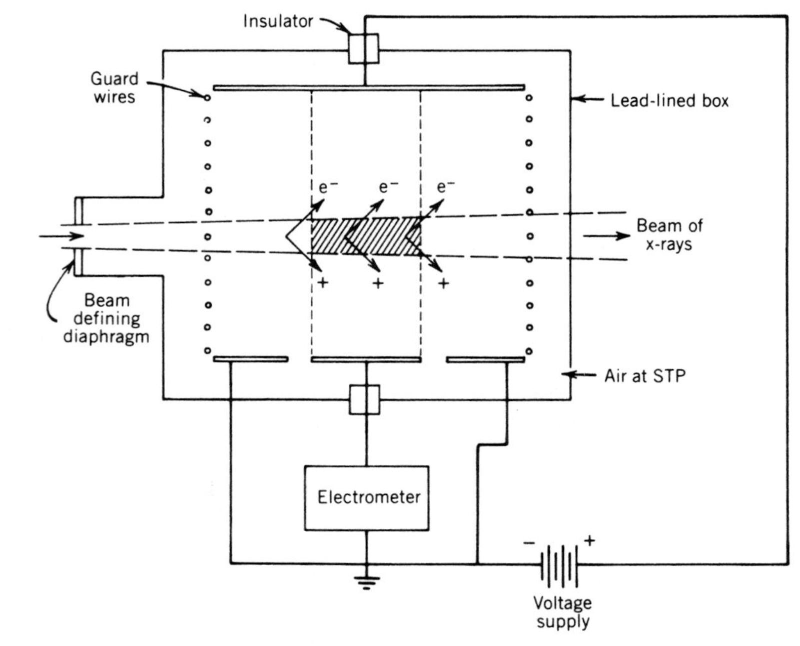

Figure 11.2 Ionization chamber for the measurement of the exposure from x-rays. The sensitive volume is the shaded area in the center. The electrometer measures the amount of electric charge passing through it. The guard wires ensure that the electric field is uniform over the sensitive volume.

exposure. One of the principles of such an ion chamber is that as many ion pairs will be scattered into the sensitive volume as will be scattered out of the volume. For practical measurements around x-ray sources that might present a health hazard, far simpler portable ionization chambers are used; these are then calibrated against a chamber of the type shown in Figure 11.2.

The concept of radiation exposure is based on the assumption that the absorbing medium is air. To relate an exposure to a given biological effect, it is more convenient to use a different measure of the radiation intensity. The absorbed dose in rads is defined as that amount of radiation of any kind that deposits 100 ergs* per gram of material. The material can be air, body tissue, or any other substance. The density of air is 0.001293 g/cm^3 at STP, and an ion pair produced in air requires on average 33.7 eV. With this information, we can calculate how many ergs are deposited in 1 g of air by 1 R.

Example 11.2

Calculate the number of ergs per gram deposited in air by an exposure of 1 R.

Solution

$$1\,\text{R} = \frac{1\ \text{esu}}{1\ \text{cm}^3} \times \frac{1\ \text{cm}^3}{0.001293\ \text{g}} \times \frac{1\ \text{C}}{3 \times 10^9\ \text{esu}} \times \frac{1\ \text{ion pair}}{1.6 \times 10^{-19}\ \text{C}}$$

$$\times \frac{33.7\ \text{eV}}{\text{ion pair}} \times 1.6 \times 10^{-12}\,\frac{\text{ergs}}{\text{eV}} = 86.9\ \text{ergs per gram of air}$$

*The erg is the energy unit of the centimeter–gram–second system of units that is frequently used in scientific calculations. An erg is equal to 10^{-7} J.

When tissue is the material of concern, the energy deposited by 1 R is 94.8 ergs/g, so that numerically 1 rad is quite close to 1 R for air or body tissue. The exposure is measured in roentgens using some device such as an ionization chamber, and the dose in rads must be calculated from the exposure for the given material. The dose in rads is the quantity of interest in health physics, radiobiology, or radiology. Although the rad is the unit commonly used, the correct metric unit has now been defined to be the gray (1 Gy = 1 J/kg = 100 rads).

When explicit biological damage is the concern, then a correction must be applied to the dose in rads to take into account the fact that various types of radiation have different biological effects for a dose of 1 rad. The unit rem (roentgen equivalent man) has been introduced to take these differences into account. Thus,

$$\text{rem} = \text{rad} \times (\text{RBE or QF})$$

where RBE is the relative biological effectiveness and QF is the quality factor. By definition, RBE is the ratio of the absorbed dose of any radiation required to produce a given biological effect to that required for 200 keV x-rays to produce the same effect. The term RBE is used most often in radiation biology where specific effects on specific tissues are being investigated. In the field of health physics, in which the overall effects of radiation on human health are of concern, RBE is often replaced by the quality factor (QF), which along with another factor (DF) that takes into account the distribution of the dose over the body, is used to multiply the absorbed dose in rads to obtain rem. For the present discussion there is no need to distinguish between RBE and QF. The reason for the dependence of the value of RBE on the energy and type of radiation will be discussed qualitatively later, but it is directly related to the energy deposited in the tissue per unit of track length. Table 11.1 lists the values of RBE for several types of radiation at various energies.

In enumerating the radiation dose encountered by people in various situations, the rem is the useful quantity since a mixture of several different kinds of radiation may be involved, each with a different value of RBE. The sievert (1 Sv = 100 rem) has recently been introduced as the proper metric (SI) unit. A summary of radiation units is given in Table 11.2.

Table 11.1 THE RELATIVE BIOLOGICAL EFFECTIVENESS (RBE), OR QUALITY FACTOR (QF), AND THE LINEAR ENERGY TRANSFER (LET) FOR VARIOUS TYPES OF RADIATION AT DIFFERENT ENERGIES

Type of Radiation	Energy	RBE (or QF)	LET (keV/μm)
X-rays, gamma rays	200 keV	1 (by definition)	0.06
Gamma rays	4 MeV	0.7	0.01
Electrons	<0.03 MeV	1.7	>2.3
Electrons	>0.03 MeV	1	0.2–2.3
Thermal neutrons	0.025 eV	3	4
Fast neutrons	5 MeV	7	10
Fast neutrons	10 MeV	6.5	45
Alpha particles	1–10 MeV	10–20	50–150
Recoil nuclei (heavy ions)	<1 MeV	20	4000–9000

Table 11.2 RADIATION UNITS

Roentgen (R)–Unit of Exposure
One roentgen is that quantity of x-ray or gamma radiation that produces, in air, one electrostatic unit of charge. The roentgen applies only to the measurement of x-ray or gamma radiation in air.
1 R = the absorption of 87 ergs per gram of air
Rad (Radiation Absorbed Dose)–Unit of Absorbed Dose
1 rad = the absorption of 100 ergs per gram of absorbing material.
Rem (Roentgen Equivalent Man)–Rem = Rad × RBE
One rem is that amount of any type of radiation that produces the same biological effect as is obtained from an absorbed dose of 1 rad of 200 keV x-rays.
RBE (Relative Biological Effectiveness)
Values given in Table 11.1.
Gray (Gy)
1 gray = 100 rad = 1 joule/kilogram
Sievert (Sv)
1 sievert = 100 rem

11.3 DIRECT EFFECTS OF LARGE DOSES OF RADIATION

The effects of radiation on human beings are divided into two major categories. The first is the direct (somatic) effects of massive short-term exposures of people. The effects of such exposures are generally felt very quickly, and death frequently results within hours or days. The second category includes effects caused by lower levels of exposure. These are frequently long-term or chronic exposures where the effects may not be manifest for many years. Genetic effects and the induction of various kinds of cancer are the major concerns with this type of exposure.

Massive or acute exposures of people to ionizing radiation have occurred in accidents involving x-ray machines, with personnel operating nuclear reactors for research purposes, or with personnel involved in the manufacture of nuclear weapons. The high-level exposure of the greatest numbers of people occurred following the detonation of the nuclear weapons at Hiroshima and Nagasaki, Japan, in 1945, at the close of World War II. The effects of acute exposure are reasonably well understood, but there are many different responses of individuals depending on their general state of health and the effectiveness of medical treatment.

If a group of people receive a total body dose of 400 rem, it can be expected that about 50% of the group will die in about 30 days. If the dose is increased to 600 rem, almost all can be expected to succumb. Often the lethal dose is stated as $LD_{50/30}$ = 400 rem, meaning that 50% of the people exposed to that short-term dose will die within 30 days. There are two ways in which we can examine the gross effects of 600 rads on a person. First, let us calculate the rise in body temperature.

Example 11.3

How much will the body temperature rise if a person receives a whole-body x-ray dose of 600 rads?

Solution

$$600 \text{ rad} = 600 \text{ rad} \times \frac{100 \text{ ergs}}{\text{g} \cdot \text{rad}} \times \frac{1 \text{ cal}}{4.2 \times 10^{7} \text{ ergs}}$$
$$= 1.4 \times 10^{-3} \text{ cal/g}$$

Each gram of body tissue receives 1.4×10^{-3} cal. Assuming the specific heat of tissue is that of water, that is, 1 cal/°C • g, then each gram of tissue will be raised 1.4×10^{-3} °C in temperature.

This is, of course, a negligible temperature increase. The effects of radiation are indeed more subtle than just a gross addition of thermal energy. Another question could be asked concerning the fraction of the total molecules in the body that are ionized.

Example 11.4

What fraction of the molecules in 1 g of tissue is ionized by an absorbed dose of 600 rads? Assume tissue has the chemical structure of water, and that it takes 34 eV to create an ion pair.

Solution

$$600 \text{ rad} \times \frac{100 \text{ ergs}}{\text{g} \cdot \text{rad}} \times \frac{1 \text{ eV}}{1.6 \times 10^{-12} \text{ ergs}} \times \frac{1 \text{ ion pair}}{34 \text{ eV}}$$
$$= 1.1 \times 10^{15} \text{ ion pairs per gram}$$

Each gram of tissue contains

$$\frac{1}{18} \times 6.02 \times 10^{23} = 3.3 \times 10^{22} \text{ molecules per gram}$$

since water has a molecular weight of 18, and 1 g molecular weight has 6.02×10^{23} molecules. The fraction affected in each gram is thus

$$1.1 \times 10^{15} \frac{\text{ion pair}}{\text{g}} \times \frac{\text{g}}{3.3 \times 10^{22} \text{ molecules}} = 3.3 \times 10^{-8} \text{ ion pairs per molecule}$$

To restate the results of this calculation in another way, if the body were made of water only one out of 3×10^{7} molecules would be affected by a lethal dose of radiation. The body, however, has in the nucleus of its cells some very complex molecules, such as deoxyribonucleic acid (DNA), that have molecular weights greater than 500,000, and these important molecules experience severe damage with much higher probability than indicated by this calculation.

Acute short-term exposures can be classified according to their effects. In this ordering each larger dose will, of course, include the effects of the lesser exposures. The information on these effects is obtained from animal studies for the larger doses and from the history of both animal and human exposure for the lower doses.

(a) Vast total body doses in the range of 100,000 rads cause the inactivation of many substances needed for the metabolic processes of cells and tissues. Death, sometimes called molecular death, generally occurs during the exposure.

(b) Doses in the range of 2000 to 15,000 rads damage the neurological and cardiovascular system as well as other organ systems. Severe sickness and disorientation occur essentially immediately. Death comes in a matter of hours.

(c) Doses in the range of about 1000 to 2000 rads result in damage to the gastrointestinal system, resulting in some cases in gangrene of the gut. Death normally occurs in 1 to 2 weeks.

(d) Doses in the range of about 200 to 600 rads primarily affect the blood-forming organs. Because of depression of the bone marrow there is a loss of mature red and white cells and platelets in the blood supply. As noted previously, doses in the range of 400 to 600 rads generally result in death within 1 to 2 months after exposure. Some changes in the blood constituents are observed for doses down to about 25 rads.

(e) Generally there are no immediate and directly observable effects for absorbed doses of less than 25 to 50 rads. There are, however, some special situations where doses in this range can be significant. If a pregnant woman receives doses as small as 10 rads, particularly during the 4th to 11th week of the gestation period, there is a significant probability that the offspring will suffer some abnormality. A pregnant woman should, according to many obstetricians, receive no more than about 0.5 rem during the entire gestation period.

There are, of course, long-term indirect carcinogenic and genetic effects that occur with some probability, however small, for doses far less than 25 rads or even 1 rad. This aspect of the effects of ionizing radiation is explored in the next section.

11.4 LONG-TERM INDIRECT EFFECTS OF IONIZING RADIATION

Unless there is a major accident involving a nuclear reactor or the explosion of nuclear weapons, the high-level direct effects of ionizing radiation are not of general significance to the human population because so few individuals are involved. What is of far greater interest at present is the exposure of large numbers of people to relatively low levels of radiation and the possibilities of long-term effects.

It has been established by direct experiment that the cellular nucleus is the part of the cell most sensitive to radiation. The outer part of the cell, the cytoplasm, can absorb thousands of rads before its functions are seriously enough impaired to jeopardize the cell's survival. On the other hand, one alpha particle can cause sufficient damage to the cell nucleus that the cell cannot survive further division.

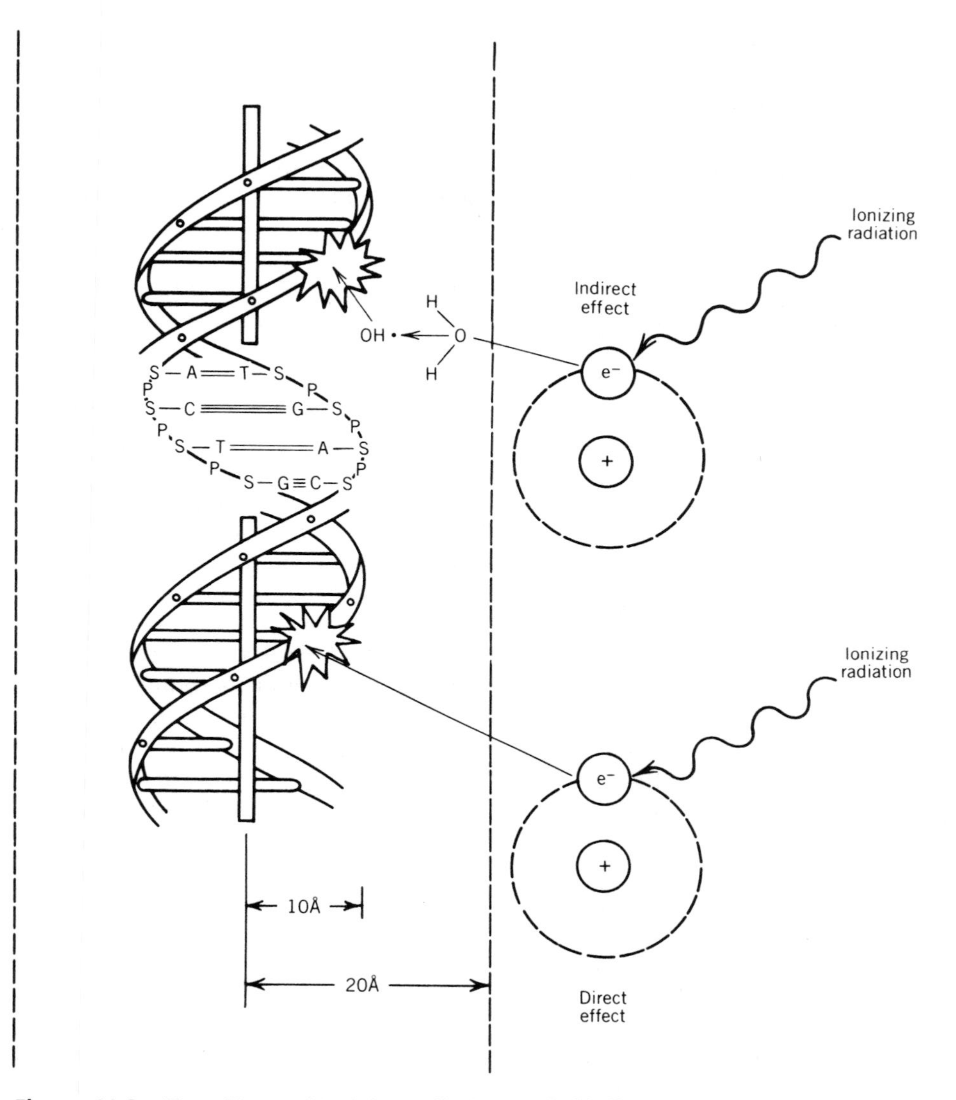

Figure 11.3 The effects of ionizing radiation on DNA. Radicals such as OH •, created by ionizing radiation interacting with water molecules within a radius of about 20 Å from the DNA, can cause breaks in the chemical bonds in the DNA. Ionizing radiation can also directly interact with the DNA structure to disrupt chemical bonds. The letters S, P, A, T, G, and C stand for sugar, phosphorus, adenine, thymine, guanine, and cytosine, respectively. (1 Å = 1 angstrom = 10^{-10} m.)

Each nucleus of a human cell except the sex cells contains 46 chromosomes. The basic constituent of each chromosome is DNA, the long, double helix molecule. Figure 11.3 shows a section of a DNA molecule with the sugar and phosphorus molecules along the strands and with the cross-linking by adenine–thymine or guanine–cytosine base pairs. The basic genetic information is contained in the ordering of the pairs of bases. Genes are sections or regions of the chromosome (or DNA) that control specific physiological or biochemical functions and contain 600 to 1800 base molecules. The number of genes in each chromosome ranges from hundreds to thousands.

Each cell goes through a cycle that lasts about 10 to 100 hours during which time it replicates very faithfully the DNA structure in each chromosome. The DNA stretches out and the two strands separate; each strand will have the appropriate missing member of the base pair added and a new strand formed. In such a way two separate DNA chains are created that are identical. After replication of the chromosome, a complete set of chromosomes gathers at the poles of the nucleus, and finally the cell undergoes mitosis and separates into two cells with identical nuclei.

Ionizing radiation can interact with DNA in two general ways. An ionizing event can occur either in the sugar–phosphorus strands or in the base pairs, or a free chemical radical such as OH • can be created by ionization in the vicinity of the DNA. Such radicals are chemically very reactive and can subsequently affect the bonding of the DNA in a fashion similar to that of the direct ionization process.

The most likely damage that radiation can cause is to break one of the hydrogen bonds in the base pairs or in the sugar molecules. This type of damage appears to be rather readily repaired, and there are no apparent long-term effects. The complete scission of one of the strands is the next most likely event. Although this can be repaired, it is also possible that some molecule, such as oxygen, will attach itself to the free broken end before the repair is accomplished, thus preventing complete restoration of the original bonds. Double strand scissions and complete breaking of the bonds of a cross-link (base pair) are less likely because more energy is required, but when they occur repair is less likely.

If the effect of the radiation is to cause a change in the ordering of the base pairs, the coded genetic information will, of course, have been altered. The gene where this alteration occurred will have a mutation, and on DNA replication followed by mitosis, both cells will carry the same error. This error will be perpetuated through further cell cycles. The chromosomes, which are generally visible with a light microscope, will show no change. If, on the other hand, there was a single or double strand scission that resulted in a permanent break in the DNA chain or a joining of the incorrect loose ends of a chain, major changes in the structure of the chromosome can result. Many gross structural changes of this kind prevent the cell from reproducing itself or surviving mitosis. Some chromosome aberrations, however, as with gene mutations, can lead to cells that can survive and reproduce themselves. Chromosomal changes are generally observable. For example, Figure 11.4 shows abnormal chromosomes that were induced by radiation. Both fragments of chromosomes can be seen as well as one chromosome that resulted from the combining of two.

If these mutations take place in the gonadal cells, there is some chance that the cell with the gene mutation will be involved in the reproduction process. Most frequently such mutations lead to the death of the fetus, but occasionally the offspring will be born with a genetic defect such as hemophilia, mongoloidism, or albinism. Genetic mutations can also be caused by various chemicals, heat, and natural mistakes made in the replication of the DNA in the cell. The mutagenic effects of x-rays were first studied extensively by H. J. Muller in his fundamental experiments with drosophila (fruit flies) in 1927. Since then extensive experiments have been carried out with mice by W. L. Russell and others at Oak Ridge National Laboratories, so that a reasonably

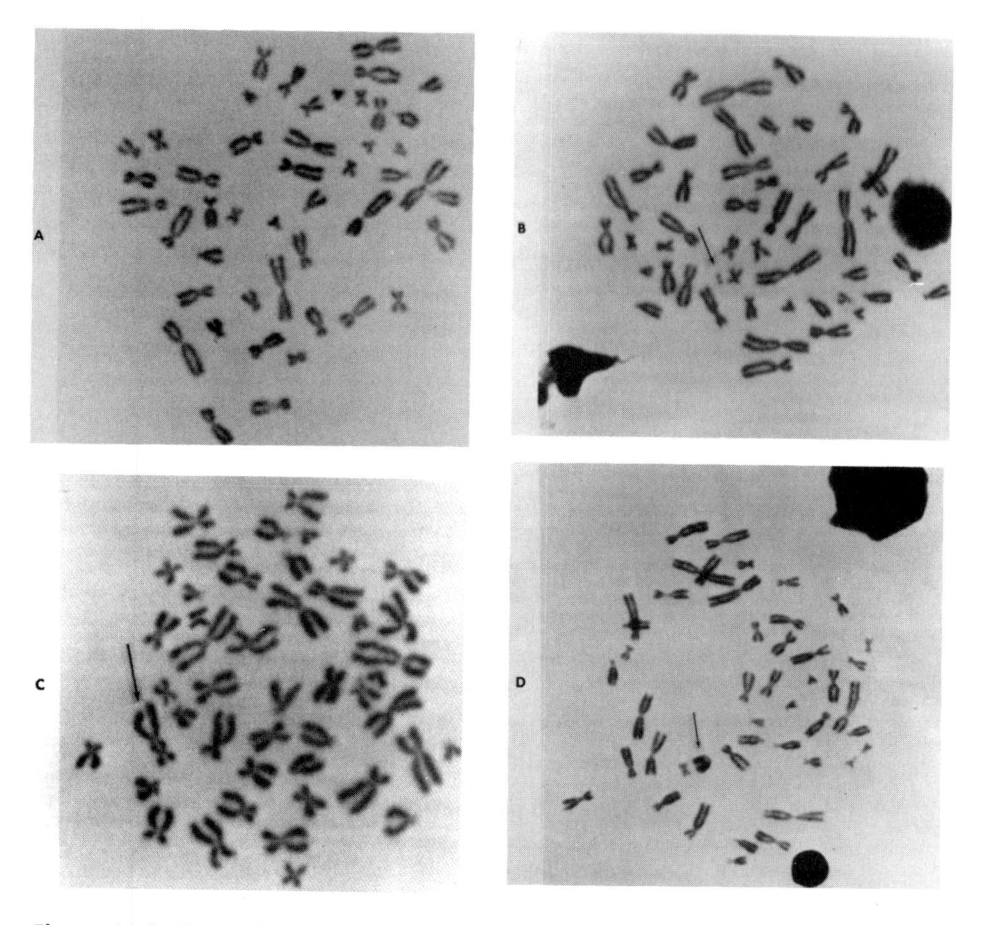

Figure 11.4 Examples of some radiation-induced aberrations in chromosomes from human cells. In (*a*) the chromosomes are normal, in (*b*) a deletion or fragment of a chromosome is shown, in (*c*) an abnormal chromosome with two centromeres is shown, and in (*d*) a ring formed from chromosome fragments is shown.

complete understanding of radiation-induced genetic effects exists. The application of these results obtained with rather large doses of radiation to the situation of humans exposed to low doses is not straightforward, however.

In addition to genetic effects, mutations and chromosomal aberrations may produce cells that demonstrate abnormal growth behavior, as in cancer. In human beings there is often a latency period of perhaps two or three decades between the creation of an abnormal cell and the appearance of cancer. There is no clear picture at present of the series of events in this latency period that finally leads to various types of cancer.

11.5 CELL SURVIVAL STUDIES

To provide a better understanding of the effects of radiation on cells, separated living cells can be irradiated in the laboratory and their survival studied. Most

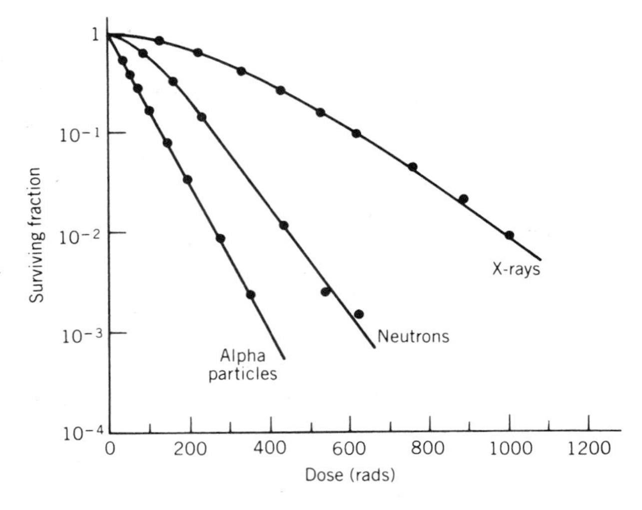

Figure 11.5 Cell survival curves for different types of ionizing radiation. The x-rays are from an x-ray tube with 250 kV on the anode, the neutrons have an energy of 15 MeV, and the alpha particles 4 MeV. As the LET increases from x-rays to alpha particles, the survival curves become steeper and show less curvature at low doses.

of the studies are carried out outside of the animal by placing a certain number of proliferating cells in a culture medium and irradiating them with a particular dose and type of radiation under controlled conditions. The percentage of cells that survives to form colonies later, compared to a control group that has not been irradiated, is determined as a function of absorbed dose. A typical set of survival curves is shown in Figure 11.5. As the dose increases, the surviving fraction decreases in an almost exponential fashion. It can be seen that the steepness of the curve is very much greater for alpha particles than for x-rays and also greater for alpha particles than for 15-MeV neutrons. As the introduction to this chapter notes, x-rays take much more absorber to attenuate them than do alpha particles. The quantity that is relevant is the linear energy transfer (LET), or the energy transferred to the medium per unit length of the particle track. For example, low-energy alpha particles have an LET of about 50 to 150 keV/μm, fast neutrons about 10 to 45 keV/μm, and x-rays or gamma rays less than 1 keV/μm. What is obvious from Figure 11.5 is that particles with higher LET have higher killing power. The RBE can be obtained numerically from the ratios of the slopes. Alpha particles are more effective in killing cells because they can create so many breaks in the DNA that there is little chance that the cell will survive. On the other hand, x-rays are sparsely ionizing and only disrupt a chemical bond every once in a while. These types of sparsely distributed breaks can frequently repair themselves. Although the argument is complex, it can be deduced from the curvature in the upper portion of the x-ray curve in Figure 11.5 that a repair mechanism is operative. With low doses of x-rays some DNA bonds are broken, but many breaks are not severe, and the bonds are restored with time. At larger doses of x-rays, multiple breaks may occur in a small region and repair is less probable. With higher LET particles, such severe damage is done that the repair process is not operative.

A number of other results have come from cell survival curves. It has been found, for example, that when oxygen is present in abundance in the vicinity of the cells being irradiated by x-rays, it takes far less absorbed dose to kill a certain fraction of cells than it does when the cells are deprived of oxygen. The ratio of the required doses for a certain survival rate without and with oxygen is called the oxygen enhancement ratio (OER). This means that the presence of oxygen enhances the effect of the radiation, not the survival of the cells. Although the value of OER is about 2.7 for x-rays, it is about 1 for low-energy alpha particles. These effects have been explained by noting that if an oxygen molecule is present when a break occurs in a DNA molecule, the oxygen molecule can attach itself to the loose end and stabilize the break, preventing the repair mechanism from operating. With alpha particles the breaks are so severe that the presence of oxygen is not needed to prevent repair. Figure 11.6 shows both OER and RBE plotted as a function of LET. Above the plots is a diagrammatic representation of low, intermediate, and very high LET radiation passing through a series of cells.

Cell survival curves have been analyzed to provide information on the size of the target in the nucleus of the cell that is needed to explain the data, and whether two hits on one target or one hit on each of two targets is involved. It has been discovered recently that the survival of cells irradiated by x-rays is considerably less when the cells are at an elevated temperature of about 42°C rather than the normal body temperature of 37°C. This temperature dependence is not completely understood. With low LET radiation, there is also a dependence on the rate at which the dose is administered.

Cancerous tumors that cannot be removed by surgery are frequently treated by ionizing radiation to kill the cells of the tumor and to reduce its size or eliminate it completely. X-rays of various energies are normally used for such irradiations because they are commonly available in treatment centers. Recently, energetic heavy charged particles such as protons, alpha particles, and heavy ions as well as neutrons have been used for cancer therapy because all of these particles have an OER lower than that of x-rays. This is important in

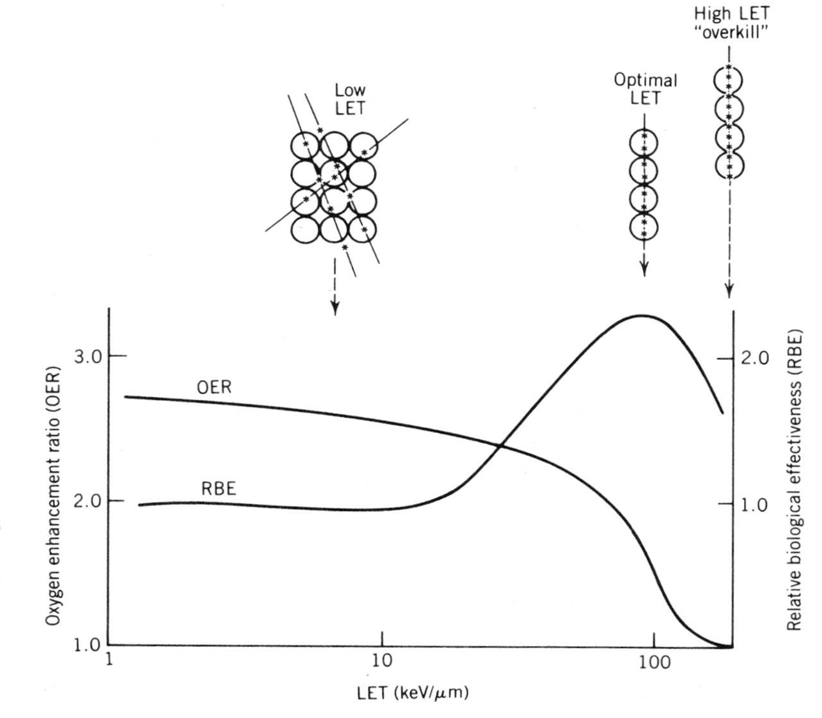

Figure 11.6 Oxygen enhancement ratio and relative biological effectiveness as observed for a specific cell population and biological effect. These curves, in general, apply to any type of ionizing radiation.

order that the radiation not unnecessarily kill the healthy cells in the region infiltrated by tumorous cells. The interiors of tumors are deficient in oxygen compared to healthy tissue; this means that high OER radiation such as x-rays is less likely to spare the healthy cells while killing the tumor cells than is the particle radiation with higher LET and lower OER.

11.6 DOSES RECEIVED FROM BACKGROUND RADIATION AND MEDICAL TREATMENT

The sources and equivalent doses from ionizing radiation received on average by the U.S. population are shown in Table 11.3. The sources of natural radiation are dominated by radon (^{222}Rn), which is a gaseous radioactive isotope in the decay chain of uranium. The uranium and thorium decay chains have as their parents ^{238}U, with a half-life of 4.5×10^9 years and ^{232}Th, with a half-life of 1.39×10^{10} years. Because of their long lifetimes, uranium and thorium are still present in the rocks and soil, where they were deposited when the earth was formed billions of years ago. The average concentration of uranium is about 1 tonne per km^2 in the first 0.3 meters of soil, and the concentration of thorium is about 3 tonnes per km^2. Both of these isotopes have gaseous radon isotopes in their decay chains, but ^{220}Rn from thorium has a half-life of only 55 seconds, and hence is not a major environmental problem. The uranium decay chain, however, includes ^{222}Rn, with a half-life of 3.8 days. This is sufficient time for the radon atom to diffuse through permeable soil a meter or so and enter a building through cracks and other openings before decaying. The decay products of radon can become attached to dust particles in the air and then inhaled. These particles then may be deposited in the lung's bronchial tubes where, upon decay, they emit an alpha particle that can damage the nearby cells. The effective dose equivalent listed in Table 11.3 has been reduced from

Table 11.3 AVERAGE ANNUAL EFFECTIVE DOSE EQUIVALENT OF IONIZING RADIATION RECEIVED BY U.S. POPULATION

Source	Dose equivalent (mrem/yr)	Effective Dose Equivalent (mrem/yr)
Natural		
Radon	2400	200
Cosmic rays	27	27
Terrestrial	28	28
Internal	39	39
Man-made		
X-ray diagnosis	39	39
Nuclear medicine	14	14
Consumer products	10	10
Occupational	0.9	0.9
Nuclear fuel cycle	<1	<1
Fallout	<1	<1
Miscellaneous	<1	<1
Total		358

Source: Adapted from BEIR V.

2400 mrem per year to 200 mrem per year, since it is a local dose to the lungs and not a whole-body dose, as are the other entries in the table.

Cosmic radiation is another important component of the background radiation we all receive. The primary cosmic rays originate in outer space and are incident on the earth's upper atmosphere. They are predominantly (about 92%) very energetic protons. The other particles are nuclei of helium (alpha particles), which constitute about 7% of the primary component, the rest being nuclei of heavier elements. The protons interact with nitrogen and oxygen in the upper atmosphere to produce a variety of secondary cosmic rays. Those reaching sea level are predominantly neutrons and muons, which irradiate all of us continuously. The muon component is very penetrating and accounts for most of the cosmic-ray radiation dose. The muon flux at sea level is about 1 per square centimeter per minute. This means that several hundred muons pass through your brain every minute and several thousand pass through your body per minute. At sea level the dose received from this source is about 27 mrem per year, but at the elevation of Denver (5300 feet), it is increased to about 52 mrem per year.

In addition to these sources of radiation, we are exposed to gamma rays from the long-lived radioisotopes ^{232}Th and ^{238}U and their daughters which are present in soil, rocks, and concrete. Outside buildings, the average dose received from these sources is about 28 mrem per year. However, the exact amount depends on the concentration of the radioisotopes in the earth, and this varies widely from one location to another. For example, the terrestrial gamma-ray dose ranges from 15 to 35 mrem per year for the Atlantic and Gulf coastal plains to 75 to 140 mrem per year for the Colorado Plateau. The doses inside stone or concrete buildings will be higher.

Several naturally occurring radioisotopes are found in our bodies. These account for the internal dose shown in Table 11.3. One of these is ^{40}K, which has a half-life of 1.28×10^9 years. It deposits a significant dose (about 20 mrem per year), since potassium is an essential element in the chemistry of the body. Radium isotopes incorporated into our bodies are in the ^{238}U and ^{232}Th decay chains; because radium is a bone seeker, we all have some of it present in our skeleton. These isotopes are taken in through our food. We receive a dose of about 7 mrem per year, mainly to the bone marrow, from that source. Nitrogen in the atmosphere can undergo a nuclear reaction from incident cosmic ray neutrons to form ^{14}C, a radioactive element with a half-life of 5700 years. We all receive a small internal dose from that source, because it is in the air we breathe and becomes incorporated into our body tissue. This isotope is the basis of the well-known carbon dating method now often used in archeology.

In addition to natural sources of radiation, there is a growing list of sources of man-made radiation. Medical x-ray exposures from various diagnostic procedures are by far the largest component. Chest and dental x-rays are the most common, but examinations for bone fractures and various internal disorders are also frequently done with x-rays. Use of faster x-ray film, image-enchancing screens, and lead aprons to protect parts of the body not involved in the examination have all helped to reduce the exposure, but it is estimated that the average dose received presently in the United States from diagnostic x-rays is about 39 mrem per year and about 14 mrem per year from radiopharmaceuticals.

Whenever a nuclear weapon is exploded above ground, many of the radioactive fission products become mixed with the atmosphere and are transported great distances before being deposited back on the earth. The atmospheric testing of nuclear weapons by the Soviet Union, the United States, and Great Britain up to the Test Ban Treaty in the early 1960s resulted in a dispersal of fission products mainly in the northern hemisphere. More recently, France, China, and India have exploded nuclear weapons. The radioactive fallout from all the above-ground bomb tests is to some extent present in the water we drink and the food we eat. The long-lived ^{137}Cs and ^{90}Sr ($T_{1/2} \approx 30$ years) fission products are of primary concern, since most of the others have shorter half-lives. If the radioactive isotope content of our bodies is examined with a large gamma ray detector in a well-shielded room, the 0.662 MeV gamma ray of ^{137}Cs is usually quite prominent. The average dose from nuclear weapon fallout is estimated to be less than 1 mrem per year.

In the category of consumer products, the main contributor is radon in domestic water supplies. Also included in the category are contributions from television sets, computer display screens, wristwatches with radium-painted numbers on the face, smoke alarm systems and coal burning. In addition, cigarette smokers are exposed to alpha particles in their lungs from smoking tobacco that has incorporated ^{210}Po from the uranium decay chain in the soil where the tobacco was grown. It is estimated that the general public receives a total of about 10 mrem per year from these consumer products.

Certain people are exposed to radiation because of their occupation. This group includes radiologists, x-ray technicians, nuclear reactor personnel, and individuals involved in the fabrication of materials for nuclear reactor fuel rods and nuclear weapons. Although some of these individuals may receive up to about 5000 mrem per year, the average occupational dose for the entire U.S. population is only about 1 mrem per year.

As discussed in Chapter 5, reactors do emit some gaseous radioisotopes under normal operation. The average dose for the general population at present from such sources is estimated to be 0.005 mrem per year and hence extremely small compared to the total dose of about 64 mrem per year from other man-made sources. In the next section we consider the consequences of these low-level doses of ionizing radiation.

11.7 BASIS FOR DETERMINING THE EFFECTS OF IONIZING RADIATION ON MAN

The effects of low levels of radiation are virtually impossible to determine through laboratory experiments with animals. Because doses below 10 rads produce so few incidents of cancer or genetic effects, the number of subjects needed to obtain statistically meaningful results becomes too large for any laboratory to manage. In addition, the time delays between the irradiation and the onset of the evidence for either cancer or genetic effects are large and quite different for humans and small animals.

It is also impossible to take directly the results of cell survival studies and predict the consequences for humans. Although an understanding of the basic interactions between ionizing radiation and cells can be gleaned from such studies, there is no way to predict the probability of cancer forming after a 20-

year period, as there are perhaps several intermediate steps that are not well understood.

The only basis we have for predicting the consequences of low doses of radiation is to take information from large doses and extrapolate down to the low-dose region. Since it is difficult to rely on animal studies done at high levels of radiation, and since experiments cannot be carried out on human beings, other information must be used.

There have been a series of accidents, use and misuse of radiation for medical treatments, occupational exposures, and, of course, irradiation of people in the detonation of the nuclear weapons at Nagasaki and Hiroshima during the war with Japan. These incidents form the basis of our knowledge of the carcinogenic and genetic effects of radiation at relatively high doses. Since the medical records of many of the people who were irradiated are very incomplete, and because reasonably accurate assessment of the doses delivered was not available, many of these incidents have provided little useful information. As outlined earlier, the information obtained at large doses must be extrapolated down to doses that range from a few millirem to a few rem.

A number of epidemiological studies of radiologists in the United States have been carried out in the last 40 years. In the early part of the study, it was determined that radiologists died from leukemia nine times more frequently than other physicians. For the more recent period, this ratio dropped down to 3.6. This qualitative information on the relationship between radiation and leukemia is useful, but it has been difficult to obtain estimates of the doses received by the radiologists, although it is known that 100 rad per year was not uncommon around 1930.

Between 1935 and 1955, 13,000 patients were treated in Great Britain with large doses of x-rays for ankylosing spondylitis, a form of rheumatoid arthritis of the spine. The doses ranged from 121 to more than 3000 rads over limited regions of the body. The number of definite cases of leukemia found so far has been 36, and two more are probable in this group, whereas the expected number for a normal population of this size is 2.9. Other forms of cancer were also noted in the patients treated in this way for ankylosing spondylitis. Additional useful case studies of external radiation with x-rays include breast cancer resulting from fluoroscopy diagnosis in Canada and the United States and thyroid cancer from treatment for an enlarged thyroid gland.

The most notable case of internal exposure from radioactivity involved the young women employed in the first half of this century to paint radium numerals on the faces of watches so the dials would glow in the dark. They would frequently put a point on the brush they used with the tip of their tongue. In so doing they ingested significant amounts of radium. From a group of 264 such persons in the Chicago area, 15 bone sarcomas and 11 tumors of the cranial structures have been diagnosed. Thirty and forty years afterward, body burdens of 0.6 to 10 microcuries of radium were still in the bones of some of the women involved.

The survivors of the bombs dropped on Hiroshima (August 6, 1945) and Nagasaki (August 9, 1945) have provided by far the greatest amount of data on the effects of ionizing radiation on human beings. The Hiroshima bomb was made of ^{235}U and was exploded in the air, causing about 60,000 civilian deaths within 2 weeks and leaving about 70,000 injured survivors. The bomb dropped on Nagasaki was made of ^{239}Pu; it resulted in 33,000 civilian deaths and about 25,000 injured survivors. Both weapons had the explosive power of roughly 20 kilotons of TNT. Of the injured survivors of both cities, nearly 24,000 had received doses of 10 rads or more.

Study of the relevant data from the Japanese atomic bomb survivors as well as other documented exposure information has been an ongoing task of the U.S. National Research Council committees on the Biological Effects of Ionizing Radiation (BEIR). Reports of this group are issued periodically, and their results and conclusions serve as an authoritative summary of the effects of ionizing radiation. BEIR III, published in 1980, was concerned with the effects on populations of low levels of ionizing radiation and was mainly based on estimates made from 1950 to 1974 of the gamma-ray and neutron doses received at various locations near ground zero in Nagasaki and Hiroshima. For a number of years these estimated doses were used in assessing the relationship between equivalent doses and the incidence of leukemia. Just recently, it has been discovered that these earlier dose calculations were subject to some significant errors. It is now believed that the neutron dose at Hiroshima was very small, whereas the gamma-ray dose was underestimated at Hiroshima and overestimated at Nagasaki. The new dose estimates have apparently made the data from the cities far more consistent. Figure 11.7 shows the effects of the new and old dose calculations.

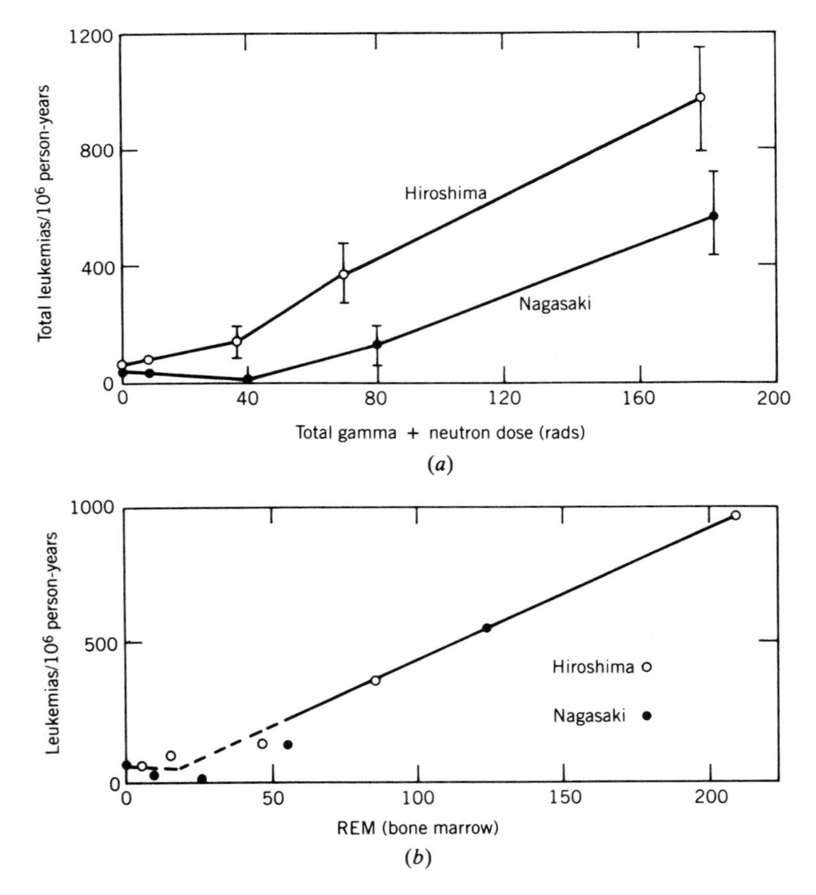

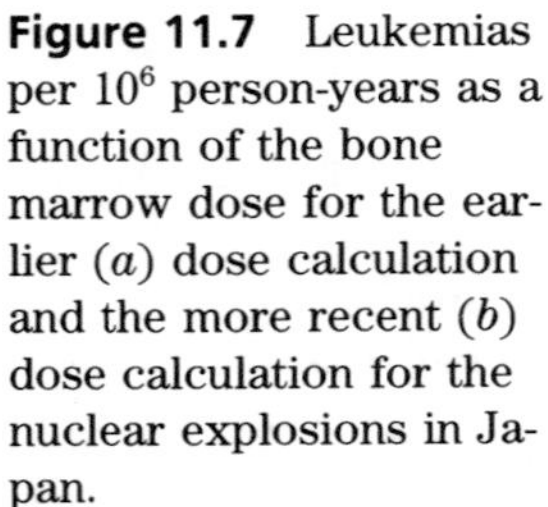
Figure 11.7 Leukemias per 10^6 person-years as a function of the bone marrow dose for the earlier (*a*) dose calculation and the more recent (*b*) dose calculation for the nuclear explosions in Japan.

The conclusions of BEIR V, published in 1990 and based on the new dosimetry information, indicated significant increases in the lifetime risk of cancer due to 10 rem of exposure relative to the older BEIR III study. There were 8714 records available on the Japanese survivors that were reasonably complete and that formed a major part of the data base.

Statistically significant excesses of various kinds of cancer have been observed for the Hiroshima and Nagasaki survivors. Leukemia was the first type of cancer found, followed by thyroid cancer and recently lung and breast tumors. The increased risk of leukemia became apparent just a few years after the explosions. The average latency period was 4 to 8 years for those less than 2000 meters from the center, and it increased to 15 to 18 years out to distances of 10,000 or more meters.

Various assumptions can be made to provide a useful scheme for extrapolating the effects of radiation at high doses to the low-dose region. Figure 11.8 shows two of the possible extrapolation procedures. The linear procedure appears to account for low LET data at high doses and high LET data at all doses, and it is the relationship that has usually been adopted by various groups assessing the effects of radiation. There appears, however, to be growing evidence that a combination of linear and quadratic terms more closely describes the data for low LET radiation at low doses, and that a linear relationship overestimates the consequences for that situation. Physically, this could come about if sparsely ionizing x-rays or gamma rays damaged the DNA, which was subsequently repaired naturally. There is no basis for assuming an absolute threshold below which no effects are observed.

The BEIR V study did more in the way of modeling of the data using a variety of statistical techniques than had been done previously. For leukemia the researchers adopted a linear–quadratic dose function, as is shown in Figure 11.8. The risk of an individual's getting leukemia, however, depends on his or her sex, attained age, age at exposure, and time since exposure. For the other forms of cancer (breast, lung, colon, urinary tract, and others), a linear dose response was adopted. A summary of the study's conclusions is shown in Table 11.4. The estimates are far less precise than it would appear from the numbers listed. For example, the total lifetime excess deaths listed for a 10-rem single exposure are 790. The 90% confidence limits, however, are about 540 to 1240.

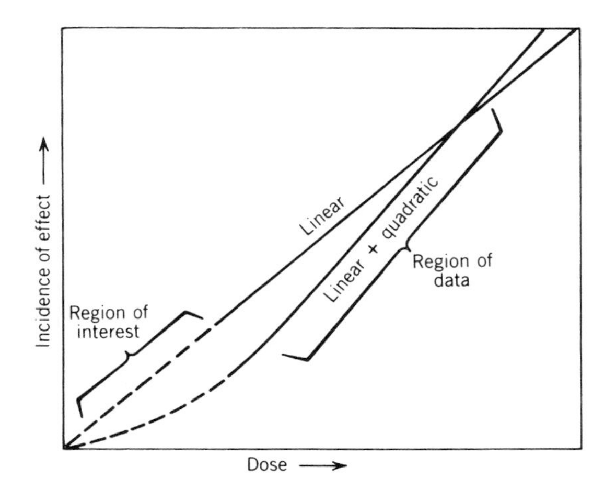

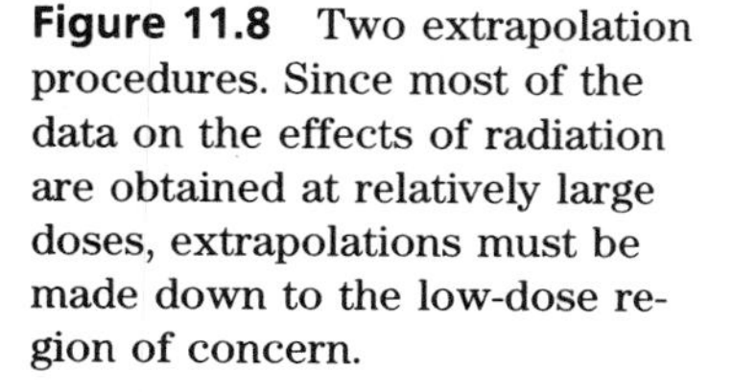

Figure 11.8 Two extrapolation procedures. Since most of the data on the effects of radiation are obtained at relatively large doses, extrapolations must be made down to the low-dose region of concern.

Table 11.4 ESTIMATES OF EXCESS CANCER MORTALITY PER 100,000 PERSONS EXPOSED

	Total	Nonleukemia	Leukemia
Single exposure to 10 rem	790	695	95
Continuous lifetime exposure to 0.1 rem/yr	560	495	65
Continuous exposure to 1 rem/yr from age 18 to 65	2,975	2,620	355
Normal expectation	19,300	18,600	700

Source: Adapted from BEIR V. It has been assumed that 50% of the 100,000 persons exposed are female and 50% are male.

These estimates by BEIR V are larger than similar estimates by BEIR III by a factor of about 4 to 18. For example, the cancer deaths estimated by BEIR III for the above case were 80 to 220.

A linear dose relationship would imply that a 1 rem single exposure would result in 79 cancer deaths for 100,000 people, and a 100 rem exposure, 7900 cancer deaths. Assuming the background and medical exposures to the 250 million people in the United States average 358 mrem per year, there would be about 5 million cancer deaths due to these types of radiation exposure. The normal expectation is 47.6 million cancer deaths per 250 million people.

There are other long-term consequences of relatively low doses of radiation beyond the excess incidence of cancer—namely, genetic effects. The BEIR V report extended the previous studies in an attempt to make current an evaluation of these effects. The children of the survivors of atomic bombs in Japan have not shown any excess genetic effects such as congenital defects, sex ratio changes, or decreased survival rates of live-born infants. This in no way implies that genetic effects are absent, but only that the statistical sample is too small to determine whether or not those effects are present. Some 78,000 children who were not yet conceived at the time of the irradiation, were born to irradiated parents involved in the study, but the average dose to the gonads of the parents was only 50 rem. The results are, in fact, quite consistent with the rate of genetic effects observed in studies of irradiated mice and with the results of the BEIR V study. There was an increased number of stillbirths among the pregnant women who were exposed to radiation in Japan, but, of course, this is not a genetic effect.

The BEIR V study took 100 rem as the dose that would be required to double the frequency of heritable mutations in human beings. This was based primarily on animal studies. The study thus makes it possible to estimate the various genetic effects that would result from some explicit dose. For example, if the natural occurrence of some genetic disease is 20,000 per million live-born infants and the mutation component is one half, then the additional cases due to 1 rem would be $1/100 \times 10{,}000$, or 100. In a similar fashion, Table 11.5 was constructed by the BEIR V committee and represents the best current estimates of genetic effects of radiation.

It may seem surprising that the genetic effects of radiation with humans have never been seen beyond the expected natural occurrence. The basic reason for this is that the normal incidence of heritable disease is very high.

Table 11.5 ESTIMATED GENETIC EFFECTS OF 1 REM PER GENERATION PER MILLION LIVE-BORN OFFSPRING

Type of Disorder	Current Incidence	Additional Cases	
		First Generation	Equilibrium
Dominant or x-linked diseases	10,400	6–35	100
Recessive diseases	2,500	<3	
Chromosomal abnormalities	4,000	<6	<1
Congenital abnormalities	20,000–30,000	10	10–100

Source: Adapted from BEIR V.

From the BEIR study it may be calculated that less than 2% of all human genetic disease can be attributed to natural background radiation.

11.8 MAXIMUM PERMISSIBLE DOSES

The exposure of people to ionizing radiation is clearly something to be avoided because of the harmful effects that have been described. It is also clear that many medical procedures involve ionizing radiation, but there are now usually ample benefits to warrant the negative effects of the exposure; this was not as true 20 or more years ago. As the harmful effects of radiation have become better known, the medical community has endeavored to minimize the exposure through improved procedures and more sensitive equipment.

Aside from medical exposures, we might try to eliminate all other sources of exposure. This, of course, is not a practical approach in an age where nuclear reactors; nuclear research; the use of radioisotopes in hundreds of applications in commercial products, science laboratories, and industry; and the use of x-rays in the study of crystals, metal castings, luggage at airports, and so forth are common. Since ionizing radiation is very much a part of a technical society, the next question concerns what doses are safe. We have seen in the preceding section that there is no safe level in the sense that there is no threshold below which effects are not expected. The general philosophy of governmental regulating bodies has always been to minimize any exposure by following prudent procedures, but then to keep the absorbed dose below some set maximum permissible levels. It is obvious that these maximum permissible levels are rather arbitrarily set to a level where, in the judgment of a commission, the effects on the individual and the community will be small enough to justify the benefits being derived from the use of the radiation.

Nationally, the Committee on the Biological Effects of Ionizing Radiation and the National Committee on Radiation Protection present information on risk estimates, as well as recommend standards, together with international groups, to the Environmental Protection Agency. The EPA then issues guidelines to various U.S. regulatory agencies such as the Nuclear Regulatory Commission.

For occupational exposure, the standards have been reduced drastically since 1925, when 1 R per week for 200-kV x-rays was set. At present, 5 rem per year after age 18 is the permitted maximum total-body dose. There are many other stipulations regarding the doses to various organs and maximum doses for a 13-week period, but the 5 rem per year or about 100 mrem per week is the basic level that has been set. Many workers mistakenly believe that no ill effects will be observed until the 5 rem level is reached. As we have seen, it is best to assume that the effects are proportional to the dose and that no dose is the best. The exposure of occupational workers should be monitored by requiring them to wear personal dosimeters that indicate the dose received over some period of time and by using various survey meters to assess the levels of radiation in the environment in which they work. In situations such as the cleanup of the reactor at Three Mile Island that require people to reach their maximum permissible levels, it makes sense for older workers to bear the brunt of the high exposures. The older workers will not be concerned about genetic effects for their offspring, and because of a 20-year or so latency period for most cancers, that fear may also be reduced.

The number of workers who are exposed to radiation at the occupational level is a very small fraction of the population. The general population should, of course, not receive doses of that magnitude. It has been determined that one tenth of the occupational dose would be acceptable for the general public where no monitoring or controls are available. The exposure of the general public is intended to include occasional exposure from various sources, but not some general source, such as fallout, that would irradiate everyone in the population somewhat uniformly. The general public maximum permissible dose of 0.5 rem per year is in addition to background radiation and any medical exposures. This limit of about 10 mrem per week would apply, for example, to the dose rate permitted at the fence line of a nuclear installation.

The maximum dose acceptable for the entire population is, of course, much lower than that for individual members of the general public; it is based on genetic defects to the entire gene pool. The BEIR II report, considering the costs and benefits to society, recommended that 5 rem per generation be set as the maximum permissible dose for the entire population. Taking a generation as 30 years means that the permissible dose is 170 mrem per year in addition to natural background and medical exposures. This dose level has been criticized as being more lenient than necessary, as there is almost no situation outside of a major nuclear war that would create doses of that magnitude. We have seen, for example, that a full-scale program of nuclear reactors for electricity generation in the United States would present a dose of only 0.5 mrem to the general population by the year 2000. It is somewhat by chance that the 170 mrem per year maximum permissible dose is about one half of the natural background and medical doses. We know, however, that man has successfully evolved in the presence of a background of ionizing radiation, so no extreme consequences of an additional 170 mrem per year can be anticipated.

Various radioactive isotopes that either occur naturally or are man-made can find their way into the air and water and possibly end up as a body burden. Using the occupational total body dose of 5 rem per year and some special maximum permissible doses to critical organs, tables have been prepared by the government for the maximum permissible body burdens of all common radioisotopes based on the absorbed dose per microcurie. Based on information on the retention of various substances from the air by the lungs and from water by the digestive system, maximum permissible concentrations (microcuries/cm^3) have also been tabulated for air and water. When the general public is involved, the occupational body burdens and concentrations must be scaled down from the occupational levels.

11.9 RADON IN HOMES

The basic nature of the radon problem was discussed in section 11.6, and it was noted that the average equivalent effective dose in the United States was 200 mrem per year, by far the largest source of background or man-made radiation. Since radon gas is invisible, odorless, and tasteless, it is a problem that can easily be overlooked, but according to the Environmental Protection Agency it is probably the worst environmental hazard we have. The estimate of 0.4% as the average lifetime risk of dying of radon-related lung cancer puts it far above such hazards as asbestos, ethylene dibromide, orange dye number 19 in lipstick, and the other concerns of the EPA. It is within the last 10 years or so that radon has received much attention. Radon concentration measurements in homes are now rather routine and are in fact required by many private and federal mortgage agencies.

Measurements of the radon level can be made rather easily in a closed environment by either an alpha particle track detector or by a charcoal adsorption collector. The measurements provide the number of picocuries per liter of air of the ^{222}Rn activity. One picocurie per liter of air is equivalent to 37 disintegrations/sec • m^3 or 37 Bq/m^3. Surveys have found an average annual concentration of 1.5 picocuries per liter of air for the living spaces in U.S. homes. The EPA has set a guideline of 4 picocuries per liter, above which remedial action is advised. Based on one survey about 7%, or 4 million houses, were found to have a concentration above 4 picocuries per liter. A few percent are over 8 picocuries per liter, and some ranged up to several hundred. An EPA survey found that 21% of the houses tested in 10 states had concentrations higher than the guidelines. Short-term measurements are frequently inconsistent, as there are variations from month to month, from room to room, and with the weather. Basements generally provide the highest readings. There is a definite geographical dependence that is related to the type of soil and uranium concentrations.

Evaluation of the risk of lung cancers from radon in houses is based mainly on the experience of uranium miners. The radon levels in the mines were sometimes no more than five times greater than those encountered in average homes. A study (BEIR IV) was made in 1988 of the internal alpha emitter problem, which included radon and miners. Based on their findings, it is estimated that in the United States between 5000 and 20,000 deaths annually are due to radon in our homes. This study was the basis of the 0.4% cancer death risk estimate. This is indeed a high risk factor and is comparable to the 2% risk of death in an auto accident.

There are, however, a growing number of questions being raised about these high radon risk factors. The data from the miners' experiences are confounded by the fact that many of the miners were smokers, and the risk of lung cancer is 10 times greater for cigarette smokers than for nonsmokers, and the effects are probably multiplicative. Another problem occurs because the atmosphere in the mines was very dusty, and dust particles can serve as a vehicle for radon daughter entry into the lungs. The risk estimates also assume that people are exposed continually for their lifetimes to the radon levels in their homes, whereas in fact they are away for a large fraction of the day and frequently move from one house to another. There are thus some major uncertainties that enter the risk estimates. Epidemiological studies so far have failed to show any relationship between lung cancer and radon concentrations in homes. It can be assumed that these risk factors will be better delineated in the future, but in the meantime the EPA guideline stays at 4 picocuries per liter of air. In terms of risk of lung cancer, this is said to be about equivalent to smoking one half a pack of cigarettes a day or having 200 chest x-rays a year.

Whether for health or economic reasons, it behooves the homeowner with radon levels over 4 picocuries per liter of air to look into measures to mitigate the problem. While sealing cracks in the walls and floor of the cellar or the concrete slab can be helpful, it has been found far more effective to put pipes under the concrete floor and use a blower connected to these pipes to draw the radon from under the floor to the outside to reduce the pressure beneath the slab. Proper venting of crawlspaces and cellars can also be effective. Most homes can be brought well below the 4 picocurie per liter level for a $1,000 investment in remediation measures.

PROBLEMS

1. How many ion pairs are made per second in a 50-cm^3 air ionization chamber if it is irradiated with 150 R/sec? To how many amperes of current does this correspond?
2. A person receives an absorbed whole-body dose of 10 rads from 200 keV gamma rays, 16 rads from thermal neutrons, and 21 rads from 10 MeV neutrons all at the same time. How many total rem did the person receive? How many sieverts?
3. Make a careful estimate of the total dose in millirem that you received from all sources last year.
4. If, as a result of the explosion of a small nuclear bomb, 100,000 people are exposed to 50 rem, about how many excess cases of leukemia would you expect to occur over their lifetime? How soon after the explosion would the leukemias appear? Assume linearity.
5. How many serious genetic defects beyond those normally expected would occur in the first generation in children born in the United States if the whole population is subjected to 33 mrem per year? Assume an annual birth rate of 15 per thousand.
6. A worker over 18 years old who is helping to clean up the radioactive spill at the Three Mile Island reactor is assigned to work in an area where she will receive 25 mrem/hr. How many hours can she work in the area in 1 year's time?
7. How many excess cancer deaths will occur over a lifetime in the 3 million people in the Rocky Mountain region who live at 5300-ft altitude rather than at sea level? Assume linearity.
8. How many excess cancer deaths will occur over a lifetime in the 250 million people in the United States because of radiation from consumer products? Assume linearity.
9. A particle of dust is deposited on the surface of a person's lung. It is radioactive and emits 5 MeV alpha particles that have a range of 25 μm in tissue. If 100 alpha

particles are emitted per second and the density of tissue is 1 g/cm^3, calculate the dose in rem received in a localized area of the lung by the person per year. Hint: Assume all of the energy of the alpha particles is deposited in a hemisphere of tissue with a radius of 25 μm, and the RBE is 15.

10. The flight crews of commercial airlines receive an additional dose of 100 mrem per year. Assuming a crew is exposed to this level from age 18 to 65, what is the risk these individuals face in dying from cancer over their lifetime due to this exposure? Assume linearity.

11. The average dose from a chest x-ray is 10 mrem. How many excess cancer deaths will be experienced if 50 million people receive such an x-ray? Assume linearity.

SUGGESTED READING AND REFERENCES

1. Cember, H. *Introduction to Health Physics.* Second Edition. Oxford, England: Pergamon Press, 1983.
2. Morgan, K. Z., and Turner, J. E., Eds. *Principles of Radiation Protection.* Huntington, N.Y.: Robert E. Krieger, 1973.
3. Schull, W. J.; Otake, M.; and Neel, J. V. "Genetic Effects of the Atomic Bombs: A Reappraisal." *Science,* **213** 4513 (September 11 1981), pp. 1220–1224.
4. Upton, A. C. "The Biological Effects of Low-Level Ionizing Radiation," *Scientific American,* **246** 2 (February 1982), pp. 41–49.
5. Wallace, B., and Dobzhansky, Th. *Radiation, Genes and Man.* New York: Holt, Rinehart & Winston, 1963.
6. National Research Council Committee on the Biological Effects of Ionizing Radiation. *The Effects on Populations of Exposure to Low Levels of Ionizing Radiation: 1980* (BEIR III). Washington, D.C.: National Academy Press, 1980.
7. National Committee on Radiation Protection. *Permissible Dose from External Sources of Ionizing Radiation.* National Bureau of Standards Handbook 59. Washington, D.C.: U.S. Government Printing Office, 1954.
8. National Research Council Committee on the Biological Effects of Ionizing Radiation. "Health Effects of Exposure to Low Levels of Ionizing Radiation" (BEIR V). Washington, D.C.: National Academy Press, 1990.
9. Nero, Anthony V. Jr. "Controlling Indoor Air Pollution." *Scientific American,* **256** 5 (May 1988), pp. 42–48.
10. Kerr, Richard A. "Indoor Radon: 'The Deadliest Pollutant.' " *Science,* **240** 4852 (April 1988), pp. 606–608.
11. Upton, Arthur C. "Health Effects of Low-Level Ionizing Radiation." *Physics Today,* **44** 8 (August 1991), pp. 34–39.

CHAPTER TWELVE

TRANSPORTATION

12.1 INTRODUCTION

Every civilization has had as one of its first goals the achievement of fast and reliable transportation. Well before recorded history, animals provided swifter and stronger land transport than did the human foot, and rafts and boats of all descriptions made waterways favorite avenues of commerce. History shows that success in trading and even in warfare has generally been governed by the quality of transportation. We now mark in our memories many ancient civilizations by the lingering artifacts of their transportation systems—canals, roadways, sea lanes, and even heavily worn footpaths.

With the technological developments of recent centuries, transportation systems have been the focus of major effort. The auto, truck, and train have evolved from the earliest wheel, which is still considered by many to be mankind's greatest invention. Our steel-hulled ocean vessels are direct descendants of the simplest watercraft, but our aircraft are almost entirely a modern development, indebted to concentrated energy sources. Although much of the advance of transportation technology is related to the exploitation of fossil fuels, that is not the whole story; we must recall that the early steam-powered trains and boats were fueled by firewood, hardly a modern fuel.

We now find that transportation is a major influence in our lives. Our cities and suburbs have been shaped by the automobile, and in a few locations by public-transit systems. We have given up 1% of our land in this country to roadways, an area equivalent to the state of West Virginia. The road area in the United States is nearly double that of all the national parks, but still only about one fortieth that of the farms. The fuel alone to power our transportation systems accounts for 25% of the total national energy budget, and the construction and maintenance of the roadways and manufacture of the vehicles accounts for a significant additional fraction. Indicating the value we place on transportation, only food and housing represent greater expenditures for the average citizen, who spends about 10% of personal income on transportation. In the United States one fourth of all retail sales are related to the automobile, nearly one fifth of our Gross National Product is accounted for by motor vehicle and allied industries, and one sixth of American workers are employed in the manufacture, distribution, service, and commercial use of motor vehicles. Our choices in mode of transportation are reflected to some degree in the amount of fuel consumed by each mode, shown in Tables 12.1 and 12.2 for both passenger and freight transport. It is clear from this and other indicators that the personal automobile is overwhelmingly the favorite, and any national pro-

Table 12.1 ENERGY USED IN U.S. TRANSPORTATION (1990)

	Btu (10^{15})	Percent
Light-duty vehicles	11.9	54
Freight trucks	5.1	23
Air transport	3.3	15
Marine	1.1	5
Rail	0.4	2
Total	22.1	(3.8×10^9 bbl petroleum)

Source: *Annual Energy Outlook*, U.S. Department of Energy, 1992.

Table 12.2 ENERGY INTENSIVENESS IN VARIOUS FORMS OF TRANSPORTATION

Passenger Transportation	Btu/Passenger-Mile
Bicycle (8 mph)	310
Walking (3 mph)	524
Bus	1,700
Train	2,620
Automobile	5,400
Airplane	7,150
Freight Transportation	**Btu/Ton-Mile**
Pipelines	450
Waterways	540
Railroads	680
Trucks	2,340
Aircraft	37,000

Source: Richard C. Dorf, *The Energy Factbook*, McGraw-Hill Inc., 1981.

gram to conserve fuel in transportation must focus on the biggest user, the automobile.

Some reductions in fuel energy used in transportation can be achieved purely through technological changes, such as by the use of diesel rather than gasoline-powered engines. Other changes may involve urban restructuring; still other improvements can be effected by electing to use alternative modes of transport such as railroads rather than trucks for intercity freight transport. Some possibilities in this direction are apparent from the data presented in Table 12.2. Table 12.1 shows that trucks consume over 10 times as much fuel in this country as do trains, even though trains are typically many times more efficient in terms of ton-miles per Btu. Other changes in transportation mode in the interest of conserving energy are suggested by the data of Table 12.2.

Several specific areas in which transportation and science are related are discussed in the following sections of this chapter, an equally large number could easily be added.

12.2 FORCE, POWER, AND ENERGY REQUIREMENTS

FORCE REQUIREMENTS

In evaluating the power requirements for any wheeled vehicle, it is first necessary to know what forces must be provided to overcome the various resistances to motion. There are various accepted parameterizations of these forces depending on specific engineering approaches and approximations. To some extent, they all are ways of formalizing what is, to a considerable degree, empirical information. The approach we shall follow is neither the most nor the least sophisticated and complete.

In general, the total motive force that must be provided for satisfactory vehicle performance may be written as the sum of four terms:

$$F = ma + msg + C_r mv + C_{ad} v^2$$

or as

$$F = F_\mathrm{a} + F_\mathrm{h} + F_\mathrm{r} + F_\mathrm{ad}$$

These terms may be considered separately:

$F_a = ma$. A force necessary to provide acceleration, a, or change of speed. The acceleration is an important variable in describing the motion of a car. If a is constant, then

$$a = \frac{v_\mathrm{f} - v_\mathrm{i}}{t}$$

where v_f is the final speed, v_i is the initial speed, and t is the time during which the speed changes. If the final velocity is greater than the initial velocity, then the acceleration is positive. If, on the other hand, v_f is less than v_i, then the acceleration is negative (deceleration). The basic law of physics $F = ma$ is a direct consequence of Newton's laws of motion. The force, F_a, is positive if the speed is increasing and negative if it is decreasing, as when braking. The force is also proportional to the mass, m, of the vehicle.

$F_h = msg$. A hill-climbing force, given by the product of the mass times the slope, s (in percent/100), times the acceleration of gravity, g (32.2 ft/sec^2). This term is positive if the vehicle is ascending the grade, negative if descending.

$F_r = C_r mv$. The force necessary to overcome rolling resistance due to the tires encountering irregularities in the road surface and, more important, to the flexing of the tires even on perfectly smooth roads, as well as to frictional losses in moving parts of the vehicle. This force requirement increases in proportion to the product of the vehicle mass times its speed. Under ordinary conditions, the energy put into overcoming rolling resistance goes predominantly into flexing, and thus heating, the tires. Only about 1 to 5% of the energy loss to rolling resistance is due to tread slippage on the road, and 1 to 3% is due to air friction with the rolling tire. The frictional losses in the vehicle drive train bearings are usually small enough to ignore, and the rest goes into the direct tire losses. These tire losses can be minimized through use of old, worn, more flexible tires, radial-ply tires, or highly inflated tires. If the vehicle mass is given in the engineering units of slugs (slugs = weight in pounds/32.2), then $C_r = 0.01$ if v is in miles per hour, and $C_r = 0.007$ if v is in feet per second. These choices of C_r will give F_r in pounds force.

$F_{ad} = C_{ad}v^2$. This term represents the force necessary to overcome aerodynamic drag on the vehicle. It is usually negligibly small at low velocities but increases rapidly as the speed increases. As a rule of thumb for most vehicles, this is the most important drag term at speeds greater than about 40 mph, and it is responsible for most of the decrease in fuel economy at higher speeds. The magnitude of the aerodynamic drag on a vehicle depends on several factors, including air density, speed, the vehicle shape, size, and even the smoothness of the outer surface. For a given road speed, this drag force is increased when the vehicle is heading into the wind and reduced when there is a tail wind. Cross-wind conditions are difficult to analyze and are beyond the scope of this discussion, but they usually increase the effective drag force. As a reasonable approximation for typical conditions with no wind present, the aerodynamic drag force is given by C_{ad} times the square of the vehicle velocity, where C_{ad} may be written as

$$C_{ad} = \frac{C_D A_f}{370}$$

In this equation C_D is the commonly tabulated aerodynamic drag coefficient, and A_f is the frontal area of the vehicle given in square feet. Then, if v is given in miles per hour,

$$F_{ad} = \frac{C_D A_f v^2}{370}$$

will be given in pounds force. The aerodynamic drag coefficient, C_D, expresses how well the vehicle is streamlined so that it will move through the air with a minimum of energy used to pushing air out of the way or creating turbulence. Small values of C_D are preferable; some representative values are shown in Table 12.3. The frontal area of the vehicle is the total area projected in the direction of motion, including projecting ski racks, mirrors, door handles, tires, and antennae. The vehicle can be thought of as cutting its frontal area out of the air just as a cookie cutter cuts a pattern out of dough. Any drag-producing components that do not add to the frontal area, such as a rear bumpers or windshield wipers, have their aerodynamic drag effects included entirely in the aerodynamic drag coefficient. The reduction of the aerodynamic drag coefficient provides great opportunity for the auto designer to improve fuel economy at no sacrifice in comfort or convenience to the driver. A series of small changes—setting the headlights flush, fairing in the rear bumpers, using flatter wheel covers, upturning the trailing edge of the rear deck, and so forth—all

Table 12.3 SOME VALUES FOR THE AERODYNAMIC DRAG COEFFICIENT (C_D)

Square flat plate (worst)	1.17
Rectangular block	1.0
Ordinary truck	0.7
1972 Dodge Polara Wagon	0.6
Streamlined truck	0.55
1981 Cadillac Eldorado	0.55
Porsche 928	0.45
Jaguar XK-E	0.40
Escort/Lynx	0.39
Camaro	0.39
Datsun 280 ZX	0.39
Porsche 924	0.34
1992 Ford Taurus	0.32
Lotus Europa	0.29
VW research vehicle	0.15
Teardrop (theoretical best)	0.03

add up to substantial gains in fuel economy, probably 10 to 20% over the cars on the market 10 years ago. These changes are usually as much a product of an empirical design art as of automotive science, with frequent testing of ideas in a trial-and-error approach. The net result on the national transportation fuel budget has been, and will continue to be, a significant part of our energy conservation effort. The drag coefficient of a particular vehicle can be determined by wind tunnel tests. There are several design changes to the traditional automobile that can reduce the drag coefficient, for instance, including a smooth underbody pan to reduce the drag of the various drive components. (Notice how smooth an airplane is on the underside.) In doing this, the designer must be careful not to impede unduly access for maintenance or the airflow that ordinarily provides engine and muffler cooling. Also, an underbody pan can introduce a fire hazard by collecting fuel and oil from leaks, which would merely drip onto the road in the absence of the pan. In spite of these possible difficulties, the bellypan is an attractive design change because it can reduce the drag coefficient by about 15%. A dam under the front bumper to divert airflow away from under the car provides some of the same beneficial effects as an underbody pan.

The designer must also be aware of the more subtle contributions to aerodynamic drag, such as the air pressure produced on the leading surface of the firewall by air admitted through the grille to cool the engine. This added pressure on the firewall, the partition between the engine compartment and the passenger compartment, normally accounts for about 5% of the aerodynamic drag. This effect calls for effort on the part of the designer to provide for streamlined airflow into and out of the engine compartment as well as over the exterior surfaces of the auto. It is also important to consider the flow of air through the passenger compartment. There is a good deal of variation from one auto to another in the drag effects produced when windows are open at high speeds. In recent years, with increasing fuel costs, various accessories to streamline truck-trailer combinations have become more popular. These devices often have a very short payback time because of the large number of miles these trucks travel at high speed in a year's time.

Example 12.1

What is the maximum speed (terminal velocity) that a car will attain when coasting down a long 10% grade if aerodynamic drag is the dominant retarding force? The car weighs 3200 lb, has a frontal area of 30 ft^2, and has a drag coefficient of 0.55.

Solution

At terminal velocity, the acceleration is zero, and for a coasting vehicle, the force provided by the engine is zero. Therefore, the force down the hill (msg) must be equal to the aerodynamic drag force ($C_D A_f v^2/370$) if the rolling resistance is ignored.

$$msg = \frac{C_D A_f v^2}{370} \quad \text{or}$$

$$v^2 = \frac{370\, msg}{C_D A_f} \quad \text{and}$$

$$v = \sqrt{\frac{370\,msg}{C_D A_f}} = \sqrt{\frac{370 \times 3200/g \times 0.1 \times g}{0.55 \times 30}}$$

$$= 84.7 \text{ mph}$$

As a check on the validity of ignoring the rolling resistance relative to the aerodynamic drag, we may calculate these forces separately at the terminal velocity just calculated:

$$F_r = 0.01\,mv = 0.01 \times \frac{3200}{32} \times 84.7 = 84.7 \text{ lb}$$

$$F_{ad} = \frac{C_D A_f v^2}{370} = \frac{0.55 \times 30 \times (84.7)^2}{370} = 320 \text{ lb}$$

Thus, we see that the aerodynamic drag is indeed much larger than the rolling resistance. A more accurate calculation, which includes both the rolling resistance and the aerodynamic drag, gives a terminal velocity of 74.2 mph. This calculation is not shown here because it involves the solution of a quadratic equation.

POWER REQUIREMENTS

Once the force required to propel a vehicle is determined, it is a simple matter to calculate the power requirement. The power expended against any force is the product of the force times the velocity with which an object moves against that force. This can be written as

$$\text{power (ft} \cdot \text{lb/sec)} = \text{force (lb)} \times \text{velocity (ft/sec)}$$

or

$$\text{power (horsepower)} = \text{force (lb)} \times \text{velocity (ft/sec)} \times \frac{1 \text{ hp}}{550 \text{ ft} \cdot \text{lb/sec}}$$

These equations may be used to determine the power necessary to overcome the total drag forces on a vehicle or any one of the individual drag forces.

For most vehicles the power available at the drive wheels is but a small fraction of the power released by the combustion of the fuel for several reasons. The various losses are shown diagrammatically in Figure 12.1. The net result is that only about 15% of the fuel energy is available for propelling a typical vehicle, with the range of actual efficiencies being about 10 to 20%.

Example 12.2 illustrates the relationship between aerodynamic drag, velocity, and power. Some useful formulas are given in Table 12.4.

Example 12.2

What is the top speed on a level road for a 3200-lb car with 40 hp available at the drive wheels if $C_D = 0.55$ and $A_f = 30$ ft^2? Assume that at top speed the dominant drag force is aerodynamic, so that rolling resistance can be ignored.

Solution

On a level road, no power goes into climbing hills, and at top speed, the acceleration is zero. Therefore, if rolling resistance can be ignored, the total delivered power goes into overcoming aerodynamic drag. Then, using an equation from Table 12.4,

$$40 \text{ hp} = \frac{C_D A_f v^3}{139{,}000}$$

$$v^3 = \frac{40 \times 139{,}000}{0.55 \times 30}$$

$$v = \left(\frac{40 \times 139{,}000}{0.55 \times 30}\right)^{1/3} = (336{,}970)^{1/3} = 69.6 \text{ mph}$$

ENERGY REQUIREMENTS

It has long been a popular myth that if certain revolutionary carburetor designs were made available, automobile fuel mileage would be dramatically increased to 100 or 200 miles per gallon or so for vehicles of ordinary dimensions. The truth of this myth can be examined by calculating the energy requirement for vehicle propulsion, starting from the heat of combustion of gasoline. As we recall from an earlier chapter, this heat of combustion is the maximum energy released on combustion of the fuel, as measured under laboratory conditions. No clever carburetion devices can increase the heat of combustion.

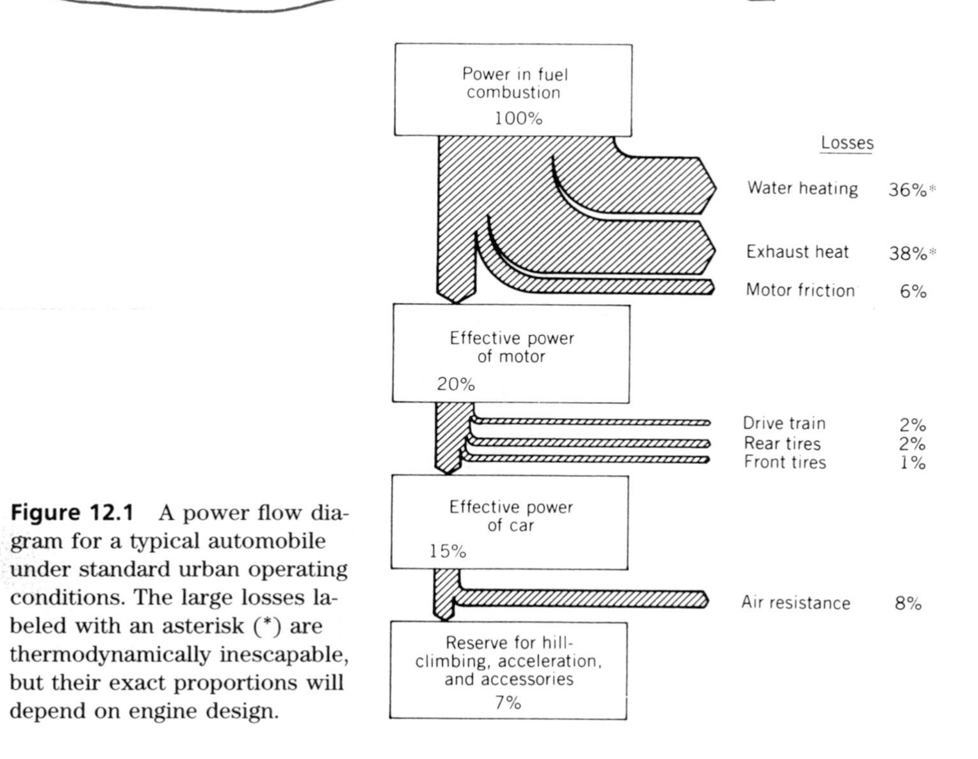

Figure 12.1 A power flow diagram for a typical automobile under standard urban operating conditions. The large losses labeled with an asterisk (*) are thermodynamically inescapable, but their exact proportions will depend on engine design.

Table 12.4 SOME USEFUL FORMULAS FOR POWER AND ENERGY REQUIREMENTS OF VEHICLES (BRITISH SYSTEM UNITS)

$F_T = F_a + F_h + F_r + F_{ad}$ (all forces in pounds)
$F_a = ma$ (m in slugs, a in ft/sec)
$F_h = msg$ (s = slope in percent/100)
$F_r = 0.01\ mv$ (v in mph)

$$F_{ad} = \frac{C_D A_f V^2}{370} \ (A_f \ in\ ft^2,\ v \text{ in mph})$$

Power = force × velocity (power in ft • lb/sec, force in lb, velocity in ft/sec)

$$P_{ad} = \frac{C_D A_f v^3}{139{,}000} \ (P_{ad} \text{ in hp}, A_f \text{ in ft}^2, v \text{ in mph})$$

$$P\,(\text{hp}) = \frac{1}{550} P\,(\text{ft} \cdot \text{lb/sec})$$

$$\text{Energy (ft} \cdot \text{lb)} = \text{force (lb)} \times \text{distance (ft)}$$
$$= \text{force (lb)} \times \text{velocity (ft/sec)} \times \text{time (sec)}$$

The energy required to move a vehicle a given distance is equivalent to work, which is given by the product of force times distance, or

$$\text{Energy (ft} \cdot \text{lb)} = \text{force (lb)} \times \text{distance (ft)}$$

The energy will also be given by the product of power times time, the power is force times velocity. Therefore

$$\text{Energy (ft} \cdot \text{lb)} = \text{power} \left(\frac{\text{ft} \cdot \text{lb}}{\text{sec}}\right) \times \text{time (sec)}$$

or

$$\text{Energy (ft} \cdot \text{lb)} = \text{force (lb)} \times \text{velocity (ft/sec)} \times \text{time (sec)}$$

Figure 12.2 Even small cars require considerable horsepower for acceleration and hill climbing.

Example 12.3 shows that the calculated fuel consumption rate is close to what is actually experienced. From this example, it is clear that the fuel consumption would be decreased if the mass, velocity, frontal area, or drag coefficient were smaller, or if the efficiency for converting power in the fuel combustion into power at the drive wheels were improved.

Example 12.3

What is the maximum mileage in miles per gallon for an auto weighing 3200 lb, $C_D = 0.55$, and $A_f = 30\ \text{ft}^2$, on a level road with no wind at a steady speed of 50 mph? Assume that the delivered energy is 15% of the fuel energy of gasoline.

Solution

First calculate the total motive force required:

$$F_T = F_r + F_{ad} = 0.01\,mv + \frac{C_D A_f V^2}{370}$$

$$= 0.01\,\frac{3200}{32} \times 50 + \frac{0.55 \times 30 \times (50)^2}{370}$$

$$= 50\ \text{lb} + 111.5\ \text{lb} = 161.5\ \text{lb}$$

Per mile, the energy expended is the product of force times distance:

$$161.5\ \text{lb} \times 1\ \text{mile} \times 5280\ \text{ft/mile} = 853{,}000\ \text{ft} \bullet \text{lb}$$

At 15% efficiency, the *input energy* must be

$$\frac{8.53 \times 10^5\ \text{ft} \bullet \text{lb/mile}}{0.15} = 5.68 \times 10^6\ \text{ft} \bullet \text{lb/mile}$$

In terms of Btus, at 1 Btu = 778 ft • lb, the input energy is

$$\frac{5.68 \times 10^6\ \text{ft} \bullet \text{lb/mile}}{778\ \text{ft} \bullet \text{lb/Btu}} = 7307\ \text{Btu/mile}$$

In terms of gasoline this is

$$\frac{1.25 \times 10^5\ \text{Btu/gal}}{7307\ \text{Btu/mile}} = 17.1\ \text{miles per gallon}$$

Note that the energy expended against each component of resistance is in proportion to the magnitude of that resistance. Thus, 50/161.5 = 31% of the delivered energy goes primarily into heating the tires, and 111.5/161.5 = 69% of the delivered energy goes into pushing air out of the way under these specific driving conditions.

The amount of fuel consumed daily in the United States for transportation went up by a factor of roughly 3 from 1948 to 1990. The increase was brought about by a growing population, a decrease in the fuel efficiency of automobiles, and a greater reliance on cars rather than other means of transportation. The oil embargo of 1973 served to remind the public that this rate of increase could not continue and that changes were in store for the wasteful driving habits of the average American.

There are several ways reductions can be made in the fuel consumed by automobiles in the United States. One obvious way is for people to drive less. This can come about in a variety of ways, such as the greater use of public transportation, carpooling, living near to one's place of employment, riding bicycles, walking, and reducing the number of shopping trips. A second way to reduce gasoline consumption is to drive with energy conservation in mind. Although most of the mass is the car itself, it is important to keep the load mass to a minimum since the mass appears in three of the four force terms. Extra cans of gasoline or bags of sand carried in the vehicle can be costly in fuel consumption. The velocity appears in two of the force terms, but as we have seen, it enters as v^2 in the important aerodynamic drag term. The 55 mph federal speed limit was mandated in 1973 because of the lower fuel consumption and the reduced rate and severity of accidents at this reduced speed.

Proper car maintenance can also play a significant role in reducing fuel consumption. Frequent car tune-ups with attention to the ignition timing, spark plugs, air filter, and the correct air–fuel mixture adjustment can easily bring about an average 10% fuel saving. Proper inflation of the tires can significantly reduce the rolling resistance force. On the other hand, it is clear that automatic transmissions and accessories such as air conditioners and luggage or ski racks can add to gasoline consumption.

Aside from reducing the use of cars and driving and maintaining them well to conserve fuel, the large remaining factor is the purchase or use of a car that has good mileage. When the automobile first became popular in the United States in the 1920s, the typical car was light, devoid of accessories and automation, and got 25 to 30 miles per gallon of gasoline. As cars became larger and heavier, with more powerful engines, more accessories, and more automatic devices, gas mileage became progressively lower. Because the cost of fuel was not a major consideration for many car owners, the gas guzzler became a common sight on American streets and highways. As Figure 12.3 shows, in 1974 the gas mileage of new cars averaged less than 14 miles per gallon. Since that time, there has been considerable improvement in gas mileage, and this trend is likely to continue. What are the steps a car manufacturer can take to increase gas mileage and what is the limit?

The obvious first step is a reduction in the mass of the car. This can be brought about by a reduction in its size and in the mass of the materials; for example, by substituting aluminum and plastics for steel. With proper design, the reduction in size does not necessarily mean a reduction in the space for passengers. In the 1960s and early 1970s, standard U.S. sedans weighed in the range of 4000 to 5000 lb, whereas a subcompact now weighs about half as much. When the mass of a car is reduced along with perhaps some of the performance characteristics, sizable reductions can also be made in the size, horsepower, and weight of the engine. It was not uncommon to have 200 hp in a new car 20 years ago, whereas today's subcompact can perform well with no more than 75 hp.

As has been discussed, there is a large potential for mileage improvement by decreasing the aerodynamic drag, which accounts for about 25% of the

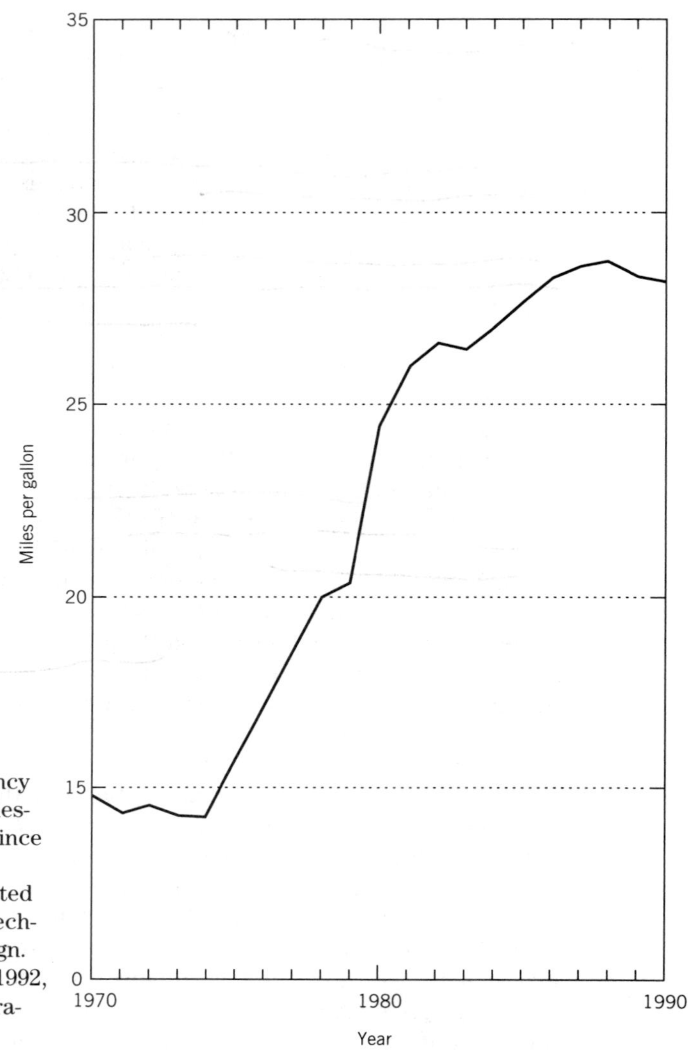

Figure 12.3 New car fuel efficiency in the United States. These are sales-weighted averages. The increase since the minimum in 1974 is due to increased fuel cost, federally mandated standards for fuel economy, and technical advances in automobile design. (Source: *Annual Energy Outlook* 1992, U.S. Energy Information Administration.)

energy used. The drag coefficient has been reduced from values around 0.5 to about 0.3 or 0.4. The smaller cars also have a smaller frontal area.

The car engine has been and will continue to be a target for improved efficiency, but the problems are intermeshed with the requirements for the reduction of air pollutants. The direct injection stratified-charge engine appears to have advantages both in efficiency and in reducing nitrogen oxide emissions. The transmission is also subject to improvement. Automatic transmissions account for about 10% of the energy used by a car. A manual four- or five-speed transmission is appreciably more effiicient.

The ultimate in gas mileage is not easily arrived at because many factors, such as performance in accelerating, space for passengers and luggage, accessories, and safety features, are involved. How far can the process go in reducing gas consumption for a usable family car? There now are cars with diesel and small conventional gasoline engines that achieve better than 40 miles per gallon, and there is certainly room for further economies. As outlined by Gray and von Hippel in their *Scientific American* article, 60 miles per gallon would appear to be achievable. If the majority of the U.S. public were driving such vehicles, significant reductions in gasoline consumption would ensue.

12.3 TRAFFIC FLOW THEORY

In designing a highway or any kind of public transit system, it is important to be able to make firm estimates of the maximum rate of traffic flow per lane of traffic. This analysis becomes very complex when a mix of different types of vehicles is involved, but it is quite tractable if certain simplifying assumptions are made.

The most important condition governing the permissible rate of traffic flow is the car-to-car spacing consistent with safety. If two identical vehicles, traveling at the same speed, one behind the other, are separated by a distance equal to the sum of the reaction distance (distance traveled between sighting a problem and applying brakes) and the braking distance (distance traveled after applying brakes), then this is a quite safe situation because the second driver can bring his car to a safe stop even if the first car suddenly stops with no advance warning given to the second driver. This is probably an unrealistically safe scenario. On the other hand, with perfectly similar vehicles separated by just the reaction distance, if the second driver begins to react as soon as the first driver hits the brakes, a collision may be avoided, but just barely. This is obviously an unrealistically hazardous situation.

We sometimes hear that vehicles should be separated by one car length for every 10 miles per hour of speed or by the distance the vehicles will travel in 2 seconds. At 55 mph (81 ft/sec), for vehicles 16 ft long, the first condition dictates a separation of 88 ft and the second a separation of 161 ft, appreciably more conservative in the direction of safety. It would seem that a satisfactory theory of traffic flow should be based on some recognition that braking distance depends on the square of the velocity, rather than linearly as is implicit in the latter two stipulations of vehicle separation.

A probably realistic proposal is that vehicles should be separated by the reaction distance plus one-half the braking distance. At 55 mph, a reaction time (t_r) of 0.5 sec and a braking deceleration of 0.6 g leads to a vehicle separation of 126 ft, not too different a value than found with the simpler prescriptions.

The *braking distance* is the distance the vehicle travels as it comes to a stop with the brakes fully applied. If the deceleration is constant during the stopping process, the braking distance is computed from a formula that relates the initial and final velocities to the acceleration and braking distance:

$$v_f^2 = v_i^2 + 2ax$$

In this case, $v_f = 0$, v_i is the initial velocity, a is negative (deceleration), and x is the braking distance.

$$x = v_i^2/2a$$

Often a is given in terms of a coefficient α times the gravitational acceleration.

In this simplified model, in which we assume that the brakes are either not in use or are fully applied, the braking distance is determined largely by the frictional forces between the tires and the road surface. The *coefficient of friction* is defined to be the ratio of the frictional drag force to the weight of an object when the object is sliding over a level surface. In the example of a vehicle skidding to a stop, the greater this drag force, the more quickly the vehicle stops, and in a shorter distance. The magnitude of the coefficient of friction depends on tire material and tread pattern, type of pavement, temperature, and even velocity. The product of the coefficient of friction (f) times the acceleration of gravity (g = 32 ft/sec^2) is the deceleration of the vehicle, or $a = f \times g$, in the direction opposite to the velocity. Although as a vehicle slows to a stop, its coefficient of friction changes slightly, in this text we treat the deceleration of the vehicle as constant during the stopping process. Some coefficients of friction are shown in Table 12.5.

If we adopt the traffic model mentioned earlier and illustrated in Figure 12.4, the vehicle separation is $x_R + \frac{1}{2} x_B$, and each vehicle in effect occupies a highway space of length (L) equal to $x_R + \frac{1}{2} x_B + l$, where l is the length of the vehicle. Now the number of vehicles passing a given point will be given by the velocity divided by the length of the vehicle "space," or L. In the form of an equation, this is

$$N\left(\frac{\text{number vehicles}}{\text{sec}}\right) = \frac{v(\text{ft/sec})}{L(\text{ft/vehicle})}$$

The vehicle space is velocity-dependent in both the reaction distance and braking distance

$$L = x_R + \frac{1}{2}x_B + l = vt_R + \frac{1}{2}\frac{v^2}{2a} + l$$

where t_R is the reaction time and a is the deceleration provided by the brakes. The distribution of driver reaction times is shown in Figure 12.5. Combining the last two equations gives

$$N = \frac{v}{vt_R + v^2/4a + l} = \frac{1}{t_R + v/4a + l/v}$$

Table 12.5 COEFFICIENTS OF FRICTION AND BRAKING DISTANCES ON A LEVEL ROAD FOR WET AND DRY CONDITIONS

Speed	Coefficient of Friction		Braking Distance on Level (ft)	
(mph)	Wet Pavement	Dry Pavement	Wet Pavement	Dry Pavement
30	0.35	0.62	87	49
40	0.32	0.60	169	90
50	0.30	0.58	280	145
60	0.28	0.56	432	216
70	—	0.55	—	299
80	—	0.53	—	406

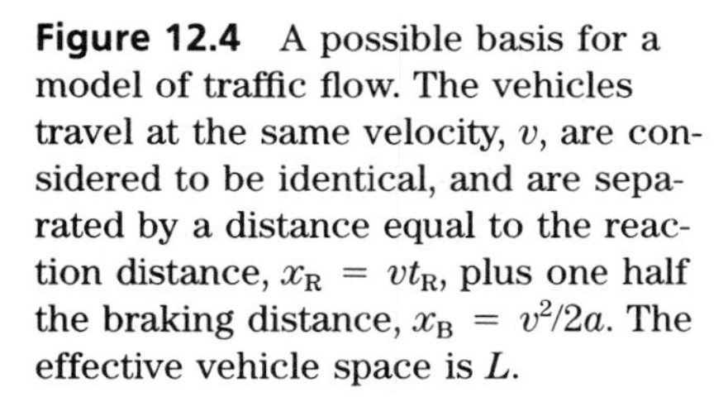

Figure 12.4 A possible basis for a model of traffic flow. The vehicles travel at the same velocity, v, are considered to be identical, and are separated by a distance equal to the reaction distance, $x_R = vt_R$, plus one half the braking distance, $x_B = v^2/2a$. The effective vehicle space is L.

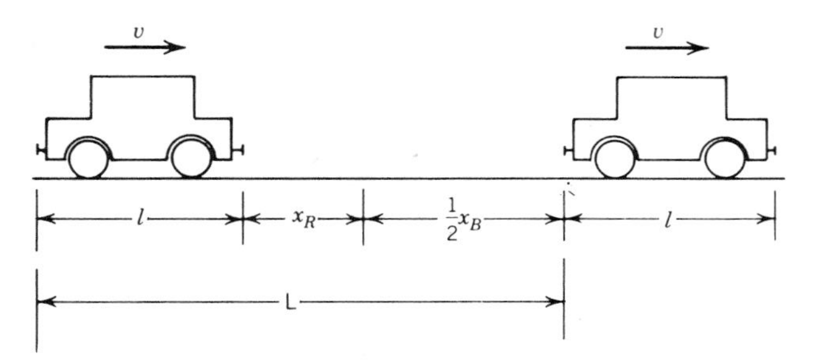

This equation shows that in the limit of zero velocity, N tends toward zero, as we would certainly guess. This comes about mathematically through the term l/v in the denominator, which tends toward infinity as v tends toward zero; the other term, $v/4a$, of course, vanishes as the velocity goes to zero. The other velocity limit, v, tending toward infinity, also results in N tending toward zero because it means that the vehicle separation also tends toward infinity. In this limit, the l/v term vanishes and the $v/4a$ term results in the denominator's tending toward infinity. The velocity dependence of N is illustrated in Figure 12.6. The point of interest on this curve is where N is at a maximum, as this is the velocity at which traffic should flow for the maximum number of vehicles per second down a single lane of traffic. This becomes important as tens of thousands of vehicles leave a stadium after a Madonna concert or football game, or during the evening rush hour when large numbers of people want to leave the central city within less than an hour's time. Note that the velocity at which N attains a maximum may empty a stadium parking lot the most rapidly, but it is not the velocity that will minimize an individual's driving time. The velocity for which N is a maximum may be determined for any particular values of l, a, and t_R by algebraically calculating N for a number of values of v and observing which value of v produces the maximum N. There is a more formal approach, however, using methods of differential calculus to determine exactly the value of v for which N is a maximum. This latter method indicates that N

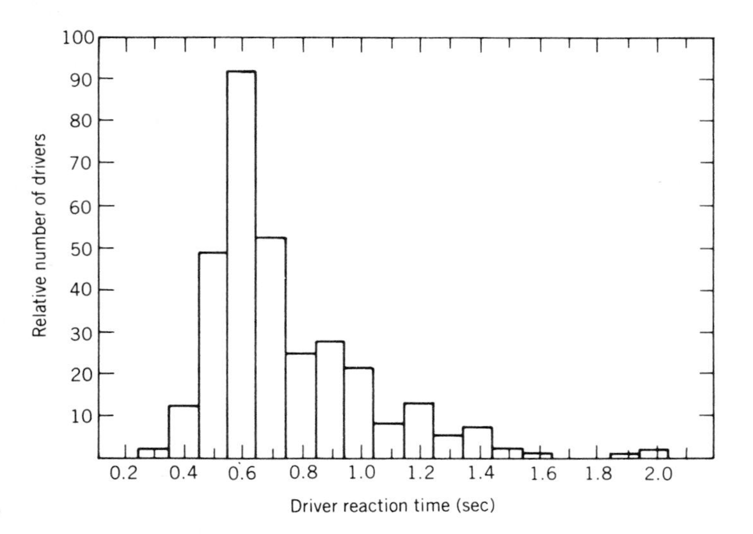

Figure 12.5 A distribution of measured driver reaction times.

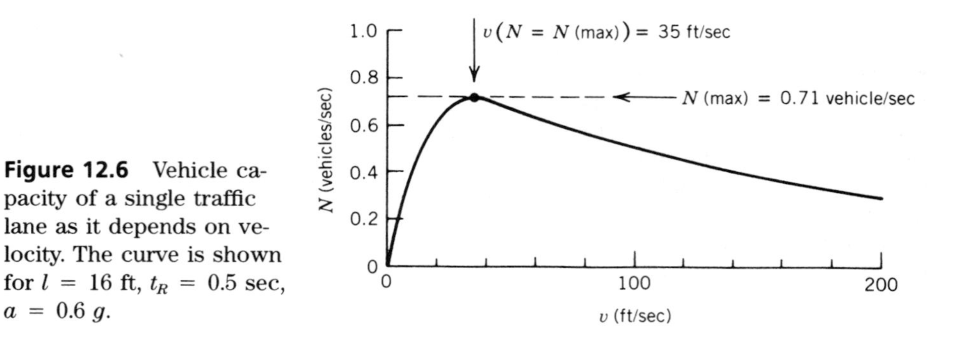

Figure 12.6 Vehicle capacity of a single traffic lane as it depends on velocity. The curve is shown for $l = 16$ ft, $t_R = 0.5$ sec, $a = 0.6\ g$.

is a maximum at the point where $v = 2\sqrt{al}$. For the conditions shown in Figure 12.6, this yields

$$v = 2\sqrt{0.6 \times 32 \times 16} = 2 \times 17.5 = 35 \text{ ft/sec} = 23.9 \text{ mph}$$

This speed may be lower than one might have first guessed, but it is consistent with the common observation that speed drops when traffic flow is heavy; perhaps drivers do unconsciously adjust their speed to maximum flow when cars are spaced as closely as safety allows. Note that $v(N = \text{max})$ depends *only* on a and l, and is independent of t_R. This means that the velocity at which N is a maximum is the same for all values of driver reaction time even though vehicle spacing and N do depend on reaction time.

Now that we know $v(N = \text{max})$, it is possible to calculate the actual maximum vehicle flow rate by using this value of v in the expression for N:

$$N(\text{max}) = \frac{1}{t_R + v/4a + l/v} = \frac{1}{0.5 + [35/(4 \times 0.6 \times 32)] + 16/35}$$

$$= \frac{1}{0.5 + 0.46 + 0.457} = \frac{1}{1.42}$$

$$= 0.71 \text{ vehicles/sec}$$

for the given t_R, l, and a. Note that in this model of traffic flow, $N(\text{max})$ increases as the braking deceleration increases, and decreases as reaction time and vehicle length become larger. The value of $N(\text{max}) = 0.71$ vehicles per second corresponds to a possible passenger flow rate of

$$0.71 \frac{\text{vehicles}}{\text{sec}} \times 3600 \frac{\text{sec}}{\text{hr}} \times 2 \frac{\text{passengers}}{\text{vehicle}} = 5112 \text{ passengers/hour}$$

which suggests that it takes four traffic lanes to move 20,000 persons per hour.

Under the conditions just considered, the space between cars would be

$$vt_R + \frac{1}{2} \bullet \frac{v^2}{2a} = 35 \times 0.5 + \frac{(35)^2}{4 \times 0.6 \times 32} = 33 \text{ ft}$$

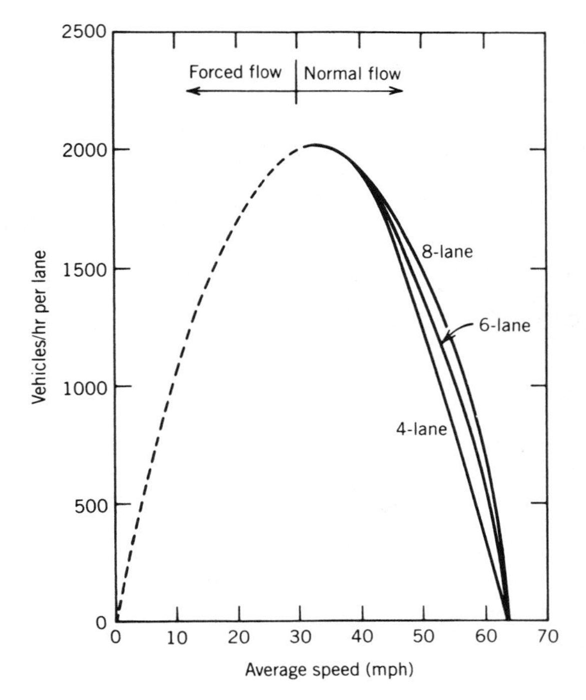

Figure 12.7 Typical measured traffic flow for multilane highways. In the normal flow conditions (above 30 mph), the lane capacity increases as the average speed is decreased. In the forced-flow condition, the vehicles become more tightly bunched and both the lane capacity and the average speed decrease. On a per-lane basis, there is rather little dependence on the number of lanes available.

which seems reasonable, as this is about two car lengths, or the distance traveled in 1 second.

Figure 12.7 shows the actual flow of passenger cars per hour for one lane of traffic as a function of the average speed in miles per hour. It is interesting to note that the maximum flow occurs at about 30 to 35 mph, which is quite close to the result of our simple model. The general features of the model of decreasing flow at low speeds and again at higher speeds are also clearly shown in Figure 12.7. There is surprisingly little dependence on the total number of lanes available. The maximum number of vehicles per hour predicted by the model is 0.71 vehicles/sec $\times$ 3600 sec/hr $=$ 2556 vehicles/hr, per lane, whereas in practice the number is about 2000 vehicles per hour. As the number of cars begins to exceed about 1900 cars per hour at 40 mph, a zone of instability is reached where the quality of flow begins to deteriorate. If the volume increases, there will be more braking and occasionally someone coming to a full stop. The average speed begins to decrease and with it the flow rate; a traffic jam develops as the flow follows the dashed curve indicated by a forced flow zone towards zero flow and zero speed.

Example 12.4

The previous discussion indicated that four traffic lanes could safely transport 20,000 persons per hour under certain stated assumptions. How many persons per hour could be transported by the same four traffic lanes in buses 50 ft long, with 60 passengers each and with brakes capable of 0.6 g, if the drivers' reaction time is 0.5 sec and if the same traffic flow model is assumed?

Solution

$$v(N = \max) = 2\sqrt{al} = 2\sqrt{0.6g \times 50} = 62 \text{ ft/sec}$$
$$= 42 \text{ mph}$$

Note that now $v(N = \max)$ is higher than for automobiles, thereby shortening the travel time for each passenger. To calculate the number of buses per second in each lane

$$N(\max) = \frac{1}{t_R + v/4a + l/v} = \frac{1}{0.5 + [62/(4 \times 0.6 \times 32)] + 50/62}$$
$$= 0.47 \text{ buses/sec} \cdot \text{lane}$$

Then

$$0.47 \frac{\text{buses}}{\text{sec} \cdot \text{lane}} \times 3600 \frac{\text{sec}}{\text{hr}} \times 60 \frac{\text{passengers}}{\text{bus}} \times 4 \text{ lanes} = 406{,}080 \text{ passengers/hour}$$

transported in four lanes.

This example demonstrates the much greater carrying capacity of a highway with the use of mass transit vehicles rather than private automombiles. It is obvious that the savings in this approach include the land, energy, and materials devoted to highway construction and maintenance as well as to the vehicles themselves. Of course, such an analysis applies exactly only to an idealized situation where mass transit vehicles travel with full passenger loads; in most situations the advantages are somewhat reduced, it is also clear that the velocities of maximum vehicle flow depend on vehicle length; thus buses do not achieve their full advantage when mixed in a traffic lane with automobiles. This is a good argument for exclusive bus lanes in areas of high traffic volume.

Although the traffic flow model just discussed does produce some insights

Figure 12.8 City expressways seldom have sufficient capacity for Friday evening rush hour traffic in the summertime.

into the factors governing patterns of traffic, it is far too simple a model to account for many of the subtleties of how drivers actually control their vehicles in traffic. A more complete model would have to take into account that each driver can often see much farther than just one car length ahead, that drivers do not drive at exactly constant velocity but tend to let their vehicle's speed fluctuate around some mean value, and that traffic tends to form into clumps of several vehicles rather than being uniformly spaced along a traffic lane. These details will have to be left to a more specialized treatment of traffic flow.

12.4 TRAFFIC SAFETY

STATISTICAL BACKGROUND

Transportation, in general, certainly ranks among the most hazardous enterprises ever undertaken by mankind. In many countries, such as the United States, even wars have not come close to traffic in terms of total deaths in this century. And like the casualties of war, traffic victims tend to be young, healthy, and male. In the past two decades in this country, we have averaged more than 20 million accidents a year, about 6 million injuries, and about 50,000 deaths. By now there have been more than 2 million traffic deaths in the United States. If the 50,000 annual toll were to occur in a single yearly event, it would exceed any disaster ever recorded in this country, and each year we would experience a single disaster exceeded only by a few events in human history. One might seriously ask whether this is the equivalent of a single every-10-years disaster that takes 500,000 lives, or whether the risk of traffic death surpasses the risk of energy-producing technologies, or whether the risk of traffic injury is not greater than that of the injurious effects of industrial pollutants in the environment.

In terms of the average life-shortening effect, traffic possibly surpasses the total of environmental pollutants because of the relative youth of its victims. This can be estimated by assuming that the average victim loses 30 years of life, and by considering the relative probability of traffic death compared with all other causes of death.

$$\begin{aligned}\text{average life-shortening effect} &= (\text{number years life lost per traffic death}) \times \\ &\quad \text{probability of traffic death}) \\ &= 30 \text{ yr} \times \left(\frac{46{,}000}{2{,}200{,}000}\right) \\ &= 0.63 \text{ yr} = 7.5 \text{ months}\end{aligned}$$

In this calculation we have used the 1989 statistics of 46,000 traffic deaths and 2.2 million total deaths in the United States.

The impact of vehicle accidents on our population and social structure is so profound that a fairly complete impression of this impact can be formed only by viewing the accident statistics in a number of ways. Several sets of comparative statistical data are presented in Tables 12.6 through 12.9 as well as in Figure 12.9. In the first of these tables, we see that vehicle occupants

Table 12.6 MOTOR VEHICLE DEATHS IN THE UNITED STATES (1989)

Vehicle occupants	34,900
Pedestrians	6,600
Motorcyclists	3,100
Bicyclists	800
Scheduled airlines	131
Bus passengers (1988)	54
Railway accidents (1988)	571
Total (death within 30 days)	46,000

Source: *Statistical Abstract of the United States,* 1991.

represent by far the largest number of victims; in the second table, motor vehicle accidents are seen to rank first among all accidental causes of death. Per passenger mile, as Table 12.9 shows, motorcycles far surpass the automobile in relative risk, but the automobile is, by an appreciable margin, more risky than other modern forms of transportation. Of the various types of automobile accidents, vehicle–vehicle collisions are shown in Table 12.9 to account for the largest number of fatalities. Figure 12.9 compares by age group the percent of registered drivers with the percent of drivers in fatal accidents. Here the younger drivers are seen to be in a disproportionately large number of fatal accidents.

However the statistics are arrayed, it is evident that traffic accidents represent a large and immediate risk to nearly all members of our population, and that improvements in traffic safety must rank among our important social goals. In the remainder of this section, our discussion of traffic safety focuses on the automobile because the applications of physics are so obvious and because automotive travel dominates all other forms of travel in the number of victims.

AUTOMOBILE COLLISIONS

The immediate causes of injury and death in traffic accidents are numerous, ranging from broken glass, to fire, to crushing, to the hurling of victims against components of the car's interior. All of these factors must be considered in designing a safer vehicle, but there is one basic factor in determining the survivability of a crash—the acceleration experienced by the passenger. This is related to the fundamental equation $F = ma$, which describes the force on any material object as being given by the product of its mass times its acceleration. In other words, the force per unit mass on any object is equal in magnitude to the acceleration and in the direction of the acceleration. If a 150-

Table 12.7 FATALITIES IN THE UNITED STATES BY TYPE OF ACCIDENT (1988)

Motor vehicle	49,000
Falls	12,000
Drowning	4,200
Fires	5,000
Electrocution	700
Lightning	100 (est.)

Source: *Statistical Abstract of the United States,* 1991.

Table 12.8 TRAFFIC FATALITIES PER PASSENGER MILE IN UNITED STATES

	Deaths/10^8 Passenger Miles
Automobiles	2.2 (1988)
Automobiles on interstate highways	1.2 (1988)
Bus	0.19 (1972)
Train	0.53 (1972)
	0.07 (1969)
Airline	0.08 (1989)
Motorcycle	22 (1968)
Horses (26×10^6 horses, 1.3×10^{10} miles, 3850 deaths)	30 (1909)

pound person experiences a horizontal acceleration of 5 g (or $5 \times 32 = 160$ ft/sec^2), that person will have an effective weight of $5 \times 150 = 750$ pounds in the horizontal direction in addition to 150 pounds vertically downward.

Since acceleration is defined as the change in velocity divided by the time during which the change took place, collisions of the type in which velocity is abruptly changed produce the most severe acceleration and, thereby, the largest forces. They are the most dangerous. An especially severe situation is presented in the case of a vehicle striking an immovable object such that the vehicle's speed is changed from its initial value to zero in a very short time. There are several ways to express the maximum tolerable acceleration in a collision. Experiments with human volunteers have shown that an acceleration of 20 g is tolerable without injury. Published safety standards are in the range of 10 to 80 g where the higher values are tolerable only if the exposure is for a shorter time. There is a specification, based on what is known as a severity index (S.I.), that takes into account both the magnitude of acceleration and the length of time during which it is experienced.

$$\text{S.I.} = \alpha^{5/2} \times t$$

where α is the acceleration in g and t is the time of exposure in seconds. A proposed specification is S.I. ≤ 1000 for the human head in a collision at 30 mph.

It is important for vehicle designers to have some idea of what values of acceleration are survivable so that they can design vehicles accordingly. It should be noted that it is not sufficient to ensure that the passenger compart-

Table 12.9 AUTO FATALITIES BY TYPE OF ACCIDENT

	Number	Percent
Other vehicles	23,000	41
Noncollision (includes run off roadway, then strike object)	18,000	32
Pedestrians	10,000	18
Railroad trains	1,600	3
Bicycles	900	1.6
Fixed objects (walls, abutments, four wheels on road)	2,500	4.6

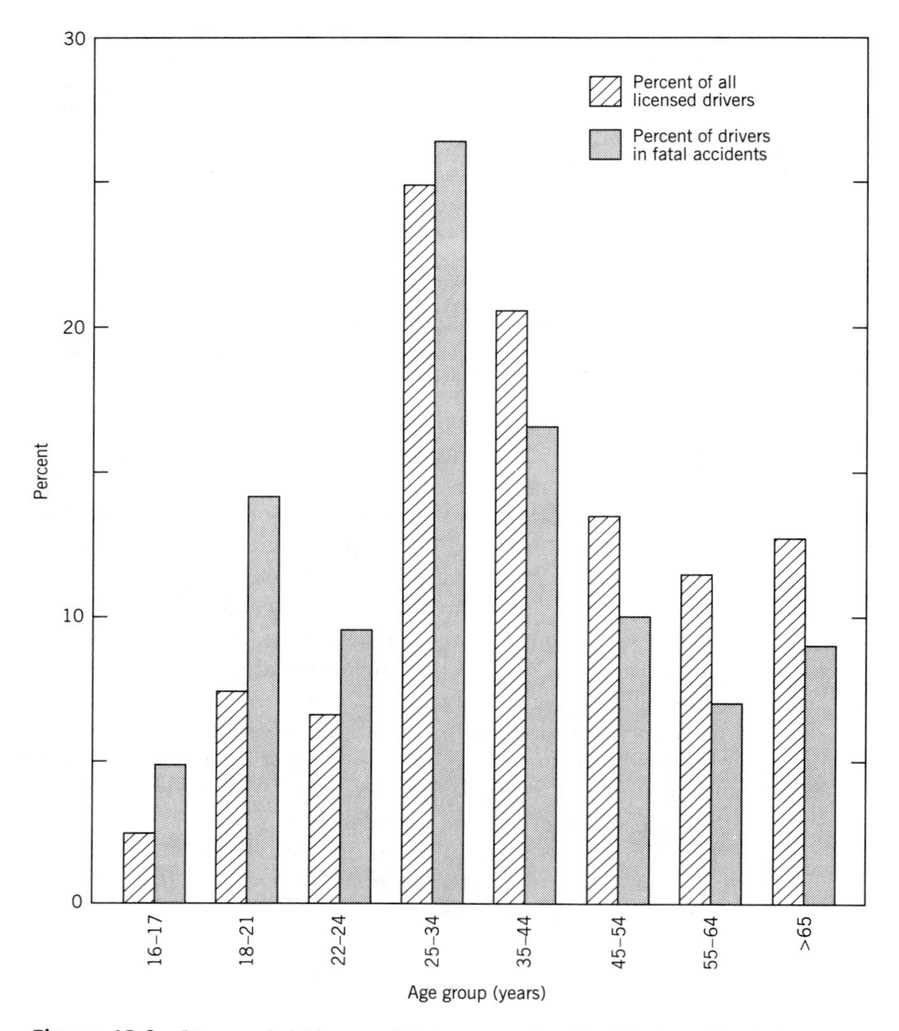

Figure 12.9 Licensed drivers and drivers involved in fatal accidents by age group in 1989. (Source: *Statistical Abstract of the United States*, 1991).

ment of a vehicle has a tolerable acceleration or severity index; these considerations must apply specifically to the bodies of the passengers, and even specifically to certain parts of the body. In some circumstances (see Example 12.5), the passenger compartment of a vehicle may undergo acceleration of a magnitude that is clearly survivable if the passengers are properly restrained by seat belts or air bags, but those passengers not so restrained will have no hope of survival. A second case in which the acceleration experienced by persons must be properly considered is that of an auto–pedestrian collision. This is usually actually two collisions; first the car hits the pedestrian, then the pedestrian hits the pavement. Even if the auto is designed to be free of protrusions and unyielding materials on the front end so that the first collision in this sequence is not lethal, the second may be.

Any attempts to reduce the death and injury toll in automobile accidents must take into account that front-end collisions are responsible for most of the injury-producing accidents (59%), followed by rollovers (21%). This suggests that a major effort must be given to designing the front ends of automobiles so that they will crumple in a way that results in a uniform acceleration over a relatively long distance, in order to keep forces on the passengers low. It is equally important for the passengers to be physically restrained within the passenger compartment so that their acceleration will be controlled and they will not be thrown from, or thrown about inside, the vehicle during a rollover. It is difficult to overestimate the benefits of good seat belt systems; a Swedish study found *no* deaths of seat-belted passengers in collisions up to 60 mph, but reported deaths in collisions at speeds as low as 12 mph for passengers not restrained by seat belts. Similar statistics were reported by a study in the state of Utah.

The major shortcoming of seat belts is that they are not used on a regular basis by most drivers and passengers, at least in the United States. As demonstrated in several countries of the world, it is possible to enforce the wearing of seat belts through laws, and thus reduce the injury and death rate, but that is still not a universal practice in this country. Air bags provide much the same benefits as seat belts for front-end collisions, even though the driver and passengers take no deliberate measures to protect themselves. For various technical and legal reasons, however, air bags are not yet standard equipment on all new cars in the United States. Objections to air bags include the arguments that they are relatively ineffective in collisions other than the front-end type, their sudden inflation may produce hearing damage, their accidental inflation could cause an accident, there is no obvious way to test their readiness, and they do not provide a uniform deceleration during the course of a collision. Nevertheless, there is growing evidence that air bags are very effective in saving lives and reducing injury. Their universal adoption in autos seems to be just a matter of time.

Example 12.5

An automobile crashes head-on into an immovable barrier when it is traveling at 55 mph. Assume a uniform rate of deceleration during the collision.

(a) How much crumple distance must be designed into the front end of the car so that the passenger compartment will not experience an acceleration of greater than 30 g?

(b) What is the duration of the crash in seconds?

(c) What is the Severity Index for this collision? Would properly seat-belted passengers be likely to survive?

(d) A careless passenger not wearing a seat belt stops in one fifteenth the distance it took to stop the passenger compartment (because the passenger flies through the air before striking the already stopped dashboard and windshield). How many g acceleration would the passenger experience? What is the Severity Index? Would the passenger be likely to survive?

Solution

(a) $$v_f^2 = v_i^2 + 2ax, \qquad \text{or} \qquad x = \frac{v_i^2}{2a} = \frac{v_i^2}{2\alpha g}, \qquad \text{when } v_f = 0$$

Given: $\alpha = 30$, $g = 32$ ft/sec^2, $v_i = 55$ mph $= 81$ ft/sec, and $a = \alpha g$. Then

$$x = \frac{(81)^2}{2\times 30 \times 32} = 3.42 \text{ ft}$$

(b) From $v_f = v_i + at$, with $v_f = 0$, and a negative, we get

$$t = \frac{v_i}{a} = \frac{v_i}{\alpha g} = \frac{81}{30 \times 32} = 0.0844 \text{ seconds}$$

(c)

$$\begin{aligned} \text{S.I.} = \alpha^{5/2} \times t &= 30^{5/2} \times 8.44 \times 10^{-2} \\ &= 30 \times 30 \times \sqrt{30} \times 8.44 \times 10^{-2} \\ &= 416 \end{aligned}$$

This is significantly less than the S.I. $\leqslant$ 1000 specification, so *properly restrained passengers would be likely to survive.*

(d) In this case the stopping distance is

$$x = 1/15 \times 3.42 \text{ ft} = 0.23 \text{ ft (2.74 in.)}$$

and the acceleration is

$$a = \frac{v_i^2}{2x} = \frac{81^2}{2 \times 0.23} = 14{,}263 \text{ ft/sec}^2$$

$$\alpha = \frac{a}{g} = \frac{14{,}263}{32} = 446$$

This large value of α indicates *the crash would not be survivable for an unrestrained passenger.* Another check on survivability is provided by the Severity Index:

$$\text{S.I.} = \alpha^{5/2}t = \alpha^{5/2}\frac{v_i}{\alpha g} = \alpha^{3/2}\frac{v_i}{g}$$

$$= 446^{3/2} \times \frac{81}{32} = 446 \times \sqrt{446} \times \frac{81}{32} = 23{,}842$$

This value of the Severity Index is much greater than the S.I. $\leqslant$ 1000 specification, clearly showing that the crash would not be survivable.

In addition to the measures indicated above, it is equally important that automobiles be designed so that the passenger compartment maintains its integrity during a collision. The engine must not move back into the passenger space. Certainly, the steering column should collapse rather than force the steering wheel back against the driver.

In recent years it has become common practice for highway departments to place arrays of certain yielding materials, such as empty oil drums, in front of unyielding bridge abutments and other solid roadside objects to reduce the acceleration experienced by vehicles that may collide with these objects. Such measures, in combination with seat belts, make it likely that collisions even at 60 mph will be survived by all the passengers in an automobile. In the absence of these simple, inexpensive, and yet very effective provisions, survival would be highly unlikely.

Of all possible types of vehicle–vehicle collisions, those of the head-on type between two vehicles can be especially severe. The probability of death or severe injury can be assessed particularly easily if the two vehicles are identical and are traveling at the same speed. In this case, once the front bumpers touch, the point of collision remains fixed, and each vehicle crunches up and comes to a stop just as if it had collided with an immovable object. This is contrary to the popular belief that such a collision between two vehicles is at least twice as severe as a collision between a vehicle and a solid object. We see that the two cases are the same by observing that if a curtain were either to cover a solid wall or simply be suspended in the air at the exact point of a vehicle–vehicle collision, the driver, or any other observer not able to see through the curtain, would not be able to distinguish which was the case. In either case, an analysis of the acceleration or the Severity Index would be the same as if the vehicle had crashed into an immovable object.

If the two vehicles colliding head-on have different mass or different speed, the situation is more complex. In this case, it is necessary to introduce another principle of physics known as *conservation of momentum.* In *any* collision, a property known as momentum is known to be conserved, just as energy must be conserved. The momentum of any moving object is given by the product of the object's mass and its velocity. In this case, it is necessary to specify velocity rather than speed because the direction of motion of each object must be considered in a consistent way. The principle involved and method of computation are illustrated by Examples 12.6 and 12.7.

Example 12.6

Consider the general case of a head-on collision between two vehicles of mass M_1 and M_2 and velocities V_1 and V_2, respectively. Assume that the collision results in the two cars remaining in contact and moving off together from the point of collision, rather than rebounding. Calculate their final velocity, V_f, after the collision.

Solution

The initial momentum is the sum of the individual momenta, or $P_i = P_1 + P_2 = M_1V_1 - M_2V_2$, where the motion to the right is taken to be positive. (Note that in this example the symbol P is used for momentum rather than power.) After the collision, the momentum is $P_a = (M_1 + M_2)V_f$. Because of momentum conser-

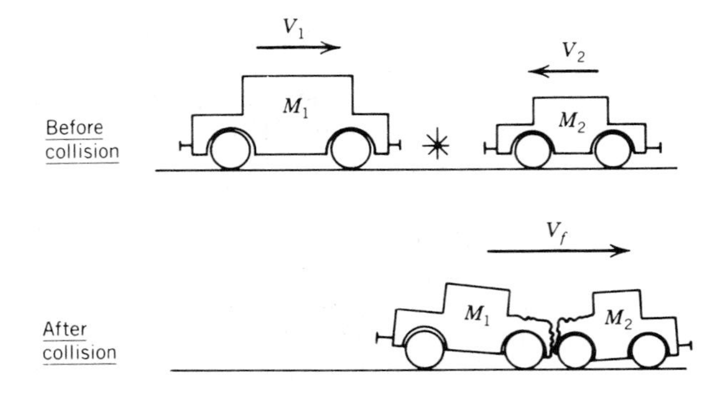

vation, we must have $P_i = P_a$, or $M_1V_1 - M_2V_2 = (M_1 + M_2)V_f$. This lets V_f be calculated as follows:

$$V_f = \frac{M_1V_1 - M_2V_2}{M_1 + M_2}$$

Note that V_f will be positive if $M_1V_1 > M_2V_2$, and negative if $M_1V_1 < M_2V_2$. If $M_1V_1 = M_2V_2$, then $V_f = 0$.

Example 12.7

Consider a head-on collision between a large auto ($M_1 = 4000$ lb) traveling at 60 mph and a smaller auto ($M_2 = 2100$ lb) traveling at 50 mph. Estimate the relative accelerations experienced by the two vehicles if they remain in contact following the collision. (Properly these masses should be expressed in slugs, but numerically it makes no difference in this example.)

Solution

Using the results of Example 12.6,

$$V_f = \frac{M_1V_1 - M_2V_2}{M_1 + M_2} = \frac{4000 \times 60 - 2100 \times 50}{4000 + 2100} = 22.1 \text{ mph}$$

The final motion is in the direction of the larger vehicle's initial velocity. The larger car experiences a *change* in velocity of $(60 - 22.1) = 37.9$ mph, and the smaller vehicle has its velocity *changed* by 72.1 mph (it is now going backward). Since the time duration of the collision is the same for the two vehicles, the accelerations experienced are proportional to the velocity changes. Therefore, the smaller vehicle experiences an acceleration $72.1/37.9 = 1.9$ times as great as does the larger vehicle. This obviously can be the difference between a fatal accident and a survivable accident.

These examples make it clear that there is definitely an increased hazard to the occupants of smaller vehicles in traffic situations where there is a mix of vehicle sizes traveling in opposite directions. This only emphasizes the need

Figure 12.10 Vehicle passengers are protected against injury in head-on collisions by an automobile front end that collapses at the proper rate.

for effective safety measures of all types, as smaller cars are now becoming more common. As the larger vehicles become fewer in number, the risk to occupants of smaller vehicles should become proportionately reduced, but at the cost of increased risk to those persons who have gone from heavier to lighter cars.

NONTECHNICAL TRAFFIC SAFETY MEASURES

An effective program of traffic safety certainly depends on more than the mere knowledge of possible means of improving the vehicles and roadways technically. Driver education and testing are obviously important, as are strict and effective measures to keep drinking drivers off the road. Traffic safety can be as much a legislative matter as it is a matter of engineering and education. Some of the impact of legislation is indicated in Figure 12.11, which suggests that a rising traffic death rate over a several-year period was reversed by the passage of the Traffic Safety Act in 1966. This act included numerous specifications for the crashworthiness of automobiles. The legal requirement for seat belts, head restraints, padded dashboards, collapsible steering columns, hazard flashers, dual braking systems, and many other safety provisions made all these features a standard part of every automobile on the market for the first time, even though they had been technically feasible for many years. It took the force of legislation to prevent traffic death and injury through these technical measures. Of course, the full effect of such measures, which apply only to new vehicles, does not show up in accident statistics until several years after the regulations are put into force. A further reduction in the highway death rate may have been brought about starting in 1973 with the lowering of the national speed limit to 55 mph, another purely regulatory action. It is certain that many thousands of lives have been saved and countless serious injuries prevented in these ways with relatively little sacrifice on the part of those using the roadways.

There is convincing evidence that these safety measures are effective. This is seen in Figure 12.11, and reinforced by the more recent statistic of only 41,000 U.S. traffic deaths in 1991, considerably lower than in 1989 and 1990, and the lowest highway death count in 30 years. The total miles driven each year in the United States have increased appreciably during this time.

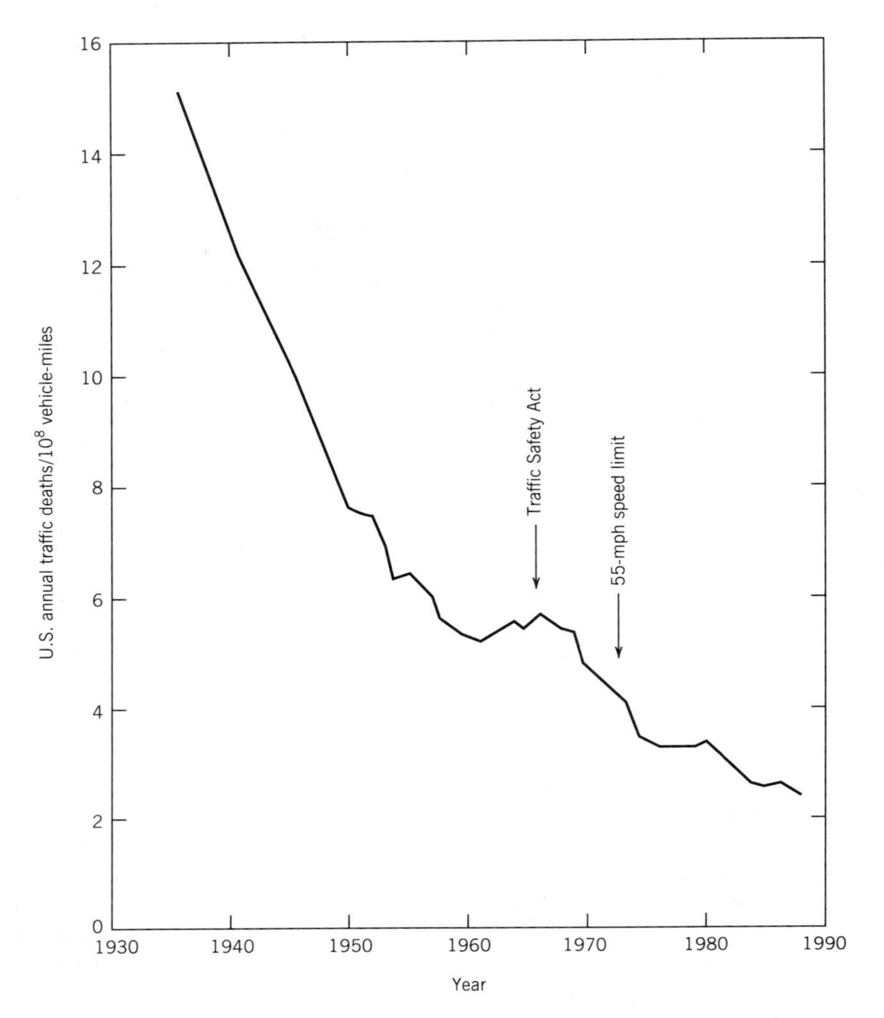

Figure 12.11 A 50-year record of U.S. traffic deaths per 100 million vehicle-miles. From 1936 to 1960, there was a fairly steady decline due to improvement of vehicles, roadways, and driver licensing. After 1960, the rate started to turn up again until the passage of the Traffic Safety Act. A further drop after 1973 is often attributed to the 55-mph speed limit. In 1988 the speed limit on some interstate highways was increased to 65 mph. Over the time period shown on this graph the number of annual vehicle-miles in the United States increased more than six times.

12.5 MASS TRANSPORTATION

Our present transportation system is emphatically centered about the private automobile. More than three quarters of our 137 milion motor vehicles are passenger-carrying automobiles. The attraction of the private auto is obvious: the reliability, privacy, luxury, and flexibility it offers in arranging people's travel are unmatched by any other known system of passenger transport. And in many situations, the cost of automotive travel is competitive with buses, trains,

Table 12.10 MEANS OF TRANSPORTATION TO WORK IN THE UNITED STATES

	Workers (× 10^3)	Percent
Drive alone		
Automobile	44,830	56.0
Truck	7,464	9.3
Carpool	15,575	19.4
Public Transportation		
Bus or streetcar	3,100	3.9
Subway or elevated train	1,179	1.5
Railroad	405	0.5
Taxicabs	141	0.2
Other means	1,067	1.3
Walk	3,778	4.7
Work at home	2,585	3.2

Source: Derived from *Statistical Abstract of the U.S.*, 1980.

planes, or any other alternative. Table 12.10 shows that two thirds of American workers drive to work alone. It seems unlikely that Americans will be enthusiastic about any changes in our transportation system that will separate them from their precious private conveyances.

Why should anyone question the superiority of our system of private vehicles when it is so clearly what the public has chosen? The reasons are several. As we discuss in Chapter 13, our cities are being increasingly plagued with problems of air pollution, primarily due to use of the private automobile. This is more than a matter of simple annoyance; human health is involved, as is the deterioration of property. There is also a strong argument on behalf of the national economy. We are now exporting many tens of billions of dollars annually to purchase petroleum so that we may enjoy our private automobiles. It is possible that we may be able to reduce these imports somewhat during the next few years, but it is also likely that the price of imported petroleum will increase as our domestic reserves diminish. Aside from petroleum, the proliferation of automobiles is enormously consumptive of other resources such as steel, aluminum, and other metals. This problem exists because our recycling system is far from complete in its recovery of these resources and because of the relatively short lifetime of automobiles (50% of them are less than 6 years old). And when we do attempt to recycle our worn-out vehicles, we create the much-unloved automobile salvage yard. As we have just concluded, the private auto is a spectacularly dangerous way to travel; all means of public conveyance are safer. Switching from private to mass transit would, on average, add several months to our lives. Another factor involves the land dedicated to the auto: a parking space at each end of each journey, a network of highways in between, and a myriad of drive-through facilities. In the section on traffic theory above, we saw how many fewer traffic lanes are needed for buses as compared to cars. At some point, this will become important as we continue to remove agricultural land from production to provide paved highways for cars. Valuable residential and business areas in many of our cities have been blighted by the construction of highways, either surface routes or elevated expressways. There seems to be no end to the automobile's appetite for more highways. As new highways are built, travel by automobile becomes easier and more automobiles appear; these in turn require still more highways, which are built in response to popular demand, and financed by a fund fattened in proportion to automobile miles traveled. The result is cities afflicted with multilane, multilevel expanses of concrete and asphalt.

Most of the problems caused by a transportation system based on a private vehicle for every two citizens can at least be alleviated by the adoption of mass transit systems. Certainly something is lost in the way of convenience and privacy, but, on the other hand, we can be freed of problems of maintenance and parking, and the time in transit can be used for pursuits other than guiding a vehicle through a frantic maze of traffic. Many of the advantages of mass transit can be quantified through a measure commonly known as passenger efficiency. This can be calculated for any vehicle as

$$\text{passenger efficiency} = \frac{\text{number of passengers} \times \text{number of miles traveled}}{\text{amount of fuel consumed}}$$

The passenger efficiency is commonly reported in terms of passenger-miles per gallon of fuel consumed. For buses and trains with most of their seats occupied, the passenger efficiency is generally better than for an automobile, at least when the automobile carries only one or two passengers. However, a highly efficient small car carrying four passengers on the open highway can achieve a passenger efficiency comparable to that of the best mass transit systems. Airplanes typically have rather poor passenger efficiency relative to wheeled vehicles.

By shifting our national transportation system to vehicles operating at high passenger efficiencies, the air pollution problems related to fuel combustion will be relieved, as will the national balance-of-payments problem. Beyond this, trains and local buses can be powered by electricity produced at central power stations equipped with emission systems far more effective than those operating on the internal combustion engine. These central power stations can even be based on hydropower or geothermal energy, or, more likely, they can be coal-burning or nuclear; but they need not be dependent on imported petroleum. Electric-powered railway systems are common in Europe and are still found in our larger cities in the form of trains, subways, and cable cars; but they are much less popular in this country now than they were in the first half of the century, before the automobile achieved its overwhelming dominance.

For a bus system or any other mass transportation system to be economically viable, it must be supported by a general public subsidy, by fares, or by some combination of the two. Very few local mass transit systems in the world are supported solely by fares. However, intercity bus systems with travel at higher speeds and with higher capacity factors have consistently earned profits based on passenger fares alone. Taking buses as an example, it is easy to see why local systems of any type seldom earn profits.

Suppose a bus costs \$100,000 and remains serviceable for 300,000 miles in an urban system. This reduces to \$0.33 per mile as a purely capital cost. The interest cost on the financing of this capital investment may add about \$0.05 per mile. With fuel at \$1.20 per gallon and a mileage rate of 4 miles per gallon, another \$0.30 per mile is added. Maintenance costs, including tires, labor, oil, parts, and cleaning can be estimated at \$0.10 per mile. For the moment, let us ignore the considerable costs of street construction and maintenance, publishing of schedules, office and garage space, supervisors, and many other overhead items. The driver's salary is typically at least \$12.00 per hour, and with an average bus speed of 20 mph (see Example 12.8), this adds \$0.60 per mile. So the operating cost is at least

Item	Dollars per Mile
Bus	\$0.33
Financing	0.05
Fuel	0.30
Maintenance	0.10
Driver salary	0.60
Total	\$1.38

Note that only about half of the total cost is related to the vehicle; the largest single item by far is the driver's salary. This suggests that only minor cost savings can be accomplished by reducing the size of the vehicle at times and in places where a full-sized bus is not needed.

Example 12.8

Consider a bus route with stops every quarter mile. The bus can accelerate at 3 ft/sec^2, it can comfortably decelerate at 8 ft/sec^2, and it achieves a maximum speed of 60 ft/sec = 41 mph between stops. It spends 10 sec at each stop. What is the average speed on this route?

Solution

The average speed is the total distance traveled between stops (¼ mile × 5280 ft/mile = 1320 ft) divided by the total time from stop to stop. This time is given by the sum of the acceleration time (t_a), the time at constant velocity (t_c), the braking time (t_b), and the stop time (10 sec).

$$t_a = \frac{v}{a} = \frac{60 \text{ ft/sec}}{3 \text{ ft/sec}^2} = 20 \text{ sec}$$

$$t_b = \frac{60 \text{ ft/sec}}{8 \text{ ft/sec}^2} = 7.5 \text{ sec}$$

To get the time at constant velocity, the distance traveled at this velocity must be known. This is the total distance minus the accelerating and braking distances, or

$$1320 - X_a - X_b,$$

where

$$X_a = \tfrac{1}{2}at^2 = \tfrac{1}{2} \times 3 \times (20)^2 = 600 \text{ ft}$$

and

$$X_b = \tfrac{1}{2} \times 8 \times (7.5)^2 = 225 \text{ ft}$$

so

$$X \text{ (at 60 ft/sec)} = 1320 - 600 - 225 = 495 \text{ ft}$$

and

$$t_c = \frac{495}{60} = 8.25 \text{ sec}$$

Now that the total distance and the total time are known, the average velocity is

$$\bar{v} = \frac{1320 \text{ ft}}{(20 + 8.25 + 7.5 + 10) \text{ sec}} = 28.85 \text{ ft/sec}$$
$$= 19.7 \text{ mph}$$

At a passenger fare of $0.50, each bus in this system needs to pick up an average of 2.7 passengers per mile of route to break even. With stops every three blocks, or one quarter mile apart, this requires a ⅔ probability of one passenger waiting at each stop. During the busy early morning and late afternoon hours, or in areas of the city with high population density, there is usually no problem achieving the necessary passenger loading. But in the typical city with workers coming in each day from the outlying suburbs, each bus and driver probably makes one profitable run into the city in the morning and one out in the late afternoon. The rest of the day the suburban routes of the system operate at a financial deficit. Any attempts to improve the financial return by increasing fares can have the opposite effect, reducing the ridership so much that the change results in a net worsening of the operating deficit. It seems clear that local mass transit systems can be self-supporting only in areas with sufficiently high population density so that there will generally be several passengers boarding per mile. Our prevalent pattern of low-density suburban sprawl certainly works against the success of any local mass transit system.

Local mass transit rail systems have the added handicap of needing to finance their own roadways rather than using existing city streets. A study in 1982 of a light rail system to serve the Denver metropolitan area reported estimated costs of $20 million to $26 million *per mile* to construct the system. These capital construction costs alone are a tremendous burden. If each mile of track were traversed 100 times a day, 365 days a year, for 20 years, the capital cost of the system per mile traveled by a light rail train can be estimated as

$$\text{capital cost per mile of train travel} = \frac{2 \times 10^7 \text{ dollars/mile}}{100 \text{ trips/day} \times 365 \text{ days/yr} \times 20 \text{ yr}}$$
$$= \$27/\text{mile}$$

By comparing this to the previous figure of less than $1 per mile for the capital costs of a bus system, it is apparent that one train must be the equivalent of a very large number of buses for the train system to be financially competitive. Of course, the train system does require fewer drivers per passenger, and thus may save on a major operating cost. In general, rail systems have higher initial costs and lower operating costs than do bus systems. And, obviously, it is more difficult to rearrange routes and schedules with rail systems.

PROBLEMS

1. A car goes from 30 mph (44 ft/sec) to 60 mph (88 ft/sec) in 20 sec. What was its acceleration? How much distance did it cover during the 20 sec?
2. How much force in pounds is needed for a 3200-lb car to accelerate at 4 ft/sec^2? Neglect all other forces such as aerodynamic drags and hills.
3. How much force in pounds is needed for the 3200-lb car to go up an 8% grade with a constant velocity? Neglect other forces.
4. How much force in pounds is needed for the 3200-lb car to overcome aerodynamic drag at 60 mph if $C_D = 0.5$ and the frontal area is 30 ft^2?
5. What is the total force needed to have the car accelerate up the hill as described in problems 2, 3, and 4, including aerodynamic drag? What is the total horsepower needed? Neglect rolling friction.
6. If a car is going 60 mph (88 ft/sec) and the driver sees an obstruction in the road 100 ft ahead, what deceleration is needed to stop in time, assuming a reaction time of 0.5 sec? Is this magnitude of deceleration possible with a typical car?
7. At 30 mph (44 ft/sec), a force of 200 lb is required to move a car on a flat stretch of road with no acceleration. Assuming an overall engine efficiency of 20%, how many Btus need to be put in the gas tank to go 1 mile (5280 ft)? How many miles per gallon does the car get?
8. Calculate the force in pounds on the side of a house 10 ft high $\times$ 30 ft long due to a 100-mph wind blowing directly onto this wall. Take $C_D = 1.0$.
9. Calculate the approximate terminal velocity for a large truck coasting down a 7% grade. Assume $C_D = 0.7$, $A_f = 96$ ft^2, and weight $= 60{,}000$ lb.
10. Estimate the terminal velocity for a bicyclist coasting down a 10% grade. Make reasonable estimates for A_f, C_D, and m.
11. Calculate the terminal free-fall velocity for a parachutist. Take $C_D = 1.0$, effective area $= 300$ ft^2, and weight $= 180$ lb.
12. Convert the equation for P_{ad} (hp) in which v is in miles per hour into a similar equation in which v is in feet per second.
13. It is reliably reported that about 50% of the U.S. traffic deaths are due to irresponsible use of alcohol. It is also estimated that about 52% of all auto-occupant deaths could be prevented by the universal use of lap and shoulder harnesses. Discuss the probable annual savings of lives in this country if alcohol were eliminated as a factor in traffic deaths *and* if all auto occupants wore lap and shoulder harnesses. How would this saving of life compare to the complete elimination of homicide from our society (about 20,000 victims annually)?
14. Describe several measures that can be taken to improve the safety of streets and highways. Do not include changes or improvements in vehicles and drivers, but only changes to the fixed elements of streets and highways.
15. Use the information given in Tables 12.1 and 12.2 to estimate how much fuel could be saved in this country each year if the workers who now commute to their jobs

by driving alone would switch to some form of public transportation. Assume that each worker commutes a total distance of 100 miles weekly. How does this savings compare to the amount of petroleum we import?

16. Show that for the traffic flow model assumed in this chapter, the maximum vehicle flow rate can be written as

$$N(\max) = \frac{1}{t_r + \sqrt{l/a}}$$

17. Using the expression given in Problem 16, calculate $N(\max)$ for $l = 18$ ft (big car), $a = 0.1\ g$ (icy), and $t_R = 1.5$ sec. Also calculate the velocity in miles per hour at which $N = N(\max)$. Under these conditions, how long would it take to empty a 20,000-seat stadium safely following a concert? Assume three passengers per vehicle and four available traffic lines.

18. If an auto is brought to a stop in a distance of 12 ft through use of a collapsible collision barrier, what is the maximum speed (in miles per hour) consistent with survival ($\alpha \leq 30$)?

SUGGESTED READING AND REFERENCES

1. Bartlett, A. A. "The Highway Explosion," *Civil Engineering*, December 1969.
2. Gray, Charles L. Jr., and von Hippel, Frank. "The Fuel Economy of Light Vehicles." *Scientific American* **241** 5 (May 1981), pp. 48–49.
3. Herman, Robert, and Gardels, Keith. "The Crashworthiness of Automobiles." *Scientific American* **228** 2 (February 1973), pp. 76–78.
4. Marston, Edwin N. *The Dynamic Environment: Water, Transportation, and Energy.* Greenwich, Conn.: Xerox College Publishing, 1975.
5. Ogelsby, Clarkson H. *Highway Engineering.* New York: John Wiley, 1975.
6. Pierce, John R. "The Fuel Consumption of Automobiles." *Scientific American* **232** 1 (January 1975), pp. 34–44.
7. Shonle, John I. *Environmental Applications of General Physics.* Reading, Mass.: Addison-Wesley, 1975.
8. Wright, Paul H., and Paquette, Radnor J. *Highway Engineering.* New York: John Wiley, 1979.
9. Ross, Marc. "Energy and Transportation in The United States." *Annual Review of Energy,* 1989, p. 131.

CHAPTER THIRTEEN

POLLUTION OF THE ATMOSPHERE

13.1 INTRODUCTION

One of the most striking products of our age is the view of our planet as seen from the moon. This view brings vividly to mind the fact that the earth is but one of nine planets of the solar system. It is neither the largest nor the smallest, but it is exceptional in that it is the only planet with an atmosphere of nitrogen and oxygen and having a temperature range compatible with life as we know it. Our sun, at a distance of 93 million miles and the gravitational center of the solar system, is but an ordinary star, one of billions in our galaxy, the Milky Way. It lies about two thirds of the way out from the galactic center and is very remote (about 24×10^{12} miles) from the nearest neighboring star. The general view is that our planet, isolated as it is, will be the only home for mankind for a very long time, most likely forever; and the resources now on the planet and available through solar radiation are all that we have for the future.

In the past, the earth's oceans and atmosphere were considered virtually infinite, and little thought was given to the polluting effects of our dumping wastes into the atmosphere or large bodies of water. These were felt to be so vast that they could absorb any conceivable amount of abuse. However, over the past decade or two, several factors have come together to bring this attitude into sharp question. Along with the realization that we are on an all too finite spaceship, the earth, the increasing population and technological base for our way of life and the increased sensitivity of our instruments for measuring pollution have all played a role in bringing about an awareness about the environment. This chapter explores the ways we are affecting the atmosphere, the consequences of this pollution, and methods and controls for lessening the damage.

By some measures the atmosphere is vast. It weighs about 5.7×10^{15} tons, about one millionth of the weight of the earth. The surface of the earth has an area of about 2×10^{8} square miles, and the atmosphere extends up about 600 miles over the entire surface, although most of the air is in a layer over the earth's surface, which has a thickness very small compared to the earth's 4000-mile radius. The density of the air at the earth's surface is 1.293 kg/m^3, where the pressure is 14.7 lb/in.^{2} (1.01×10^5 N/m^2). The pressure and density gradually lessen as the altitude increases. Half of the air is below 18,000 feet altitude. By 50,000 feet, the pressure is only 1.6 lb/in.^{2}, and by 600 miles, the pressure is essentially zero. The atmosphere is structured with altitude in that properties such as the temperature gradient vary from one region to another. Figure 13.1 shows the major stratification and how the temperature varies with altitude. We are most concerned, of course, with the troposphere, because that layer affects us most directly. We shall see, however, that all regions of the atmosphere are important, as we can with various pollutants affect processes that take place right up to the mesosphere. Pollutants at high altitudes can affect conditions down in the troposphere where we live.

The major permanent constituents of the atmosphere are listed in Table 13.1. In addition to these permanent gases, there are a number of others, such as water vapor, carbon dioxide, methane, carbon monoxide, ozone, and ammonia, that fluctuate with time, altitude, and location. Water vapor is present to 0.7% by volume, on average, but its concentration depends on a number of factors. Likewise, carbon dioxide is present, on average, to 0.036%, but its

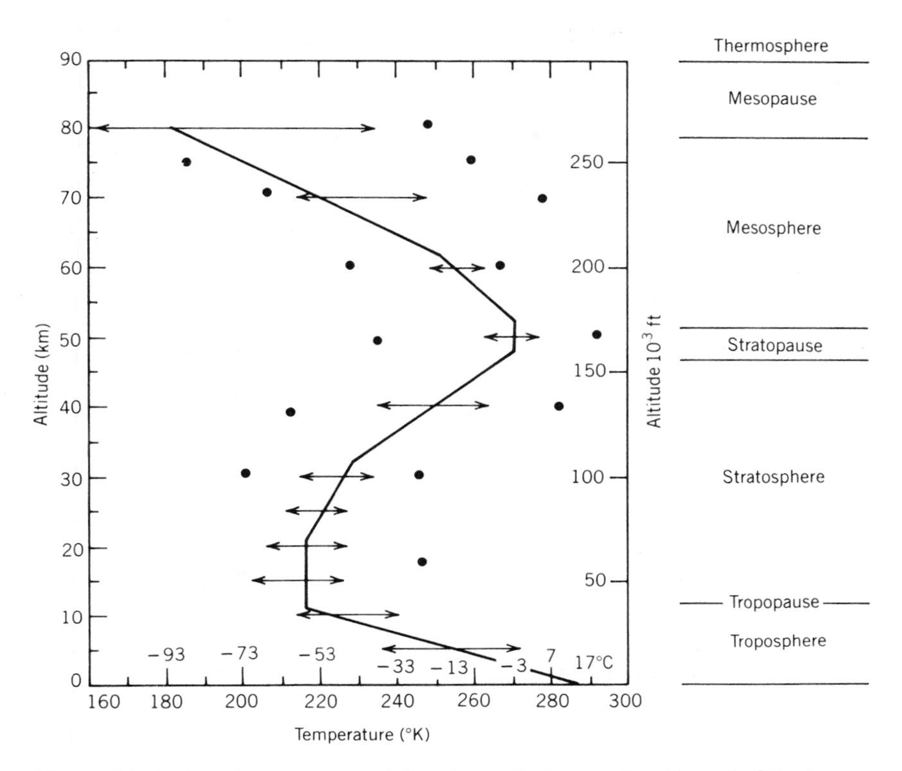

Figure 13.1 The temperature of the atmosphere as a function of altitude. The arrows indicate the normal range of temperature variation and the dots the extreme observed values. The names given to the various regions of the atmosphere are shown on the right.

concentration varies with the time of year and location. Some of the gases that are present in relatively small amounts in the atmosphere play vital roles in absorption of certain parts of the spectrum of solar radiation, in the complex chemical interactions that take place in the upper atmosphere, and in the heat balance of the earth. Man is changing the concentrations of some of these gases, and the results of these changes are not always known. Thus, we are in some instances doing giant experiments in the atmosphere with our pollutants; the results will not be known for many years, and the processes may not be reversible.

Table 13.1 MAJOR PERMANENT CONSTITUENTS OF THE ATMOSPHERE

Element	Percent by Volume	Parts per Million
Nitrogen	78.08	
Oxygen	20.95	
Argon	0.93	
Neon		18.2
Helium		5.2
Krypton		1.1
Hydrogen		0.5

13.2 THERMAL INVERSIONS

The saying that "the solution to pollution is dilution" has some basic truth in it. Although the real long-term solution to pollution problems is clearly to reduce the emission of the pollutants, the dispersal of pollutants in the atmosphere is important to the health of many city dwellers. The purpose of the tall smokestacks associated with electric power plants and some industries is to emit the pollutants at an elevation where their chance of dispersal by being carried aloft, rather than hovering near the ground, is greatly enhanced.

The temperature of the atmosphere normally decreases as the altitude increases up to about 10,000 meters. As Figure 13.1 shows, the average temperature decreases from roughly 20°C at ground level to about −60°C at 10,000 meters. Above the troposphere, the temperature gradually increases through the stratosphere until about 50,000 meters, where it starts to decrease again. There is a great deal of variation in the temperature profile with the time of year and location, but normally a negative temperature gradient exists near the earth, and this has important consequences for the dispersal of pollutants. It is commonly recognized that warm air will rise in the presence of surrounding cooler air. This observation derives from the fact that as air temperature increases, its density decreases, requiring a parcel of warm air to float upward in an ambient atmosphere of colder, more dense, air. If a parcel of warm, polluted air, for instance, from a smokestack, is released into the lower levels of the atmosphere under normal meteorological conditions, it will rise in the atmosphere to as much as 10,000 meters, at which height it usually presents no immediate problems.

Not all meteorological conditions, however, are conducive to this upward motion of the warmed polluted air. There is a relationship between temperature and altitude that is determined by the basic thermodynamic laws governing the behavior of a gas. The generally prevalent temperature–altitude relationship in the lower atmosphere is known as the *adiabatic lapse rate* (ALR). In thermodynamics, the word *adiabatic* describes any process in which no heat energy is either gained or lost by some defined volume of gas. The word *lapse* indicates that temperature decreases with increasing altitude. If a given parcel of air, warmer than its surroundings, starts to rise in the atmosphere, and if it can be considered to do so without exchanging heat energy with the neighboring air, it will expand and cool at the adiabatic lapse rate. Expansion occurs because as the air rises, the air parcel is under less pressure, due to the thinner burden and lesser weight of the overlying air. The adiabatic lapse rate for dry air is −1°C/100 m. If the air is moist, the ALR is less; it can be as low as −0.35°C/100 m. An approximate average ALR is −0.65°C/100 m. The ALR, in simple terms, states the rate at which the temperature of a volume of air will naturally tend to decrease as altitude increases or increase as altitude decreases.

If the prevailing temperature profile, because of unusual meteorological circumstances, is such that the atmospheric temperature decreases more rapidly with the altitude than given by the ALR, then any parcel of air released near ground level and warmer than its surroundings will rise indefinitely into the upper atmosphere. This is because as it cools at the ALR, it will always be warmer, and thereby less dense, than the surrounding air. This unstable condition is obviously desirable because it leads to good vertical mixing and a relatively pollution-free lower atmosphere.

If the ALR and the existing temperature profile happen to be the same, there will be a neutral condition that neither forces the warm air upward nor traps it near the earth. On the other hand, the existing temperature profile may be such that the temperature decreases more slowly with altitude than indicated by the ALR. Then a volume of warm air released near ground level will rise in the ambient cooler air, cooling at the ALR as it rises, until at some level it is no longer warmer than its surroundings and it ceases to rise. An extreme condition exists when the temperature actually increases with altitude. This corresponds to a very stable condition where any polluted warm air released near the ground will be trapped and not be vertically dispersed to any degree. This condition is known as a thermal inversion. There are several causes of thermal inversions, and their duration varies widely from a few hours to many days. Their frequent occurrence in cities such as Los Angeles and Denver is a major contribution to the pollution problems of these cities. Figure 13.2 illustrates these conditions of stability.

There have been several notable incidents of thermal inversions in the past in the Meuse Valley in Belgium and in Donora, Pennsylvania, as well as in New York City and London, England, where the human deaths caused by trapped air pollutants numbered in the hundreds to thousands. On December 5, 1952, a thermal inversion developed in London, enveloping the city in fog with essentially no vertical movement of air above 150 ft. The sulfur dioxide and particulates from burning coal, in addition to the other air pollutants of a big city, accumulated for four days. After 12 hours of the inversion, people began coughing and complaining of respiratory ailments. During the next four days, an estimated 4000 deaths occurred beyond those normally experienced for a four-day period. The majority, but not all, of the people who succumbed were over 55 years old. Two similar, but not as drastic, episodes occurred in London in 1956 and 1962. Cities such as London and New York have been plagued for many years with this type of classic smog, primarily SO_2 and particulates. In addition to causing various respiratory problems, this type of smog puts a

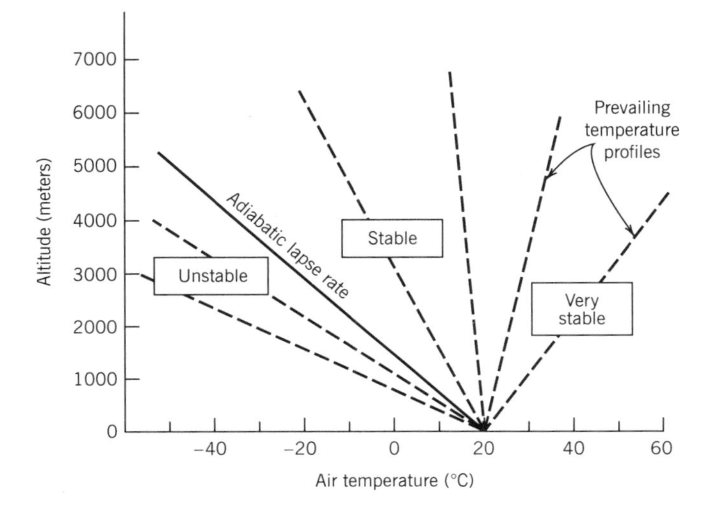

Figure 13.2 A variety of possible temperature profiles. The region to the left of the adiabatic lapse rate is unstable and will lead to appreciable vertical mixing of polluted air released near the ground. The region to the right represents stable conditions and air stagnation.

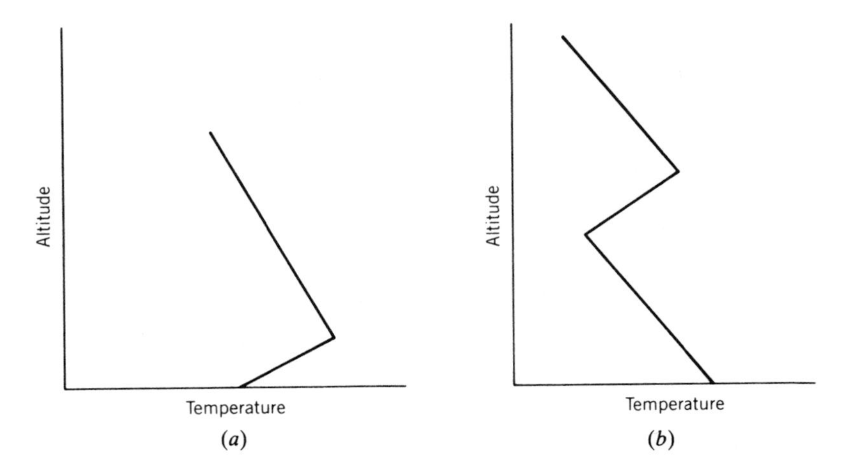

Figure 13.3 Two types of thermal inversions. In (*a*), a radiative inversion: Because of radiative cooling during the night, the air near the ground surface is cooler than the air above it. In (*b*), a high-pressure subsidence: A high-pressure mass of air subsides toward the earth and is compressed and heated in the process, causing a thermal inversion layer some distance above the ground.

coating of soot over the people and buildings. The problems of cities like Los Angeles and Denver differ from those of London in that radiation from the sun interacts with air pollutants largely associated with automobiles and trucks, creating what is called photochemical smog; however, the major pollution incidents in these cities have also been associated with thermal inversions. In general, all major air pollution incidents that have led to documented elevated levels of human mortality have occurred during periods of thermal inversions.

Many of the causes of thermal inversions are now well understood, and their occurrence can often be predicted by meteorologists. One such cause is a high pressure subsidence. When a high pressure region of the atmosphere subsides, or moves downward toward the earth where the pressure is greater, the air mass will be compressed and its temperature will rise. This relatively dense warm air will continue to move toward the earth until it meets the higher density air near the surface. A temperature profile will result such as shown in Figure 13.3*b*; the air tends to get cooler with increasing elevation above the ground, up to a point where the high pressure air mass is sitting, and then the temperature gradient reverses and the air becomes warmer with altitude. Some mixing of the air close to the earth's surface will occur, but none of the warm polluted air can break through the lid provided by the thermal inversion. It was this type of thermal inversion that caused the problems in London, Donora, New York, and the Meuse Valley.

Example 13.1

Find the height to which polluted air 10°C warmer than its surroundings will rise if the prevailing air temperature profile is +5°C/100 m (a thermal inversion).

Solution

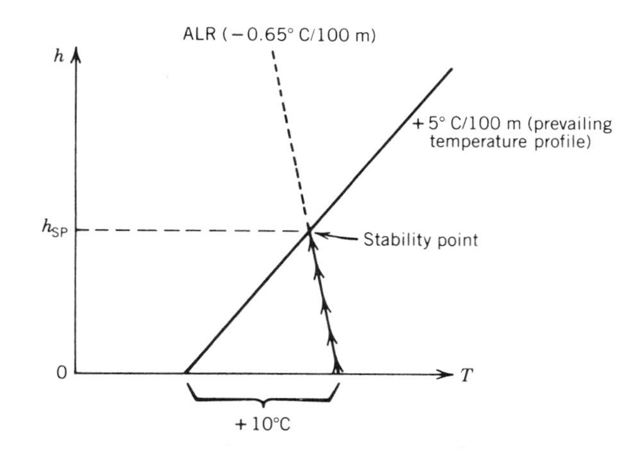

For each 1 m height, the lines converge by $(0.05 + 0.0065)°$ C/m. To reach the stability point, they must converge by 10°C. This will occur at a height of

$$\frac{10°\text{C}}{0.0565°\text{C/m}} = 177 \text{ m}$$

The polluted air will rise to this height and accumulate there unless dispersed by winds.

Radiative thermal inversions are much more frequent but are less troublesome than those caused by high pressure air masses. On a clear night, the earth's surface will radiate thermal energy into space and cool both the earth's surface and the air near the surface. The energy being radiated was provided by the sun on the previous day. After a night of cooling, the air near the surface the next morning will be cooler than air above it, and thus a thermal inversion, such as shown in Figure 13.3*a*, will result. Normally the sun, as the morning progresses, will warm the surface of the ground and the lower atmosphere, and the thermal inversion will disappear by mid-afternoon. Such radiative inversions are almost a daily occurrence in cities such as Denver.

Los Angeles has a reputation for air pollution that is not to be envied. Later we discuss in some detail the rather complicated reactions and pollutants that are involved, but the problem originates just as much in the geography and weather patterns experienced by that city. A large part of the problem stems from global air circulation patterns, two of which meet at about the latitude of Los Angeles, sending air down toward the earth. This air is warmed by compression and forms a lasting thermal inversion. The San Gabriel Mountains to the east of Los Angeles act to deter winds that would help move the air from the area. In addition to these effects, there is the usual land–shore air movement. During the daytime the heated air over the area tends to rise, and the cool air from the sea comes in to replace it. At night, however, the sea does not cool as fast as the land, and the direction is reversed and the polluted air returns.

There are other meteorological causes of thermal inversions, but those already mentioned are the most significant. It would be tempting to blame air pollution on the weather, but this would be self-defeating. The real culprit is us. In the next sections we learn what the major pollutants are, how they come about, and what can be done to reduce their concentration in the atmosphere.

13.3 CARBON MONOXIDE

Most of our air pollution is produced either directly or indirectly by the combustion of fuels, or, more exactly, by the oxidizing of the carbon content of the fuels. In addition to the ideal combustion process that combines carbon (C) and oxygen (O_2) to form carbon dioxide (CO_2), there can also be incomplete combustion of carbonaceous materials that leads to the formation of carbon monoxide:

$$2C + O_2 \rightarrow 2CO$$

The formation of carbon monoxide takes place when insufficient oxygen is present during combustion to form carbon dioxide. A prime source of CO is the internal combustion engine, where the burning of gasoline takes place at high pressures. Table 13.2 lists the amounts of CO that are estimated to be emitted in the United States from various sources. We see that gasoline-powered motor vehicles alone account for about 54% of the carbon monoxide.

Carbon monoxide is an odorless, colorless gas that is toxic at high con-

Table 13.2 MAJOR U.S. AIR POLLUTANTS (10^6 TONNES/YEAR)

Source	CO	Particulates	SO_x	HC	NO_x
Motor vehicles (gasoline)	53.5	0.5	0.2	13.8	6.0
Motor vehicles (diesel)	0.2	0.3	0.1	0.4	0.5
Aircraft	2.4	0.0	0.0	0.3	0.0
Railroads and others	2.0	0.4	0.5	0.6	0.8
Total transportation	58.1	1.2	0.8	15.1	7.3
Coal	0.7	7.4	18.3	0.2	3.6
Fuel oil	0.1	0.3	3.9	0.1	0.9
Natural gas	0.0	0.2	0.0	0.0	4.1
Wood	0.9	0.2	0.0	0.4	0.2
Total fuel combustion	1.7	8.1	22.2	0.6	9.1
Industrial processes	8.8	6.8	6.6	4.2	0.2
Solid waste disposal	7.1	1.0	0.1	1.5	0.5
Forest fires	6.5	6.1	0.0	2.0	1.1
Agricultural burning	7.5	2.2	0.0	1.5	0.3
Coal refuse burning	1.1	0.4	0.5	0.2	0.2
Structural fires	0.2	0.1	0.0	0.1	0.0
Total miscellaneous	15.3	8.8	0.5	7.7	1.5
Total	91.0	25.9	30.2	25.3	18.4

Source: National Air Pollution Control Administration.

centrations. Its main toxicity stems from its ability to form a stable compound with hemoglobin called carboxyhemoglobin. Hemoglobin is the substance in the red blood cells that carries oxygen to the tissues. Carbon monoxide has an affinity for hemoglobin 200 times as great as that of oxygen. Because of this, CO tends to block the normal distribution of O_2 in the body and leads eventually to suffocation. The effect of carbon monoxide on people is a function of the concentration and duration of exposure. A concentration of 100 parts per million (ppm) in air for 10 hours of exposure will lead to headaches and a reduced ability to think clearly. Concentration of 300 ppm for 10 hours leads to nausea and possibly loss of consciousness. At 600 ppm after a similar length of time, death can result. At 1000 ppm, unconsciousness occurs in 1 hour and death in 4 hours. It has not been determined if there are chronic low-level effects. There are, however, a number of people with respiratory diseases or anemia who are susceptible to even relatively low concentrations.

Carbon monoxide is generally considered the most serious air pollutant in cities. In Los Angeles County, where there are over 3 million cars, more than 8000 tons of CO are emitted every day from internal combustion engines. This amounts to about 5 lb per vehicle per day. The average carbon monoxide levels in cities range from a few parts per million to more than 100, depending on the traffic volume, car speeds, and various aspects of the weather such as the temperature profile discussed previously. The concentration of carbon monoxide has been of concern in U.S. cities because of the increasing number of cars. The CO level in downtown areas, parking garages, and on freeways during rush hours has reached a point where the general health of those exposed is endangered. The present National Ambient Air Quality Standard (NAAQS) maximum permitted concentrations are 9 ppm for an 8-hour period and 35 ppm for 1 hour. These limits should not be exceeded more than once a year, but they are exceeded much more frequently than that in cities such as Los Angeles, Denver, Cincinnati, and Detroit.

Table 13.3 lists the NAAQS for carbon monoxide as well as the other major air pollutants. It should be noted that the standards are stated in terms of the number of micrograms (μg) or milligrams (mg) per cubic meter (m^3); this is the mass of the pollutant in every cubic meter of air. The standards are also stated in parts per million (ppm) by volume or, equivalently, by number of molecules. To understand the equivalence of the mass and volume measures of concentration, it is useful to recall that a volume of 22.4 liters at standard temperature and pressure will contain 1 gram molecular weight and Avogadro's number of molecules of any gas.

Example 13.2

If carbon monoxide is present in air at a concentration of 1 ppm, what are the number of CO molecules in 1 m^3, and what is the mass of CO in 1 m^3?

Solution

Air at STP has Avogadro's number (6.02×10^{23}) of molecules per mole (29 gm), which occupies 22.4 liters or 0.0224 m^3. Therefore the molecular density of air is:

$$\frac{6.02 \times 10^{23}\ \text{molecules}}{0.0224\ \text{m}^3} = 2.68 \times 10^{25}\ \text{molecules/m}^3$$

At 1 ppm there would be $1 \times 10^{-6} \times 2.68 \times 10^{25} = 2.68 \times 10^{19}$ molecules of CO per cubic meter. Since CO has a molecular weight of 28, the corresponding mass density would be

$$2.68 \times 10^{19} \frac{\text{molecules}}{\text{m}^3} \times \frac{28 \text{ g/mole}}{6.02 \times 10^{23} \text{ molecules/mole}} \times 10^3 \frac{\text{mg}}{\text{g}}$$
$$= 1.25 \text{ mg/m}^3$$

Figure 13.4 shows the daily variation of carbon monoxide at a downtown location in Denver. We see that the levels of CO encountered greatly exceed the 8-hour standard of 9 ppm. The two major peaks occur somewhat after the main commuter traffic at the beginning and end of the working day. Because of its high altitude, Denver has a severe problem in meeting the national standards; the automotive emission controls are primarily set for sea level, and they are not as effective at 5000-ft altitude.

The seriousness of the problem is shown in Figure 13.5, where the second highest maximum of the 8-hour CO average in parts per million is plotted versus time. Up until 1986 there was very little improvement, in spite of the federal emission standards for cars, since there was an increase in the number of vehicle miles. Since then the State of Colorado's oxygenated gasoline program for the winter months, the annual inspection required for auto emissions, technical improvements in automobile engines, the improvement of public transportation, and controls on fixed-source emissions have helped bring the number of days where the 8-hour standard was exceeded down from 65 in 1975 to 3 in 1990.

Table 13.3 SUMMARY OF NATIONAL AMBIENT AIR QUALITY STANDARDS

Pollutant	Averaging Time	Primary Standards
Particulate matter	Annual (geometric mean)	75 μg/m³
	24 hours[a]	260 μg/m³
Sulfur oxides	Annual (arithmetic mean)	80 μg/m³ (0.03 ppm)
	24 hours[a]	365 μg/m³ (0.14 ppm)
Carbon monoxide	8 hours[a]	10 mg/m³ (9 ppm)
	1 hour[a]	40 mg/m³ (35 ppm)
Nitrogen dioxide	Annual (arithmetic mean)	100 μg/m³ (0.05 ppm)
Photochemical oxidants	1 hour[a]	160 μg/m³ (0.08 ppm)
Hydrocarbons	3 hours[a] (6 to 9 A.M.)	160 μg/m³ (0.24 ppm)

[a] Not to be exceeded more than once a year.

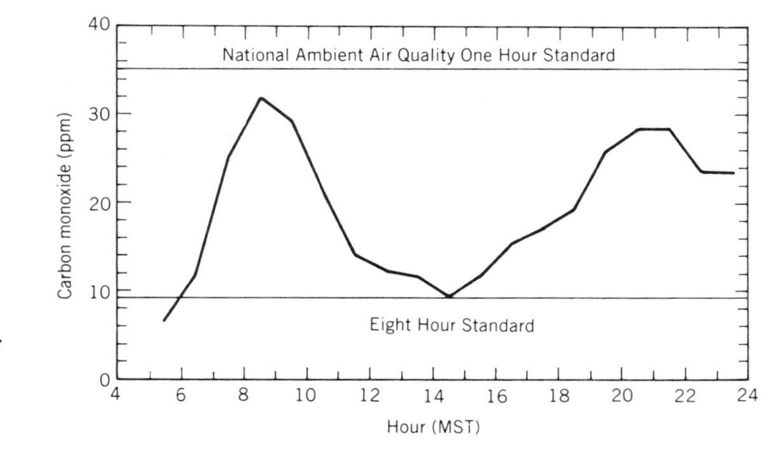

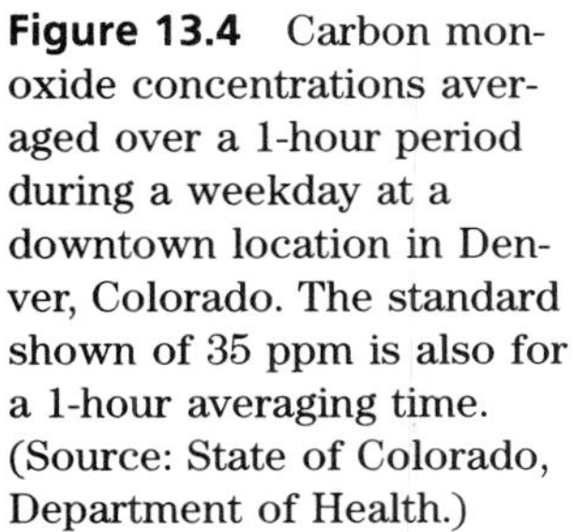

Figure 13.4 Carbon monoxide concentrations averaged over a 1-hour period during a weekday at a downtown location in Denver, Colorado. The standard shown of 35 ppm is also for a 1-hour averaging time. (Source: State of Colorado, Department of Health.)

Carbon monoxide is relatively stable in the atmosphere. It has an estimated half-life of 0.2 year and is probably converted to CO_2 by interaction with OH molecules in the tropopause. About 290×10^6 tons of CO are injected into the world's atmosphere ever year, the majority in the northern hemisphere. Even though the emission of carbon monoxide by natural processes is a factor of 15 or so greater than this, there should be an increase of about 0.03 ppm per year in the average atmospheric concentrations owing to man-made contributions. For reasons that are not completely understood, the average atmospheric concentration of about 0.08 ppm does not seem to be increasing.

The problems of carbon monoxide are basically local problems in cities where, because of short-term high traffic volume and weather patterns, excessively high concentrations are reached only for relatively short periods of time. There are two apparent types of solutions to the problem. The replacement of the personal car with mass transportation systems, particularly those that use electric power, is an option that many cities are pursuing. Because of the capital costs of such systems and because of the great convenience many people find in the use of their own cars, this approach is not being widely accepted. The other major approach is to use technical measures for reducing carbon monoxide emissions from automobiles. Since automotive emission control involves more air pollutants than just CO, we discuss that topic after learning about some of the other ways we are fouling the air.

13.4 THE OXIDES OF NITROGEN

If a nitrogen–oxygen mixture such as air is heated to over 1100°C, the nitrogen and oxygen will combine to form nitrogen oxide (NO). If the cooling process is slow, the reaction is reversible, and the NO will decompose into N_2 and O_2. The following equation summarizes the process:

$$N_2 + O_2 \rightleftarrows 2NO$$

However, if the cooling takes place rapidly, as is the case for most internal combustion engines, then the nitrogen oxide does not have a chance to decom-

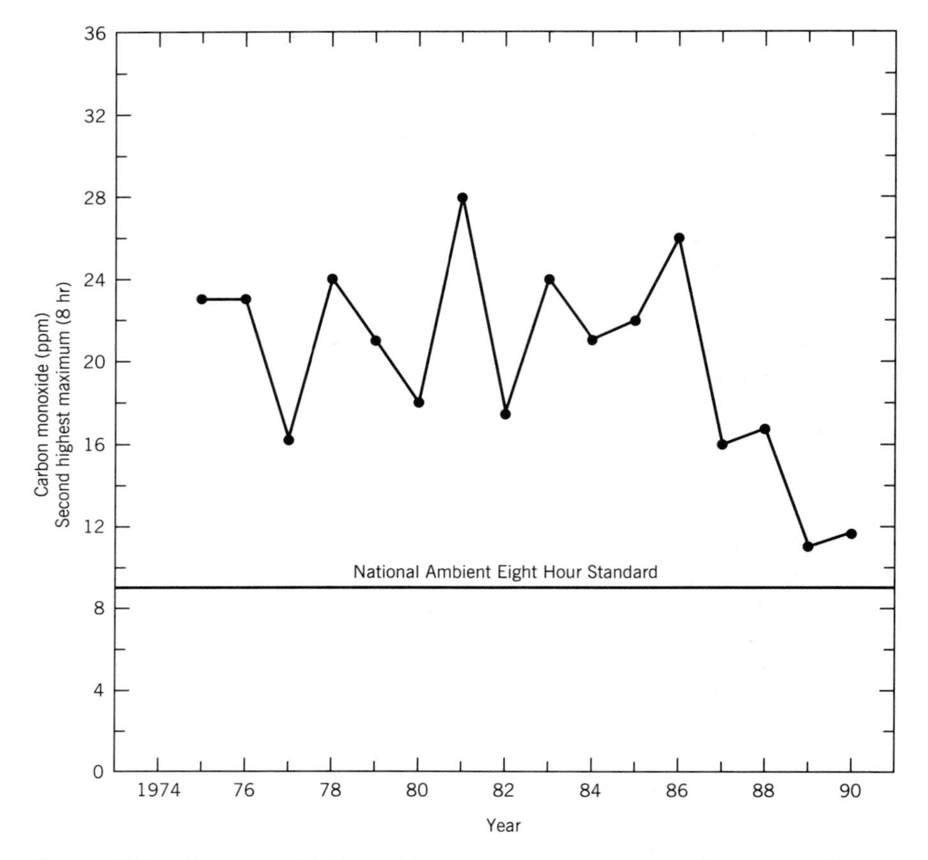

Figure 13.5 The second highest 8-hour average maximum carbon monoxide concentration observed each year at a downtown location in Denver, Colorado, from 1975 to 1990. The 8-hour standard is 9 ppm as a second maximum for the year. (Source: State of Colorado, Department of Health.)

pose and will go off as NO. It should be noted that the reaction that forms NO does not directly involve the fuel used in the combustion process but only the nitrogen and oxygen that are the major constituents of air. An internal combustion engine at full speed will emit about 4000 ppm of NO in the exhaust and a coal-fired steam generator 200 to 1200 ppm; but both of these figures depend on the details of the treatment of the exhaust gases. Nitrogen oxide is a colorless gas that is toxic in sufficiently high concentrations, but its toxicity is generally considered to be minor (about 20%) compared to nitrogen dioxide (NO_2).

Nitrogen dioxide is produced in the same manner as nitrogen oxide, although for most combustion processes, NO is much more prevalent. Nitrogen oxide, however, when released to the atmosphere will go to NO_2, partially by reacting with ozone (O_3). After about 10 hours, 50% of the NO will have been converted to NO_2. Because of its greater toxicity compared with NO, NO_2 is of much greater importance environmentally, although frequently the nitrogen oxides are treated together as NO_x. Nitrogen dioxide is a reddish-brown gas that contributes the brownish color to the familiar smog in cities. The major sources of NO_x in the United States are listed in Table 13.2. Although motor vehicles are important, there are several other major sources, such as combustion of coal, fuel oil, and natural gas.

At about 0.1 ppm in air, NO_2 can be smelled, and at 5 ppm it begins to affect the respiratory system. At concentrations of 20 to 50 ppm, there is a strong odor, one's eyes begin to become irritated, and damage to lungs, liver, and heart has been observed. At 150 ppm, serious lung problems occur with 3- to 8-hour exposures. Many feel that chronic lung damage will occur with concentrations as low as 5 ppm if a person is exposed all day. The National Air Quality Standard is 0.05 ppm annual arithmetic mean (see Table 13.3). There are alerts in cities when the concentration reaches 4 ppm. In a city such as Los Angeles, about 750 tons of NO_x are put into the atmosphere every day, about 500 tons from internal combustion engines and 250 tons from electric power plants. The major effects of the nitrogen oxides, however, are indirect. In the presence of water vapor in the atmosphere, they are partially converted to nitric acid (HNO_3). The consequences of the formation of the acid are considered in chapter 14 when acid rain is discussed. Another indirect consequence of NO_x in the atmosphere is the important role it plays in photochemical reactions and the formation of smog.

Under solar radiation the following reaction occurs:

$$NO_2 \xrightarrow{\text{light}} NO + O$$

The sunlight that is absorbed is in the ultraviolet and blue (light with wavelengths less than 4200 Å), hence the reddish-brown color of nitrogen dioxide. The consequences of this, however, are much more than reduced visibility, as the atomic oxygen can react in the presence of any third molecule, M, with molecular oxygen (O_2) to form ozone (O_3), a very strong oxidant.

$$O + O_2 + M \rightarrow O_3 + M$$

Among other consequences, the ozone can react with NO as follows:

$$O_3 + NO \longrightarrow NO_2 + O_2$$

and the original nitrogen oxide is converted back to nitrogen dioxide, and the photochemical process can go on again. More ingredients are needed, however, to explain fully the photochemical smog that engulfs our cities.

13.5 HYDROCARBON EMISSIONS AND PHOTOCHEMICAL SMOG

In about 1943, Los Angeles began to experience a new kind of air pollution. The new form of pollution damaged plants, caused eye irritations, cracked stressed rubber, and generally decreased visibility. After an intensive period of research, in the early 1950s A. J. Haagen-Smit and his colleagues arrived at an understanding of the basic processes that were responsible. Although the major reactions have been identified by Haagen-Smit and other subsequent investigators, the entire process is complicated, involving many substances and chemical reactions, and research continues at the present time.

Figure 13.6 Visibility in downtown Los Angeles is often restricted by smog.

We have seen that a combination of nitrogen dioxide and sunlight leads to the formation of ozone. There are a number of other reactions involving various hydrocarbons that also result in strong oxidants such as ozone. Thus, the basic ingredients of photochemical smog are sunlight, NO_2, and hydrocarbons. Another necessity is a meteorological condition in which the ingredients have time to interact before being dispersed. Most of the nitrogen dioxide and hydrocarbons are related to automobile emissions. A city such as Los Angeles, with its sunlight, traffic, and thermal inversions, has all of the necessary ingredients of photochemical smog in great abundance. The problems in Los Angeles have increased with the number of people and automobiles. In 1962, 80% of the people in Los Angeles County were bothered by photochemical smog. Although local automotive emission controls began in 1966 and the federal Clean Air Act became law in 1970, the effects of the controls have been largely offset by the continued increase in vehicle miles traveled.

As the name implies, hydrocarbon molecules involve atoms of hydrogen and carbon, but atoms such as oxygen and chlorine can also be involved. In a study in Los Angeles in 1970, 56 different species of gaseous hydrocarbons were identified in the air; but apparently the number observed is limited only by the sensitivity of the analytical techniques used.

The alkanes (or paraffins) are the simplest type of hydrocarbon. As listed in Table 2.4, the group includes methane, ethane, propane, and so forth. Methane (CH_4, the basic ingredient of natural gas) is emitted in great abundance by natural sources such as swamps and is also a major man-made pollutant. Fortunately, methane plays no important role in photochemical smog. Some of the other types of hydrocarbons are alkenes (also known as alkylenes or olefins), aromatics (with the basic unit of a benzene ring), and aldehydes or ketones (involving one or more oxygen atoms, for example, formaldehyde).

These various hydrocarbons enter the atmosphere from a number of different sources (see Table 13.2), but some of the most important are:

(a) Auto exhaust and partially burned gasoline

(b) Gasoline evaporated in various steps in production, refining, and handling

(c) Organic solvents used in manufacturing, dry cleaning fluids, inks and paints

(d) Chemical manufacturing

(e) Incineration of various materials, industrial driers, and ovens

The general role that hydrocarbons play in photochemical smog can be summarized in the following equation:

$$R + O \longrightarrow R'O\bullet + R''$$

As discussed, the free oxygen atom can come from the interaction of light on NO_2 producing O and NO. The original hydrocarbon is designated R in the above equation, and R′ and R″ are resulting hydrocarbons that may differ or be the same as the original hydrocarbons. The notation R′O• indicates an oxygen-bearing free radical, which in general is very reactive and is at the heart of the destructive properties of photochemical smog. Another similar set of reactions that can take place involves ozone:

$$R + O_3 \rightarrow R'O\bullet + R''$$

Some of the oxygenated hydrocarbons that are produced are the aldehydes, formaldehyde, and acrolein. At the start of the smog process, the photochemical reactions are limited to NO_2 since there is not enough light energy to break down the hydrocarbons directly. After the smog has been building for a while, however, and some amount of oxygenated hydrocarbons is in the air, photochemical reactions can take place directly on these hydrocarbons producing more free radicals. One of the common radicals is peroxy (a radical containing O_2). Peroxy radicals are thought to offer a competing mechanism for the conversion of NO into NO_2. When the concentration of NO_2 is sufficiently great, these same peroxy radicals can interact with NO_2 to form peroxyacyl nitrate (PAN).

$$RCOOO\bullet + NO_2 \longrightarrow RCOOONO_2$$

where the OOO notation indicates that the oxygen atoms are bonded individually and not as an oxygen molecule. A large number of PANs can be formed depending on the exact hydrocarbon, R, that is involved. The PANs are felt to be an important product of the various reactions that take place and, because of their strong oxidizing ability, are responsible for many of the harmful properties of smog.

The measure of the intensity of photochemical smog is the total oxidant concentration. The standards (see Table 13.3) relate to ozone and the equivalent oxidizing power of the various hydrocarbons and free radicals just discussed. The standards now call for 0.24 ppm (160 $\mu g/m^3$) based on a 3-hour average not to be exceeded more than once a year. In other words, it is the same as for carbon monoxide in that it is the second highest maximum that counts, as the first highest maximum is forgiven. U.S. cities (Los Angeles in particular) have a very difficult time meeting this standard. Figure 13.7 shows the concentrations of hydrocarbons (HC), NO, NO_2, and O_3 (actually all oxidants) as a function of time during a weekday. At the beginning of the day, NO and HC show a large increase due to commuter traffic and automotive emissions. After

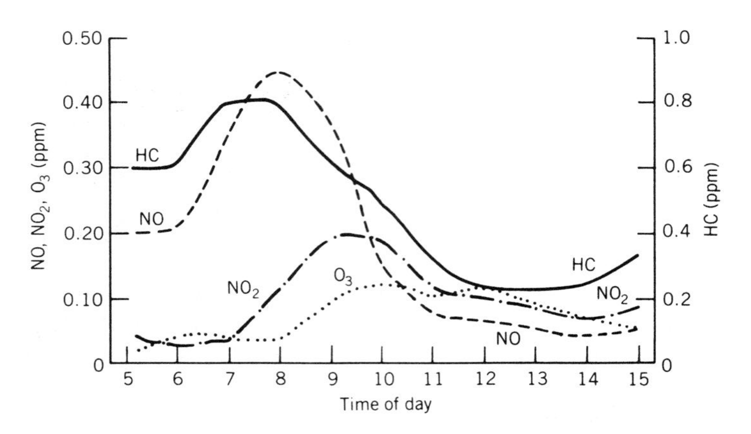

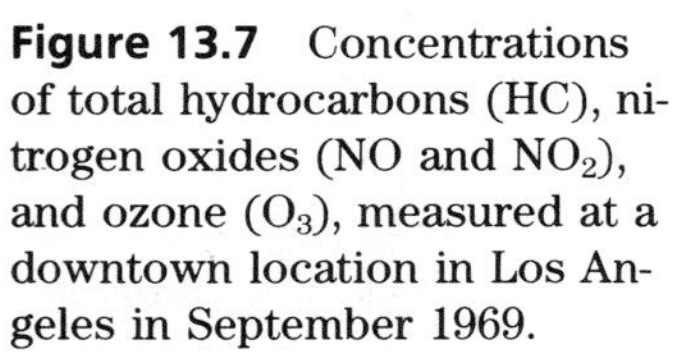

Figure 13.7 Concentrations of total hydrocarbons (HC), nitrogen oxides (NO and NO_2), and ozone (O_3), measured at a downtown location in Los Angeles in September 1969.

an hour or so, the NO has been converted to NO_2, and the photochemical reactions can now begin with the strong morning sun. The resulting oxidants peak about an hour later and then diminish during the day. It can be seen by comparing Figure 13.7 and the National Ambient Air Quality Standards (Table 13.3) that Los Angeles has not always met those standards.

What are the harmful effects of smog? If one asks a typical resident, he or she would certainly complain about the eye irritations that affect about three fourths of the population. The eyes smart, and then tears form to help wash away the irritants. It is the oxidants, particularly the PANs, aromatic olefins, aldehydes, formaldehyde, and acrolein that cause the problems, but interestingly enough, ozone does not seem to be involved. Another feature of photochemical smog widely experienced is the odor, which is largely caused by ozone. The threshold for detection of the ozone smell is 0.02 ppm, and it is readily noticed at 0.2 ppm or more. Single exposures of a few hours to O_3 in the range of 80 to 400 parts per billion (ppb) have been found to have noticeable effects even on young people. Increased coughs, reduced athletic performance and a series of respiratory problems have been documented. With continued exposure, many residents apparently become less sensitive to the ozone odor.

Beyond these irritations, researchers feel there are undesirable consequences for the respiratory system. So far, there have been no definitive epidemiological studies that directly associate photochemical smog with respiratory disease and death, as was the case in London and Donora with classic smog. It certainly is known that conditions such as sinus trouble, hayfever, bronchitis, asthma, and other respiratory problems grow worse in an atmosphere such as Los Angeles presents. It is also suspected that photochemical smog contributes to lung cancer and chronic pulmonary disease. In animal studies, it has been shown that 1 ppm of O_3 can cause lung diseases.

There appear to be two distinct types of plant damage—smog injury and grape stipple (or weather fleck). Smog injury, mostly from the PANs, refers to a collapse of the cells on the under part of the leaf that causes the leaf to appear water-soaked. Grape stipple refers to the blotched or stippled appearance of affected leaves such as tobacco or grape. Even low concentrations of ozone appear to be capable of producing such effects. The loss of productive crops due to the effects of smog on plants is a documented problem with serious economic consequences.

The deterioration of materials exposed to photochemical smog is another

Figure 13.8 Automobile exhaust emissions contribute in a major way to atmospheric pollution even though they may not be as visible as the exhaust from a diesel truck.

expensive and inconvenient result. The cracking and disintegration of stretched rubber was mentioned above as one of the initial manifestations of smog. These effects are generally serious for materials such as paints and fabrics, and they eventually lose their strength and color from the oxidants in the air.

The visibility problem that plagues many cities is mainly due to aerosols and particulates, and we shall postpone discussion of these for the moment. The brown cloud, however, is the NO_2 undergoing photochemical reactions, and it is a direct component of photochemical smog.

Overall, it is difficult to assess quantitatively the damage in terms of human misery and the direct economic consequences of photochemical smog and carbon monoxide. It is obvious, from any point of view, that our atmosphere is too finite to absorb the pollutants that are being dumped into it and still let us pursue a reasonably healthful existence. In the next section we discuss the role the automobile plays in generating the pollutants and summarize the steps that can be taken to reduce the harmful emissions.

13.6 REDUCTION OF AUTOMOBILE EMISSIONS

The internal combustion engine, and, hence, the automobile, is the major source of carbon monoxide, nitrogen oxides, and hydrocarbons in many cities. It has also been a prime source of lead as an environmental pollutant as a result of the burning of gasoline with lead compounds as additives. Figure 13.9 shows the main components of a car engine. Also shown are the main avenues through which emissions of various kinds reach the atmosphere. Evaporation of gasoline leads to hydrocarbons escaping from the tank (*A*), the carburetor (*B*), and the crankcase (*C*). Because of the high pressure in the cylinder above the piston during the combustion process, some of the unburned gasoline and other organic compounds find their way past the piston rings into the crankcase. Before emission controls were mandated, the crankcase was vented, as shown in Figure 13.9, to the atmosphere, and this accounted for about 25% of hydrocarbon emissions. All new cars now have blowby devices that recycle the gases from the crankcase back to the engine intake manifold. Similarly, the evaporative losses of gasoline from the carburetor and fuel tank, which accounted for about 20% of the hydrocarbon emissions, have been largely eliminated.

Of the remaining emissions, almost all now come from the tailpipe (*D*).

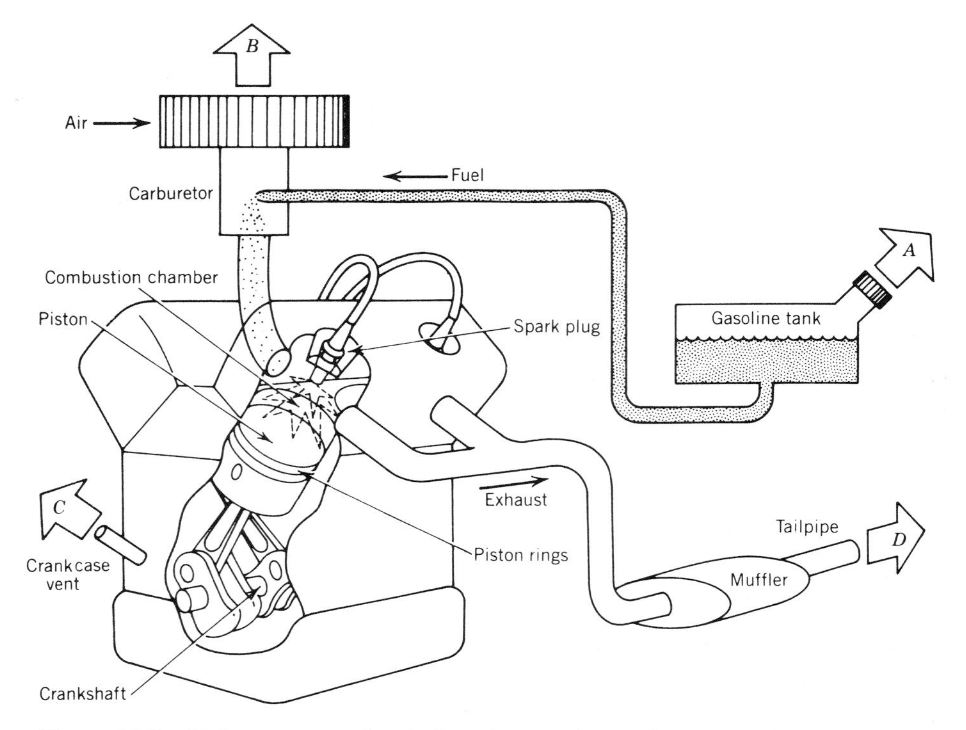

Figure 13.9 Major sources of emissions from an internal combustion engine. There are evaporative losses of hydrocarbons from the fuel tank (*A*) and the carburetor (*B*). Some combustion gases get into the crankcase and out of a vent (*C*); they are emitted in large quantity through the exhaust and tailpipe system (*D*).

There are two general approaches to reducing automotive emissions. The first is to improve the combustion process; the second is to provide an exhaust system that traps the pollutants or converts them into harmless gases.

The Clean Air Act in 1968 and 1970, in addition to providing standards for air pollutants, also provided restrictive standards on exhaust emissions that must be met by newly manufactured autombiles. Before that, in 1966 the State of California had imposed restrictions on the emissions of new cars sold in that state. Both sets of regulations specify the number of grams of hydrocarbons, CO, and NO_x that can be emitted per mile. It is interesting to note that the size of the car is not specified, only the pollutants emitted per mile by each automobile. The driving tests used in determining the emissions involve a test cycle that includes acceleration, cruising, deceleration, and idling. Table 13.4 lists some of the California and U.S. regulations. These regulations have been the focus of much attention because the operation and costs of automobiles are very much involved, as is, to some extent, the profit margin of the manufacturers. Because fundamental changes in car designs cannot be made rapidly, the Clean Air Act of 1970 specified that 1975 model cars must have no more emissions than those listed in Table 13.4. The Environmental Protection Agency (EPA) was the governmental body entrusted with enforcing the regulations and providing test procedures.

The relationship of the details of internal combustion engine operation to

Table 13.4 AUTOMOBILE EXHAUST EMISSION IN GRAMS PER MILE

	HC	CO	NO_x
Actual uncontrolled emissions on pre-1966 cars	17.0	124.0	5.4
1966 California standards	3.4	34	—
1968 California "low emission vehicle"	0.5	11	0.75
1970 U.S. standards	2.2	23	—
U.S. standard for 1990	0.41	3.4	1.0
Tier I[a]	0.25[b]	3.4	0.4
Tier II[a]	0.125[b]	1.7	0.2
Chrysler–Esso engines			
Manifold reactor	1.5	20	1.3
Catalytic reactor	1.7	12	1.0
Wankel engine (with controls)	1.8	23	2.2
Natural gas fueled internal combustion engine	1.5	6	1.5
Gas turbine	0.5–1.2	3.0–7.0	1.3–5.2
Stirling-electric hybrid (GM)	0.006	0.3	2.2
Stratified-charge engine (Ford)	0.37	0.93	0.33

Source: Adapted from Laurent Hodges, *Environmental Pollution.*
[a] 1990 Clean Air Act amendments (special future standards).
[b] Nonmethane organic gases.

the pollutants emitted is rather complicated. There are dependencies on the air–fuel ratio, ignition timing, intake manifold vacuum, compression ratio, engine speed, coolant temperature, and other factors. The effect of variation of the air–fuel ratio on emissions of NO_x, HC, and CO is shown in Figure 13.10. The overall improvement in engine performance by better design and use of controlled fuel injection and ignition timing has been very impressive. The availability of inexpensive microprocessors has aided the complex control of the combustion functions.

At the same time that engine improvements have been made, there has been a concentrated effort to process the exhaust fumes in various ways to meet the standards. The device that is the most widely used in the United States at present is the catalytic converter. One of the objects of the device is to convert NO into N_2 and O_2. This decomposition proceeds very slowly at room temperature, but at higher temperatures, in the presence of a catalyst such as platinum, the process can go very much faster. The catalytic converter is usually located near the muffler. Other converters, containing mixtures of oxides of chromium, iron, copper, and so forth, are also used to promote the oxidation of hydrocarbons and carbon monoxide. It is very difficult to reduce NO, HC, and CO in the same converter. A converter in wide use in this country uses platinum, which is subject to being poisoned by the lead contained in the antiknock additive for gasoline. For this reason, and because of the toxic effects of lead, most cars have been designed to accept only lead-free gasoline since 1974.

The 1990 Clean Air Act amendments called for three alternative fuel programs. The first is a summer reformulated gasoline program, which will be

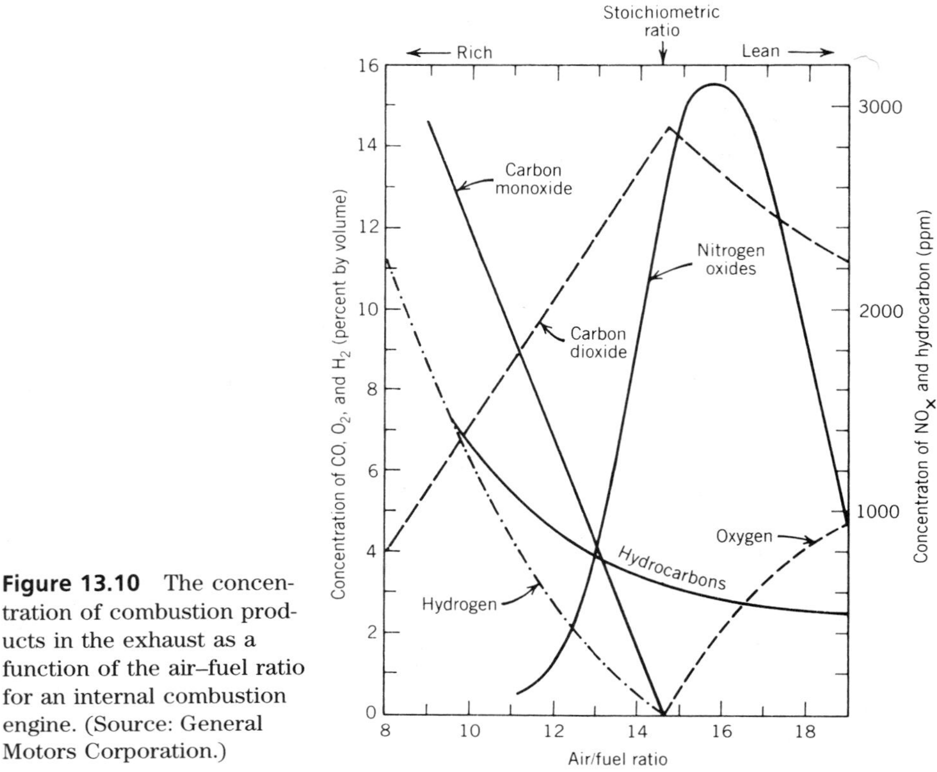

Figure 13.10 The concentration of combustion products in the exhaust as a function of the air–fuel ratio for an internal combustion engine. (Source: General Motors Corporation.)

required after 1995 in the nine worst ozone areas. The second is a provision for a winter oxygenated fuel program for areas not in compliance with the CO standards. The third is a clean-fuel fleet program to start in 1998 in the 20 current ozone noncompliance areas. This means that some fraction of the new car fleet must operate on low-emission alternative fuels.

13.7 SULFUR DIOXIDES IN THE ATMOSPHERE

Since sulfur is present to some extent in almost all fossil fuels, and since the United States and the rest of the world are dependent on fossil fuels as a source of most of our energy, it should not be surprising to learn that sulfur dioxide (SO_2) is an important atmospheric pollutant. When fossil fuel is burned, the sulfur goes off as SO_2, a colorless and nonflammable gas. Various sulfur compounds in the fossil fuels are converted to SO_2 in the process of burning the fuel. Table 13.2 lists the major sources of SO_2, and we see that in this case the automobile is not a major source. Although a relatively small amount of sulfur is emitted by vehicles that burn gasoline and diesel oil, the major SO_2 sources are stationary, particularly coal- and fuel oil–burning electric plants. The next most important sources are various industrial processes, with the smelting of copper, zinc, and lead being the leading contributors. The sulfur content of coal and petroleum varies widely, but is generally in the range of 0.3 to 5% by weight.

About one third of the total sulfur compounds put into the atmosphere come from man-made sources, with about 93% going into the northern hemisphere. The natural sources of sulfur are mainly the decay of terrestrial and marine organic matter, where the sulfur is in the form of hydrogen sulfide (H_2S). Of comparable importance are the sulfates released in the form of aerosols in ocean spray. The H_2S is converted into SO_2 in a day or two, presumably by interaction with the ozone in the troposphere. There are, of course, various avenues by which SO_2 is removed from the atmosphere. Dry deposition, precipitation, and plant uptake are the major ones. It is estimated that the total SO_2 in the atmosphere is about 11×10^6 metric tons. As can be seen from Table 13.2, just the U.S. annual contribution of man-made SO_2 to the atmosphere is many times this residual amount. The problems of SO_2 are related more to its regional rather than its global effects. As discussed earlier, sulfur dioxide and particulate matter are the essential ingredients of the classic smog that has plagued cities such as London and New York.

It is difficult to separate the effects of SO_2 from those of the H_2SO_4 (sulfuric acid) that results from SO_2. When sulfur dioxide enters the atmosphere, it is oxidized to SO_3 in a relatively short time (a few days), and the SO_3 can then combine with moisture to form H_2SO_4 or sulfate salts. The problem presented by acid rain is discussed separately in the next chapter. The local effects of SO_2 are somewhat intermixed with those of H_2SO_4, but both substances are known to be irritants of the respiratory system. In combination with particulates, measurable increases in the rate of illness and death have been observed in cities where the SO_2 concentration is around 0.1 or 0.2 ppm. Concentrations of 0.01 ppm for a year, ranging to concentrations of a few ppm for 30 seconds, are felt to affect the health of people, particularly through the respiratory and cardiovascular systems. The experience in the Netherlands has been that there is a definite increase in the mortality rate from lung cancer and bronchitis for SO_2 concentrations of 0.04 ppm. The effects directly due to sulfur dioxide are hard to isolate, however, because particulates and moisture are usually present in the same environment and play a synergistic role.

Building materials such as marble, limestone, and mortar are severely affected by sulfur dioxide because the carbonates present are, to some extent, exchanged for sulfates originating from SO_2. The sulfates are soluble in water and washed away with time by rain. Buildings and statues of great historic and aesthetic value throughout the industrial world are suffering under such deterioration. The oxide coatings on metals (rust, for example) partially lose their ability to protect the metal in an environment of SO_2 and moisture. Organic materials and paints are also weakened and discolored by the effects of sulfur dioxide. Plants, too, are affected by SO_2. Various crops and trees suffer appreciable damage to their leaves and internal cells at exposures of 0.01 ppm for 1 year and to 1 ppm for 1 hour.

Example 13.3

Consider a 1000-MW_e coal-burning power plant operating 24 hours a day at an efficiency of 33%.

(a) How much coal (in tons) is burned daily?

Solution

$$24\ \frac{\text{hr}}{\text{day}} \times 10^6\ \text{kW} \times \frac{1}{0.33} \times \frac{1\ \text{ton coal}}{7800\ \text{kW}\bullet\text{hr}} = 9324\ \text{tons coal a day}$$

(b) If the coal has 1% sulfur content, how many tons of SO_2 are released daily by this plant?

Solution

0.01 × 9324 tons = 93 tons sulfur released a day, and this results in 186 tons SO_2 because SO_2 by weight is half sulfur (atomic weight = 32) and half oxygen (2 × at. wt. = 2 × 16 = 32). In units of micrograms (using the approximation 1 ton ≈1 tonne) this is

$$186\ \text{tons} \times 10^3\ \text{kg/ton} \times 10^3\ \text{g/kg} \times 10^6\ \mu\text{g/g}$$
$$= 1.9 \times 10^{14}\ \mu\text{g}\ SO_2\ \text{released daily}$$

(c) Now consider a thermal inversion episode with no horizontal winds so that the emitted SO_2 is confined to a volume 1 km high × 10 km × 10 km. What SO_2 concentration, in $\mu g/m^3$, will be achieved in 1 day?

Solution

$$\text{Air volume} = 10\ \text{km} \times 10\ \text{km} \times 1\ \text{km}$$
$$= 10^4\ \text{m} \times 10^4\ \text{m} \times 10^3\ \text{m} = 10^{11}\ \text{m}^3$$

Therefore,

$$SO_2\ \text{concentration} = \frac{1.9 \times 10^{14}\ \mu\text{g}}{10^{11}\ \text{m}^3} = 1900\ \mu\text{g/m}^3$$

achieved in 1 day. This is 24 times the U.S. Primary Air Quality Standard of 80 $\mu g/m^3$, about 5 times the annual maximum standard of 365 $\mu g/m^3$, and more than half the concentration experienced during the famous 1952 London smog episode that caused 4000 excess deaths in a few days.

To accomplish a reduction in SO_2 emissions, one must look principally to the coal-burning power plants. There are at least three useful directions to pursue: burning coal with less sulfur content, removing the sulfur before burning the coal, and removing the SO_2 from the stack gases. There is, in fact, a vast amount of low-sulfur coal (0 to 1% sulfur content) in the United States, 90% of which is in the West. This coal is largely located in the Rocky Mountain region, in areas such as in the Powder River Basin in Wyoming and Montana. Although this coal has significantly less sulfur than eastern coal, which has as much as 3 to 5% sulfur content, the western coal is largely subbituminous and lignite, which have a lower energy content per ton. The eastern coal is, in general, hard coal (bituminous and some anthracite). For a given energy output of a power plant, it is not always certain that western coal will result in lower total SO_2 emissions. There is also the problem that western coal is located at relatively large distances from the eastern cities and, hence, a significantly larger transportation cost, in both dollars and energy, is involved. Another related problem is that the efficiency of fly ash precipitators is reduced when low-sulfur coal is burned.

Another attack on the problem is to remove the sulfur before burning the coal. With high sulfur coals, FeS_2 is the most prevalent sulfur compound present. After the coal is crushed to a fine powder, it is possible to remove most of the FeS_2 by washing the coal with water and taking advantage of the density of FeS_2, which is about four times greater than that of pure coal. Hydrogenation can also be used, particularly when the coal is to undergo liquefaction. These processes, however, are expensive, and in many cases are not effective because of the numerous sulfur compounds involved.

There are a variety of processes for removing SO_2 from the stack gases. One of the most commonly used processes is to have the SO_2 combine chemically in a flue gas scrubber with some alkaline substance such as limestone, lime, or dolomite along with air to form $CaSO_4$. In this particular process, the resulting material is just thrown away. In other processes the by-products are processed to recover the sulfur for resale. There is a vigorous research effort to refine the existing techniques of SO_2 removal and to provide new and better processes, as there are technical and economic problems that discourage routine use of the existing removal processes.

Title IV of the 1990 Clean Air Act amendments provides a two-phase program for further control of SO_2 emissions. In Phase I the act imposes an annual emission limit of 2.5 lb/10^6 Btu on 110 utility plants. In Phase II a lower emission standard of 1.2 lb/10^6 Btu would be applied to utility plants that were in operation with more than 25 MW before 1990. These new requirements are calculated to reduce the SO_2 emissions by 1×10^7 tons from the 1980 emission levels.

13.8 PARTICULATES AS POLLUTANTS

Particulates are quite different in nature from the other pollutants we have discussed inasmuch as they are not gases. They are either solids or liquids and, as such, can have a certain size as well as a chemical constituency. Aerosol is a term often used to describe either solid or liquid matter suspended in the atmosphere; however, we shall use the equivalent term *particulates.* There are a number of natural sources of particulates, such as salt from ocean spray, dust from fields, volcanic ash, and forest fires. Worldwide, these natural sources are about 14 times as prevalent as the man-made sources. Although we tend to focus on man-made pollutants as sources of our atmospheric problems, the recent eruptions of Mount St. Helens in Washington state, El Chichon in Mexico, and Mount Pinatubo in the Philippines were dramatic reminders that nature, on occasion, can outdo us by many orders of magnitude.

One of the major problems of man-made particulates is that they are frequently emitted in areas where the population density is very high, and, thus, they tend to have more serious health consequences than do particulates from a forest fire in Montana. The greatest contribution to man-made particulates is the fly ash from coal combustion. Petroleum combustion yields roughly one twentieth the particulate emissions from coal burning. Iron and steel mills,

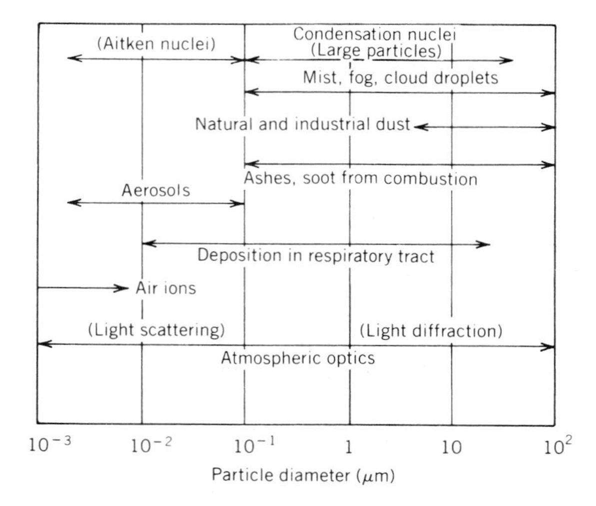

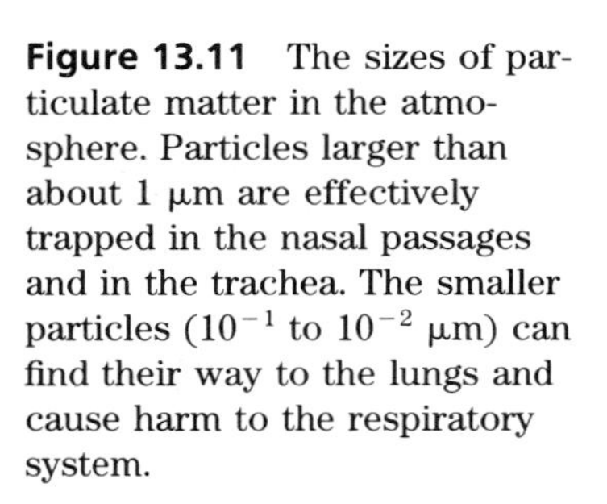

Figure 13.11 The sizes of particulate matter in the atmosphere. Particles larger than about 1 μm are effectively trapped in the nasal passages and in the trachea. The smaller particles (10^{-1} to 10^{-2} μm) can find their way to the lungs and cause harm to the respiratory system.

cement manufacturing, and the burning of wood and other materials are all important sources, as are agricultural processes. Table 13.2 indicates the sources of man-made particulate matter in the United States and shows that the total emissions are 26×10^6 tonnes/year.

Figure 13.11 displays some of the different types of particulates and their size. One of the main threats to health presented by particulates results from their deposition in the lungs. There are various mechanisms that prevent this from happening, such as the filtering of large particles in the nose, the trapping of particles in the mucous membrane of the airways, and sneezing and coughing to remove particles. There are, however, some small particles, in the range of 10^{-2} to 10 microns diameter, that can get through these traps to the sensitive air sacs (the alveoli) that are important in the proper functioning of the lung. The particles reaching the inner lung can directly interfere with the respiratory system, or the particles may be themselves toxic or carry a toxic substance with them. Some particulates are believed to transport to the lungs free chemical radicals, which have effects similar to those of ionizing radiation. Lung cancer is a known occupational hazard of chimney sweeps and coal miners. Epidemiological studies have associated the presence of particulates with the occurrence of bronchitis and emphysema. General deterioration of health appears for long-term exposures to concentrations of about 80 μg/m^3.

Example 13.4

A published epidemiological study has indicated that in the United States from 1960 to 1961 the excess death rate due to particulates in the atmosphere was 0.04 × the particulate concentration in μg/m^3 and that due to sulfur oxides was 0.07 × the sulfur oxide concentration in μg/m^3, both expressed as annual deaths per 10,000 population. Apply these results to the 1990 population to predict the corresponding annual excess death rates in the United States, *if* the average particulate concentration is 100 μg/m^3 and the average SO_2 concentration is 40 μg/m^3.

Solution

Particulates

$$2.5 \times 10^8 \text{ persons} \times 4 \times 10^{-2} \frac{\text{deaths}}{10^4 \text{ persons} \cdot \mu\text{g/m}^3} \times 10^2 \ \mu\text{g/m}^3$$

$$= 100{,}000 \text{ excess deaths per year}$$

SO_2

$$2.5 \times 10^8 \text{ persons} \times 7 \times 10^{-2} \frac{\text{deaths}}{10^4 \text{ persons} \cdot \mu\text{g/m}^3} \times 40 \ \mu\text{g/m}^3$$

$$= 70{,}000 \text{ excess deaths per year}$$

In addition to the direct effects on human health, particulates in the atmosphere can scatter and absorb an appreciable amount of sunlight. It is known that the volcanic ash put into the atmosphere by the eruption of Tambora in the Dutch East Indies in 1815 resulted in a general lowering of the global temperature for several years afterward. On a more local level, cities receive about 20% less sunlight because of atmospheric particulates than do the areas with less industry and fewer power plants. Coupled with this is a general reduction in visibility and the enhancement of fog formation. In addition to the soiling of clothing and buildings with the deposition of soot, corrosion of metals and degradation of other materials is caused by particulates that have had sulfuric acid and other corrosive liquids condensed on them. Paint, masonry, electrical contacts, and textiles are all affected.

The measure of the amount of particulates in the atmosphere is in terms of the number of micrograms per cubic meter ($\mu g/m^3$). The present standards (see Table 13.3) are that the annual mean should not exceed 75 $\mu g/m^3$, and the 24-hour maximum that should not be exceeded more than once a year is 260 $\mu g/m^3$. The present standards do not put a stipulation on the size of the particles or their chemical composition. The cities of the United States have an average of about 100 $\mu g/m^3$, which is considerably in excess of the standard, and heavily polluted areas are at times as high as 2000 $\mu g/m^3$. When the deadly classical smog episode occurred in London in 1952, the particulate concentration reached a maximum of 1700 $\mu g/m^3$ and the SO_2 about 0.70 ppm. U.S. suburban areas with a lower population density have an average of about 45 $\mu g/m^3$. Very remote areas (wilderness areas, deserts, national forests, etc.) average 21 $\mu g/m^3$. Large cities such as New York, Chicago, Pittsburgh, and Philadelphia are among the worst areas, having particulate concentrations in the range of 155 to 180 $\mu g/m^3$.

There are natural processes that tend to remove the particulates put into the air by either man-made or natural sources. The gravitational force acts on all particles and tends to move them toward the earth. This sedimentation, or gravitational settling process, is really only effective for large particles that have a radius larger than 20 μm. Particles smaller than this can be removed to some extent when they are near the ground by impact on various objects such as trees and buildings. In the higher atmosphere, small particles can coalesce and form nucleation centers for raindrops and are thus brought to earth. There is also a wash-out process where precipitation intercepts some of the larger particles in the air beneath clouds and thus brings them to earth. The natural residence time for particles in the atmosphere ranges from days to years. Unfortunately, we cannot rely on nature to clean up our air because of the rate at which we are fouling it.

As Table 13.2 shows, there are a number of different sources of particulates, and the controls or devices for reducing emissions from these various sources differ. Since coal-burning power plants are the largest source, we shall discuss the control devices used by this industry. After coal is burned, the gaseous effluents go off; along with them, small particulates or fly ash go up a stack and are released to the atmosphere. The larger residue, or bottom ash, left from combustion of the coal must be physically removed from the collection areas at the bottom of the boilers and disposed of separately.

Power plants use a number of different types of devices to remove the fly ash. Generally, more than 99% of the fly ash must be removed, so it takes a very extensive system, usually involving two separate types of devices to bring the effluent up to standards. The 1979 Federal New Source Performance Standards (NSPS) require that 90% of the sulfur and all but 0.03 lb of the particulates per 10^6 Btu of thermal energy should be removed. It was expected that these standards would be strengthened to 95% and 0.02 lb by the mid-1980s. Even 0.03 or 0.02 pounds of particulates per 10^6 Btu can represent the emission of enormous numbers of smaller particles, and these smaller particles tend to be more damaging to human health than particles of larger size.

The exact devices used depend on the type of coal being burned and other factors, but filters are used very commonly. A porous woven fabric material is usually used in a bag house, arranged so that the filters can automatically be shaken and air blown through them in the reverse direction to clean them. The removed fly ash collects at the bottom of the bag house and is carted off for disposal. One version of a bag house is shown in Figure 13.12. Devices of this kind have typical removal efficiencies (in terms of $\mu g/m^3$) greater than 99% and are quite effective for particles smaller than 1 μm; however, high-temperature gases must be cooled before they can pass through filters. Sometimes a set of

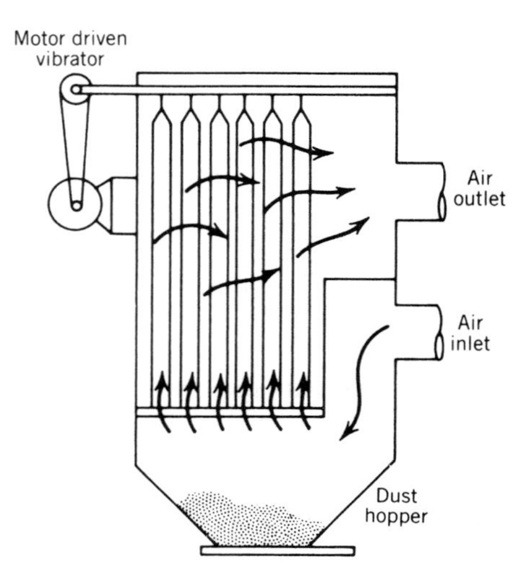

Figure 13.12 A bag house with a motor-driven vibrator. Vibration of the filter bags frees the fabric of the particulates, which then collect at the bottom.

filters is preceded by a settling chamber, which lets the heavier particles settle out by gravity prior to entering the bag house.

Electrostatic precipitators are also widely used control devices because they have high removal efficiencies and no moving parts. The stack gases are passed through a chamber where, in a region of high electric fields, some of the gas molecules are ionized. These ions go on, and some of them become attached to the fly ash particles. The particles thus become charged and are collected on metal plates maintained at a high electrostatic potential.

The cyclone separator is arranged to have the stack gases spiral upward with a circular motion. The heavy particles hit the walls; they settle out and are collected in the bottom. Cyclone separators are not very efficient for submicron particles, but they can be used for high-temperature moisture-laden gases. There is also a family of devices called scrubbers, or wet collectors, which remove particles by having them come into contact with water. In the simplest version, the stack gases are passed directly through water. Spraying water in fine droplets into the stack gases is a more effective arrangement. The wastewater must, of course, be treated to remove the particulates, and this is rather costly. One such device is shown in Figure 13.13.

13.9 THE GREENHOUSE EFFECT AND WORLD CLIMATE CHANGES

As discussed briefly in Chapter 6, the term "greenhouse effect" follows from the fact that solar radiation penetrates the glass coverings of an ordinary greenhouse rather efficiently, but the infrared radiation from objects inside the greenhouse does not. The result is an entrapment of the energy of solar radiation, which leads to heating the greenhouse. A similar phenomenon takes place

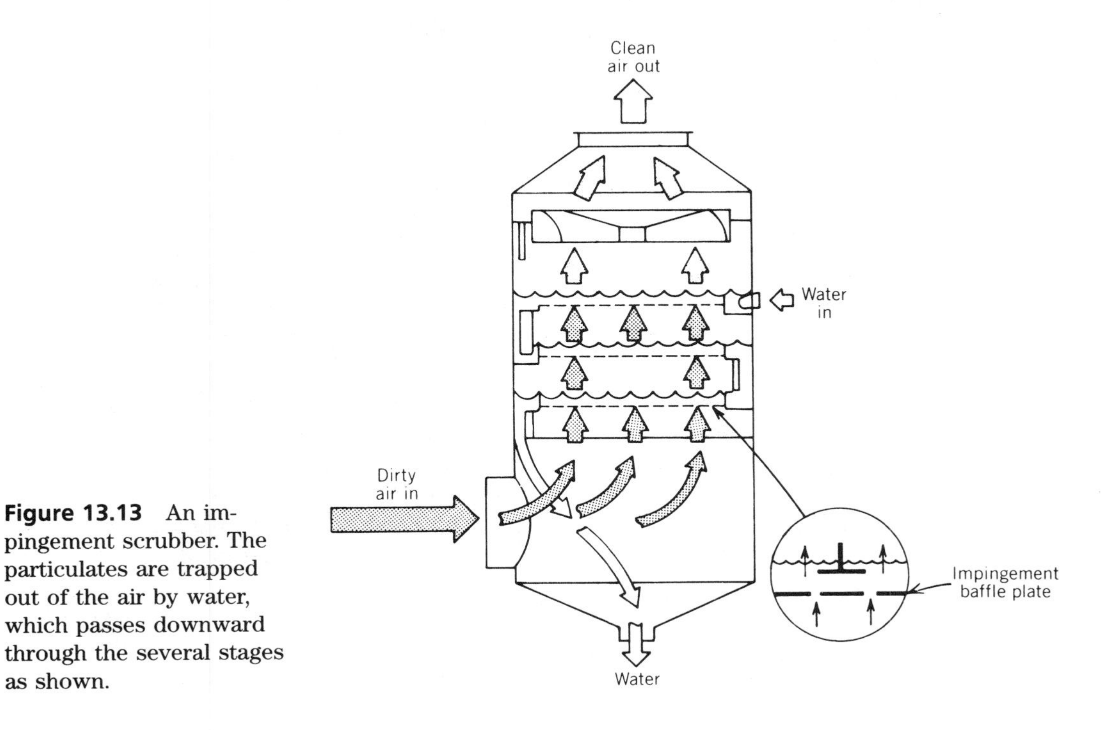

Figure 13.13 An impingement scrubber. The particulates are trapped out of the air by water, which passes downward through the several stages as shown.

in the earth's atmosphere. Certain gases, now called the greenhouse gases, act like the glass roof does in greenhouses to warm the atmosphere. The effect is related to the fact that the sun's radiation temperature is very high compared to the earth's surface temperature. The wavelength of the radiated thermal energy from the earth's surface is in the infrared region (about 4 to 20 μm). Certain gases absorb electromagnetic radiation very effectively in this region, and if they are present in the atmosphere, will trap the infrared radiation moving upward from the earth's surface. This contributes to a warming of the earth's atmosphere. The most important greenhouse gases are carbon dioxide (CO_2), methane (CH_4), nitrous oxide (N_2O), chlorofluorocarbons (CFCs), and tropospheric ozone (O_3). Carbon dioxide now accounts for about two thirds of the greenhouse effect, methane about 25%, and CFCs about 10%. There is growing evidence that various human activities are leading to an increase in the concentration of these gases in the atmosphere, and hence to the possibility of an increase in the temperature of the earth's atmosphere. This may result in significant climate changes on a global scale.

Carbon dioxide cannot be classified as an air pollutant, as it is naturally reasonably abundant in the atmosphere (0.036%, or 358 ppm). Any burning of carbon-based fuels leads to the formation of CO_2. Although the chemical reaction in the combustion of the various fossil fuels is not always simple, the following equation summarizes the general process:

$$C + O \rightarrow CO_2$$

Whether wood, coal, petroleum, or natural gas is burned, carbon dioxide will result. As discussed in Chapter 10, CO_2 plays a vital role in photosynthesis, and when organic matter decays, CO_2 is released to the atmosphere. The problem is that the concentration of CO_2 in the atmosphere is increasing at the rate of 0.4% annually, and human activities, especially the ever-increasing consumption of fossil fuels and the reduction of forests, are the likely cause of the increase.

Methane is far less abundant in the atmosphere (1.65 ppm) than CO_2, but it is molecule-for-molecule 20 to 30 times more effective in absorbing infrared radiation than CO_2, and it is increasing at the rate of 1% annually. Anaerobic bacteria in rice fields, the digestive tract of cattle, release from handling fossil fuels, termite colonies, sewage treatment plants, and landfills are thought to be the main sources of atmospheric CH_4. Nitrous oxide is present in the atmosphere to 305 ppb, and is increasing at the rate of 0.2% annually. The main sources are fossil fuel combustion and nitrogen fertilizers. The CFCs (CCl_2F and CCl_3F) are present only in trace amounts [11 and 19 parts per trillion (ppt)] but are increasing at the rate of 5% annually. Their importance is related to the fact that a CFC molecule is about 10,000 times as effective in trapping heat as is a CO_2 molecule. The CFC problem is outlined in detail in the following section. Tropospheric ozone is present to about 35 ppb and is increasing at 1% annually. Carbon dioxide is the main greenhouse gas discussed here, as it is the most important, but the other gases just enumerated play a significant role in the global heating problems. Table 13.5 summarizes the information on the various greenhouse gases.

The question of the effect of man-made CO_2 and other greenhouse gases on the earth's temperature and climate is a complex one involving some processes that are not completely understood. On the other hand, some aspects

Table 13.5 GREENHOUSE GASES IN THE ATMOSPHERE

Greenhouse Gas	Present Concentration	Annual Rate of Increase	Approximate Importance to Greenhouse Effect	Relative Infrared Absorption per Molecule
Carbon dioxide (CO_2)	360 ppm	0.4%	65%	1
Methane (CH_4)	1.65 ppm	1%	25%	20–30
Nitrous oxide (N_2O)	305 ppb	0.2%		
Chlorofluorocarbon (CFC)	10–20 ppt	5%	10%	10,000
Ozone (tropospheric) (O_3)	35 ppb	1%		

of the problem are well established. It is abundantly clear that we are injecting large amounts of CO_2 into the atmosphere. Since the amount of coal and other fuels that are burned each year is approximately known, reasonable estimates can be made of the CO_2 released. This amount has increased steadily from about 0.5×10^9 tonnes/yr in 1860 to about 5×10^9 tonnes/yr at present. It can be expected that atmospheric CO_2 will increase with time as more fossil fuels are burned, until the total recoverable fossil fuels are consumed. The total recoverable resource of fossil fuels is about 2×10^{12} tonnes.

An aspect of the problem that is not open to speculation is the amount of CO_2 present in the atmosphere. Since 1958 accurate measurements of the CO_2 concentration have been carried out at an observatory on Mauna Loa. The results of these measurements are shown in Figure 13.14. There is a general

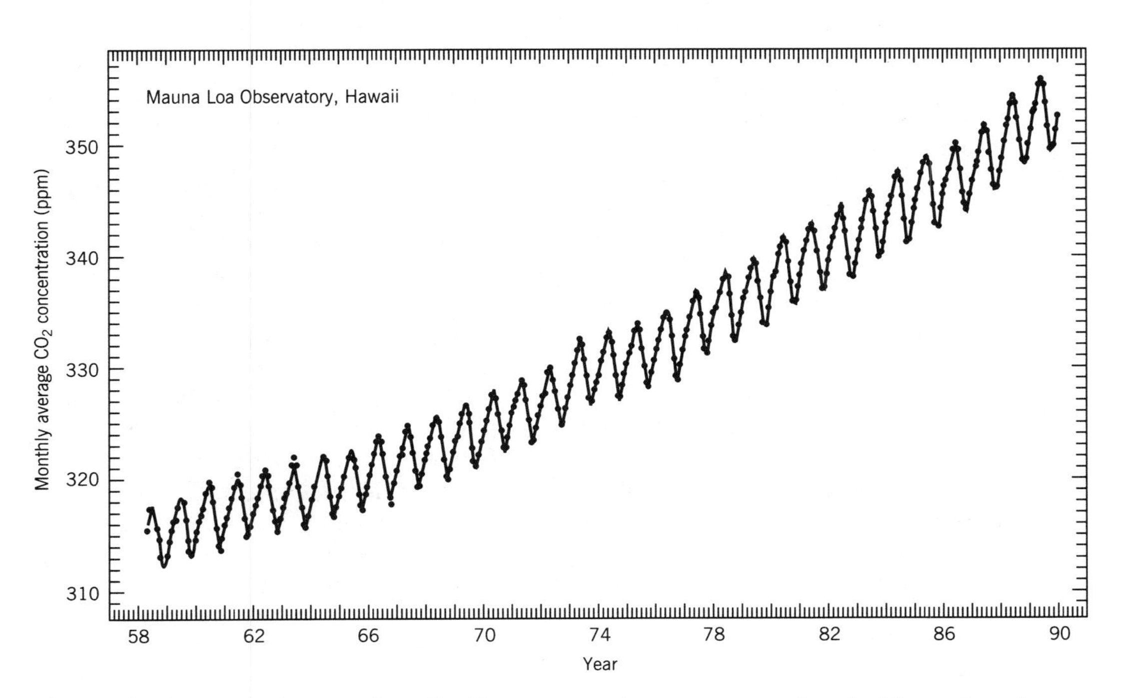

Figure 13.14 Atmospheric carbon dioxide concentrations as measured at the Mauna Loa observatory on the island of Hawaii. The data are monthly averages in parts per million. (Source: U.S. National Oceanic and Atmospheric Administration.)

trend upward with time that amounts to about 1.4 ppm per year, or an annual increase of about 0.4%. The interesting seasonal oscillation in the data is caused by the removal of CO_2 by photosynthesis during the growing season in the northern hemisphere. A similar general increase of CO_2 was measured at the South Pole. It can be calculated from these increases that only 60% of the CO_2 being added by fossil fuel consumption each year is being retained in the atmosphere. Based on the measured trends and various models, it has been estimated that by the year 2000 the CO_2 concentration will increase from its value in 1992 of about 358 ppm to about 370 ppm. It has been estimated that the carbon dioxide level was about 280 ppm in 1860. If the present growth rate continues, by the year 2075 the CO_2 concentration will be double the preindustrial level. The question of the exact source of this CO_2 is harder to answer.

In addition to the 700×10^9 tonnes of carbon in the form of carbon dioxide in the atmosphere, there are extensive pools in the biota, the organic matter in the soil, and the oceans. It is estimated that the biota hold about 700×10^9 tonnes and the organic matter in the soil 1×10^{12} to 3×10^{12} tonnes. It is further known that the photosynthetic process removes about 110×10^9 tonnes of carbon from the atmosphere each year, and the amount of CO_2 added by respiration and the decay of organic matter is about the same. The deep layers of the ocean hold a vast amount of carbon in the form of dissolved CO_2 (about 40×10^{15} tonnes); the surface layers (the top 100 m), where carbon dioxide mixes with the atmosphere, hold very much less, about 600×10^9 tonnes. Although mixing can occur between the deep and surface layers of the ocean, this is believed to happen over a long time span. The surface layers of the oceans appear to be the sinks for about 3×10^9 tonnes of CO_2 per year.

The basic question seems to be whether the observed increase in atmospheric CO_2 is due largely to a reduction of the biomass. There has been a reduction in the size of the tropical rain forests, which are excellent pools of CO_2, and arguments have been made that this is the likely source of the increased atmospheric CO_2. More recent calculations and models, however, indicate that the net biotic pool of CO_2 has probably not changed enough to account for the carbon dioxide increase. Although there has been cutting of forests, there has also been intensive agriculture established by irrigation of land that was previously not very productive. Although there are large uncertainties, it appears that the combustion of fossil fuels contributes about 5×10^9 tonnes/yr of CO_2, and deforestation 1 to 2×10^9 tonnes/yr.

The next question relates to the effect on the world's climate of these carbon dioxide increases. The temperature of the earth has changed over geologic time. During the periods of glaciation, it was about 5°C cooler and in the interglacial periods about 5°C warmer than now. The last glaciation period was 10,000 years ago. Since then there have been fluctuations of a few degrees, the period of 1430 to 1850 A.D. being one of particularly low temperatures in Europe. The mean annual global temperature from 1854 to 1989 is shown in Figure 13.15. Although there are large fluctuations from year to year, it seems evident that there has been about 0.5°C average temperature increase in the last 135 years. Eight of the warmest years in the 110-year record of land surface temperature have all occurred in the past 12 years. The years 1989, 1990, and 1991 were the warmest years on record. There are at least two effects that tend to lower the earth's temperature. Volcanoes spew forth a very fine ash that can stay aloft for several years and form a layer of sun-blocking haze.

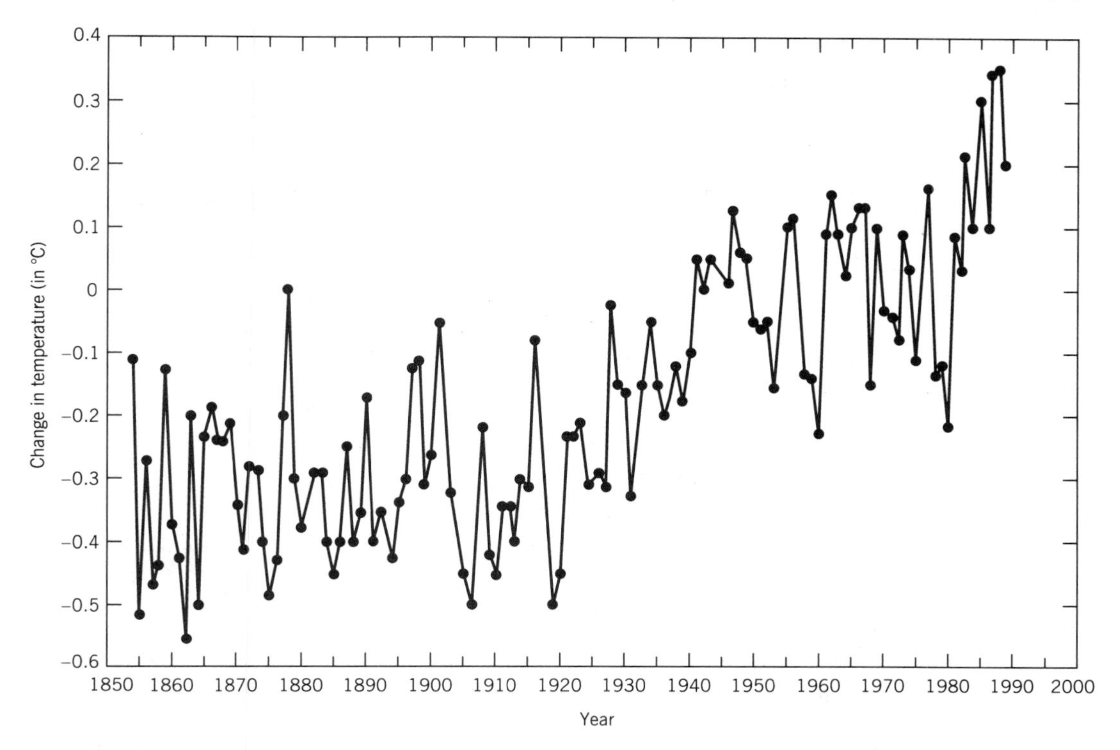

Figure 13.15 The mean annual global relative temperatures from 1854 to 1989. (Source: Data are from T. H. L. Wigley, P. B. Wright, and P. D. Jones, Hadley Center, University of East Anglia, Norwich, England.)

Industrialized countries contribute a variety of aerosols to the atmosphere from smokestacks, cars, tilled fields, and burning forests. Sulfur dioxide, for example, can become sulfate aerosols that form a haze and tend to reflect solar radiation. There is concern that these temperature-lowering effects have masked the temperature increases caused by the greenhouse gases and that we will eventually be faced with greater greenhouse increases than are indicated by the recent temperature records.

It is also known that the average sea level is currently rising at 1.6 to 3.3 cm per decade. About a fourth of this rise can be attributed to the thermal expansion of the oceans due to the temperature increase, and the rest is due to a melting of the glaciers and permanent snowfields.

What can be said about the future world climate? The problem has several aspects. The first is the prediction of the increase in greenhouse gases in the atmosphere. Second is an understanding of the effects of the increased concentration on the global temperature, and third is the effect of the increased temperature on the global climate. The global average temperature over the past 160,000 years is correlated rather closely with the CO_2 concentration. On a finer scale, however, the correlation over the last 100 years is less convincing. The majority of the concerned scientific community feels that there is a high probability that the 0.5°C temperature increase in the last 100 years is related to the increase in CO_2 concentrations.

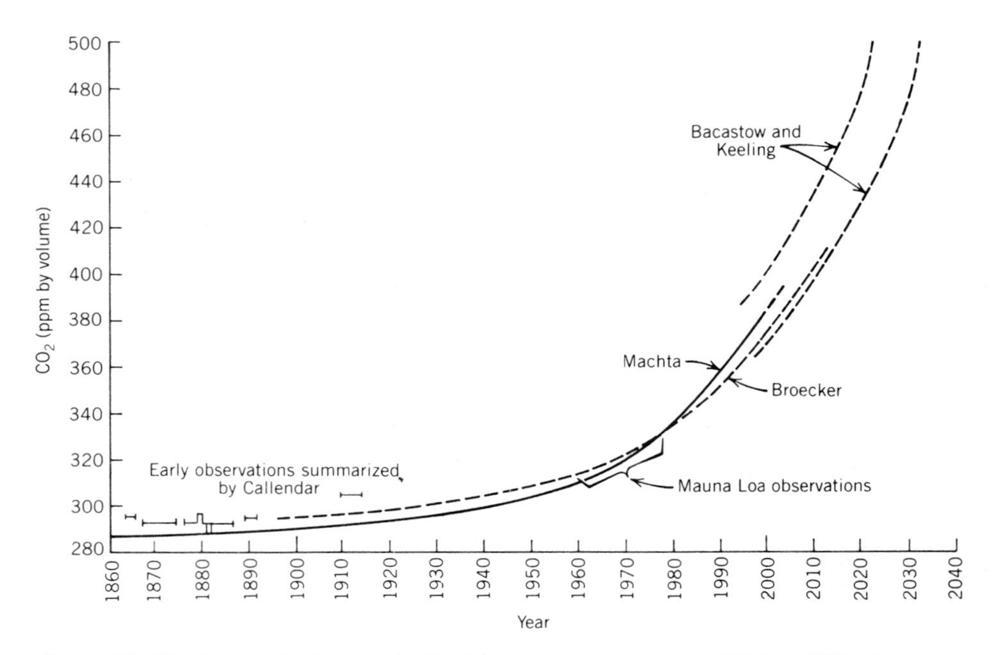

Figure 13.16 Atmospheric carbon dioxide concentration from 1860 to 1975. Several projections or estimates of future trends are also shown. (Source: W. W. Kellogg, *Effects of Human Activities on Global Climate*, Geneva, Switzerland: World Meteorological Organization, Tech. Note 156, 1977.)

Models of the world's climate have been run on very large computers, and scientists have used these models to vary parameters in order to ascertain the effect of a doubling of the concentration of CO_2 or the equivalent composite greenhouse gases. These models tend to predict a 3 to 5.5°C increase in the global temperature. Other estimates have been in the range of 1.5 to 4.5°C. Such predictions are very uncertain because of the crude treatment of various meteorological phenomena. There is no reason to think, however, that the chances of the predictions being high are larger than the chances of their being low. Figure 13.17 shows one projection of the global temperature along with the uncertainties associated with the projections.

An increase of 3°C over the next several decades, as estimated in Figure 13.17, would be many times larger than any such change ever experienced in such a short time. The consequences of such global warming could be severe. Precipitation patterns could be changed; some arid regions could receive more rain, and the important agricultural areas such as the U.S. Midwest could become arid. Many coastal areas would be inundated with the rising sea level.

What can be done to avoid such extreme global warming? Some of the greenhouse gases such as the CFCs hopefully will no longer be manufactured and used, and the injection into the atmosphere of methane, nitrous oxide, and ozone could be reduced by reasonable air pollution control measures. The reduction of carbon dioxide emissions, however, would require less burning of fossil fuels, but there is a world tendency to increase use of fossil fuels. Switching from oil or coal to methane would be helpful to some degree because

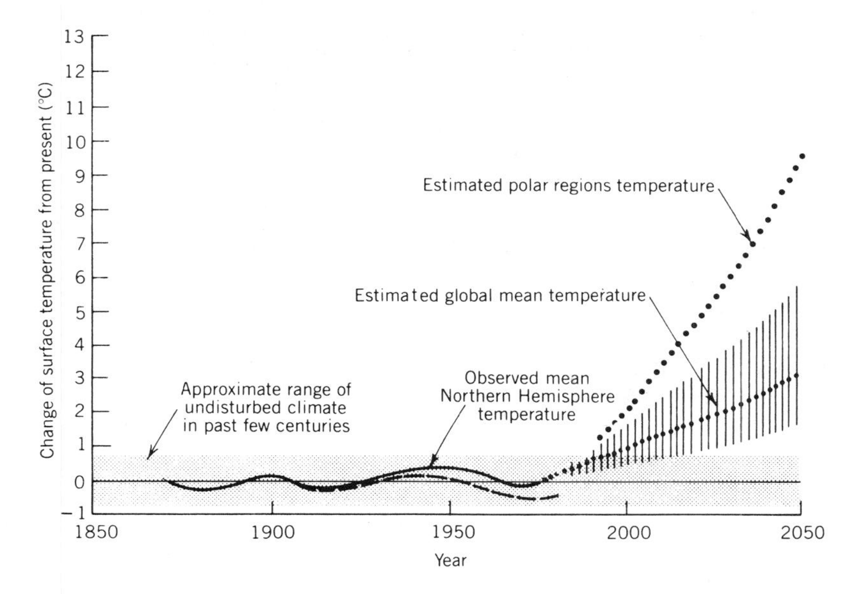

Figure 13.17 The mean surface temperature for the northern hemisphere from 1850 to 1977. Also shown are estimates of the temperatures of the world and polar regions up to 2050. The dashed line is an estimate of what the temperature of the northern hemisphere might have been had carbon dioxide not been added to the atmosphere by human activities. (Source: W. W. Kellogg, *Effects of Human Activities on Global Climate*, Geneva, Switzerland: World Meteorological Organization, Tech. Note 156, 1977.)

less CO_2 and other pollutants are emitted per Btu of energy released by burning methane. Solar, geothermal, and nuclear energy avoid the carbon dioxide problem entirely. Energy conservation in all its forms would certainly be a useful step in reducing the rate of global warming and other forms of pollution.

The United States now accounts for about 25% of the world's fossil fuel–derived CO_2 emissions, the countries of the former Soviet Union about 25%, Western Europe about 15%, China about 10%, and Japan about 6%. The developing countries (except China) contribute about 15%, but it is feared that they will dominate the CO_2 emissions in 50 or 75 years. Various conferences such as the United Nations Intergovernmental Panel on Climate Changes in 1991 and the United Nations Conference on Environment and Development held in Rio de Janiero in June 1992 have tried to establish agreements on greenhouse gas reductions. While many countries have endorsed significant reductions, the Bush administration has been reluctant to do so because of economic implications and questions of the reality of global warming due to greenhouse gases.

At the Earth Summit in Rio de Janiero the United States was mainly responsible for pushing through a treaty on CO_2 reductions that had no specific targets or timetables. It is estimated that a 60% reduction of greenhouse gas emissions due to human activities would be needed to stabilize their concentration at today's levels.

13.10 OZONE DEPLETION IN THE STRATOSPHERE

One of the most dramatic examples of human interference with the earth's atmosphere is the reduction of ozone in the stratosphere, particularly over the Antarctic in the spring months. This ozone hole is a well-measured and reasonably well-understood problem, whose causes are directly linked to a class of widely used industrial gases.

Ozone in the troposphere is a pollutant and, as we have seen, plays an important role in the formation of smog. In addition, it is one of the greenhouse gases. Its presence in the stratosphere at altitudes of 10 to 30 km, however, is vital, as it is a very strong absorber of ultraviolet radiation. If it were not present, excess amounts of such radiation would reach the earth's surface, causing a rise in the incidence of skin cancer and damaging the ocean phytoplankton, which is basic to the food chain.

In 1985, British scientists working in Halley Bay, Antarctica, announced a startling drop in the ozone that they had been measuring in a vertical column extending upward from their ground-based instrument. From 1957 to 1968 the concentration of ozone was relatively constant at about 300 Dobson units (DU, a milliatmosphere-centimeter of ozone), but by 1984 it had dropped to 200 DU and in October of 1991 it was less than 150 DU. A DU is directly related to the amount of ozone in a vertical column. These initial discoveries have since been verified by measurements made from satellites and high-altitude aircraft by groups from various countries. The ozone depletion has since been noted in the Arctic and mid-latitude regions. Overall it represents about a 3% reduction in the global stratospheric ozone but a 50% reduction in the Antarctic ozone hole, and the reduction continues.

Ozone is formed in the stratosphere when ultraviolet radiation dissociates O_2 into two free oxygen atoms. One of the oxygens combines with O_2 to form O_3. Ozone itself is dissociated by the ultraviolet, but a net balance remains from the rate of formation and destruction. In a series of studies, the main cause of the ozone depletion was found to be the injection into the atmosphere of man-made CFCs. Two of the more widely used CFCs are Freon 11 ($CFCl_3$) and Freon 12 ($CFCl_2$). These are used in air conditioners, refrigerators, building insulation, some plastic foam products, solvents used in the electronics industry, and propellants in spray cans. Measurements in Antarctica and elsewhere showed that the amount of CFCs in the atmosphere had gone up by a factor of 3 from 1970 to 1980. The CFCs can be transported into the stratosphere, where again ultraviolet radiation dissociates the molecule into free chlorine atoms and other molecular fragments. Free chlorine can then act as a catalyst in dissociating O_3 in the following way:

$$Cl + O_3 \rightarrow ClO + O_2$$
$$ClO + O \rightarrow Cl + O_2$$

The chlorine atoms are then able to repeat the process and remove more ozone.

It was in fact the detection of ClO over the Antarctic in amounts about 100 times more than expected that led to man-made CFCs being tagged as the main cause of the ozone depletion. The complete process is somewhat more complicated than indicated in the previous two chemical equations. For example, bromine is thought to play a role in these reactions, leading to about 25% of the ozone destruction. It is known that extensive clouds form over the Antarctic in the wintertime. The clouds include aerosols containing nitric acid. The chemical reactions leading to the ozone reduction apparently take place on the surface of these nitric acid particles. It is for this reason that the greatest ozone depletion takes place in Antarctica.

It is likely that ozone depletion is a global problem that can be solved only by limiting the use of CFCs. Their use as a propellant in spray cans and in many foam plastics has been largely discontinued in the United States, and there has been some success in using other solvents for cleaning electronic circuits. The replacement of Freon 11 and Freon 12 in refrigeration and air conditioning appliances has been held back by the lack of a good substitute. Some large chemical companies are investing substantial amounts of money in trying to develop other refrigerants. At this time hydrofluorocarbons (HFCs) seem to offer the greatest promise and are gradually being introduced as a substitute.

In 1987 a world conference was held in Montreal to reach an agreement to reduce the use of CFCs. The Montreal Protocol, which was signed by most nations, mandated a 50% reduction in CFC production by the year 1998. Since that time, because of the deepening of the ozone hole, the protocol has been modified to phase out CFC production entirely by 2000. Some developing countries, such as China and India, did not sign the protocol because they were concerned about the economic implications for their countries. At the present time, global ozone is decreasing by 2.3% per decade. The EPA has estimated that if this continues, 200,000 additional deaths from skin cancer could result in the next 50 years in the United States alone. Since the CFCs are also greenhouse gases, there is an additional incentive for phasing them out. In early 1992 it was found that the ozone hole was spreading to more densely populated areas in the northern hemisphere at a rather alarming rate. Because of this, the United States has now called for elimination of the use of CFCs by December 1995 and the phasing out of HFCs by 2005.

PROBLEMS

1. As a result of a thermal inversion, the prevailing air temperature profile increases 1°C/100 m above the ground level. To what maximum height will polluted air rise if it is released at a ground level 15°C warmer than the atmosphere? The adiabatic lapse rate is -0.65°C/100 m.
2. What are the basic differences between the smog responsible for the London disaster and that found in Los Angeles?
3. How many molecules are there in 1 m^3 of air at standard temperature and pressure?
4. A concentration of 1 ppm of NO corresponds to how many $\mu g/m^3$? The atomic weight of nitrogen (N) is 14 and oxygen (O) is 16. How many molecules of NO at 1 ppm are there in 1 m^3 of air?
5. The primary standard for SO_2 in the atmosphere is 365 $\mu g/m^3$ averaged over a 24-hour period. How many parts per million by number of molecules is this?
6. In the year 2000, it is conceivable that 2×10^{10} W of electricity could be continuously generated in a single Western state by coal-burning power plants.
   (a) Assuming a 38% efficiency in generating the power, how many tons of coal would be burned a year?

(b) If the sulfur content of the coal averages 1%, how many tons of SO_2 are released a year?

(c) If a standard of 0.1 pounds of particulates per 10^6 Btu is adhered to, how many tons of particulates are released a year?

(d) How many tons of CO_2 are released a year assuming coal is pure carbon?

7. In Problem 6(b), assume that the correct result is 1.07×10^6 tons of SO_2 released per year. If a thermal inversion persists for a 24-hour period over the whole state and the released SO_2 can rise no higher than 1 km, what concentration of SO_2 ($\mu g/m^3$) will be present in the air? Assume that the state is 500×500 km in area. Will the concentration of SO_2 be above the 24-hour standard?

8. Assume that the total coal resources of the world are eventually burned and that the resulting CO_2 is added to the atmosphere. Estimate the resulting increase in concentration of CO_2 (in parts per million) in the atmosphere of the world. See Table 2.6 for coal resources, and section 13.1 for the mass of the atmosphere. Assume the coal is 100% carbon.

9. Estimate the load-lifting capacity in kilograms of a helium-filled balloon having a volume of 1000 m^3. One mole of helium has a mass of 4 g.

10. If the rate of increase of CO_2 in the atmosphere is 0.4% per year, in what year would you expect the concentration to double from the preindustrial level in 1860?

11. How many pounds of SO_2 could be released if 1 ton of coal is burned under Phase I of the 1990 Clean Air Act amendments? Phase II? Assume 2.7×10^7 Btu/ton for the coal.

12. Propose a program to reduce by 60% the amount of CO_2 released through the burning of fossil fuels by the year 2000.

SUGGESTED READING AND REFERENCES

1. Hodges, L. *Environmental Pollution.* New York: Holt, Rinehart and Winston, 1973.
2. Seinfeld, J. H. *Air Pollution, Physical and Chemical Fundamentals.* New York: McGraw-Hill, 1975.
3. Turiel, I. *Physics, the Environment and Man.* Englewood Cliffs, N.J.: Prentice-Hall, 1975.
4. Williamson, S. J. *Fundamentals of Air Pollution.* Reading, Mass.: Addison-Wesley, 1973.
5. Hamil, P., and Owen, B. T. "Polar Stratospheric Clouds and the Ozone Hole." *Physics Today,* **44** 12 (December 1991), pp. 34–42.
6. Graedel, T. E., and Crutzen, P. J. "The Changing Atmosphere." *Scientific American* **261** 3 (September 1989), pp. 58–68.
7. Schneider, S. H. "The Changing Climate." *Scientific American* **261** 3 (September 1989), pp. 70–78.
8. Mintzer, I. M. "Energy, Greenhouse Gases and Climate Change." *Annual Review of Energy,* **15** (1990), p. 513.
9. Ramanathan, V. "The Greenhouse Theory of Climate Change: A Test by an Inadvertent Global Experiment." *Science,* **240** 4850 (April 15, 1988), pp. 293–299.
10. Chang, T. Y.; Hammerle, R. H.; Japar, S. M.; and Salmeen, I. T. "Alternative Transportation Fuels and Air Quality." *Environmental Science and Technology,* **25,** 7 (July 1991), pp. 1190–1197.
11. Maibodi, M. "Implications of the Clean Air Act Acid Rain Title on Industrial Boilers." *Environmental Progress,* **10** 4 (November 1991), pp. 307–312.

12. Jones, Phillip D., and Wigley, Tom M. L. "Global Warming Trends." *Scientific American* **263** 2 (August 1990), pp. 84–91.

13. White, Robert M. 'The Great Climate Debate." *Scientific American* **261** 3 (July 1990), pp. 36–43.

14. Houghton, Robert A., and Woodwell, George M. "Global Climate." *Scientific American* **260** 4 (April 1989), pp. 36–44.

15. Stolarski, Richard S. "The Antarctic Ozone Climate Change." *Scientific American* **258** 1 (January 1988), pp. 30–37.

16. Balzhiser, Richard E., and Yeager, Kurt E. "Coal Fired Power Plants for the Future." *Scientific American* **257** 3 (September 1987), pp. 100–107.

17. Alpert, S. B. "Clean Coal Technology and Advanced Coal-Based Power Plants." *Annual Review of Energy and the Environment,* **16** (1991), pp. 1–23.

18. Spencer, D. F. "A Preliminary Assessment of Carbon Dioxide Mitigation Options." *Annual Review of Energy and the Environment,* **16** (1991), pp. 259–273.

CHAPTER FOURTEEN

WATER—THE RESOURCE AND ITS POLLUTION

14.1 THE SPECIAL PROPERTIES OF WATER

Our planet is unique among those of the solar system in that it alone has abundant amounts of water in a form usable by organisms. The biosystem that has evolved on earth has been strongly influenced by the rather special chemical and physical properties of water. Hence, all animals and plants are in some way highly dependent on the availability of this colorless, odorless, and tasteless substance. Water is a precious resource, but because of its relative abundance in some areas, it has been abused and has served more as a dumping ground for wastes rather than being valued for its life-giving properties. Although the attitudes responsible for this misuse are slowly changing in the United States, the increased demands for the limited water, particularly in the western part of the country, and the high cost of water purification have proved to be barriers to sensible water policies that can endure. Before discussing water on a national or global scale, it is instructive to examine the molecule itself.

The chemical symbol for water, H_2O, is no doubt better known to the general public than any other. Beyond the fact that two hydrogen atoms and one oxygen atom are contained in the water molecule, the properties of water are rather complex. The water vapor molecule has strong covalent bonds between the hydrogen and oxygen atoms with an O–H bonding energy of 110.2 kcal per mole. The angle between the two lines connecting the hydrogen atoms to the oxygen atoms is about 104°.

The average separation between molecules is reduced as the liquid state is formed from a vapor because there are weak forces that bond the hydrogen atoms of one molecule with the oxygen atom of another. In a liquid, the volume of a water molecule is 29.7 $Å^3$ (1 angstrom $= 10^{-10}$ m). At a pressure of 1 atmosphere, it takes 539 cal/g to transform the liquid into a vapor (i.e., the heat of vaporization equals 539 cal/g). The boiling point at a pressure of 1 atm is by definition 100°C (212°F).

As the temperature of water is lowered to the freezing point (0°C or 32°F), 80 cal/g of energy (the heat of fusion) must be removed to transform it to the solid, or ice, stage. In ice each oxygen atom is bonded to four others by hydrogen bonds forming a tetrahedral configuration. The volume occupied by a water molecule in ice is appreciably greater (32.3 $Å^3$) than that occupied in the liquid state. Upon freezing, water increases in volume by one eleventh. This means, of course, that ice is less dense than liquid water, and this is the reason ice floats. This property is important for the survival of aquatic life in regions where the temperatures go below freezing. The density of water increases as temperatures decreases until the density reaches a maxiumum near 4°C. As water is cooled below 4°C, it expands until it reaches the freezing point. In a body of water, the cooler (down to 4°C) and, hence, more dense water will go toward the bottom.

Water, also by definition, has a specific heat of 1.0 cal/g °C. As this specific heat is considerably higher than that of most common materials, water has an exceptional ability to store heat. Large bodies of water, hence, have a stabilizing effect on the surrounding temperature, since it takes a relatively large change in stored thermal energy to raise or lower the temperature of water. The fact that sand has a specific heat of only 0.2 cal/g °C partially explains why deserts cool off so drastically at night. The relatively high heats of fusion and vaporization of water thus act as temperature stabilizing features, and they serve to protect the organisms that live in it.

Water is an excellent solvent of inorganic compounds such as ammonia, salt, potash, and phosphates. It can, in fact, dissolve more substances than any other liquid. It is also important that water is an inert solvent; it does not change the chemical structure of the dissolved materials. Because of these properties, it can transport nutrients in a biological system in a very effective fashion.

Water adheres to biological materials or wets them very readily. This is vitally important for the transport of giant molecules such as proteins that are not dissolved, but, because of the wetting action of water, are moved along with the water. Another important and related property of water is its surface tension. Except for mercury, water has the highest surface tension of any liquid. The formation of spherical water droplets is one manifestation of this property. The rise of liquids in fine-bore tubes is another consequence of surface tension. This phenomenon, called capillary action, is important for the movement of water up through the soil and through the roots of plants.

It is clear that water has some very special properties that are intimately related to the ways the plants and animals of the earth have evolved. Human beings are 71% water by weight. We can lose nearly all our body fat and half our protein and survive, but if we lose more than 10% of the water in our body, the consequence is death.

The water problems of the United States and the world are serious. Some regard the problems as more significant for the general well-being of the people of the world than the problems of energy. As we shall see, there are very limited quantities of available fresh water, and there are no easy solutions to this shortage.

The problems of water resources and energy are interconnected in many ways. As we saw in Chapter 2, one of the limitations on the production of shale oil in the Green River Formation is the availability of water, since 3 barrels of water are needed to produce 1 barrel of shale oil. The oil shale occurs in the semiarid regions of Colorado, Wyoming, and Utah, where a barrel of water is a very precious item badly needed for agriculture. Another example of this relationship between water and energy is the use of water in transporting coal in a slurry form in a pipeline. Although the process is economically sensible, in general most of the western coal is found in regions of Wyoming, eastern Montana, and New Mexico, where there were water shortages before coal slurry pipelines were ever considered.

The subject of water is complex and diverse. It would be impossible to cover the field in anything like a complete fashion in one chapter. Because of this we have selected for discussion a few topics that involve water as a resource and a few involving water pollution. Many important aspects of both water resources and pollution are not mentioned, and it is hoped that the interested reader will find material in the references at the end of the chapter that covers the entire field more adequately.

14.2 WATER RESOURCES OF THE WORLD

At the time of the earth's formation, about 4½ billion years ago, all of the planet's water was tied up in minerals. As the gravitational forces compacted and heated the earth, some of this water was released into the atmosphere as vapor, but the surface of the earth was too hot for it to condense into the liquid state. Eventually, the earth cooled and the vapor condensed. This led to the formation of the primitive oceans. With time, the oceans and land masses as we know them were formed. A similar process of vapor formation is thought to have occurred on the moon after its formation. The gravitational field of the moon, however, is not sufficiently strong to keep the water vapor molecules from escaping into space. The moon is thus devoid not only of any surface water but also of any atmosphere. Table 14.1 lists the various water resources of the earth. The oceans dominate with 97.2% of the water. Unfortunately, because of their salinity, the oceans cannot serve directly to meet most of our water needs, such as for agricultural irrigation. The runoff from the land contains dissolved sodium chloride (salt) and other minerals from the exposed rocks, and this mineral-laden water has been carried to the sea. Subsequent evaporation left the salt and other minerals behind, so the salinity of the oceans has continually increased with time, but at a very slow rate.

A schematic view of the hydrological cycle is shown in Figure 14.1. The main process is straightforward in that solar energy evaporates water from the oceans, lakes, and rivers, and water is transpired from plants. A certain amount of moisture is retained in the atmosphere (34×10^{15} gal). Under an equilibrium condition, the same amount of water as was evaporated must return to the earth as rainfall. Because the moisture in the atmosphere is subject to global weather patterns, somewhat more rainfall is received by the land mass than is evaporated from the land, and this provides a net world runoff of 9.8×10^{15} gal/yr. There are several complicating factors to this simple view of the hydrological cycle. The first is that the water stored in the icecaps and glaciers is appreciable (7700×10^{15} gal), and the amount increases or decreases depending on the prevailing temperatures in the area. The second complication occurs because the water stored in the ground or in the soil as moisture (vadose water) is subject to change depending on the climate and human interference. These factors do not normally contribute or withdraw much from the runoff

Table 14.1 DISTRIBUTION OF THE EARTH'S WATER RESOURCES

Location	Water Volume (10^{15} gal)	Percentage of Total Water
Oceans	348,700	97.2
Icecaps and glaciers	7,700	2.15
Atmosphere	34.1	0.01
Freshwater lakes	33	0.009
Saline lakes and inland seas	27.5	0.008
Average in stream channels	0.3	0.0001
Soil moisture	17.6	0.005
Ground water within depth of half a mile	1,100	0.31
Ground water–deep lying	1,100	0.31

Source: Adapted from Brian J. Skinner, *Earth Resources*, Foundation of Earth Science Series, Englewood Cliffs, N.J.: Prentice-Hall, 1969.

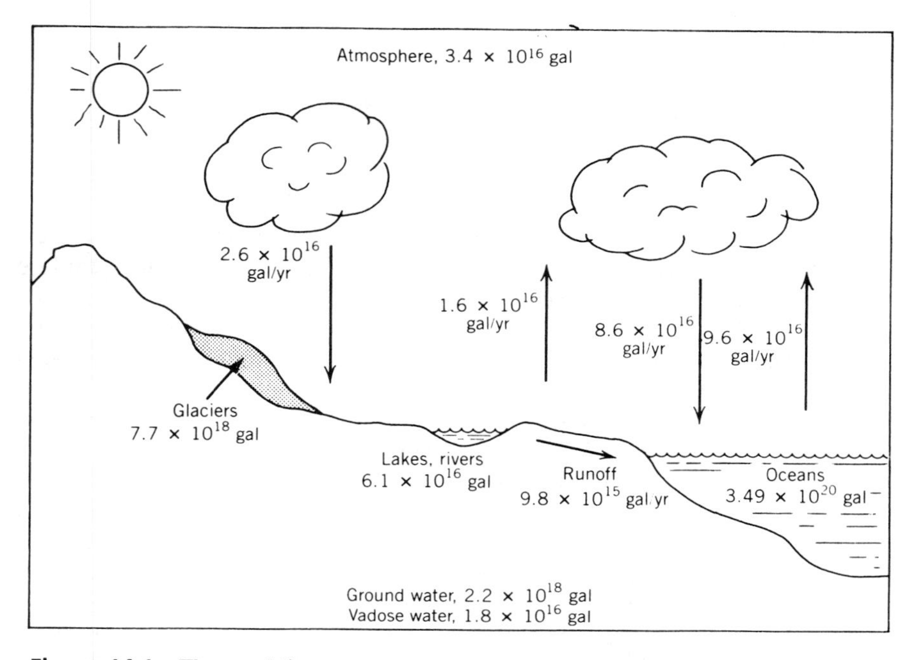

Figure 14.1 The world's water resources and hydrological cycle.

over a time span of a few years. But on a time scale of thousands of years, the amount of water stored in the ground or icecaps can change appreciably.

Of the water resources listed in Table 14.1 or shown in Figure 14.1, only the water in freshwater lakes, stream channels, and ground water is directly available for mankind's needs. Do these resources provide enough water for the daily use of the present population? The answer depends, of course, on style of living, state of industrial development, and so forth, but we shall see later that in an industrialized country such as the United States, about 186 gal are used per person a day for drinking, washing, household needs, cooking, and irrigation of lawns and gardens. Of course, the 186 gal/day does not include the water used to irrigate the crops grown for food, nor does it include the water used in mining, manufacturing, petroleum refining, and so forth. In round numbers, it is estimated that the agricultural sector adds 570 gal/day and the industrial sector about 894 gal/day for a total need per person of 1650 gal/day.

The quantities listed in Table 14.1 are not immediately useful in discussing needs because it would be difficult to assign a time for the consumption of all of the water resources in lakes or in the ground. Unless we are willing to reduce the amount of stored water, the water available to us is the annual runoff from the land.

Example 14.1

Using the estimated rate of water usage per person per day for the United States, calculate the total number of people in the world that could be furnished water if the entire world runoff per year were used once.

Solution

$$9.8 \times 10^{15}\,\frac{\text{gal}}{\text{yr}} \times \frac{1\ \text{yr}}{365\ \text{days}} \times \frac{1\ \text{person-day}}{1650\ \text{gal}} = 1.6 \times 10^{10}\ \text{persons}$$

A number of unrealistic assumptions are made in the above calculation. Perhaps the most glaring is related to our inability to utilize the complete runoff from the world's rivers. First, there are huge rivers in the world, such as the Mississippi, Columbia, Amazon, Zaire, Yangtze, and Nile, where, because of their sheer size and importance for transportation and other uses, only a small fraction of the flow can be consumed. There are rivers in areas where, because of the climate and low population density, there is very little incentive to utilize much of its flow. The Mackenzie River in Canada is one example. There is also a major problem with the distribution in time of the runoff. Because heavy rains and snow melt usually occur at certain times of the year, the flow of most rivers varies a great deal during the year, and this fact makes it difficult to utilize the river's flow efficiently.

Another difficulty in estimating the number of people that can be supported with the world's runoff is the repeated use of any given volume of water. When a ton of coal or a barrel of oil is burned, the chemical energy is extracted and the process cannot be repeated. Water, on the other hand, can theoretically be used an infinite number of times as long as the molecular structure is not disturbed or the water is not tied up in some compound. For example, water from a river may be purified and distributed as the water supply of a city. Aside from evaporation and irrigation, most water can be returned to the river after being used for drinking, washing, waste disposal, and so forth, and then being treated at a wastewater treatment plant. It is hoped that the water returned to the river will be pure enough to be used by another community downstream. The water might then be used as a coolant for a power plant. The water from the plant's heat exchangers will be returned to the river at a slightly elevated temperature, but it will still be usable water. This process can go on with water serving many users until the river empties into the ocean. Water used for irrigation is generally not directly reusable, as it is mostly evaporated with some going into plants, where transpiration takes place, and some going into ground water. Water used by some cities and industries is sometimes returned in a polluted form that prevents its reuse.

There are thus many complications to be considered in estimating the number of people in the world that can use water at the rate of 1650 gal/day. The stable runoff that can be utilized varies for the different continents, but is generally in the range of 32 to 42% of the total runoff. Ignoring some of the complexities just discussed, there would appear to be sufficient water for a population of about 37% of 1.3×10^{10} people, or 4.8×10^{9} people, which is just about equal to the present population of the world. The water availability per capita for Asia is lower than for any other continent.

14.3 THE USE OF WATER IN THE UNITED STATES

Table 14.2 shows the approximate water flow rates for the United States. The stream flow not used is in the rivers that flow to the oceans or to Mexico and

Table 14.2 WATER FLOW RATES FOR THE UNITED STATES (10^{12} gal/yr)

Precipitation	1552
Evaporation and transpiration	1086
Runoff	466
Withdrawn and used	124
Not withdrawn	342

Canada. The part of the stream flow that is used (124×10^{12} gal/yr) corresponds to about 1500 gal/person-day.

Of the 1086×10^{12} gal/yr indicated for the annual volume of water evaporated and transpired, 44%, or 481×10^{12} gal/yr, has no economic value. It is estimated that 33% (357×10^{12} gal/yr) is involved with crops and pastures and 23% (248×10^{12} gal/yr) is from forests and browsing vegetation.

Table 14.3 lists the water used in the United States since 1940. The amount of water used per capita per day in 1985, when the population was 246 million people, was 186 gal for domestic use, 570 gal for irrigation, and 894 gal for industrial use and power generation, for a total of 1650 gal. The total amount of water used, as well as the amount used per capita, has increased considerably since 1940. This increase is due to a greater personal use of water in the home, as well as to more extensive irrigation and a greater industrial use of water. Water is involved in all of the items we use in our daily lives. Producing a single newspaper requires 20 gal of water; the steel for a car, 52,000 gal of water; 1 gal of gasoline, 18 gal of water; and a can of lima beans, 250 gal of water. With time we have been using more of these manufactured goods as well as relying on automatic dishwashers and clothes washers in the home. In Europe the use of water is considerably less. The domestic use of water, for example, is only 40 gal a day per person compared to 186 gal in the United States.

It is interesting to note in Table 14.3 that the total use of water in 1985 went down. This reflects the greater reuse of treated water from city wastewater treatment facilities, industries, and power plant effluents.

The overall statistics on water flow in the United States do not reflect either the magnitude or the nature of the severe water problems facing the

Table 14.3 USE OF WATER IN THE UNITED STATES (10^9 gal/day)

Year	Total	Irrigation	Public Water Utilities	Rural Domestic	Industrial and Miscellaneous	Steam Electric Utilities	Per Capita (gal./day)
1940	140	71	10	3	29	23	1027
1950	180	89	14	4	37	40	1185
1960	270	110	21	4	38	100	1454
1970	370	120	27	5	47	170	1815
1975	420	140	29	5	45	200	1972
1980	450	150	34	6	45	210	1953
1985	400	140	38	8	30	190	1650

Source: Statistical Abstract of the United States, 1991.

Figure 14.2 Irrigation sprinklers in Teton county, Idaho. Snow melt from the Teton mountains irrigates a field of alfalfa through a gravity flow system.

country. For example, the distribution of the runoff is far from uniform across the country. The area west of the Mississippi River, which is considerably larger than the eastern area, receives only 35% of the precipitation (excluding Alaska and Hawaii). Within the western area there are also great differences, with cities such as Seattle in the Northwest averaging about 39 in. of rain a year, and cities in the Southwest such as Phoenix only 7 in. a year.

To provide some insight into the multifaceted water problems of the United States, two examples will be discussed that point up quite different aspects of the water problem.

THE COLORADO RIVER

The uneven distribution of water in the United States would not be such a major problem if people lived where the water is. Unfortunately, people are going where the sun is and where there is little water. Not only are people flocking to the Southwest (Arizona and southern California, in particular) to enjoy the sunny, mild winters, but farmers are also trying to take advantage of the long growing seasons, and water is needed to make the desert bloom. The Colorado River, with its headwaters high in the snow-packed Rocky Mountains, has the potential to provide water for the parched Southwest; however, there are many needs that must be served by the river as it winds its way some 1450 miles to the Gulf of California. In fact, the river's waters are already oversubscribed, and all of the lawsuits and countersuits going through the courts are not going to increase the water available by 1 gallon.

The Colorado River drains a large part of the Southwest, an area of 250,000 square miles that includes the western parts of Colorado and New Mexico, Utah, and Arizona, along with southern Nevada and California. As with all great river systems, it is fed by a number of rivers as it works its way toward the sea. The Green River in Utah is a major contributor along with the San Juan and Little Colorado in northern Arizona. Beyond the Grand Canyon, the Virgin

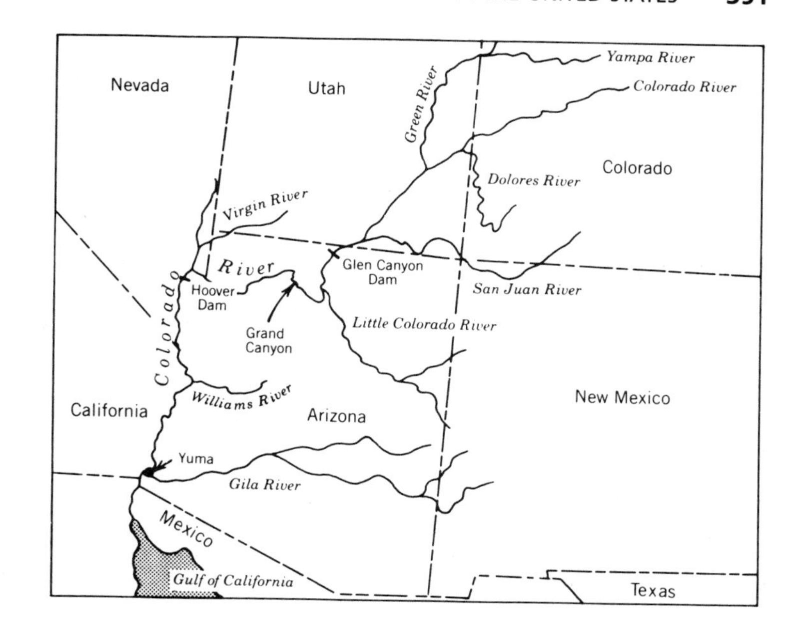

Figure 14.3 The Colorado River system.

River in Nevada joins and then the Williams and Gila rivers join in southern Arizona. A map of the course of the river is shown in Figure 14.3.

The city of Denver is partially supplied by water from the Colorado; this water is taken near the river's origin on the western slope of the Rocky Mountains and brought through a tunnel to the eastern side of the mountains. Downstream to the west near Glenwood Springs and also near Grand Junction, there are dams to produce electricity and to furnish water to an extensive agricultural area. The Glen Canyon Dam backs up river water to form Lake Powell, 185 miles long, in southern Utah. Although such dams offer electric power, flood control, and some recreational possibilities, they also flood sandstone canyons that offer a unique and primitive beauty, lead to greater evaporation of water from the large surface area of the lake, and eventually will present some major problems because of the build-up of silt. The Hoover Dam, with its Lake Mead, situated downstream from the Grand Canyon on the Nevada border, has the same set of problems and benefits as the Glen Canyon Dam, but here a major amount of water is diverted to southern California to furnish most of the water used by the city of San Diego and about a fifth of the water used by Los Angeles. Near Yuma, Arizona, more water is removed for irrigation in the Imperial Valley and for the Coachella and Palo Verde irrigation districts in California. After leaving Arizona, the Colorado River flows 90 miles through Mexico and then empties into the Gulf of California. A treaty signed in 1944 stipulates that Mexico shall receive 12% of the water in the Colorado. Unfortunately, by the time the now sluggish and depleted river reaches Mexico its salinity is so high (about one seventh the salinity of seawater) that it is of only marginal use for agriculture, and the Mexicans are quite unhappy.

The Colorado River water is completely used up. Various legal battles have been fought under the complicated western water right laws to obtain more water for party A at the expense of party B. Finally, the U.S. Supreme Court in 1963 rejected the appropriation and beneficial use doctrine (the basis of western water right laws) and divided up the flow. Assuming 15×10^6 acre-ft/yr, the upper basin states (Colorado, Utah, New Mexico, and Wyoming) were given

Figure 14.4 The Grand Canyon of the Colorado at river level. This is the major river of the southwestern United States.

half and the lower basin states (Nevada, Arizona, and California) half, with 0.3 $\times$ 10^6 given to Nevada, 2.8 $\times$ 10^6 to Arizona, and 4.4 $\times$ 10^6 to California. An acre-foot of water is the amount of water in 1 acre (43,560 ft^2) covered to a depth of 1 ft. It is equivalent to 325,851 gal. Unfortunately, the average flow at Lee's Ferry, Arizona, just above the Grand Canyon, is only 13.2 $\times$ 10^6 acre-ft/yr, and it varies by a factor of 2 from year to year. California has been getting 5.1 $\times$ 10^6 acre-ft/yr, and if the upper basin states take their full share, there is going to be a major problem in southern California. A severe problem will occur in the near future when the Central Arizona Project begins diverting 1 $\times$ 10^6 acre-ft/yr to Phoenix and Tucson. California will have to surrender much of the water to make up this amount. In the meantime, a dozen Indian tribes are in court with the Central Arizona Project, claiming that they should have one fourth of the amount to be diverted.

The basic problem is that there is not enough water to provide the millions of people who have moved into the Southwest with sufficient domestic, industrial, and agricultural water. The situation is worsened by the losses of water by evaporation from the large lakes that have been created and by water sinking into the soil and growing vast amounts of weeds in the numerous canals and irrigation ditches that have been created. A secondary problem occurs when extensive irrigation is carried out year after year on the same fields, and the salt and other minerals in the river water are then left behind as the water evaporates from the soil. Some areas have become so salted by this process that farming is no longer possible, and the fields must be abandoned.

THE OGALLALA AQUIFER

Underlying a large portion of the midwestern United States is a huge aquifer that extends from the western portion of Texas through the northern part of Nebraska. It is named Ogallala after the Indian tribe that once hunted buffalo over the region. It is one of the largest aquifers in the world, with a freshwater capacity of about 2 $\times$ 10^9 acre-ft. If one flies over this area during the growing season, the circular patterns one sees on the earth are half-mile-diameter green circles of vegetation. These circles are created as water from the Ogallala aquifer is pumped to the surface and then distributed to crops by the quarter-mile-long arms of a center-pivot irrigation system sweeping out huge circles. The patterns are interesting, and certainly the corn and other crops produced are important for the farmers' income and as a source of food to the nation and the world. Unfortunately, the water being mined is a finite resource, and it is being depleted at an alarming rate.

The Ogallala aquifer contains water trapped in subterranean beds of sand, gravel, and silt. This material was carried down from the Rocky Mountains during the Pliocene period. The water-saturated material eventually became trapped between a layer of impermeable shale on the bottom and a cap rock on the top. The Ogallala aquifer is but one particular source of groundwater. The total amount of groundwater beneath the earth's surface is very large, but much of it occurs in rock having porosity so low that flow rates into wells are prohibitively slow for significant recovery. To have a good flow rate a suitable aquifer or water carrier is needed. As with all ground water, aquifers can be recharged by rainfall, but the process is slow because most of the rainwater runs off in rivers. If all of the ground water in the United States to a depth of 2500 ft were removed, it would take an average of 150 years for the various areas to be recharged.

Water is being removed from the Ogallala aquifer at a rate considerably higher than the recharge rate. The water table levels at various locations in this aquifer are receding from 6 in. to 3 ft a year. Each year more water is taken from the Ogallala aquifer than the entire annual flow of the Colorado River (13.5×10^6 acre-ft). At the present rate of use, it is estimated that there are about 30 years of useful life remaining. Also, as the water level drops, the amount of energy needed for pumping from the greater depth increases steadily. This is vitally important in a time of increasing energy cost and decreasing energy reserves. Some areas are experiencing water shortages already, with irrigated acreage declining in five of the six states drawing from the Ogallala aquifer. There is no conservation program mandated by federal policy, and the pumping continues. The withdrawing of water in one region eventually affects the availability of water in other areas far removed, and water right laws are inadequate to bring about a sensible, controlled use of the resource at an equilibrium level. When the Ogallala aquifer is finally depleted, the consequences are going to be felt not only by the farmers, but also by the nation's economy and in the world's food supply.

14.4 WATER POLLUTION AND PURIFICATION IN THE UNITED STATES

Water pollution in the United States is a complex problem with many different aspects. Some of the major areas of concern are:

1. Cities and towns with no sewage treatment facilities or inadequate facilities (leaky, too small, or combined with storm sewers).
2. Industrial wastes that are inadequately treated or are put into municipal systems not prepared to treat water contaminated by various chemical pollutants.

3. The water drained from mines that is often so acidic that it should be specially treated before being put into streams or lakes.
4. Drainage from various agricultural endeavors such as feedlots frequently entering streams directly without treatment.
5. Oil from various sources (wells, ships, spills, and natural sources).
6. Natural leaching of minerals from the soil and rocks into the runoff.
7. Natural organic matter, such as leaves and animal droppings, entering the streams.
8. Sanitary systems of ships not provided with adequate holding tanks.
9. Special pollutants such as detergents, toxic chemicals, trace metals, and insecticides.

To discuss fully all of these different aspects of the water pollution problem would be beyond the scope of this chapter. Hence, we focus our attention on the treatment of sewage in municipal sewage systems. Acid rain, in a sense, is also a water pollution problem, but because of its special nature and current importance, it will be treated separately.

The history of sewage systems is amazing in that the Romans developed a fairly extensive and useful sewage system, and then for 1800 years or so, civilization reverted to foul, unsanitary practices that resulted in the deaths of millions of people. The Romans had toilets that were flushed with water. Some of the rural houses had cesspools, and in Rome sewage systems such as the Cloaca Maxima in the sixth century B.C. carried the sanitary waste away from the city. With the fall of the Roman Empire these water-collecting and sanitation systems were abandoned.

As long as the population density is low, as it is in rural farming areas, sanitary sewage disposal can usually be handled without jeopardizing the public health. When people congregate in cities, however, a sanitary disposal system is needed or people will suffer and die from the proliferation of diseases. During the Middle Ages people in European cities frequently threw the human waste out of the window onto the street. The open gutters in the streets were intended to carry away all refuse to a nearby river or other body of water. There was no understanding that disease would follow if drinking water was taken from that same body of water. The Black Death of 1348, in which one fourth of the population of Europe died, can be largely attributed to such unsanitary practices.

Toward the end of the 18th century, some improvements were made in sanitary systems, such as the valved water closet and later the water trap in drainage lines. The basic problems of combining the sanitary system with storm sewers and inadequately sized cesspools or leaching fields have persisted up to the present time. Because of a lack of understanding of the role of bacteria in causing disease, the relationship of drinking water to sewage was not appreciated, even in the more developed countries, until the last half of the 19th century. In the less developed countries, these problems still persist, and almost 20% of the hospital beds in these countries are still occupied by patients with cholera and typhoid.

What is in sewage and what can be done to purify it? Average sewage is about 99.9% water and about 0.1% sand, grit, and organic matter. In more detail, the makeup of sewage and its method of treatment are listed below.

1. Floating debris is usually organic, but because of its size it cannot be reduced by biochemical action. Removal is rather readily accomplished with screens and skimmers at the beginning of wastewater treatment, and the material removed is handled as solid refuse.
2. Suspended inorganic matter, such as grit and sand, is also removed in the primary treatment phase, largely by gravitational settling in basins.
3. Dissolved inorganic materials, such as salts and other chemicals, can pass through the primary and secondary treatment stages unchanged. It is only in the tertiary treatment stage that some special chemical processes are used to remove compounds that impair the water quality. Certain phosphates and nitrates are particularly troublesome.
4. Suspended colloidal and dissolved organic matter is subject to biochemical action (oxidation, digestion, and decomposition) involving the removal of oxygen from the water. This type of material is the major cause of water pollution by sewage if it is not adequately removed. In a treatment plant the secondary stages have various devices to hasten the removal of organic matter by bacterial action.
5. Bacteria and other disease-carrying microorganisms that are potentially dangerous are not removed directly in either the primary or secondary processes. Frequently a sewage treatment plant will directly add a disinfectant, such as chlorine, to the effluent before it is discharged to a river to reduce the bacteria count.

All sewage treatment plants have at least a primary stage in which various mechanical processes are used to remove the materials listed in (1) and (2) above. These mechanical devices and processes include surface skimmers, filters and screens, grinding, sedimentation, and flocculation. Flocculation is the process of agitating the sewage water to try to have small suspended particles collect to form larger particles that settle out more readily.

Secondary treatment is also now included in most sewage facilities. In this stage the organic material, (4) in the above list, undergoes decomposition by bacteria in large tanks. The process usually takes place in water in which there is no free oxygen and is, hence, an anaerobic process. The oxygen needed comes from the water. In the case of sugars, the process is fermentation, and in the case of proteins, it is putrefaction. The following generalized equation indicates what takes place in putrefaction.

$$C_xH_yO_zN_2S + H_2O \rightarrow NH_4^+ + CO_2 + CH_4 + H_2S + 368 \text{ cal/g of protein}$$

The equation is not balanced because the numbers of carbon atoms, x, and so forth are not explicitly stated. The anaerobic digestion leads to 50 to 80% volatile products (carbon dioxide, ammonium ion, methane, and hydrogen sulfide) and 20 to 50% sludge. Methane accounts for about 72% of the volatile products and is frequently a usable by-product of a sewage treatment plant. Sometimes the sludge can be dried and burned for heat, and sometimes it is used as a fertilizer. In a secondary treatment facility, the anaerobic process is helped by the use of trickling filters or activated sludge. In a trickling filter the sewage is sprayed over a bed of crushed rocks to increase the supply of free oxygen. As the water goes through the rocks the organic matter gets absorbed by the slime on the rocks where the bacteria can then decompose it. Thus, at the surface the aerobic process is important, and on the rocks the anaerobic process dominates.

The activated sludge method involves the pumping of sludge from a clarification (or sedimentation) tank that follows the secondary treatment stage into the sewage water. This sludge helps to clarify the water by absorbing colloidal and suspended solids, and it promotes the bacterial action by introducing the necessary bacteria. A schematic diagram of a typical sewage treatment system is shown in Figure 14.5.

If sewage does not have its organic matter removed, there is said to be a biochemical oxygen demand (BOD) present. This is measured by the amount of oxygen needed to oxidize the organic material in a volume of sewage, and it is frequently stated as the amount of oxygen used up (in grams) in 1 liter of sewage in 5-day incubation period where the temperature is kept at 20°C. One mg/liter is equivalent to 1 part per million by weight. In a typical sewage treatment facility, 90% of the BOD is removed in the primary and secondary stages. The BOD of drinking water should be no higher than about 5 mg/liter, and raw sewage runs about 140 mg/liter.

If raw sewage with a higher value of BOD is dumped into fresh water, oxygen will be removed from the water as the bacterial action on the organic matter proceeds. After some time the aquatic life, such as fish that depend on free oxygen, will no longer be able to exist in the water. As the higher life forms die or leave the contaminated area, algae will grow in abundance, having received a great deal of nourishment from the organic matter present. As the process continues, the bacteria increase in number until a septic stage is reached; a smelly condition exists with a lot of algae and bacteria but no higher life forms. If no further sewage is introduced, the organic matter will eventually be decomposed and the original clear condition of the water will be reached again. In a stream the natural clear water will reappear some miles downstream from the place where the sewage was introduced. Figure 14.6 illustrates these transitions from clear water to septic stage and back to clear water after introduction of organic pollutants.

Although the situation is improving, there have been a number of lakes that received so much sewage that they were continually in the septic stage. Lake Erie, for example, was classified in the past as a dead lake in the sense that the BOD and chemical pollution were so high that no desirable aquatic life could be maintained. It has since been partially restored.

The tertiary or advanced sewage treatment stage varies considerably with the immediate problem being solved. The general purpose of this third stage is to remove more completely the organic material left from the primary and secondary treatment and to remove some dissolved inorganics such as phosphates and nitrates. Many chemical techniques are available, such as activated carbon, filtration, electrodialysis, and reverse osmosis, that can be useful in tertiary treatment, depending on the need. Because of the relatively high cost, only about 18 million people in the United States are presently served by tertiary treatment plants. It is quite possible to have the effluents of a tertiary plant of sufficient quality that the discharged water can be directly used for household needs and drinking. Closed-cycle water systems are available for mountain homes and some municipal water systems that directly utilize their purified water effluent after chlorination.

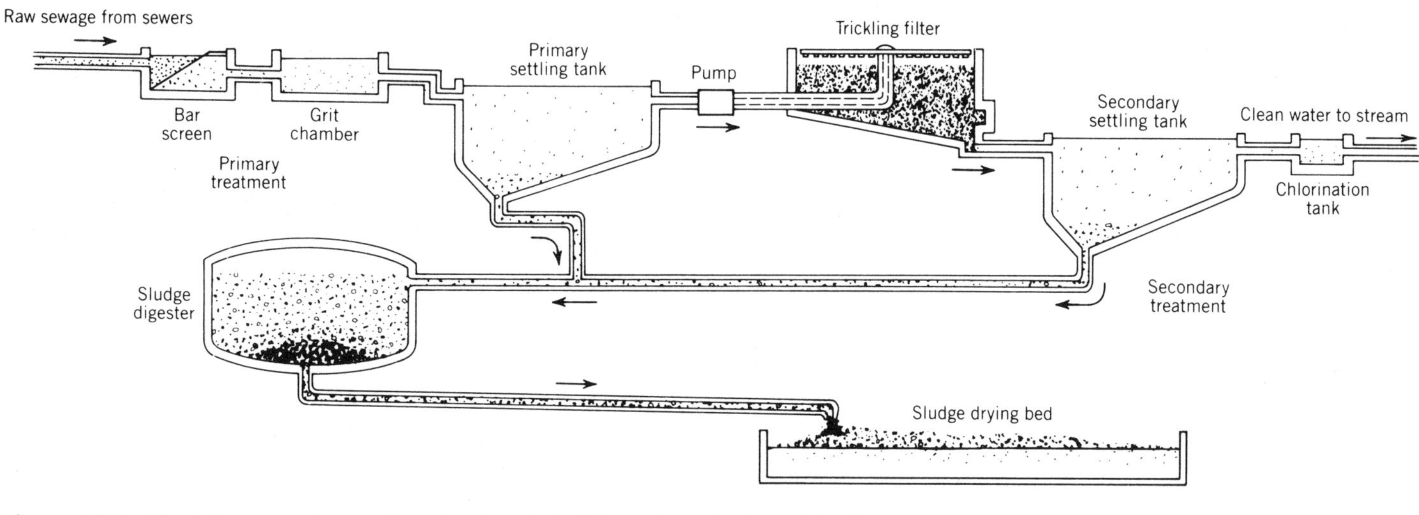

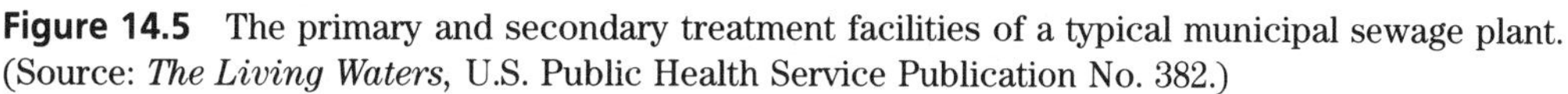

Figure 14.5 The primary and secondary treatment facilities of a typical municipal sewage plant. (Source: *The Living Waters,* U.S. Public Health Service Publication No. 382.)

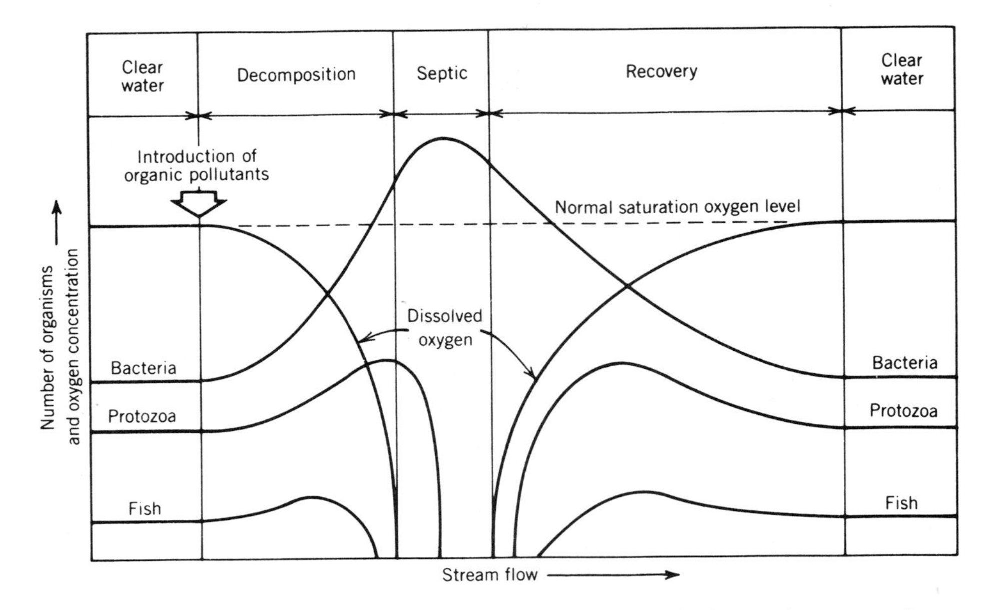

Figure 14.6 The transition from normal clear water through the septic stage and back to normal clear water after the introduction of organic pollutants into a stream.

The Water Pollution Control Act of 1966 has had a major influence on the construction of industrial and municipal sewage treatment facilities in the United States. Under the provisions of the act, the federal government will pay for 60 to 70% of the cost of approved sewage treatment facilities. The amount of pollutants that will be permitted in water discharges was to be tightened in three stages, until in 1985 virtually all pollution was to be eliminated. This goal, which was felt to be unrealistic by some, has seen the construction of new sewage treatment facilities and vast improvements in those already in existence. The data in Table 14.4 indicate the drastic reduction that has taken place in the number of people served by untreated sewage facilities with a corresponding increase in the number of people served by treated facilities. Most of these municipal facilities have both primary and secondary, but not tertiary, stages of treatment.

Although it may be technically possible for tertiary wastewater treatment systems to provide effluent water of a quality suitable for domestic use, this is not ordinarily achieved in practice. Virtually every municipality finds it necessary to provide some means of purifying its drinking water, whatever the source, so that its population will not be exposed to harmful wastes and contaminants. This need for purification would supposedly be greatly reduced if we could succeed in collecting and treating our wastewater to drinking water quality before it is returned to the environment. It has been proposed, only half seriously, that impetus in this direction could be provided by a legal requirement that every community returning treated wastewater to a river must have its drinking water inlet located downstream from its wastewater outlet. In isolated cases, such as single-family residences served by deep well water, there is no need for treatment of domestic water. But even if the wastes of society were not a problem, most domestic users would still want their water treated to

Table 14.4 POPULATION OF THE UNITED STATES SERVED BY SEWAGE SYSTEMS (IN MILLIONS OF PEOPLE)

	1960	1970	1978
Not served by sewers	70	58	60
(Percent)	(39)	(29)	(30)
Served by sewers	110	145	154
Not treated	70	59	2
Treated	40	86	152

Source: Statistical Abstract of the United States, 1980.

remove minerals and sediment and perhaps by the addition of fluoride. Water purification is now nearly a universal practice in this country, and the benefits are impressive. For example, the annual death rate from typhoid fever in our large cities was reduced from nearly 100 per 100,000 population in the late 1800s to about half that number in the early decades of this century; since about 1945, deaths from typhoid fever have become almost unknown in this country.

Cities use either surface water from lakes and rivers or ground water from wells, with reservoirs being used in either case to store the water and equalize rates of flow into the system. Before being distributed to households, the water is treated by some combination, depending on the circumstances, of sedimentation, filtration, aeration, disinfection, softening, and fluoride addition. The sedimentation, in a settling basin, removes the heavier suspended solids. A thick bed of fine sand serves as a filter to remove finer contaminants, including bacteria, from the water passing through it. Mechanical filters can be used rather than sand beds. Sometimes the sedimentation and filtering processes are aided by the addition of coagulants and by flocculation. Aeration oxidizes iron and manganese (which add an undesirable color to the water) so that they will precipitate out of solution. Disinfection, to prevent the passage of pathogenic organisms, is usually accomplished by the addition of chlorine (0.1 ppm) or in some places, mostly in Europe, by treatment with ozone. Although it is the more expensive process, ozone treatment is considered to be more effective in eliminating undesirable tastes and odors. Where necessary, lime, $Ca(OH)_2$, and soda ash, Na_2CO_3, are added to soften the water by converting calcium and magnesium salts to insoluble precipitates. Fluoride is added to help reduce tooth decay. Its concentration is typically about 1 g/m^3 or 1 ppm.

Although significant water purification problems remain in the United States, the water problems in developing nations are of a much greater magnitude, where only about half the people have access to safe drinking water, and some 1.4 billion people have no facilities for sanitary waste disposal. It has been estimated that about 80% of all human disease is related to unsafe water and poor sanitation and hygiene. Contaminated water is mainly responsible for trachoma, blindness, elephantiasis, cholera, typhoid, infectious hepatitis, poliomyelitis, and intestinal worms. Diarrhea from contaminated water seriously afflicts many in Asia, Africa, the Caribbean, and Latin America, and is a major contributor to malnutrition, particularly in children. Although progress is being made through programs such as those sponsored by the United Nations, much remains to be done, especially in education concerning hygiene and sanitation.

14.5 DESALINATION

It is apparent from the size of the oceans that there is no water shortage, only a shortage of fresh water. The vast amount of water in the oceans (349×10^{18} gal) would solve our water problems if the salt could be removed easily. It has been observed that it is far easier to put a pinch of salt in a glass of water than it is to take it out (think of this in terms of entropy changes). The core of the problem of desalination is the cost, and for most of the processes that will be discussed, it is the cost of the energy involved.

The average salt (NaCl) concentration of the oceans is 3.5% by weight. Stated in another way, this is 35 kg of salt for 1 m^3 of water, or 35×10^6 tonnes of salt in 1 km^3 of seawater. Although more than 50 known chemical compounds are found in seawater, NaCl accounts for about 85% of the dissolved material. Seawater can be used directly for some applications that require water, such as cooling for power plants and some sanitary systems. However, the basic life support that fresh water provides to animals and plants cannot be met by seawater because of its salinity. The possibility of providing abundant fresh water by desalination of the oceans has been of such immense importance that a great scientific and engineering effort has been made over the years, and a number of ingenious devices have been designed.

DISTILLATION

It was stated earlier that 539 cal are required to vaporize 1 g of water. One could imagine a simple still, fashioned from a boiler, heated by burning a fossil fuel and the vaporized water (steam) condensed to provide fresh water. The cost of providing fresh water from such a simple device would be entirely prohibitive. The next step in making the process more efficient would be to use not merely a single still or stage but a multistage distillation process. An obvious improvement on the one-stage device would be to use incoming salt water to cool the condenser where the steam is converted to a liquid, thereby preheating the seawater. The most efficient distillation devices use multistage flash distillation, in addition to having many stages in tandem. In flash distillation, the water vapor instantly evaporates off from hot brine when the pressure is reduced, so that boiling occurs at temperatures lower than 100°C. Plants of this type have been constructed that require only 40 cal/g of fresh water. The amount of power required to provide water for one person (about 2000 gal a day) can be calculated on this basis. The result is 14.5 kW, assuming continuous operation, 24 hours a day. This is roughly 50% greater than the total rate of energy used per person for all other needs in the United States.

Example 14.2

Assume an electrically heated multistage still that requires 40 cal to produce 1 g of fresh water. Calculate the energy in kW • hr to obtain 1000 gal of water. Using the present costs of electricity, what would be the cost of 1000 gal of water produced in this fashion?

Solution

(a)

$$\frac{40 \text{ cal}}{1 \text{ g}} \times \frac{3.785 \times 10^3 \text{ g}}{1 \text{ gal}} \times \frac{4.184 \text{ J}}{1 \text{ cal}} \times \frac{1 \text{ kW} \bullet \text{hr}}{3.6 \times 10^6 \text{ J}}$$

$$= 0.18 \text{ kW} \bullet \text{hr/gal}$$

$$= 180 \text{ kW} \bullet \text{hr/1000 gal}$$

(b) At a typical rate for household electric energy of \$0.06/kW • hr, the cost of 1000 gal of water would be

$$\frac{180 \text{ kW} \bullet \text{hr}}{1000 \text{ gal}} \times \frac{\$0.06}{\text{kW} \bullet \text{hr}} = \$10.80/1000 \text{ gal}$$

The present cost of domestic water in the United States varies greatly, but prices of \$0.50/1000 gal are not uncommon, and water for irrigating crops is far less. It is, thus, amply clear that even the most efficient seawater distillation system is far from competitive economically. There are additional problems caused by the mineral deposits, such as calcium carbonate, forming a coating inside the boiler and plumbing. With all desalination systems, only communities reasonably near to coasts can benefit because of the costs of transportation.

Although present technology has brought the energy cost down to 40 cal/g for a distillation system, this is not the theoretical minimum energy. It can be shown on the basis of thermodynamics alone that the absolute lower limit for any system is 0.66 cal/g for the first gram of fresh water from a volume of seawater. If the process continues until half the original volume has been extracted as fresh water, the limit is 0.90 cal/g. In practice, distillation systems usually do not reduce seawater beyond such a half-way point. It should be stressed that 0.9 cal/g is a theoretical limit, and it is unlikely that much improvement in a practical working system can be made beyond 40 cal/g.

The use of electric energy to distill water in Example 14.2 was a choice made for convience of the calculation. In practice, a less expensive source of heat energy, such as coal, would be used. There are also interesting possibilities of using the heat rejected from the turbines of an electric power plant to run a distillation plant. Proposals have also been made to use a specially built nuclear reactor to furnish heat to desalinate the Colorado River water before it enters Mexico.

SOLAR STILLS

An interesting application of solar energy can be made to provide fresh water from salt water. Figure 14.7 shows the general scheme. Seawater is pumped to area *B* where it is warmed by sunlight coming through sloping glass panes labeled *A*. The vapor condenses on the inside of the glass panels and runs down to be collected in the channels labeled *C*. The channels lead to a storage basin for the fresh water. By arranging channels *D* on the outside of the glass,

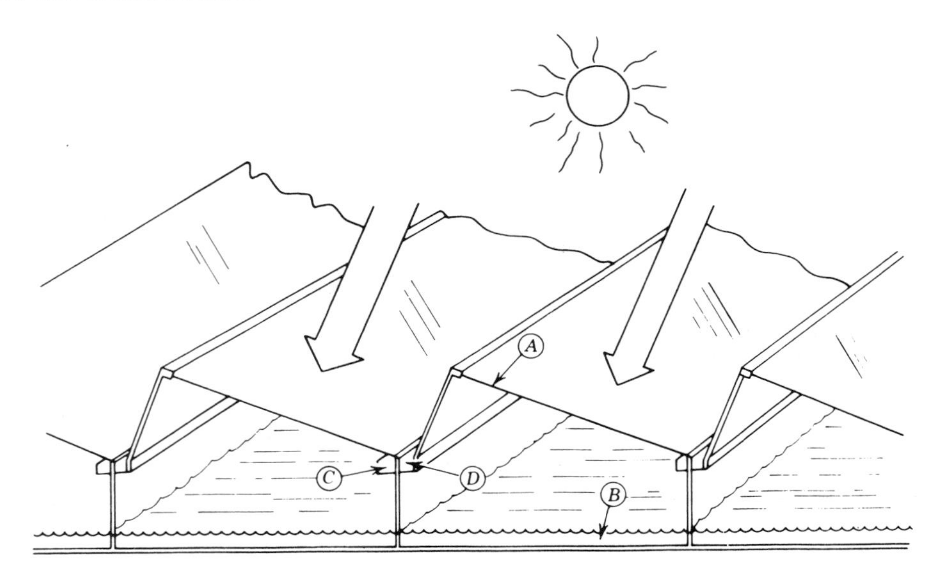

Figure 14.7 A cross-sectional view of a combination solar still and rainwater collector. Solar radiation enters through the glass panels (*A*) to heat seawater (*B*). Fresh water then condenses on the underside of the glass panes and runs down to be collected in the interior channel (*C*). Rainwater is collected in the outside channel (*D*).

any rainwater that falls on the outer surface is also collected. A solar still of this kind has been constructed on the Greek island of Patmos where, with an area of 9000 m^2, about 27 m^3 of water is collected daily. About the same amount is collected through rainfall. The seawater is retained in the still for 48 hours before it is pumped out and a fresh charge of seawater is pumped in. The costs of the electricity for the pumps and the periodic cleaning of the channels are the main expenses. It appears that the use of this type of solar still is limited to situations where land is readily available, the amount of water needed is not too great, and the insolation is relatively high.

Electrodialysis

When sodium chloride is dissolved in water, a positively charged ion of Na cation and negatively charged ion of Cl (anion) are formed. If positive (anode) and negative (cathode) electrodes are immersed in the solution, the cations will flow to the cathode and the anions to the anode. Various types of cation and anion membrane filters exist that will permit the passage of one or the other of these ions. If these filters are interposed between the anode and cathode, there will be regions where the salinity will increase and regions where it will decrease. Figure 14.8 shows a schematic view of the process: saltwater enters from the left, fresh water is taken from two of the areas on the right, and high-salinity water is taken from the other two areas. Electrodialysis appears to offer one of the most inexpensive ways to obtain fresh water from brackish water with 1% or less salt concentration. Because of this and the fact that there is much brackish water in the world, small commercial plants are using this process.

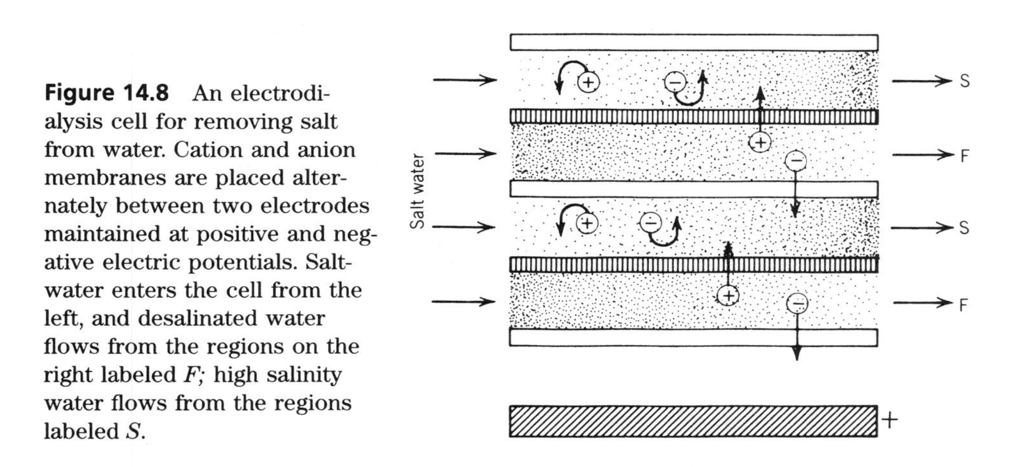

Figure 14.8 An electrodialysis cell for removing salt from water. Cation and anion membranes are placed alternately between two electrodes maintained at positive and negative electric potentials. Saltwater enters the cell from the left, and desalinated water flows from the regions on the right labeled *F;* high salinity water flows from the regions labeled *S*.

Reverse Osmosis

There are semipermeable membranes that permit the passage of fresh water but not saltwater. If saltwater and fresh water are separated by such a membrane, there is an effective pressure, osmotic pressure, that tends to push the fresh water through the membrane into the saltwater area. If a pressure greater than the osmotic pressure is mechanically created in the saltwater area, the process is reversed so that fresh water will be pushed from the salt solution across the membrane. Such a process of reverse osmosis produces fresh water as the salt is left behind. When the salt concentration is very high as in seawater, a very high reverse pressure is required. Research is now being carried out to develop improved membranes. Small-scale reverse osmosis plants indicate that the process could be the least expensive of those discussed so far, particularly for brackish water.

Other Methods

There are a number of additional schemes for the desalination of seawater. Several of these processes involve lowering the temperature of the seawater until ice crystals are formed from fresh water, leaving behind brine. If the ice is removed, fresh water can be obtained by melting the crystals. The process is actually considerably more efficient if the freezing takes place at a reduced pressure where the boiling point of water coincides in temperature with the freezing point. This occurs at a pressure of 4.6 mm of mercury compared to atmospheric pressure of 760 mm of mercury. This vacuum freezing process has been developed in Israel to a point where it is, along with reverse osmosis, one of the least expensive ways to desalinate water.

In summary, the cost of desalinating water is generally still so high that it is considered a possibility only in special situations where fresh water is very expensive. A number of ships find it less expensive to operate a desalination facility than to transport water needed during the voyage. There are isolated communities such as Coalinga, California, Buckeye, Arizona, and Port Mansfield, Texas, that have found it less expensive to desalinate seawater than to import other fresh water for domestic use.

From 1986 to 1991, there was a severe drought in southern California. With its water resources nearly gone, the city of Santa Barbara has had built, at a cost of $30 million, the largest reverse osmosis desalination plant in the United States. The plant uses 7 MW of electric power to produce 6.7×10^6 gal/day (7500 acre-ft/year) of potable water. The cost of the desalinated water is $5.85/1000 gal compared to $0.71/1000 gal for surface water from its local reservoir. While the salinity of the desalinated water is several times higher than that of normal surface or ground water, most people cannot taste the difference, and the hardness (calcium and magnesium salts) is far less.

At the national level, the Office of Saline Water was created in the Department of the Interior in 1966 to help develop desalination systems and to aid communities with brackish water to reduce the salinity. However, the use of desalinated seawater for the irrigation of crops would not appear to be at all feasible on a large scale.

14.6 ACID RAIN

Chapter 13 discusses in some detail the emission into the atmosphere of sulfur dioxides and nitrogen oxides from various man-made sources. The formation of acids, primarily H_2SO_4 and HNO_3, from these pollutants and the resulting damage caused by the acidic rain formed is a story of growing importance and interest. In terms of global atmospheric problems, many regard the overall ramifications of the acid rain problem second only to carbon dioxide and the greenhouse effect. State or country boundaries are in no way barriers to the flow of the pollutants that travel many hundreds and even thousands of miles before returning to earth as acid rain.

Before relating some of the interesting and important details of the acid rain story, it will prove useful to review the definition of pH as a measure of acidity. The pH scale ranges from 0 to 14, with the midpoint taken as the neutral point. Values of pH less than 7 represent the presence of hydrogen ions and, hence, acidity. Values of pH above 7 represent alkaline (or basic) soils or liquids. The pH scale is logarithmic. Hence, a change of 1 on the pH scale corresponds to a change by a factor of 10 in the hydrogen ion (H^+) concentration, which is generally measured in microequivalents per liter. Vinegar, for example, has a pH of 2.2, and it is, as we know, quite acidic to the taste. Milk of magnesia, with a pH of 10.5, is very alkaline.

Rainwater that is not polluted has a pH of around 5.6, which is somewhat lower than 7, which represents a completely neutral solution. This slight acidity comes about mainly from the formation of carbonic acid (H_2CO_3) from the carbon dioxide present in the atmosphere. The acid rain problem arises because of the further reduction in the pH by acids formed from SO_x and NO_x that originate primarily from fossil fuel burning, smelters, and other industrial processes. The name *acid rain* is somewhat of a misnomer, as 10 to 20%, and at times more than 50%, of the acidity comes from the dry deposition of the SO_x and NO_x particles. Some of these particles are so small that they do not fall quickly out of the atmosphere but remain suspended for a very long time. The air in which they are suspended interacts with the surface of the earth, and these particles can be caught by surface objects and contribute to the acidity by mixing with surface water. The dry deposition of acid rain is just as harmful as acids brought to the surface by precipitation.

Prior to the 1950s, no effective effort was made to monitor the acidity of the rain and the lakes in the United States; however what evidence there is indicates that the pH of the rain or snow was about 5.6 or higher. At this level of acidity, the fish and plants in the lakes do not seem to be adversely affected.

Since that time a general increase in the acidity of rain and snow has been documented by a number of investigators, and, particularly in Western Europe and the northeastern section of North America, alarming increases have been noted. These increases appear to be directly related to increased sulfur dioxide and nitrogen oxide emissions. Of the two pollutants, SO_2 is associated with about 70% of the hydrogen ion concentration. In North America, the SO_2 is primarily from coal-burning power plants in Ohio, Indiana, Missouri, Pennsylvania, and Illinois, among other states in the region. The United States is the largest emitter of SO_2 in the world, with about 25.7×10^6 tonnes a year. In Europe the primary sources of SO_2 are the Ruhr Valley of Germany, England, France, and the Low Countries. The important factors involved are the sulfur concentration in the coal burned and the prevailing wind directions and precipitation patterns. In the case of Europe, Norwegian and Swedish lakes have been the areas where the effects have been felt most. In North America, the lakes in the Adirondack Mountains of New York, upper New England, and the provinces of Ontario and Quebec, Canada, have been most adversely affected. Figure 14.9 shows the approximate values of the pH of surface waters in the United States and Canada. These values are the most extreme ones, which are generally reached in the spring. There are, of course, many variations within the regions shown because of local sources, varying rock formations, and weather patterns.

After acid rain is received, the effects on the ecosystem depend on the type of rock and soil. If there is a limestone rock base in the soil or water, the resulting acidity can be largely neutralized. On the other hand, if the bedrock is granite, gneiss, quartzite, or quartz sandstone, there is very little buffering action since the waters are soft and contain few dissolved minerals. Large sections of the northeastern United States and eastern Canada have bedrock of these types, and hence are very vulnerable to acid rain.

If one looks at a lake that has been affected by acid rain, it appears to be clear, fresh, unpolluted water. Unfortunately, when the pH of a lake goes below 5 or so, the effects on the plants and fish are very real even if the water appears clear. Effects on the reproduction process of fish are among the first consequences of a lowering pH level. The newly hatched fish, fish fry, will generally not survive a pH level in the region of 4.5 to 5.0, or if they do survive, they are frequently deformed. The acid also tends to upset the body chemistry of the fish and causes a calcium deficiency that leads to deformed, humpbacked, or dwarfed fish. Another effect of below normal pH is the release of aluminum from the soil surrounding a lake, which leads to a clogging of fish gills and gradual suffocation.

There are similar effects on frogs, salamanders, clams, crayfish, and many of the aquatic insects. Aquatic plants are also adversely affected by acidity. Eventually, after the fish and plant life of a lake have disappeared because of the increase in acidity, the lake can only support a thick mat of algae, moss, or

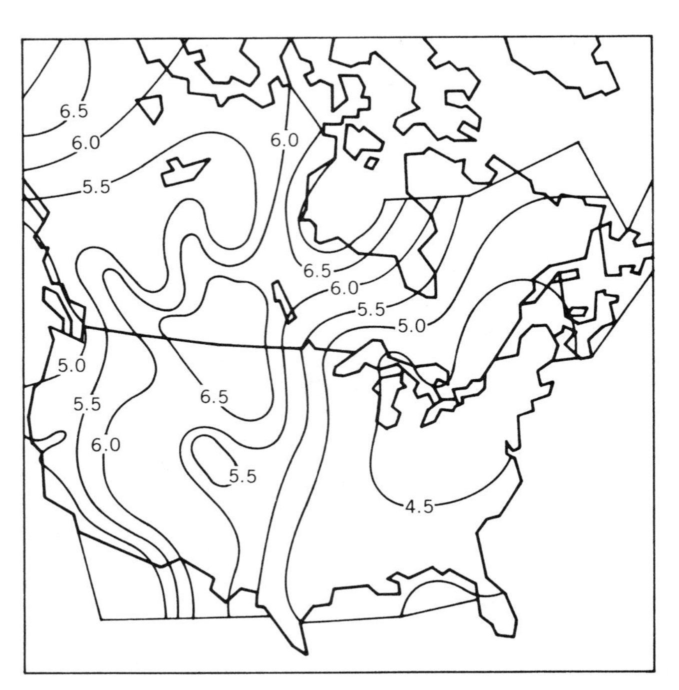

Figure 14.9 Approximate pH values measured in the spring of the year in the surface waters of various regions of the United States and Canada.

fungus at the bottom of the lake. It has been noted that the bottom of a dead lake looks as if it has been covered with "astroturf."

In the Adirondack Mountains of New York, 51% of the lakes at elevations above about 2000 ft were found in the early 1970s to have a pH lower than 5. In the period of 1920 to 1937 only 4% of the same lakes were below a pH of 5. It was estimated in 1970 that about 200 lakes were dead, and by now the number must be considerably higher. A similar series of dead lakes was found in the La Cloche Mountains near Sudbury, Ontario, where the International Nickel Company has huge smelters.

During the winter months, the snow with its acid content accumulates as snow packs in the mountains. When the spring runoff comes, this acid is released to the lakes and rivers, and the pH can suddenly be decreased in some bodies of water by as much as 2 units in the span of a few weeks. This chemical shock is even more devastating to the flora and fauna than is a steady concentration of the same low pH.

In 1980 the U.S. government established the National Acid Precipitation Assessment program (NAPAP). The goal of this program was to evaluate the causes and effects of acid rain and to propose ways to control it. It was a massive 10-year study involving 2000 scientists and costing $570 million. Many measurements of water acidity were made and mathematical models created, for example, for acid deposition on a regional basis. Unfortunately, in spite of the program's being scientifically sound, there have been many delays in issuing the final report and political controversy with the administration of the program, which have reduced the usefulness of the effort in providing guidance for legislation on control. The 1990 Clean Air Act amendments, which reduce SO_2 emissions (see Chapter 13), were drawn up without appreciable benefit from NAPAP.

The 6000-page final NAPAP report presents a picture that is perhaps not quite as disastrous as some of the earlier reports, but there is no question that a significant problem exists. In brief, the report states that acid rain has adversely affected aquatic life in about 10% of the eastern lakes and streams. The decline of the red spruce trees at high elevations is associated with acid rain, and acid rain has contributed to corrosion of buildings and materials. Acid rain has also reduced visibility in the Northeast and parts of the West.

There is evidence that progress is being made, as the concentration of sulfur compounds in the air in the Northeast has decreased and the rate at which lakes in the area are acidifying has slowed. The progress is mainly due to control measures of the Clean Air Act of 1975 and some additional controls imposed by states.

There is far greater agreement that there is a problem than there is on some definite solution. Because of the long-range transportation of the air pollutants, the people affected are generally not the people who are generating the pollutants. The reduction of acid rain in the Adirondack Mountains and eastern Canada can come about by a requirement that the SO_2 emissions be reduced from coal-burning plants in the U.S. Midwest. The cost in increased electric power rates to the residents of the Midwest to install flue-gas desulfurization units will be appreciable, and such programs have not been enthusiastically received. The Environmental Protection Agency (EPA) has ruled that 70 to 90% of the gaseous sulfur must be removed from all new coal-based power plants. Unfortunately, most of the sulfur emissions come from old power plants that will not be affected by the EPA's requirement. The 1990 amendments to the Clean Air Act along with some innovative clean-coal technologies that are being developed should help solve some part of the problem.

PROBLEMS

1. (a) Calculate in calories the energy needed to evaporate the 1.1×10^{17} gal of water that fall as rain each year worldwide.
   (b) Compare the answer from (a) with the total solar energy reaching the earth's surface and the amount of solar energy utilized in photosynthesis. See Chapter 10 for necessary information.
2. (a) Assuming that a state with an area of 100,000 mi^2 has an annual rainfall of 15 in., calculate the annual runoff in gallons, assuming 20% of the rainfall ends up as runoff.
   (b) If the per capita needs are 1600 gal a day, how many people can be provided this amount of water in the state, assuming all of the runoff is used once.
3. Calculate the number of gallons in an acre-foot of water.
4. Calculate the maximum amount of runoff needed to serve as cooling water for electric power plants in the United States, assuming the electric energy produced each year is 1.3×10^{12} kW • hr. Also assume that the efficiency of the plants averages 35%, the water temperature is increased by 12°C and the water is used only once.
5. Calculate the cost of producing 1000 gal of water with a still that requires 40 cal to produce 1 g of water. Assume that oil is burned with 100% efficiency and a 42 gal barrel of oil costs $35.
6. If the concentration of hydrogen ions (microequivalents per liter) is 300 for a lake in which the pH is 3.5, what is the hydrogen ion concentration when the pH is 4.5?

7. (a) What are some useful by-products of a municipal sewage treatment plant in addition to the purified water output?
   (b) Why is the combining of sewage disposal and storm sewers such a poor practice?

8. The greatest peak demand on a municipal water system is presented by the need for extinguishing fires. A formula used to estimate the water supply necessary to provide for adequate fire protection in a city is

$$\text{gallons/minute} = 1000 \times \sqrt{P}$$

where P is the population in thousands. For a city of 1 million people, calculate the number of gallons per minute and per hour necessary for fire protection. For a fire of 1 hr duration, how many gallons of water would be expended, per capita, in this city? Can you explain why the requirement is proportional only to the square root of the population rather than directly to the population?

9. Water conservation around the home could make significant contributions to easing the water shortages being experienced in various states. List six practical conservation practices that could be initiated.

10. Based on the number of gallons of water used per person per day in the United States in 1985, calculate the capacity in gallons per day of a desalination plant needed for a city with a population of 100,000 if all the city's water needs were to come from that plant.

11. If all of the water allotted to California from the Colorado River were used to furnish water for the residents of that state, how many people could be served? What is the present population of California?

SUGGESTED READING AND REFERENCES

1. Camp, Thomas R. *Water and Its Impurities.* New York: Reinhold Publishing, 1963.
2. Reisner, Marc. *Cadillac Desert.* New York: Penguin Books, 1986.
3. Harte, John, and Socolow, Robert H. *Patient Earth.* New York: Holt, Rinehart & Winston, 1969.
4. Hodges, Laurent. *Environmental Pollution.* New York: Holt, Rinehart and Winston, 1973.
5. Likens, G. E.; Wright, R. F.; Galoway, J. N.; and Butler, T. J. "Acid Rain." *Scientific American* **241** 4 (October 1979), pp. 43–54.
6. Marston, Edwin H. *The Dynamic Environment.* Lexington, Mass.: Xerox College Publishing, 1975.
7. McCaull, Julian, and Crossland, Janice. *Water Pollution.* New York: Harcourt, Brace, Jovanovich, 1974.
8. Corson, Walter H., Ed. "Fresh Water." *Global Ecology Handbook.* Boston, Mass.: Beacon Press, 1990.
9. Splinter, W. S. "Center-Pivot Irrigation." *Scientific American* **234** 6 (June 1976), pp. 90–99.
10. Maurits la Rivière, J. W. "Threats to the World's Water." *Scientific American* **261** 3 (September 1989), pp. 80–94.
11. Mohnen, Volker A. "The Challenge of Acid Rain." *Scientific American* **259** 2 (August 1988), pp. 30–48.
12. Balzhiser, Richard E., and Yeager, Kurt E. "Coal Fired Power Plants for the Future." *Scientific American* **257** 3 (September 1987), pp. 100–107.

CHAPTER FIFTEEN

NOISE

15.1 INTRODUCTION

Noise has become a form of pollution in our industrial society because sound levels have tended to increase with higher population density, and more mechanical devices (lawn mowers, jack hammers, automobiles, jet airplanes, sirens, air conditioners, etc.), and because of our ability to amplify sound levels with electronic devices (radio, TV, rock concerts, etc.). Over the years these sound levels have crept up over those experienced by a more rural, less technically oriented society, until not only is the sensitivity of our hearing systems being impaired, but our mental state is also being adversely affected. The average urban noise intensity is now twice what it was in 1955.

Numerous studies have shown that in general the human response to excessive noise is tension and, in some cases, anxiety neurosis, hypertension, vertigo, stomach ulcers, allergies, and a series of other undesirable conditions. Of course, not everyone exposed to high levels of noise experiences all of these symptoms, and there has been a tendency not to recognize noise as the source of many psychological and physiological problems. The increasing severity of the noise problem is now finally being recognized, and steps are being taken either to lower the noise levels or provide protection for the individual hearing system.

Noise is sometimes defined as unwanted sound. To understand the noise problem, one certainly must have some understanding of the physics of sound. In addition, some knowledge is needed of how the ear functions, how sound levels are measured, how sound can be attenuated and controlled, and what laws have been and can be enacted to protect the public.

Sound travels in a gas by means of longitudinal mechanical waves. Sound is a completely different phenomenon from light, a transverse electromagnetic wave. Light propagates at 3×10^8 m/sec and requires no medium. Sound requires some propagating medium, such as air or water, and the velocities of propagation are very much less than for light. The speed of sound in air is typically 331 m/sec.

Example 15.1

A lightning strike is seen and 4 seconds later the sound of the thunder from that strike reaches an observer's ears. What was the distance between the lightning strike and the observer?

Solution

The time for the light to reach the observer is negligible since it is traveling at 3×10^8 m/sec. If it takes 4 seconds for the sound wave to reach the observer, the distance is then

$$d = vt = (331 \text{ m/sec}) \cdot (4 \text{ sec})$$
$$= 1324 \text{ m}$$

or somewhat less than a mile.

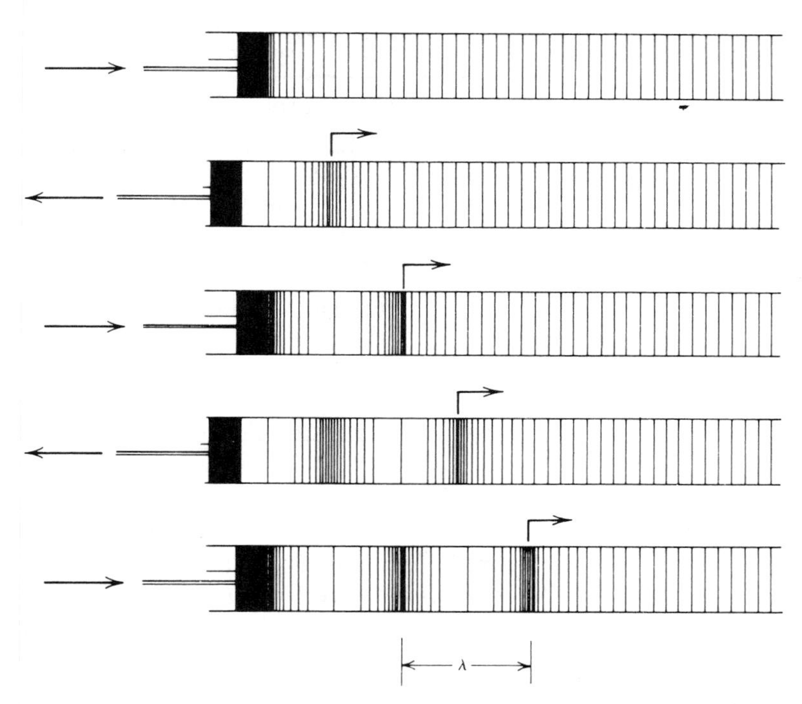

Figure 15.1 The generation of sound waves in a tube filled with a compressible gas by the cyclical back-and-forth motion of a piston or a loudspeaker cone.

Figure 15.1 illustrates the manner in which sound is propagated in a gas. If the cone of a loudspeaker or piston moves into a tube containing a gas, the gas just in front of the cone will be compressed. That is, there will be an increase in the gas pressure and density in that region. The gas next to this region of increased pressure will, in a short time, also experience a pressure increase, even as the speaker cone now goes in the reverse direction, creating a low-pressure area (area of rarefaction) in front of the cone. The region of high pressure will then propagate down the tube, followed some distance afterward by a low-pressure region, then a high-pressure region, and so forth. The distance from one region of high pressure to the next is called the wavelength, λ, and it is entirely similar to the definition of wavelength already used for electromagnetic radiation. The wave is longitudinal because the disturbance is in the direction that the sound is traveling rather than at right angles to it (transverse). The atoms of the propagating medium oscillate about their equilibrium positions as they transmit the longitudinal sound wave, but there is no net motion of the medium in the direction of the sound propagation.

The speed of sound varies considerably with the kind of medium in which it is traveling. As can be seen in Table 15.1, sound tends to travel faster in water and various solids than it does in air. For a given gas the speed is proportional to the square root of the absolute temperature. For example, the speed of sound in air at 0°C is shown in Table 15.1 as 331.3 m/sec, but at 20°C it is 344 m/sec.

The frequency of a sound wave, f, is determined by the characteristics of

Table 15.1 THE SPEED OF SOUND IN VARIOUS MEDIA

Medium	Temperature (°C)	Speed	
		m/sec	ft/sec
Air	0	331.3	1,087
Hydrogen	0	1,286	4,220
Oxygen	0	317.2	1,041
Water	15	1,450	4,760
Lead	20	1,230	4,030
Aluminum	20	5,100	16,700
Copper	20	3,560	11,700
Iron	20	5,130	16,800
Granite	—	6,000	19,700
Vulcanized rubber	0	54	177

the device that is the origin of the sound. In the case of the loudspeaker in Figure 15.1, an electronic oscillator is driving the loudspeaker at a particular frequency. The frequency is measured in hertz (cycles/sec) and is related to the wavelength, λ, and speed, v, by $\lambda f = v$. This is the same relationship that was used for electromagnetic waves except that there the velocity was c, the velocity of light.

Example 15.2

An electronic oscillator drives a loudspeaker at a frequency of 10,000 hertz (Hz). What is the wavelength of the sound in air at 0°C?

Solution

The wavelength is given by

$$\lambda = \frac{v}{f} = \frac{331 \text{ m/sec}}{10 \times 10^3/\text{sec}} = 3.31 \times 10^{-2} \text{ m}$$
$$= 3.31 \text{ cm}$$

15.2 THE INTENSITY OF SOUND

To discuss the effects of sound on people in a quantitative way, there must be an appropriately defined intensity and some way to measure it. As with other wave phenomena, the intensity of sound is the power transferred through a unit area perpendicular to the direction of propagation. It is frequently measured in watts per square meter. The intensity can be shown to be proportional to the square of the maximum displacement of the atoms disturbed by the sound wave or proportional to the square of the maximum pressure created by the sound.

The human ear is a remarkable organ. It can hear sound intensities over an enormous range. For example, the loudest sound that can be tolerated by a normal ear corresponds to a maximum displacement of 1.1×10^{-2} millimeter.

Table 15.2 VARIOUS SOUND INTENSITIES

Sound	Intensity	
	W/m^2	dB
Threshold of hearing	1×10^{-12}	0
Rustle of leaves	1×10^{-11}	10
Whisper at 4 ft	1×10^{-10}	20
Quiet room at home	1×10^{-8}	40
Normal conversation	$1 \times 10^{-6} - 1 \times 10^{-5}$	60–70
Freeway traffic	1×10^{-5}	70
Office with noisy machines	1×10^{-4}	80
Inside car in traffic	$1 \times 10^{-4} - 1 \times 10^{-3}$	80–90
Riveting	3.2×10^{-3}	95
Noisy factory	1×10^{-2}	100
Blaring radio	1×10^{-1}	110
Jet takeoff at 60 m	1	120
Threshold of pain	1	120
In midst of 75-piece orchestra	10	130
Jet takeoff at 25 m	100	140
Moon rocket lift-off	1×10^{8}	200

The softest sound that can be heard corresponds to a maximum displacement of the eardrum of only 8×10^{-9} mm, which is one tenth the radius of the hydrogen atom. The way in which the ear can sense such very small displacements of the air molecules will be discussed shortly. In terms of intensity, the maximum sound level that the ear can tolerate is about 1 W/m^2, and the hearing threshold (the softest sound that can be heard) is 1×10^{-12} W/m^2. Because of this very large range of intensities, it has been found convenient to introduce a special logarithmic scale for sound intensities. The unit for relative sound intensities is called the decibel, dB. *Deci* is simply a prefix meaning one tenth and *bel* is a unit named after Alexander Graham Bell. It is defined by

$$\text{dB} = 10 \log \frac{I}{I_0}$$

In this equation I is the sound intensity in question and I_0 is a reference intensity equal to 10^{-12} W/m^2, the hearing threshold intensity at 1000 Hz. Table 15.2 lists various sound intensities in watts per square meter and the corresponding values in decibels. The values given are only approximate because the devices and situations vary so markedly. The threshold for pain also varies with individuals, but is in the range of 120 to 140 dB. With the decibel scale ranging from 0 to about 200, the intensities in watts per square meter change by a factor of 10^{20}.

Example 15.3

The intensity of a sound wave is determined to be 3.2×10^{-5} W/m^2. What is the intensity on the decibel scale?

Solution

$$\text{dB} = 10 \log \frac{I}{I_0}$$

$$= 10 \log \frac{3.2 \times 10^{-5}}{1 \times 10^{-12}} = 10 \log 3.2 \times 10^{7}$$

$$= 10(7.5) = 75 \text{ dB}$$

Several important features of the decibel scale are sometimes confusing. Suppose there are two sources of sound, one with an intensity of 100 dB and another 90 dB. What is the intensity in decibels of both together? The answer is not 190 dB. We must first add the intensities in watts per square meter and then convert the result to the decibel scale.

Example 15.4

The intensities of two sound waves, A and B, on the decibel scale are 100 dB and 90 dB. What is the combined sound level of A and B in decibels?

Solution

To add the intensities we first calculate the intensities in watts per square meter.

$$100 \text{ dB} = 10 \log \frac{I_A}{1 \times 10^{-12}}$$

$$\frac{I_A}{1 \times 10^{-12}} = 10^{10}$$

$$I_A = 1 \times 10^{-2} \text{ W/m}^2$$

And, similarly,

$$\frac{I_B}{1 \times 10^{-12}} = 10^{9}$$

$$I_B = 1 \times 10^{-3} \text{ W/m}^2$$

Thus, the total intensity is $I_A + I_B$, or 1.1×10^{-2} W/m^2. On the decibel scale this is

$$\text{dB} = 10 \log \frac{1.1 \times 10^{-2}}{1.0 \times 10^{-12}} = 10 \log 1.1 \times 10^{10}$$

$$= 10(10.04) = 100.4 \text{ dB}$$

This 10% increase in sound power in going from 100 dB (1×10^{-2} W/m^2) to 100.4 dB (1.1×10^{-2} W/m^2) does not necessarily represent a 10% increase in perceived loudness. Loudness is a subjective measure of sound. The average

Figure 15.2 A jet aircraft overhead near Logan Airport, Boston, presents a common distraction.

judgment is that for sound levels lower than about 40 dB, a sound level increased by 10 dB sounds about twice as loud.

As we shall see, people near airports where jet airplanes are taking off experience a continuing noise problem. There has been widespread misinformation concerning the sound levels when one jet takes off versus, for example, three taking off at the same time.

Example 15.5

If one jet causes a sound level of 120 dB on takeoff, what would be the sound level of three such jets operating simultaneously?

Solution

For one jet

$$120 \text{ dB} = 10 \log \frac{I}{1 \times 10^{-12}}$$

$$\frac{I}{1 \times 10^{-12}} = 10^{12}$$

$$I = 1 \text{ W/m}^2$$

For three jets

$$I_T = 3 \text{ W/m}^2$$

$$\text{dB} = 10 \log \frac{3}{1 \times 10^{-12}} = 124.8 \text{ dB}$$

Although some sources of sound have a directional output, there are many that produce sound waves approximately equally in all directions. For such sources one can calculate the intensity at any distance. If the total power of a sound source is S watts, the intensity in watts per square meter at a distance d meters away is

$$I = \frac{S}{4\pi d^2}$$

The denominator of the term on the right is the surface area of a sphere of radius d centered at the source. The same total power will go out through any sphere around the source no matter what the radius, but the intensity (W/m^2) will decrease by $1/d^2$ as the sphere becomes larger. Sound intensity obeys the inverse square law. In attempting to reduce exposure to high noise levels, it is well to remember, for example, that going away from the source four times the distance will decrease the intensity by a factor of 16. The total sound power radiated, S (in watts), is shown in Table 15.3 for some sources of noise.

An important property of sound waves is frequency or pitch. Most sources of sound produce more than one frequency. The difference between a pleasing note on the violin and the harshness of a single frequency from an oscillator is that the violin produces not only a fundamental tone, or pitch, but also a myriad of other harmonics and overtones.

The human ear responds quite differently to different frequencies. A person with good hearing can generally hear sound with a frequency as low as 20 Hz and as high as 20 kHz, but the sensitivity of the ear falls off for frequencies lower than 1 kHz and higher than 4 kHz. Figure 15.3 shows how the thresholds of audibility and pain vary with frequency for an average person. The definition of the reference intensity I_0 as 1×10^{-12} W/m^2 is taken from the threshold of audibility at 1000 Hz.

15.3 THE HUMAN EAR

The ear is a remarkable sense organ in many ways. As has been discussed, it can distinguish intensities over a range of 10^{12}, and it can distinguish rather precisely one frequency from another. On the other hand, the ear is not so sensitive that we hear the random motion of air molecules against the eardrum, nor do we hear the many low-level sounds from our bodies such as the sounds of muscles expanding and contracting. These noises would be very disturbing, but we do not hear them because our hearing threshold increases very rapidly as the frequency decreases below 100 Hz.

Table 15.3 TOTAL SOUND POWER RADIATED BY SOME SOURCES

Source	Power (W)
Small electric clock	10^{-8}
Ordinary speech	10^{-5}
Auto at 45 mph	10^{-1}
Pneumatic hammer	1
Jet engine	10^4
Saturn V rocket at takeoff	10^8

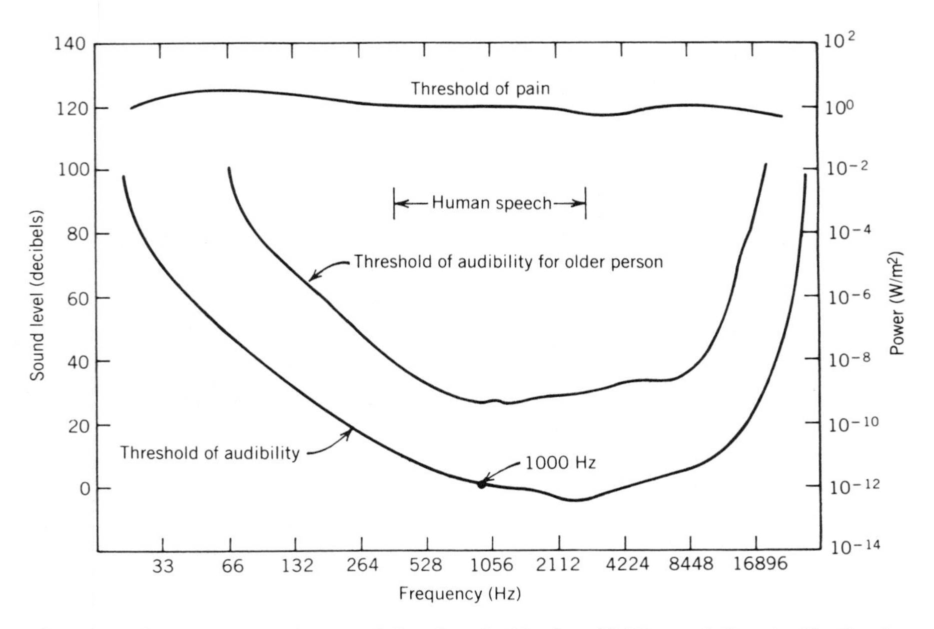

Figure 15.3 The dependence of the threshold of audibility and threshold of pain on frequency for human beings. The decibel scale is defined to be 0 at an intensity of 1×10^{-12}W/m^2 at 1000 Hz.

A diagram of the ear is shown in Figure 15.4. Although it appears quite complex, the basic functions can be understood rather readily. The outer ear (auricle) directs the sound waves down the ear or auditory canal to the eardrum (tympanic membrane). Sound waves coming down the auditory canal set in motion small vibrations of the eardrum. On the inside of the eardrum, there are three small bones (ossicles) in a region called the middle ear. The first bone, the hammer (malleus), is attached to the inside of the eardrum and moves back and forth with the sound pressure on the eardrum. This motion is coupled by the anvil (incus) to the stirrup (stapes), which is attached to the oval window of the inner ear. These three bones of the middle ear act as an acoustical transformer, coupling the motion of the eardrum to the oval window. Because of the mechanical advantage of the bone system and the smaller size of the oval window, about 20 times as much pressure is exerted on the oval window as there is on the eardrum. The middle ear and the nasal cavity are connected by the eustachian tube. This tube allows the pressure on both sides of the eardrum to be equalized. If for some reason, such as a head cold, the eustachian tubes become blocked, the resulting change of pressure, such as that experienced in an airplane, can be painful.

The inner ear takes mechanical vibrations present on the oval window and transforms them into signals that the brain can interpret as sound. The inner ear is a rather complex system of tubes known as the labyrinth. There are three semicircular canals, which are three sets of mutually perpendicular tubes. The main function of these three tubes is to provide the individual with the ability to balance. Hearing is accomplished primarily in the cochlea, which as its name suggests, has the spiral shape of a snail. It is, in fact, a coiled tube filled with a fluid called perilymph. It is divided lengthwise into two sections

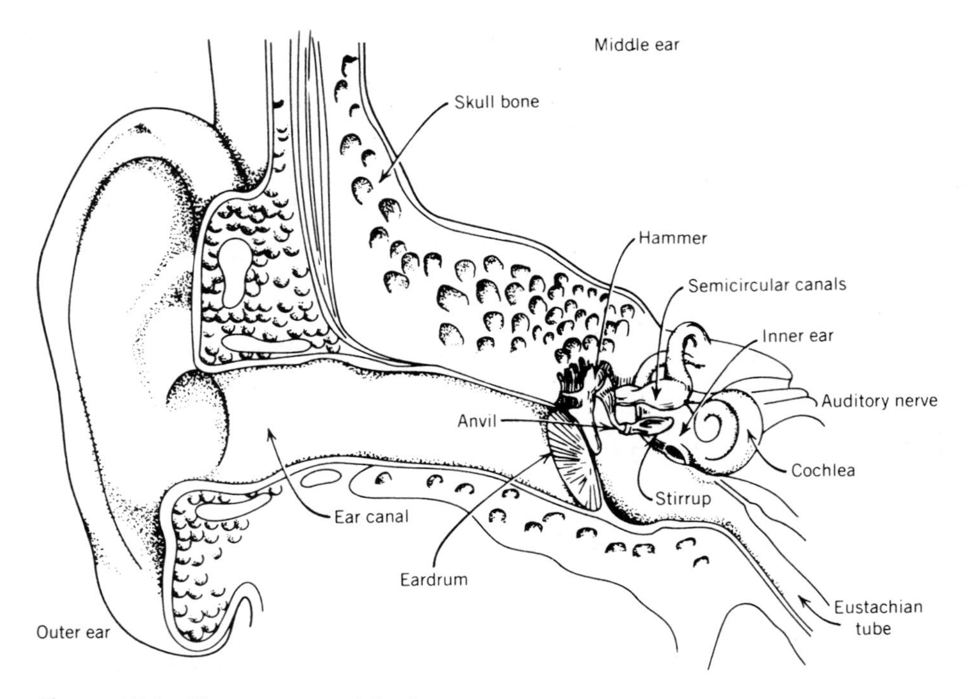

Figure 15.4 The structure of the human ear.

by the basilar membrane. The oval window is connected to one of these sections, and the sound window is connected to the other section. When the stirrup moves the oval window, vibrations occur in the fluid. Running along the basilar membrane is the organ of Corti, which contains about 24,000 fibers or hair cells. These hair cells are attached to nerve endings. Thus, when the fluid vibrates, the hair cells respond and signal the brain through the nerves. The precise mechanism by which these hair cells respond to the amplitude and the frequency of the sound wave is not completely understood. One rather widely accepted theory is that the basilar membrane is deflected at different points along its length for different frequencies. The organ of Corti and the brain cells at that point are excited and send signals proportional to the amplitude of the wave, while the position of the hair cells provides information on the frequency.

There are two other paths that sound waves can take to vibrate the fluid in the cochlea, in addition to the normal route involving the eardrum and the mechanical linkage to the oval window. Sound can pass directly in air from the eardrum, or from a hole in the eardrum, to the sound window of the cochlea. This entry to the inner ear also leads to vibration of the perilymph fluid. Sound can also be transmitted through the bones of the skull that surround the ear. The vibrations in the bone lead eventually to vibration of the cochlear fluid.

15.4 SOUND MEASUREMENTS

Sound can be measured by determining the power in watts per square meter at each frequency. A spectrum of these data can be plotted on a graph, as shown in Figure 15.3. The sound power can be determined by using a microphone, amplifier, and some sound power indicator (meter, strip chart recorder, etc.), where the system has been properly calibrated to read watts per square meter directly. This procedure is not only time-consuming and troublesome, but it is also difficult to interpret the spectrum in terms of human perception of noise. For most applications, it is far more convenient to have a meter that characterizes the noise level by indicating with one number the sound intensity that would be perceived by the average human ear. Unfortunately, because the frequency response of our ear changes somewhat with the level of the sound power, three compensating networks are incorporated into the meters, *A*, *B*, and *C*, which are used for weighting the various frequencies. Figure 15.5 shows the frequency response of the three networks. Response *A* is for power levels of about 40 dB and *B* for about 70 dB. Response curve *C* is much flatter and is appropriate for sound levels in excess of 85 dB.

In specifying measured noise levels, it is customary to indicate the response curve used, for example, 40 dB(*A*). When no letter designation is used, it is sometimes implied that response curve *A* was used, as it is the most common; this practice can be assumed for the rest of the discussion in this chapter. When determining the effects of the attenuation of various media or barriers, it is also important to specify the response curve. For example, a sound power meter could measure with the *A* network the sound level from a furnace motor at some point in the room as 75 dB(*A*). If sound-insulating material is put completely around the unit and the level is now measured as 70 dB(*A*), the attenuation of the insulating material can be said to be 5 dB on the *A* scale. It could be quite different on the *B* or *C* scale.

15.5 HEARING IMPAIRMENT

Acoustic trauma, the sudden loss of hearing from a blast or an explosion, frequently involves the rupture of the eardrum. Although the loss of hearing

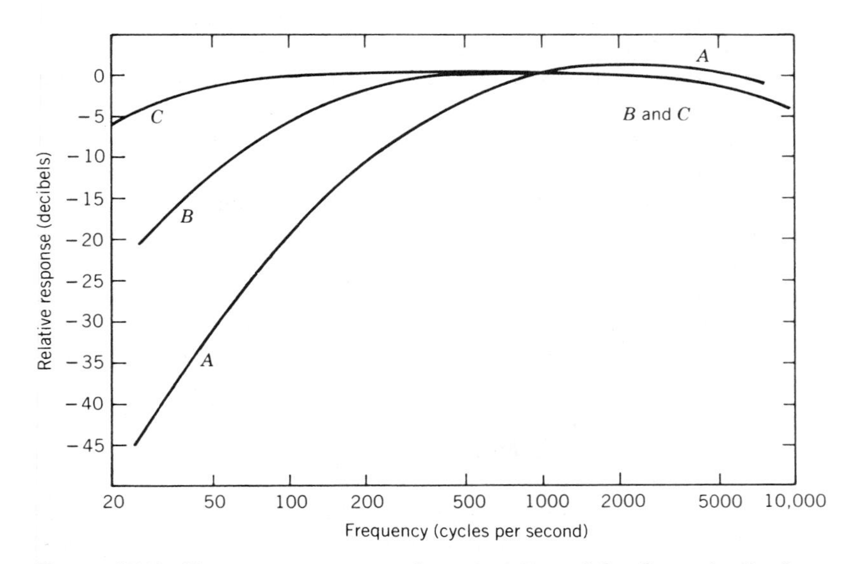

Figure 15.5 Frequency response characteristics of the three standardized American networks used in sound power meters.

from such a cause is serious, only a relatively few people are so affected. The same is true of the mechanical puncture of the eardrum by a foreign object or the loss of hearing from certain diseases. Hearing impairment, however, is experienced by a much larger number of people, generally through the continued exposure to sound at excessively high levels of intensity. An estimated 4 million to 5 million people in the United States are suffering from significant noise-induced hearing loss. In the early stages, the loss of hearing occurs primarily in the 4000- to 6000-Hz frequency range. If the exposure to excessive noise continues, the loss is extended to both higher and lower frequencies. The loss of hearing sensitivity at these relatively high frequencies means that the ability to appreciate music, to hear sounds such as telephones ringing, and to understand conversations is significantly reduced. A comparative study of American males and African males showed that between the ages of 16 and 75 the Americans had a hearing loss of 50 dB at 4000 Hz and 6000 Hz, whereas the Africans experienced only a 10 dB loss. It is striking to note that a loss of 50 dB corresponds to a factor of 100,000 in the minimum sound intensity that can be heard. In one respect, this might be regarded as a loss of 99.999% of an individual's hearing sensitivity. Continued exposure to the noises of industrial society takes its toll, compared with the quiet offered by the undeveloped and agricultural environment of Africa.

If a person is exposed to sound levels in the range of 95 to 100 dB during the 40-hour week, by Friday this person will have experienced, perhaps without realizing it, a 10- to 20-dB decrease in sensitivity in the range of 3000 to 6000 Hz. After a quiet weekend the sensitivity will generally be restored to its normal levels. This temporary threshold shift (TTS) is a well-documented consequence of exposure to sound levels that are regarded as excessive. If the exposure lasts for a shorter length of time, the time required to recover is less. If, for example, a person is exposed to 100 dB for 10 minutes, the TTS will be over in perhaps 30 minutes. The shift is usually greatest at about 4000 Hz, without any direct relationship to the frequency of the noise that induced it.

If the exposure to excessive noise levels is infrequent, then the effects will be minimal. If, however, a person is exposed week after week, as is a weaver in a textile factory or an operator of a jack hammer on a road repair crew, the restoration process stops functioning as well, and the temporary shift eventually becomes a permanent shift. With continued overexposure, the hearing loss becomes greater (the threshold is raised to more decibels and the band of frequencies involved becomes wider) until the individual has a difficult time understanding others when they speak and, perhaps later, total deafness occurs. The TTS has been ascribed to fatigue of the hair cells in the cochlea. With repeated abuse, the hair cells lose their ability to regain their sensitivity and are permanently desensitized.

The amount of noise that can be tolerated without risking hearing impairment depends on the frequency band involved. For a broad-band noise, 120 dB can be tolerated for a minute or two, and the tolerance level ranges down to 85 dB for an 8-hour day. If pure tones or a very narrow band of frequencies are involved, the tolerance levels are reduced 5 dB to 115 dB and 80 dB, respectively.

A study conducted in a large machine shop over an extended period of time compared 270 machinists working in a noise level of 80 dB and 290 office workers working in a much quieter environment. After 16 years there were only minimal differences, but after 25 years the machinists had suffered a 25-dB loss at a frequency of 4000 Hz compared with the office workers.

Various standards for noise levels have been set by various agencies over the years. The U.S. Air Force, for example, has set 85 dB over the frequency range of 300 to 4800 Hz as the limit for an 8-hour daily exposure. Above 85 dB, protective ear coverings should be used, and above 95 dB they must be used.

There are three general ways that noise can be controlled: at the source, by inserting absorbing media and barriers between the source and the individual, and by providing personal protection for the individual. With a greater realization of the harmful effects of noise and the greater number of noise sources, all three methods are being utilized to reduce the incidence of hearing impairment. Since 1908, loss of hearing (acoustic trauma) has been included in Worker's Compensation, but to have a claim recognized one previously had to show that loss of earnings resulted. It was thus extremely difficult to base claims on chronic effects and hearing losses that did not involve total deafness. Fortunately, around 1950 court cases established that even without loss of working time, a gradual loss of hearing beyond aging effects entitled the worker to compensation. These rulings have been instrumental in improving noise conditions in industries such as textile mills, for aircraft maintenance personnel, and for street repair crews.

Industrial problems still exist in the United States. The home environment, with power lawn mowers and loud stereos, and the constant increase in airplane flights and automotive traffic all tend to keep noise pollution a continuing problem.

Some cities have enacted noise ordinances, and specially trained members of the police department are available to make the necessary measurements. The city of Boulder, Colorado, for example, has an ordinance prohibiting a noise level of more than 80 dB(A) measured 25 ft from a right-of-way or from a property line. The only exception to this is for trucks of over 10,000 gross pounds on certain streets during the week and in the daytime. The limit for these trucks is 88 dB(A). Permits may also be obtained from the city for temporary activities that might be in violation of the law, such as parades, rock concerts, and fireworks displays. Generally speaking, the noise ordinances have

Figure 15.6 The use of powerful audio systems at rock concerts has reached a point where hearing defects can be expected.

been difficult to enforce, but such efforts will certainly bring greater public awareness of the seriousness of excessive noise.

15.6 SONIC BOOMS

A special type of noise problem occurs when aircraft fly faster than the speed of sound. The passengers in the airplane are not affected, but the people on the ground experience a very sharp boom when a supersonic plane passes overhead. The intensity of the boom depends on the altitude of the plane, its shape, and its speed. It is difficult to ascribe a sound level on the decibel scale to a short-term crack or pressure change of the kind that results from supersonic flight, but it is said to be equivalent to a sound intensity of about 130 dB with mostly very low frequencies. The booms may not result in hearing deficiencies, but almost everyone who experiences them is bothered by them, and they clearly can add to the other noises present in our industrial society to produce nervous tension and related problems.

In 1964 in Oklahoma City, a six-month test was carried out to assess the effect of sonic booms on a population group. The city was subjected to 1253 supersonic flights during the test program with no flights at night. After experiencing an average of about seven sonic booms a day, 27% of the people said that they could not live with sonic booms and 4900 filed damage claims. Not only are people affected, but damage to buildings can also occur. The prehistoric Indian ruins at Canyon de Chelly, Arizona, experienced severe damage in 1966 from supersonic flights.

A number of years ago there was a spirited national debate in the United States on the pros and cons of developing supersonic aircraft (SST) such as the Concorde. For several reasons, including sonic booms, the Congress decided not to fund development of such aircraft. Under a modest program, there would have been perhaps 150 SSTs in operation by 1990, and 50 million people would have been subjected to about 30 sonic booms a day.

When any object moves through the air, there is an increase in pressure just in front of the object. This pressure increase propagates as a wave at the velocity of sound. If the object or airplane is traveling at subsonic speeds, the wave will move out in all directions but will, of course, precede the advancing airplane. However, if the airplane is traveling faster than the velocity of sound, there is no opportunity for the wave to advance ahead of the airplane. As Figure 15.7 shows, the disturbance will form a conically shaped shock wave that moves out from the plane's path. The part of the shock wave that moves toward the ground will cause a sharp increase in the air pressure at ground level of about 1 to 3 lb/ft^2, depending on the type of airplane and the altitude. Any part of the airplane can generate a shock wave of its own. Because the sonic booms from the various parts of the airplane, such as the nose and wings, come in such quick succession, only one boom is generally heard. The tail of the airplane creates a decrease in pressure of about the same amount as the increase produced by the nose, and these two pressure changes are separated in time at the ground by about 0.1 sec. Figure 15.8 shows the shock wave as it hits the ground with the change in pressure associated with the nose and tail of the airplane.

The sonic booms are heard on the ground in a path about 40 miles wide

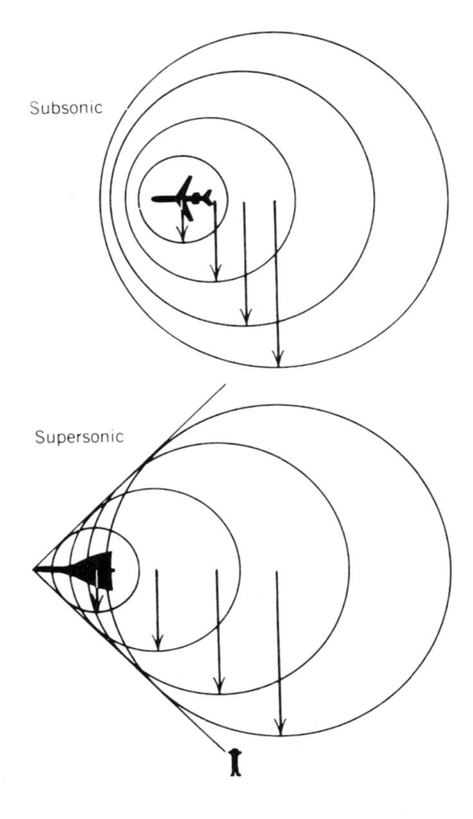

Figure 15.7 Air pressure waves from aircraft at subsonic and supersonic speeds.

directly under the plane. There is no way to avoid completely producing sonic booms with supersonic aircraft, although some changes in the shape of the airplane, such as a very sharp pointed nose, can help reduce the intensity of booms. The only obvious way to avoid having bothersome sonic booms is to restrict the flights of supersonic planes over populated land areas. The British–French Concorde SST can fly at supersonic speeds over the ocean but must reduce its speed below the velocity of sound when over the United States.

15.7 NOISE AND ARCHITECTURE

As the noise levels increase in our complex society, more attention must be given to the design of buildings, particularly homes, so that better attenuation of outside noises is achieved. Automobile traffic, the number of jet airplanes, neighbors with loud stereos, the use of gasoline-powered lawn mowers, and so forth all seem to be increasing. It is important, for a number of reasons mentioned in the introduction, to have a house or apartment one can retreat to, where these noises are reduced to a tolerable level. The noise levels at one's place of employment are also important.

Unfortunately, good acoustical design has been largely ignored in the design of apartments and houses. To some extent this is a result of architects and others not understanding the fundamentals of sound attenuation, but mainly it results from the desire to keep building costs down. Frequently, noise problems

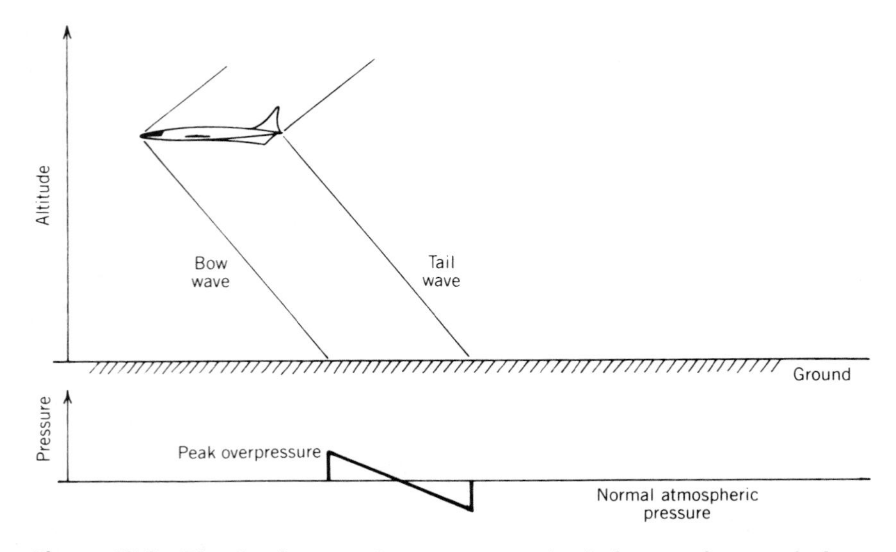

Figure 15.8 The shock waves from a supersonic airplane as they reach the ground cause an abrupt change in pressure from the nose and then from the tail of the airplane.

are not appreciated until the purchase contract has been signed and the house occupied.

There are a few principles of acoustics that are worth bearing in mind in discussing the attenuation of noise. If sound is incident on a heavy stone or concrete wall, about 99% of its intensity will be reflected. Because of the mass and rigidity of the wall, the air molecules are unable to transmit much energy to the wall. The heavier and stiffer the wall, the better the reflectivity. Although such stone walls will prevent the intrusion of outside noises into a home, they are very expensive, and they create an interior that suffers from echoes and an uncomfortably "live" feeling. These exterior walls will not prevent the noises generated in one part of a house from penetrating to the other parts.

Fibrous materials, such as drapes or rugs, are poor reflectors, that is, about 10 to 20% of the incident sound energy is reflected and the rest transmitted. For a porous material to be an effective absorber, it must have a thickness of not less than about 1 wavelength of the sound under consideration.

Example 15.6

(a) How thick should a sound barrier made of cloth be to attenuate effectively transmitted sound with a frequency of 100 Hz?

(b) 20,000 Hz?

Solution

(a)

$$\lambda = \frac{331 \text{ m/sec}}{100/\text{sec}} = 3.31 \text{ m}$$

(b)

$$\lambda = \frac{331 \text{ m/sec}}{2 \times 10^4/\text{sec}} = 1.7 \times 10^{-2} \text{ m} = 1.7 \text{ cm}$$

Thus, we see that even for the very high frequencies, the thickness of the material needed is considerable. A very heavy blanket about 2 cm thick will provide only about 3 dB attenuation for a typical noise spectrum, and hence is essentially useless for sound attenuation.

It is well known that when air is blown across the top of a bottle, it will sing at its natural resonant frequency. The frequency depends on the shape and volume of the bottle. If sound with a range of frequencies is incident on such a resonator, those sound waves corresponding to the resonant frequency will be preferentially absorbed because of the damping by viscous forces. Through proper shaping of the resonator, a moderately wide band of frequencies can be absorbed. In some auditoriums slots are cut in the cement blocks or other structural materials on the side and rear walls to create resonators that will absorb the unwanted frequencies and provide an acoustically good hall. Another technique is to provide fibrous ceiling tiles with a random assortment of holes and spacings. The holes with interior fibers to help the damping constitute a number of different resonators with different characteristic frequencies. In such a fashion, the reflectivity can be decreased and the attenuation increased. Acoustic ceiling tile is a generally effective and inexpensive way to provide sound treatment in the interior of a room, although its performance for very low and very high frequencies is not good.

The question of how to provide sound insulation in the walls to attenuate outside noises or noises from adjacent rooms or apartments is an important one. Generally, good sound insulation will be provided by panels that have high mass, low stiffness, and high damping. A sheet of lead is an example of a material that has all of these properties. Typically, one achieves an additional 6 dB for every doubling of mass per unit area. Single panels of lead, however, are expensive and not appropriate for typical wall construction. Table 15.4 indicates the attenuation achieved by some typical construction materials. A normal stud wall, which is very commonly used for both individual homes and apartments in the United States, does not provide the necessary attenuation to bring traffic noises, which are generally in the range of 70 to 90 dB, down to the 30 or 40 dB desired in one's home.

A sophisticated stud wall is shown in Figure 15.9. The mass per unit area is increased over a normal stud wall, and the sound-absorbing material in the void provides damping. Not having the panels on each side (plywood or plasterboard, for example) nailed to the same studs reduces the stiffness and prevents the vibration of one panel directly coupling to the other panel. There are a number of other modifications of conventional building practices, such

Table 15.4 TYPICAL SOUND ATTENUATION VALUES FOR COMMON WALLS AND MATERIALS

Material	Attenuation (dB)
18-in. brick, plastered	55
9-in. brick, plastered	50
Sophisticated stud wall	50
4-in. dense concrete	45
Normal stud wall	31
¼-in. cement asbestos sheet	25
$^3/_{16}$-in. glass	20

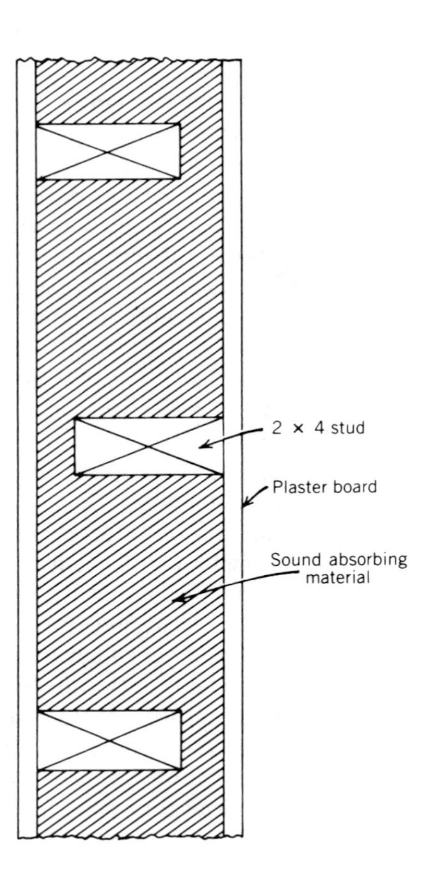

Figure 15.9 A cross-sectional view of a sophisticated stud wall as seen from above. The wall is supported by vertical 2 × 4 studs located on 16-in. centers, with each side of the wall having its own set of studs. The void is filled with sound-absorbing material.

as attaching the panels to the studs with spring clips, that can improve the acoustic properties of a house.

As can be seen in Table 15-4, single-paned windows offer a rather low attenuation of sound. Storm windows or double-paned windows with a dead air space between them can increase the attenuation to the range of 33 to 43 dB. It is interesting to note that most of the measures that can be taken to increase the thermal insulation of a house also increase the sound insulation.

Doors are frequently a major problem in noise transmission. The commonly used hollow core interior doors have an attenuation of about 18 dB. A heavier solid door with a layer of ⅛-in. felt down the middle can improve this figure by 20 dB. Gaskets or felt stripping around the edge of the door also help. Techniques for increasing the sound insulation of floors and ceilings are similar to those described for walls. Typical construction with wooden flooring and 2 × 10 in. joists on 16-in. centers has an attenuation of 35 dB, which is quite marginal.

Air ducts are frequently the cause of noise problems, as they can readily conduct sound between rooms. Lining the ducts with a sound-absorbing material helps, as does flexible coupling at the joints. Much can be done by homeowners to help reduce noise. Rugs, soft furniture, and drapes will reduce the reverberations, but will not attenuate the sound from outside by very much. Landscaping is also important; trees, stone walls, and earthen berms can all act as barriers to sound.

PROBLEMS

1. What is the wavelength in meters of a sound wave with a frequency of 2000 Hz that is traveling in air? In water?
2. A lightning flash is seen and 10.5 seconds later thunder is heard. How far away was the lightning strike from the observer?
3. What is the power per unit area in watts per square meter if the sound intensity on the decibel scale is 65?
4. The sound intensities from two automobiles are 63 dB and 68 dB. What is the resulting intensity in decibels of the two together?
5. A point source of sound has an intensity of 95 dB at 25 m. What is the intensity in decibels at 50 m?
6. A saxophone player generates a sound (noise?) intensity of 90 dB in an adjacent apartment. What attenuation in decibels of the party wall is needed to bring the sound level down to a tolerable level?
7. What are the five most intense sources of noise that you experience in a typical day? From Table 15.2 make a rough estimate of the intensities in decibels of these sound levels.
8. A loudspeaker at a rock concert is delivering 100 W of sound power in all directions. What is the sound intensity in decibels 25 ft from such a source?
9. The human ear has an effective area of about 2 cm $\times$ 2 cm. The threshold of hearing is $I_0 = 10^{-12}$ W/m^2. What is the power (in watts) detectable by the human ear? Estimate also the minimum *energy* in joules detectable by the human ear.
10. Show that:
    Increasing by 10 dB increases intensity (W/m^2) by 10 times.
    Increasing by 20 dB increases intensity (W/m^2) by 100 times.
    Increasing by 30 dB increases intensity (W/m^2) by 1000 times.
    (Hint: $\log [A \times B] = \log A + \log B$.)
11. Show that:
    To add two identical sources together, add 3.01 to decibel scale.
    To add four identical sources together, add 6.02 to decibel scale.
    To add eight identical sources together, add 9.03 to decibel scale.
    (Hint: $\log [A \times B] = \log A + \log B$.)
12. At its lowest resonant frequency, the air column in a pipe closed at one end (such as in a pipe organ) vibrates at a frequency such that the length (l) of the pipe is equal to one quarter the wavelength of the sound ($\lambda/4$). If the human auditory (ear) canal is about 4 cm long, what would be the lowest resonant frequency, considering it as a closed tube? How might this frequency relate to the frequency at which the ear is most sensitive?
13. Consider a bat, the little furry mammal that flies at night and locates flying insects, supposedly by reflected high-frequency sound. Calculate the frequency at which these sound waves must be emitted given that bugs are about 5 mm long, $v = 348$ m/sec, and λ should be no greater than the size of the object being sought.
14. An industrial worker suffers a hearing loss of 25 dB at 4000 Hz after several years of exposure to a noisy work environment. To what factor in terms of intensity (W/m^2) does this 25-dB loss correspond? What percentage of the hearing sensitivity has been lost?
15. If the speed of sound in the air is 331.3 m/sec at 0°C, what is it at 30°C and -30°C?

SUGGESTED READING AND REFERENCES

1. Bragdon, Clifford R. *Noise Pollution.* Philadelphia: University of Pennsylvania Press, 1971.
2. Hudspeth, A. J. "The Hair Cells of the Inner Ear." *Scientific American* **248** 1 (January 1983), pp. 54–64.
3. Littler, T. S. *The Physics of the Ear.* Oxford, England: Pergamon Press, 1965.
4. Meyer, Erwin, and Neumann, Ernst-Georg. *Physical and Applied Acoustics.* New York: Academic Press, 1972.
5. Peterson, Arnold P. G., and Gross, Ervin E., Jr. *Handbook of Noise Measurement.* West Concord, Mass.: General Radio Company, 1963.
6. Taylor, Rupert. *Noise.* Middlesex, England: Penguin Books, 1970.
7. Turiel, Isaac. *Physics the Environment and Man.* Englewood Cliffs, N.J.: Prentice-Hall, 1975.
8. Loeb, Gerald F. "The Functional Replacement of the Ear." *Scientific American* **252** 2 (February 1985), pp. 104–111.
9. Borg, Eric, and Counter, S. Allen. "The Middle-Ear Muscles." *Scientific American* **261** 2 (August 1989), pp. 74–80.

CHAPTER SIXTEEN

NUCLEAR WEAPONS

16.1 INTRODUCTION

The evidence for the earliest self-sustaining nuclear fission reactions taking place on earth comes from the Oklo fossil reactor site in Africa. In this area, some 1.8 billion years ago, there occurred naturally the right combination of a rich uranium deposit, water as a moderator, and surrounding geological materials to serve as neutron reflectors, so that a nuclear fission chain reaction started spontaneously and continued for hundreds of thousands of years. Such a natural reaction would be impossible today, because the 700-million-year half-life of the fissionable ^{235}U has reduced its proportion relative to the 4.5-billion-year half-life of ^{238}U, so that a chain reaction could not now be sustained under natural conditions. The Oklo reactor was discovered in 1972 when workers in the French nuclear energy program detected anomalously low abundances of ^{235}U in ores obtained from the Oklo region. Further investigation showed conclusively that the concentration had been depleted when the fissile ^{235}U was consumed in a naturally occurring nuclear fission chain reaction.

It was not until relatively recently, in 1942, that a second example of a nuclear chain reaction is known to have taken place on earth. This time it was in Chicago, during the early stages of the Manhattan Project, which culminated in the development of a nuclear fission weapon, first known as the atomic bomb. In this achievement the nuclear scientists in Chicago had only succeeded in imitating an earlier accomplishment of nature, although at that time theirs was thought to be the first-ever terrestrial nuclear reactor. In the nearly 2 billion intervening years, the earth slowly evolved from a barren, hot planet into one richly populated with the many life forms we now know. But it was only 3 years from the first man-made chain reaction (1942) to the first military use of a nuclear weapon (1945), and only 7 more years (1952) until the first demonstration of the vastly more powerful nuclear fusion weapon, commonly known as the hydrogen bomb.

In barely a decade, our weapons builders created a means of threatening all of earth's life forms in a brief exchange of bombs. The rapidity of this weapons development was unprecedented; previous advances in weaponry had been much more gradual and had never resulted in so much destructive power being so instantly available. It remains to be seen how mankind can adjust to this capability, and if we shall make the adjustment in a manner that will allow the continuance of our civilization.

We now all live in a state of suspended nuclear annihilation; our destruction can be brought about within a mere 30 minutes after any nuclear-capable adversary decides to make us a target. Even if not targeted directly, each of us who might survive the initial strike would face a highly uncertain future amid the nuclear devastation. This nuclear threat is the product of many technologies. Certainly, the nuclear explosive is at the heart of it, but the sophisticated weapons delivery systems contribute to the imminence of the threat. Advances in aerospace vehicles, submarines, computers, chemical rocket fuels, communications, electronics, satellites, and global navigation all play a role in the weapons delivery system. Taken together, these advances have resulted in the world's leading military powers now having poised for strike thousands of intercontinental nuclear missiles linked to their targets by ballistic trajectories of only tens of minutes' duration.

Even though nuclear weapons have now been available for nearly 50 years without being used in warfare since 1945, it seems obvious to many that the likelihood of their further use is continually increasing as the nuclear weapon capability spreads. Considering that in the past 50 years a nuclear capability to destroy every large city on earth has emerged, it is apparent that the next 50 years will be critical to the survival of humanity. Weapons of global range are becoming available to more and more nations. Any complacency we may feel in our daily lives should certainly be tempered through knowledge and awareness of the potential effects of nuclear weapons. This chapter focuses mostly on technical factors concerning the nuclear arms race. Much of the background for this discussion was covered in Chapters 4, 5, and 11, and will not be repeated here.

16.2 FISSION WEAPONS

The basis for energy release in a nuclear explosion is a natural extension of the energy-producing mechanism in a nuclear power reactor. A chain reaction results when neutrons from a single fission event induce further fission events in either uranium or plutonium. As we saw in Chapter 4, the products of a nuclear fission event have less mass than that of the fissioning nucleus.

The design progression in going from the concept of a controlled chain reaction in a reactor to the uncontrolled explosion of a nuclear bomb involves a number of steps. It was by no means clear during the early days of the Manhattan Project that a nuclear explosion could be accomplished, even after the first reactor became operational in 1942. Some of the initial doubts centered around whether the nuclear assembly would blow itself apart, thus extinguishing the explosion, well before a reasonable fraction of the fissionable material had undergone fission.

A reactor produces energy at a steady rate because the density of neutrons in the core is controlled by capturing the excess number in control rods, as well as by other technical measures. For this control process to be effective, the reactor must be operating in a mode slightly short of "prompt critical." In the prompt critical condition, the reactor sustains a chain reaction with fissions induced only by those neutrons emitted promptly during each fission event. In fact, reactors are designed such that they are not quite prompt critical, but require the delayed neutrons from the radioactive fission products to sustain a chain reaction. These delayed neutrons (0.75% of all emitted neutrons for ^{235}U, 0.25% for ^{239}Pu) are emitted, on the average, about 14 seconds after fission. This time delay allows control rods to be moved in if the reaction rate is tending to increase, and moved out if the reaction rate is tending to decrease. In this way stable operation is achieved. If the reactor were prompt critical, once started, it would nearly instantly increase in power output to a dangerous level, ultimately overheating or blowing apart the critical assembly. This is what accidentally occurred in Chernobyl in 1986. Although this type of an explosion would damage the reactor and could release a lot of radioactivity, it would not be nearly as powerful as the explosion of a nuclear weapon.

Even though a reactor assembly contains many times more fissionable material than any nuclear weapon, in a reactor the fissionable nuclei are sufficiently dispersed among the nonfissionable materials of the fuel rods, and the fuel rods are interspersed with coolant, that it is not possible for a reactor to explode with anything like the yield of a nuclear weapon. In a nuclear weapon, the assembly must suddenly be made to operate far in excess of the threshold of prompt criticality for the mass of fissionable material to be bathed in an intense flux of neutrons well before it is thrown apart by the force of its own explosion. Whereas in a uranium-fueled reactor the enrichment of ^{235}U is typically about 3%, in a bomb it is generally more like 90%. This high enrichment ensures that in the bomb a densely packed structure of fissionable nuclei can be assembled, so that criticality is achieved in a compact structure. In a compact structure, neutrons emitted by one fission are substantially more likely to encounter other fissionable nuclei than if they were released into a less dense structure, where the neutrons would have a larger probability of escaping to the surface before encountering nuclei and inducing further fission. For an example of how a densely packed assembly of fissionable nuclei results in a higher probability for one of the nuclei to be struck by a neutron emitted from within the volume, consider spots of a fixed size, say 1-in. diameter, lying on a spherical elastic surface, such as a balloon. As the balloon is expanded in size with the spot diameters remaining fixed, the spots grow farther apart and cover less of the spherical surface as seen from inside the sphere. If the spots may be taken to represent the nuclei of a certain size arrayed around the site of neutron emission, it is clear that as the structure is compressed, the neutron is less likely to escape from the volume occupied by the nuclei without causing a reaction. As the structure is made more dense, there is simply less open space through which the neutron can escape.

Example 16.1

Calculate the energy release in tons of TNT corresponding to 1 g of matter being converted entirely into energy. (Given: 1 ton TNT $= 4.3 \times 10^9$ J.)

Solution

The equation

$$E = mc^2$$

gives E in joules if m is in kilograms and c is in meters per second. The energy released by 1 g of matter being entirely converted to energy is then

$$E = (1 \times 10^{-3}\ \text{kg}) \times (3 \times 10^8\ \text{m/sec})^2 = 9 \times 10^{13}\ \text{J}$$

This is equivalent to

$$\frac{9 \times 10^{13}\ \text{J/g}}{4.3 \times 10^9\ \text{J/ton TNT}} = 20{,}930\ \text{tons TNT/g}$$

This is about the yield of each of the three weapons exploded in 1945.

Example 16.2

It is known (see Chapter 4) that each nuclear fission releases about 200 MeV of energy. If 1 kg of ^{235}U completely fissions, what is the equivalent energy release—in terms of tons of TNT exploded? (Given: 1 ton TNT = 1200 kW • hr; 1 kW • hr = 2.3 × 10^{19} MeV.)

Solution

Energy released per kilogram ^{235}U is equal to

$$\frac{6 \times 10^{23} \text{ nuclei}}{235 \text{ g } ^{235}\text{U}} \times 10^3 \text{ g/kg} \times 2 \times 10^2 \text{ MeV/nucleus}$$
$$= 5.1 \times 10^{26} \text{ MeV/kg } ^{235}\text{U}$$

The energy per ton exploding TNT:

$$1.2 \times 10^3 \text{ kW} \cdot \text{hr/ton TNT} \times 2.3 \times 10^{19} \text{ MeV/kW} \cdot \text{hr}$$
$$= 2.8 \times 10^{22} \text{ MeV/ton TNT}$$

Then

$$\frac{5.1 \times 10^{26} \text{ MeV/kg } ^{235}\text{U}}{2.8 \times 10^{22} \text{ MeV/ton TNT}} = 18{,}000 \text{ ton TNT/kg } ^{235}\text{U}$$

As discussed in Chapter 5, there is a minimum mass of fissionable material needed to sustain a chain reaction. This is called the critical mass. If the mass of ^{235}U or ^{239}Pu is too small, the surface-to-volume ratio is too large and too many neutrons escape. As the mass and size get larger this ratio goes down. The exact critical mass depends on the geometry, the enrichment of the material, and the surroundings, but for ^{239}Pu it is known that a sphere of 10 kg (4.9 cm in radius) is more than enough for a critical mass.

There are at least two obvious ways of assembling a critical mass quickly. One approach is to use a gun-type apparatus to fire one subcritical mass into close proximity to another; the other is to use an implosion-type arrangement to increase the density of a mass of fissionable material that is nearly critical at its normal density.

There are numerous conceivable variations on the gun-type system. For example, one could have two hemispheres, each of 0.9 critical mass, initially located at opposite ends of a cylindrical tube closed at both ends, driven together by a charge of gunpowder. Or a core of uranium can be driven by an explosive charge into an open cylinder of the same material, much as if one were reinserting the core into an apple. These possibilities are illustrated in Figure 16.1. Because of details of its spontaneous decay, which could lead to premature initiation of the chain reaction, plutonium is not a suitable fuel for a gun-type assembly. In the gun type of weapon, it is desirable to ensure that abundant neutron flux is present at the instant of assembly so that the reaction can be reliably initiated, rather than relying on the presence of an initial neutron from a spontaneous fission event or from cosmic radiation. A copious flood of

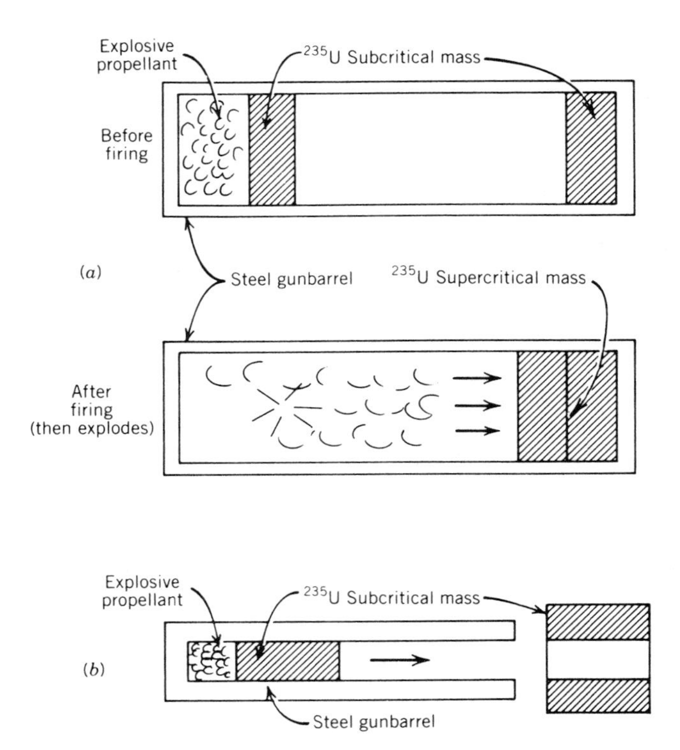

Figure 16.1 Two possible gun-type fission bombs. In (*a*) two subcritical cylinders of ^{235}U are driven into contact by a chemical explosive. In (*b*), the "putting the core back into the apple" approach, a subcritical cylinder of ^{235}U is driven into another subcritical mass of ^{235}U in the form of a hollow cylinder. The two pieces together constitute a supercritical mass.

neutrons can be provided in a number of ways to the critical assembly at the time the subcritical masses are brought together. One way would be to provide a radioactive source of ^{226}Ra, which naturally emits alpha particles, on the front surface of one of the subcritical masses facing a piece of beryllium metal mounted on the other subcritical mass. When the two masses are driven together into each other, the alpha particles from the radium impinge upon the beryllium nuclei, causing the well-known reaction

$$^{9}\mathrm{Be}(^{4}\mathrm{He,n})^{12}\mathrm{C}$$

which can also be written as

$$^{9}\mathrm{Be} + {}^{4}\mathrm{He} \rightarrow {}^{12}\mathrm{C} + \text{neutron}$$

This reaction will provide an abundance of neutrons, if the radium source is sufficiently intense. It would also be possible to produce a flood of neutrons with a small accelerator that would cause a beam of deuterons to impinge on a tritium target, resulting in neutron production by the D–T reaction, as discussed in Chapter 4. In this case the accelerator would be timed to produce a burst of neutrons just at the instant the critical assembly comes together. Either of these two approaches should ensure that the chain reaction starts reliably at the instant of critical mass assembly. Either the source arrangement or the accelerator arrangement is called the initiator.

Although both uranium and plutonium are very dense metals, with densities more than 50% greater than that of lead, these densities can be increased even further if the metals are subjected to enormous pressures such as can be

Figure 16.2 A nuclear fission weapon of the "Little Boy" type that was detonated over Hiroshima in World War II. It was the first deliverable nuclear weapon ever detonated. The bomb was 28 inches in diameter and 10 feet long. A gun-type uranium bomb, it weighed about 4.5 tons and had a yield of about 13,000 tons of TNT.

obtained from chemical explosives. If a slightly subcritical sphere of fissionable uranium or plutonium is surrounded uniformly by an arrangement of explosive charges detonated simultaneously, as shown in Figure 16.3, it can be compressed in an extremely short time to the point where a critical mass is created and a nuclear explosion results. An initiator of some sort can provide an initial burst of neutrons. All modern nuclear weapons are of this implosion type, as was the first weapon ever detonated, the plutonium bomb tested at the Trinity site in New Mexico in July 1945. The bomb exploded over Nagasaki was also of the implosion type, with plutonium as the fuel.

It might seem that the force of a nuclear explosion would blow apart the fissionable material before a significant fraction underwent fission, but this is obviously not so. The answer lies in the inertial reaction of the fissionable material itself and other material that may be situated around the nuclear explosive to serve as a tamper. The tamper is a heavy surrounding mass that resists rapid acceleration outward, thus retarding the expansion of the explosive mass. The same principle has long been a technique used to direct the force of conventional explosives. The heavy surrounding material that serves as a tamper, and other surrounding materials such as the bomb casing, also serve as neutron reflectors; they reflect some of the outgoing neutrons from

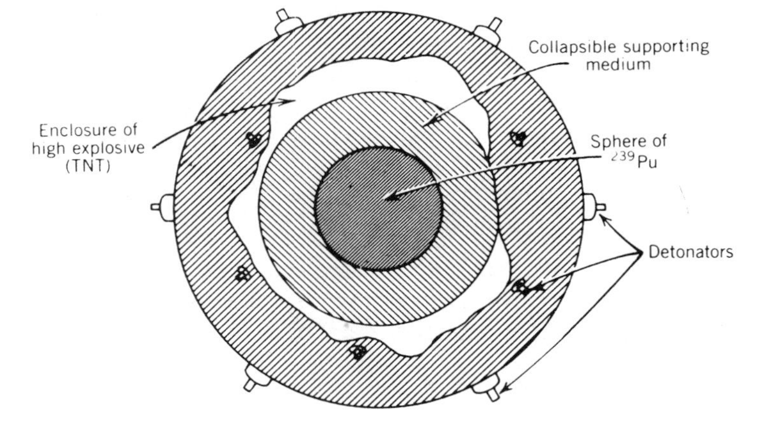

Figure 16.3 An implosion-type nuclear weapon. The initially subcritical sphere of ^{239}Pu at the center may be either a hollow shell or solid. The detonators are fired simultaneously so that the shock waves from the high explosive will converge uniformly toward the center.

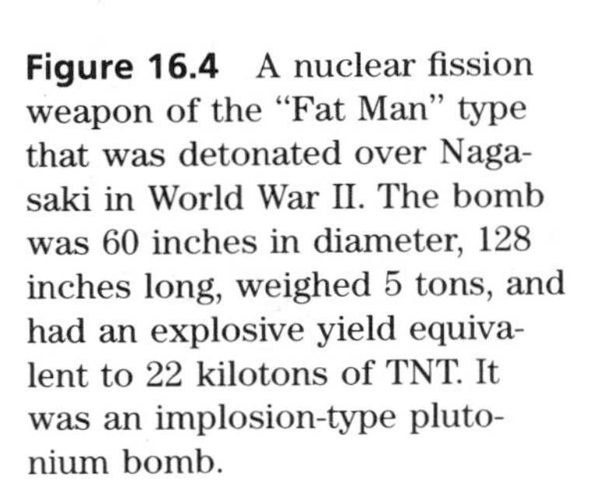

Figure 16.4 A nuclear fission weapon of the "Fat Man" type that was detonated over Nagasaki in World War II. The bomb was 60 inches in diameter, 128 inches long, weighed 5 tons, and had an explosive yield equivalent to 22 kilotons of TNT. It was an implosion-type plutonium bomb.

the initial stages of the nuclear explosion back into the fissionable material, thus increasing the yield. The confinement of the nuclear mass need not last very long, as about 10% of the fissionable material will be consumed in less than the first 10^{-6} second after the reaction is initiated. During this short time the temperature at the center of the explosive mass will rise to more than 10^8 °K, adequate to initiate nuclear fusion, as discussed in Chapter 4.

There is a definite upper limit, related to the size of a critical mass, to the yield of a fission weapon. The initial configuration, be it gun-type or implosion, must be composed of safely subcritical masses of fissionable material. In practice, this means that no more than something like 50 kg are present prior to detonation. With a yield of 20%, this amounts to no more than 200 kilotons TNT equivalent. Here, as in all discussions of nuclear weaponry, when one speaks of equivalent tons of TNT (as in kilotons, or megatons), this refers to the energy released by the explosion of that amount of TNT. This is illustrated by Example 6.2. About 250 kilotons is usually considered the practical upper limit for a fission weapon. Nuclear fission weapons with yields in the range of 1 ton to 500 kilotons have been tested, but the extreme limits are considered unusual cases.

The discussion of nuclear fission weapons would not be complete without some mention of the fact that nature has just barely allowed us to have nuclear weapons. Of the hundreds of atomic nuclei found naturally present on earth, only one, ^{235}U, is capable of sustaining a chain reaction. And this isotope of uranium is less than 1% of all uranium. Once we have ^{235}U, it serves as a fuel for reactors to provide ^{239}Pu, and then there is a choice of two nuclear fuels for either reactors or fission weapons. With fission explosives to serve as triggers, thermonuclear fusion is added to our nuclear explosives technology. If the particular ways in which 92 protons and 143 neutrons interact when assembled together were only slightly different, the course of our future would unfold ever so differently.

16.3 THE HYDROGEN BOMB

Discussion began in 1942 of creating an even more powerful weapon than the fission device that was the main goal of the Manhattan Project. This more advanced weapon would use the fusion of hydrogen nuclei to release massive amounts of energy according to the reactions discussed in detail in Chapter 4. It would be virtually unlimited in size, and the yield of radioactive debris could be substantially less than for a fission explosion of the same magnitude. By 1952 this weapon was a reality.

Of the various possible fusion reactions that lead to the release of considerable amounts of energy, the D–T reaction is among the most important. This reaction combines a deuteron and a triton, both isotopes of hydrogen, to form an alpha particle plus a neutron and to release 17.6 MeV of energy in the process. The reaction may be written as

$$^{2}_{1}H_{1} + {}^{3}_{1}H_{2} \rightarrow {}^{4}_{2}He_{2} + n + 17.6 \text{ MeV}$$

or

$$D + T \rightarrow \alpha + n + 17.6 \text{ MeV}$$

Note that the concept of a critical nuclear mass does not apply to this reaction; any amount of hydrogen, whatever its form, may be assembled with no possibility of a sustained nuclear reaction of any sort being initiated at ordinary temperatures. This means that the limit imposed on the size of a fission weapon, because of the maximum amount of fissionable material that can initially be in a single device without exceeding the critical mass, is not a constraint for a fusion weapon. There is no upper limit on the amount of hydrogen that can be safely assembled and, thus, in a sense, no upper limit on the size of a fusion bomb. Fusion bombs of over 50 megatons, or 2500 times the yield of the World War II fission bombs, have been detonated in tests. The direct products of the fusion reactions do not include radioactive nuclei.

It will be recalled from Chapter 4 that the D–T reaction requires a very high ignition temperature, on the order of 40 million °K, and a fairly high density of deuterons and tritons. One of the few known ways to achieve the requisite ignition temperature is through use of a nuclear fission explosive; and for the required density, the hydrogen must be in the form of a liquid or solid. Once the fusion reaction is initiated, it will be self-sustaining and increase in rate under its own heat generation, until the nuclear fuel is either consumed or dispersed by the force of the explosion. The trigger for every fusion weapon is a uranium or plutonium fission bomb.

The first fusion weapon, detonated at a test site in the South Pacific, used liquid hydrogen as the nuclear fuel. If this were the only practical way of achieving the high density of hydrogen isotopes necessary, it would be unlikely that we would now have our large arsenals of hydrogen-bomb tipped rockets on continual standby. The reason is that the liquid hydrogen must be maintained at a very low temperature, about 20°K, and this requires a bulky apparatus and constant maintenance. It has been estimated that this first demonstration device weighed in excess of 10 tons, hardly the type of device that lends itself to multiple-warheaded missiles.

A breakthrough in nuclear weapons design came about in the early 1950s with the recognition, in both the United States and the Soviet Union, that nature had provided an astonishingly appropriate substance for the weapons builders. This substance is a compound of lithium and hydrogen, lithium hydride, or LiH, that has a density 82% that of water. It is a solid at temperatures to nearly

Figure 16.5 The first thermonuclear explosion at Eniwetok in the South Pacific, Oct. 31, 1952. This explosion had a yield of about 10 megatons and introduced the era of fusion weaponry.

700°C. The remarkable property of this compound is that the lithium not only serves to bind the hydrogen atoms into a fairly dense solid structure, but also that the lithium nuclei add to the nuclear fusion reactions through a reaction sequence outlined previously (see Chapter 4).

If the lithium is made up of compounds of ^{6}Li, which is one of the two stable isotopes of lithium, combined with ^{2}H (deuterium) for part of the solid lithium hydride, and ^{3}H (tritium) for the other part, then there will be present in the assembly abundant amounts of ^{6}Li, ^{2}H, and ^{3}H. When this material is exposed to the 10^8 °K temperature and intense neutron flux emanating from the fission trigger, the following reactions will occur.

$$^6_3\text{Li}_3 + \text{n} \rightarrow {}^4_2\text{He}_2 + {}^3_1\text{H}_2 + 4.8 \text{ MeV}$$

and

$$^2_1\text{H}_1 + {}^3_1\text{H}_2 \rightarrow {}^4_2\text{He}_2 + \text{n} + 17.6 \text{ MeV}$$

Note that, technically, the first reaction is a nuclear fission reaction induced by a neutron. However, such reactions are usually not classified with the fission reactions of heavy nuclei because the lack of emitted neutrons makes chain reactions impossible. Once initiated, the first reaction can be considered as a source of tritium for the second, and the second reaction a source of neutrons for the first. So, in an ideal sense, it would not even have been necessary to have provided tritium. However, to ensure that the D–T reaction proceeds at a good rate, it may be reasonable to assume that both tritium and deuterium are present in the initial assembly. In fact, the 12-year radioactive half-life of tritium limits the maximum time during which fusion weapons remain effective. After a certain number of years they must be reprocessed and the tritium supply replenished. The tritium is produced by fission reactors operated for this purpose. In summary form, the two reactions can be combined to indicate that (in the presence of incident neutrons and at sufficiently high temperature) the compound lithium deuteride yields two alpha particles plus 22.4 MeV of energy per molecule:

$$^{6}\text{LiD} \rightarrow 2\ ^{4}\text{He} + 22.4\ \text{MeV}$$

An added feature of lithium deuteride as a fission weapon fuel is that the tritium produced, following neutron irradiation of the lithium deuteride, is released within the volume, indeed even within the molecule, of lithium deuteride. This gives the triton a high probability of reacting with a deuteron.

Example 16.3

In a fusion reaction sequence, each molecule of lithium deuteride can release 22.4 MeV. What is the equivalent energy release in terms of tons of TNT if 1 kg of lithium deuteride undergoes fusion? (Given: 1 ton TNT = 1200 kW • hr; 1 kW • hr = 2.3×10^{19} MeV.)

Solution

^{6}LiD has a molecular weight of 8, so

$$\frac{6 \times 10^{23}\ \text{molecules}}{8\ \text{g}\ ^{6}\text{LiD}} \times 10^{3}\ \text{g/kg} \times 22.4\ \text{MeV/molecule}$$
$$= 1.68 \times 10^{27}\ \text{MeV/kg}\ ^{6}\text{LiD}$$

and energy per ton exploding TNT is equal to

$$1.2 \times 10^{3}\ \text{kW} \cdot \text{hr/ton TNT} \times 2.3 \times 10^{16}\ \text{MeV/kW} \cdot \text{hr}$$
$$= 2.8 \times 10^{22}\ \text{MeV/ton TNT}$$

Then

$$\frac{1.68 \times 10^{27}\ \text{MeV/kg}\ ^{6}\text{LiD}}{2.8 \times 10^{22}\ \text{MeV/ton TNT}} = 60{,}000\ \text{ton TNT/kg}\ ^{6}\text{LiD}$$

This means that a 1-megaton bomb must have only about 17 kg of ^{6}LiD if it can all be considered to undergo fusion. In practice, of course, the yield is much less than 100%.

Once the weapons designers created a fusion explosive ignited by a fission explosive, the next logical step was to surround this fission–fusion bomb with a layer of natural uranium that, being a dense material, serves very well as a tamper and, moreover, enhances the weapon's yield by a large factor. The natural uranium is abundantly available and fissions under bombardment by the relatively fast neutrons from the fusion reaction, releasing 200 MeV per fission event. Instead of using natural uranium as the outer fission jacket, one might choose to use depleted uranium, which is the nearly pure ^{238}U fraction remaining after ^{235}U is separated out for weapons or reactor use. The depleted uranium is, of course, now vastly more abundant than enriched ^{235}U. As we

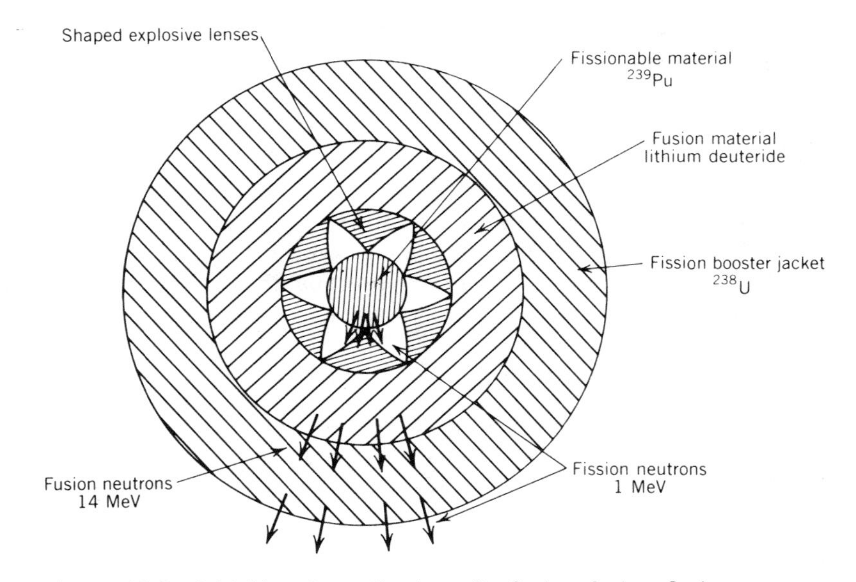

Figure 16.6 A highly schematic view of a fission–fusion–fission weapon. An enhanced radiation weapon would lack the ^{238}U booster jacket, thereby having reduced yield but allowing the energetic fusion neutrons to escape. Both types of weapons have the fission trigger detonated by a surrounding implosion array of shaped high-explosive lenses.

saw earlier in Figure 4.4, ^{238}U requires neutrons of several MeV to fission with reasonable probability. Uranium-238 does not fission sufficiently readily for it to sustain a fission chain reaction, which produces neutrons of about 1 MeV, but it will fission readily under bombardment by the neutrons from D–T fission, which average about 14 MeV. Because ^{238}U cannot sustain a chain reaction, there is no such thing as a critical mass for this isotope; thus, an arbitrarily large amount may be safely arranged around a fusion weapon, substantially increasing the energy released in a simple and inexpensive way. This type of fission–fusion–fission bomb, shown schematically in Figure 16.6, typically may have half its yield from fusion and half from fission. Such a weapon does produce more radioactive fallout per kiloton of yield than a fission–fusion weapon because of the increased amount of fission fragments.

The neutron bomb is nothing more than a small fission–fusion device, lacking the surrounding jacket of ^{238}U; the 14 MeV neutrons from fusion can, therefore, escape readily, rather than being absorbed in the uranium, and irradiate anything within their range. These 14 MeV neutrons have greater range in air than do fission neutrons. And, in terms of number, fusion releases 10

Table 16.1 NUCLEAR BOMB ENERGY RELEASE

	Typical Fission Bomb	Neutron Bomb (½ fission, ½ fusion)
Blast	50%	40%
Thermal radiation	35%	25%
Prompt nuclear radiation	5%	30%
Delayed nuclear radiation	10%	5%

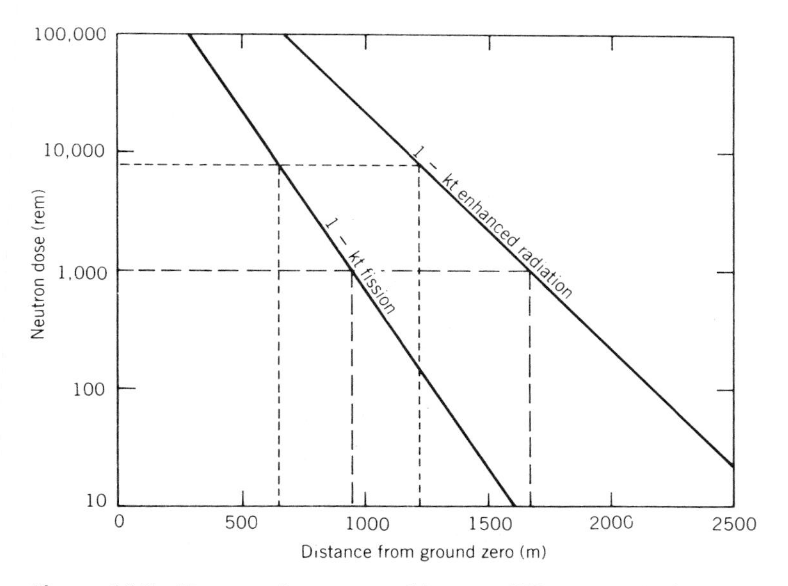

Figure 16.7 Neutron doses caused by two different tactical nuclear weapons with the same explosive yield. The ranges corresponding to doses of 8000 rem (which will cause immediate incapacitation) and 1000 rem (which ensures death within 6 weeks) are shown. For the 1-kiloton enhanced-radiation weapon, these ranges are about 80% greater than for a 1-kiloton fission weapon, corresponding to more than three times the ground area covered with lethal doses of radiation.

times more neutrons per kiloton than does fission. This results in much higher neutron doses being delivered at longer range than for fission weapons of similar explosive yield (See Figure 16.7). As Table 16.1 shows, the neutron bomb releases about 6 times more of its energy as prompt nuclear radiation than does a fission bomb of the same total yield. These fission–fusion weapons are known either as neutron bombs or enhanced-radiation weapons. The neutron bomb, which has as its purpose the exposure of enemy troops to massive lethal doses of radiation without causing excessive collateral damage to buildings and other objects by direct blast effects, must necessarily be of low yield. This is because, as we shall see later in more detail, only for the smaller weapons does the lethal range of the prompt radiation exceed that of the blast effects. Also, a small weapon in the few-kiloton range will be deliberately designed to be an inefficient weapon in the sense that only a small fraction of the fissionable material will be consumed, with the rest of the material being dispersed by the explosion. If the fission trigger has 0.5-kiloton yield, this corresponds to the fissioning of only about 30 g of plutonium, leaving the rest of the critical mass (several kilograms minimum) to be spread about by the explosion.

16.4 THE IMMEDIATE EFFECTS OF NUCLEAR WEAPONS

The long-term effects of a major nuclear war would be profound. The survival of our industries and agricultural system after such an event remains a matter of doubt. And there would certainly be potential for modification of the earth's climate, at least in the short term, possibly extending over a many-year period, through changes in the transparency of the atmosphere. Sunlight might be partially blocked because of the massive clouds of smoke and dust carried upward. Calculations indicate that nuclear explosions anywhere in the atmosphere could produce sufficient oxides of nitrogen to deplete chemically the ozone naturally present in the atmosphere. It is this ozone layer that protects all forms of life from the ultraviolet light normally present in the sunlight incident on earth. Although the ozone would probably be renewed by natural processes over a period of a few decades, its destruction would endanger many forms of life, including human, on earth. Another long-term effect might be created by the distribution over the earth's surface of thousands of kilograms of ^{239}Pu, with a half-life of 24,000 years, which would follow the detonation of a thousand nuclear weapons. This plutonium would be from the unexpended 80 to 90% of the fissionable material in the fission triggers for the hydrogen bombs. Many workers in radiation biology, however, believe that the effects of the plutonium would be small compared to the effects of the shorter-lived ^{137}Cs and ^{90}Sr fission products. It is also true that the unexpended plutonium, in even a thousand weapons, is only a few times more than the amount that would be released if a single large power reactor were vaporized by a direct hit with a 1-megaton weapon.

Although it is generally agreed that there would be significant and serious long-term effects following a nuclear war of any magnitude, these delayed effects are more difficult to predict, and, consequently, there is less agreement concerning them than there is about the immediate effects of nuclear explosions. Also, no doubt, there would be global effects of an extensive nuclear war that have not yet been thought of, since there has been no way to obtain experimental data on such a catastrophic event. The possibility of some of them being irreversible is particularly disturbing. For these reasons the discussion in this section focuses on the immediate effects of nuclear weapons, an area in which the background information can be considered to be fairly well-established. The immediate effects can be categorized, as in Table 16.2, into the five areas of prompt nuclear radiation, thermal radiation, blast wave, fallout, and electromagnetic pulse. The effective lethal radii associated with some of these effects are shown in Figure 16.8. From this figure it is seen that even a 1-kiloton bomb has a lethal range of about one half mile, sufficient to destroy a university campus and its population. The damage radius progresses rapidly with increasing explosive yield to the point where a single 1-megaton bomb would destroy a very large city.

PROMPT NUCLEAR RADIATION

From our discussion of fission and fusion reactions, it is obvious that energetic neutrons are produced copiously and instantly by either type of reaction. It is probably less obvious that such reactions are also sources of prompt, intense gamma radiation. These gamma rays are emitted as the fission fragments, which may have been formed in excited nuclear states at high energy, decay to their lowest energy states; gamma rays can also result from neutron interactions with all surrounding matter. As the neutrons encounter either weapons materials or material in the ground, or even the nitrogen of the air, they will interact with these nuclei, either by scattering or by being captured. Either process

Table 16.2 THE IMMEDIATE EFFECTS OF NUCLEAR WEAPONS

Effect	Source	Radius
Prompt nuclear radiation	Gamma rays and neutrons from nuclear reactions during explosion	Lethal radius in air less than that of blast for larger weapons, more for very small weapons
Thermal radiation	Surface of the hot fireball	Will ignite fires up to tens of miles radius; radius reduced by clouds, smog, or dampness
Blast wave	Rapid expansion of superheated material produces a high-pressure shock wave and winds; not a factor for explosions above 30,000 m altitude	Lethal radius greater than that of prompt nuclear radiation for all but the smallest weapons
Fallout	Mostly radioactive fission products, condensed on vaporized material	Fallout is locally important for ground bursts, not a serious local factor for air bursts
Electromagnetic pulse	Electrons moving rapidly outward from explosion	Damage to electrical equipment up to hundreds of miles away

results in prompt gamma ray emission. The gamma rays that follow at a later time from radioactivity induced by the neutrons are usually considered to be a part of the delayed radiation, and are treated separately from prompt nuclear radiation.

By far the most intense part of the prompt nuclear radiation is emitted during the first microsecond of the explosion, but all neutron and gamma radiation produced during the first minute will be considered as part of the prompt nuclear radiation. The radiation doses result from a mixture of gamma rays and neutrons, with the exact proportion depending on many factors, such as bomb size and altitude of burst. For high-yield weapons the gamma-ray dose exceeds that from neutrons.

The general subject of the interaction of radiation with living beings was covered in detail in Chapter 11, and the definitions of the units of radiation measurement are also given there. For the present, it is adequate to recall that a whole-body exposure of 100 rem will produce only slight immediate effects, that an exposure of 450 rem corresponds to a 50% probability of death within a few weeks whether or not medical attention is available, that 1000 rem ensures death within weeks, and that 8000 rem produces immediate incapacitation. Some of these exposure levels are indicated in Figures 16.7 and 16.8.

Although the prompt nuclear radiation is one of the more spectacular products of a nuclear explosion, it is, surprisingly, relatively unimportant in assessing the immediate lethality of large nuclear explosions. The reason is that the intensity of the radiation drops off as the inverse square of the distance in the absence of an absorbing medium such as air; but in air the effective range of the radiation is also limited by absorption. For the larger weapons, the lethal radius of the blast is so much greater than that of radiation that there

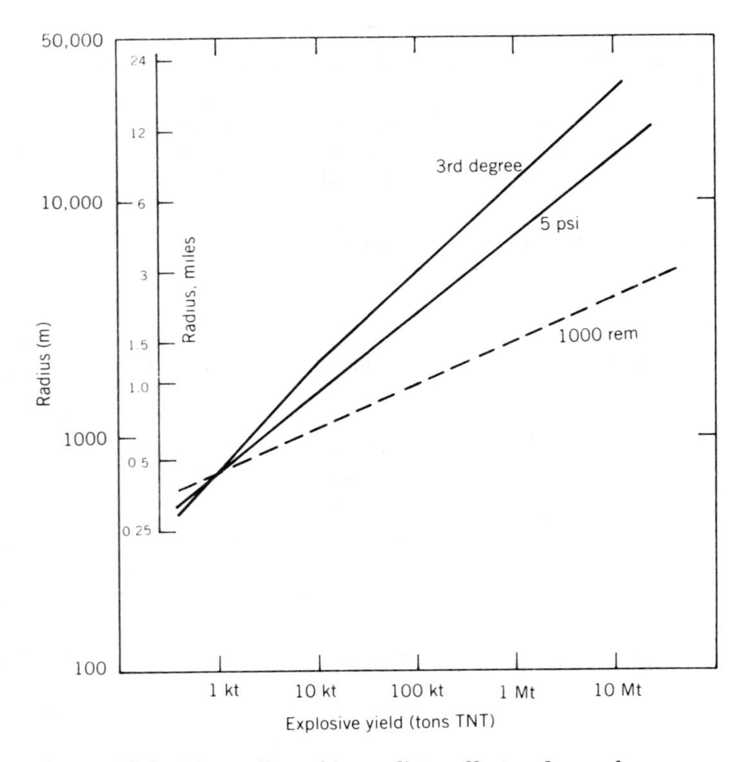

Figure 16.8 The radius of immediate effects of a nuclear explosion at the optimum burst height above ground level. The curves are: 1000 rem, the prompt nuclear radiation at a level that will ensure death within 6 weeks to all exposed persons; 5 psi, (lbs/in.^{2}), the maximum overpressure of 5 psi and a wind velocity of 160 mph; 3rd degree, third-degree burns due to thermal radiation from the fireball. All effects are shown for unshielded exposure under clear weather conditions and with a sea-level atmosphere. (Source: "Nuclear Bomb Effects Computer," rev. ed. 1977, in *The Effects of Nuclear Weapons*, U.S. Department of Defense and Department of Energy, 1977.)

is little possibility of anyone surviving the blast and suffering a lethal dose of prompt radiation. If shelter is available sufficient to protect against the blast, it will also provide at least partial protection against the radiation. This argument does not apply to the small fusion weapons, or neutron bombs. In that case, the bombs are deliberately limited in size, so that the effective radius of the prompt nuclear radiation exceeds the radius of heavy blast damage. For weapons of this smaller size, it is true that lifesaving protection from prompt radiation can be provided by shelters with walls of 1 or 2 ft of earth or concrete.

THERMAL RADIATION

The transport of heat energy by radiation was presented in Chapter 6. In that discussion we saw that the rate at which heat energy is emitted from a surface is proportional to the fourth power of the absolute temperature. For a nuclear explosion, the temperature of the bomb material and that of the fireball that forms subsequent to the explosion as the surrounding air becomes superheated is so high that prodigious amounts of energy are radiated as electromagnetic energy in the visible and nearby parts of the spectrum. In fact, more than a third of the total energy release of a nuclear weapon is through the emission of this thermal radiation. In general, as noted in Table 16.1, about 50% of the total energy release is through blast and shock waves, about 35% by thermal radiation, about 5% by prompt nuclear radiation, and the remaining 10% in the form of delayed nuclear radiation.

The thermal radiation causes injury and death by burns to any exposed individuals and also ignites flammable material, thus introducing the threat of secondary burns. As seen in Figure 16.8, the lethal range of the thermal radiation extends beyond 10 km for air bursts of the larger weapons under clear weather conditions. The effective range of the thermal radiation is more limited under conditions of smog or moisture-laden air, or if the explosion takes place above a heavy cloud layer. However, the intensity will be enhanced by reflection from cloud layers above the burst height or from snow-covered ground. Although thermal radiation has potentially the greatest lethal range of any of the prompt effects, almost any opaque surface will shield individuals against its effects. Even a layer of clothing, especially if light colored, will provide significant protection.

The number of burn casualties to be expected from a 1-megaton surface burst on the city of Detroit, Michigan, has been estimated. These estimates are shown in Table 16.3. The table gives estimates of casualties to be expected on a winter night when only 1% of the population might be exposed to the line of sight from the fireball, and for a summer weekend afternoon when 25% of the population might be exposed. In the latter case, especially, it is evident that the

Table 16.3 THERMAL RADIATION BURN CASUALTY ESTIMATES (1 MEGATON ON DETROIT)

Distance from Blast (miles)	No. of Blast Survivors	Fatalities (eventual)		Injuries	
		2-Mile Visibility	10-Mile Visibility	2-Mile Visibility	10-Mile Visibility
1% of population exposed to line of sight from fireball					
0–1.7	0	0	0	0	0
1.7–2.7	120,000	1,200	1,200	0	0
2.7–4.7	380,000	0	3,800	500	0
4.7–7.4	600,000	0	2,600	0	3,000
Total (rounded)		1,000	8,000	500	3,000
25% of population exposed to line of sight from fireball					
0–1.7	0	0	0	0	0
1.7–2.7	120,000	30,000	30,000	0	0
2.7–4.7	380,000	0	95,000	11,000	0
4.7–7.4	600,000	0	66,000	0	75,000
Total (rounded)		30,000	190,000	11,000	75,000

Note: These calculations arbitrarily assume that exposure to more than 6.7 cal/cm^2 produces eventual death, and exposure to more than 3.4 cal/cm^2 produces a significant injury, requiring specialized medical treatment.

Source: *The Effects of Nuclear War*, OTA, 1979.

number of burn injuries from just this one bomb would completely overwhelm the medical resources not only in the area, but even in the nation.

It is unlikely that a person would be able to respond in any way to obtain complete protection from thermal radiation, since it travels at the speed of light and would probably be the first signal that a weapon had exploded. For a 1-megaton burst at a height of 50,000 ft, anyone looking toward the blast would suffer permanent eye damage, from retinal burns, out to a distance of 45 miles on a clear day.

BLAST WAVE

Because of the high temperatures generated at the center of a nuclear explosion, all material objects are instantly vaporized into a high-temperature gas. Since this occurs so quickly, the vaporized material cannot expand until after tremendous pressures are developed. This pressure is as much as a million times atmospheric pressure or more. The air surrounding the high-pressure region is forced outward with great velocity, far exceeding that of sound. This shock wave, or a sudden rise in pressure, moves outward in a spherical shell centered at the point of the explosion, losing its dynamic pressure with the inverse square of the distance. The arrival of the shock front signals the beginning of winds of very high velocity blowing outward from the explosion. The time of arrival of the shock wave and the duration of the winds depend on the size of the weapon, distance, altitude of burst, and density of air. For large weapons, it is the blast effect that produces the most casualties; the lethal blast range exceeds the range of prompt nuclear radiation for weapons of larger than a kiloton or so for either air bursts or ground bursts. For explosions at very high altitudes, the blast wave is not an important factor. At lower altitudes the air communicates the blast wave but attenuates the prompt nuclear radiation. The casualties from the blast wave occur as people are hurled about at high velocity, as they are struck by flying objects, as they are crushed by collapsing buildings, as they are blown out of high-rise office buildings, or as they suffer damage due to the direct compressive effects of the high pressure.

The vulnerability of human beings to overpressure is shown in Figure 16.9, and the ranges corresponding to these overpressures are indicated in Table

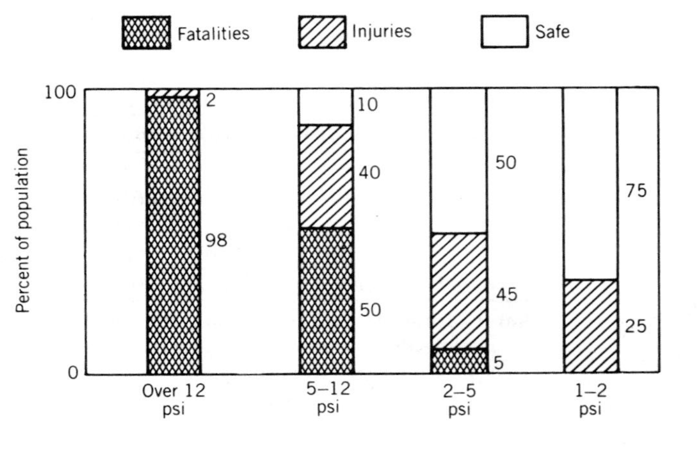

Figure 16.9 Vulnerability of population to death and injury in different overpressure zones due to the blast from a nuclear weapon (psi—lbs/in.^{2}). (Source: *The Effects of Nuclear War*, U.S. Office of Technology Assessment, U.S. Government Printing Office.)

Table 16.4 BLAST EFFECTS OF A 1-MEGATON EXPLOSION 8000 FT ABOVE THE EARTH'S SURFACE

Distance from Ground Zero		Peak Overpressure	Peak Wind Velocity	
miles	**km**	**(psi)**	**(mph)**	**Typical Blast Effects**
0.8	1.3	20	470	Reinforced concrete structures are leveled
3.0	4.8	10	290	Most factories and commercial buildings are collapsed; small wood-frame and brick residences destroyed and distributed as debris
5.9	9.5	3	95	Walls of typical steel-frame buildings are blown away; severe damage to residences; winds sufficient to kill people in the open
11.6	18.6	1	35	Damage to structures; people endangered by flying glass and debris

Source: *The Effects of Nuclear War*, OTA, 1979.

16.4. It is evident that few will survive inside a radius of about 3 miles from a 1-megaton burst. The type of damage expected from 5 psi overpressure is shown in Figure 16.10. It seems improbable that inhabitants of this house would have survived.

FALLOUT

As we learned in Chapter 4, the radioactive waste from nuclear power reactors resides primarily in the radioactive fission products. These radioactive nuclei are produced at the rate of two per fission event, and they and their radioactive daughter nuclei have half-lives ranging from much less than 1 second up to more than 100 years. In nuclear weapons debris, as in reactor waste, these fission fragments and their daughters constitute the major radioactivity. But the residual radioactivity also includes unexpended uranium or plutonium fuel and tritium, in addition to materials made radioactive by irradiation with neutrons released by the weapon. The amount of this secondary induced radioactivity depends strongly on whether the weapon was exploded as an air burst or in close proximity to the ground, as well as on the material included in the bomb assembly. The mix of the fission fragment nuclei is complex; hundreds of isotopes of more than 30 elements have been identified. Each of these isotopes decays at a different rate. The combined decay rate of the fission fragments remaining following a nuclear explosion is indicated in Figure 16.11, in which it can be seen that most of the decay occurs in the first hours following the explosion. Until about 1 year after the explosion, the activity drops with time at a rate proportional to $1/t^{1.2}$; after 1 year the decline of activity is more rapid than this. During the first year, a 7-fold increase in time results in a 10-fold decrease in activity (because $7^{1.2} \approx 10$).

Figure 16.10 A brick house before and after being hit by a blast wave having 5 psi overpressure. This is the type of damage to be expected at 7 km from ground zero for a 1-megaton air burst at 2400 m elevation.

Example 16.4

Calculate the number of radioactive fission fragment nuclei created per kiloton of fission energy field.

Solution

From Example 16.2 we know that $\frac{1}{18}$ kg of fissionable nuclei is equivalent to 1 kiloton of TNT. This is

$$\frac{1}{18}\,(1000\text{ g}) = 56\text{ g of fissioning nuclei per kiloton yield}$$

or

$$\frac{56\text{ g}}{235\text{ g/mole}} \times 6.02 \times 10^{23}\,\frac{\text{nuclei}}{\text{mole}} = 1.43 \times 10^{23}\text{ fissioning nuclei per kiloton yield}$$

$$2 \times 1.43 \times 10^{23} = 2.9 \times 10^{23}\text{ fission product nuclei per kiloton yield}$$

For a fusion weapon, the fallout radioactivity is due mostly to the products of the fission trigger, fission in the ^{238}U booster jacket, and the activities produced by neutron irradiation of surrounding materials. Either a fusion or fission weapon creates a fallout pattern that depends strongly on the height of

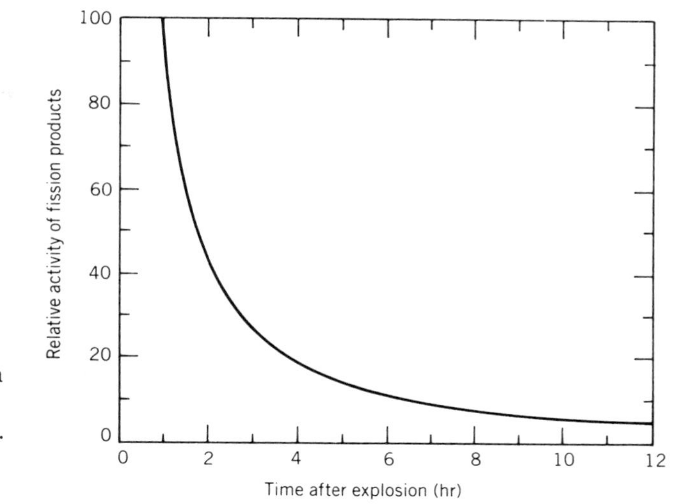

Figure 16.11 Rate of decay of the fission products in the early hours following a nuclear explosion. At 7 hours, the activity is about one tenth of what it was at 1 hour. (Source: *The Effects of Nuclear Weapons*, U.S. Department of Defense and Department of Energy, 1977.)

the burst above the ground and on weather conditions. For a high-altitude air burst, the radioactive debris from the weapon's assembly is usually carried further aloft by rising air currents and eventually dispersed over a large area, possibly over a region extending entirely around the globe. Because of the long residence time of this material in the atmosphere, much of the radioactivity decays long before the material settles to earth, thus diminishing the total radioactive fallout. In such a case the local fallout pattern is generally not a major consequence. On the other hand, for a ground burst, or any burst in which the fireball contacts the ground, an enormous amount of particulate matter is drawn up by the rising heated air column, and intensely radioactive particles are formed as the vaporized fission products condense and coalesce on the particles. These particles are sufficiently large and heavy that they cannot remain suspended for long in the atmosphere, but fall to the earth rather quickly in a fairly local region near the burst site. Depending on wind velocity, the region of intense local fallout may extend for a distance of 100 miles or more. Some indication of the fallout patterns to be expected is given in Figures 16.12, 16.13, and 16.14.

Example 16.5

After a nearby nuclear explosion, the radiation level due to fallout is observed to fall as follows:

Time (hr)	Intensity (rad/hr)
10	1000
20	435
30	268
40	189
50	145
60	116
70	97

Estimate the radiation intensity (rads per hour) to be expected at 140, 280, and 490 hours after the explosion. Assume that the intensity drops as $t^{-1.2}$.

Solution

If the intensity drops as $t^{-1.2}$, it is reduced by a factor of 10 for every sevenfold increase in time. Therefore, at any later time, the intensity will be one-tenth that at one-seventh the time. So the intensity at 140, 280, and 490 hours is obtained by scaling from 20, 40, and 70, respectively.

Time (hr)	**Intensity (rad/hr)**
140	43.5
280	18.9
490	9.7

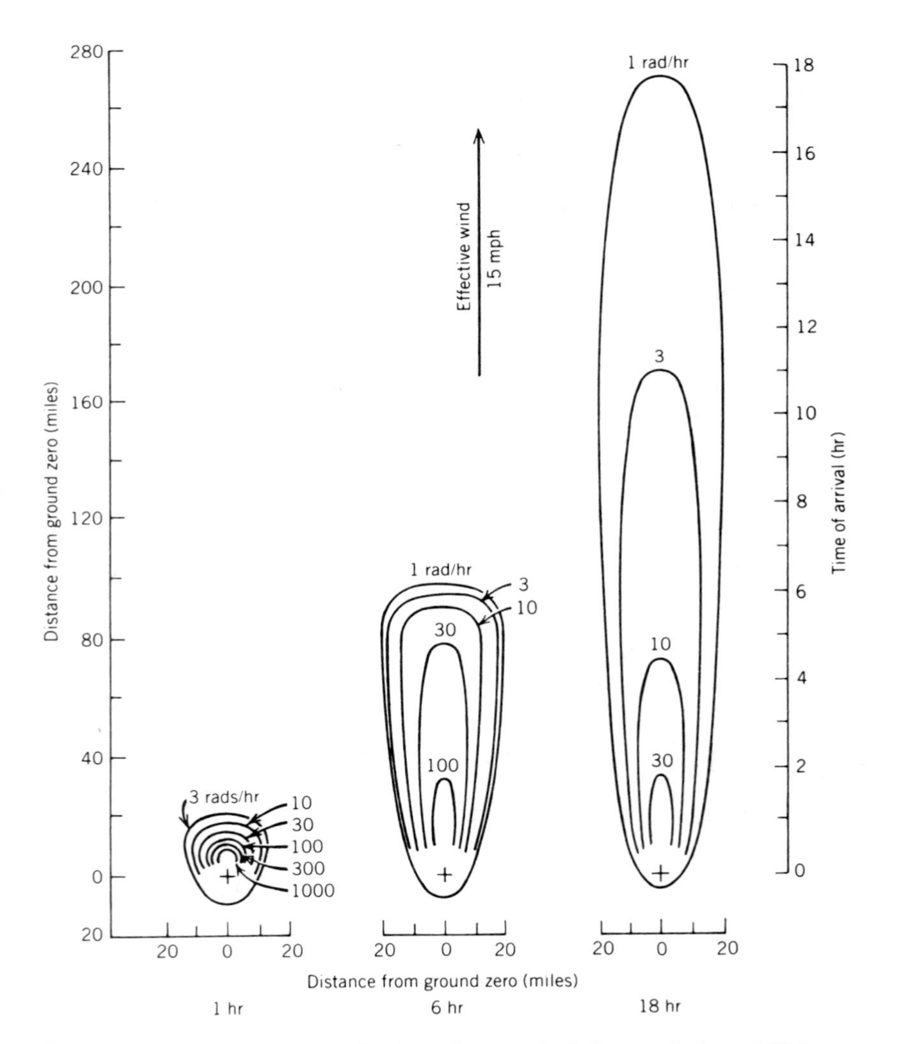

Figure 16.12 Dose *rate* distributions from early fallout at 1, 6, and 18 hours after a surface burst of a 2-megaton weapon. The distributions shown are for a steady 15 mph wind blowing in a constant direction. Note that the 30 rad/hr contour, for instance, reaches further out at 6 hours than at 18 hours. (Source: *The Effects of Nuclear Weapons*, U.S. Department of Defense and Department of Energy, 1977.)

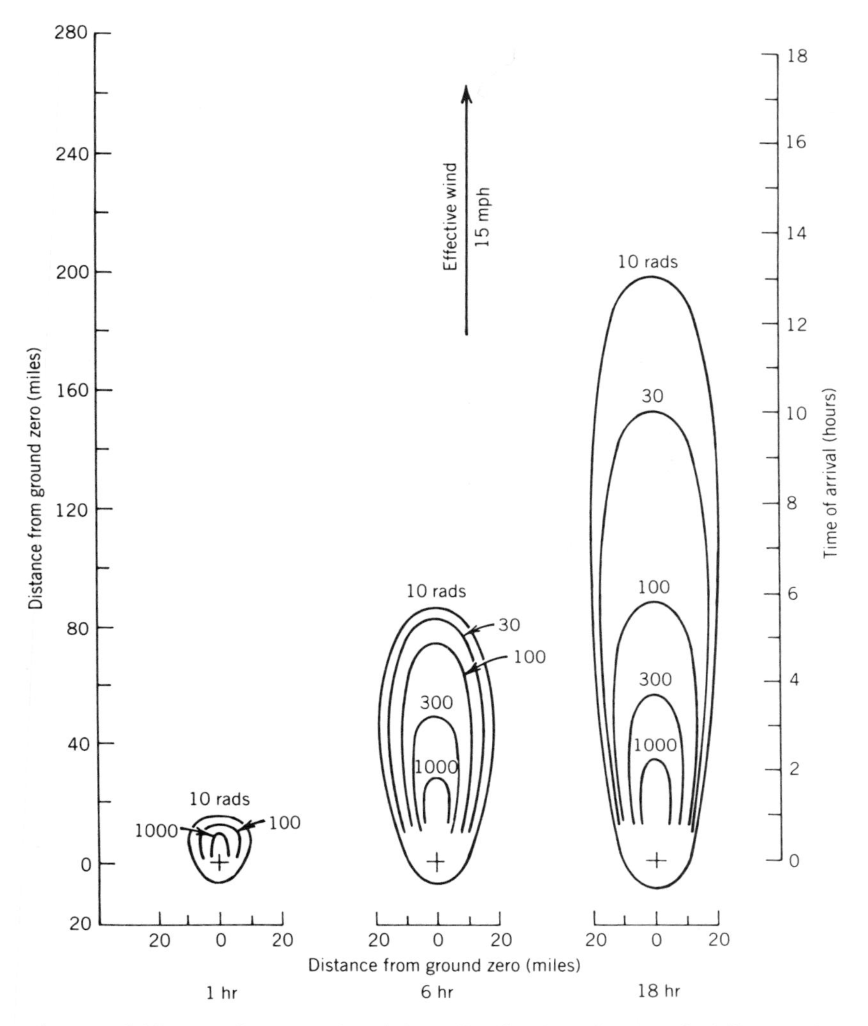

Figure 16.13 Total *accumulated dose* distributions from early fallout at 1, 6, and 18 hours after a surface burst of a 2-megaton weapon. The distributions shown are for a steady 15 mph wind blowing in a constant direction. (Source: *The Effects of Nuclear Weapons*, U.S. Department of Defense and Department of Energy, 1977.)

The radioactive fallout consists mostly of beta and gamma emitters. These isotopes can irradiate persons either as external or internal sources of radiation. The external radiation comes from isotopes in the air and on the ground, clothing, skin, rooftops, and all other surfaces. It is primarily gamma radiation, which is not absorbed appreciably in air over distances of several meters, but which is attenuated by a factor of 10 to 100 by 1 or 2 ft of earth or concrete. A shelter of this thickness will reduce the exposures shown in Figures 16.12, 16.13, and 16.14 correspondingly (see Table 16.5). Internal radiation is likely to result from ingestion of radioisotopes into the respiratory or digestive systems. This can be minimized through use of air filters on shelter ventilating systems and by washing food that may have the fallout dust deposited on it. A longer

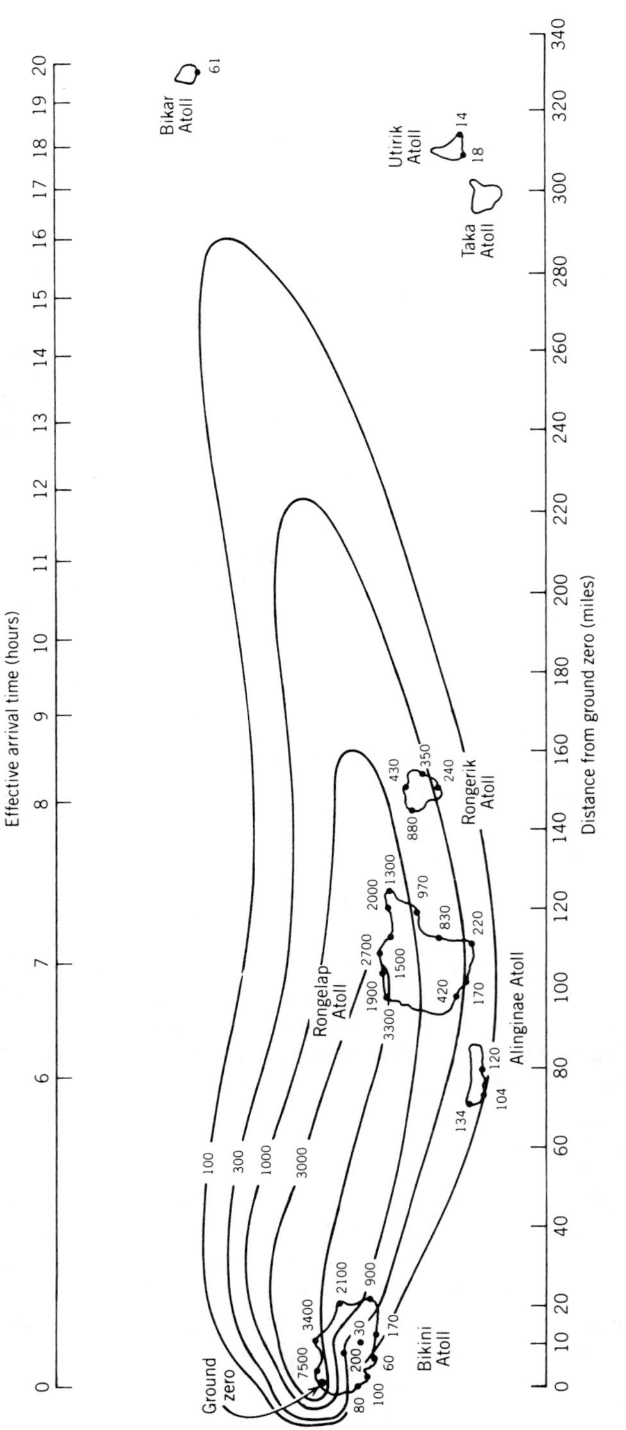

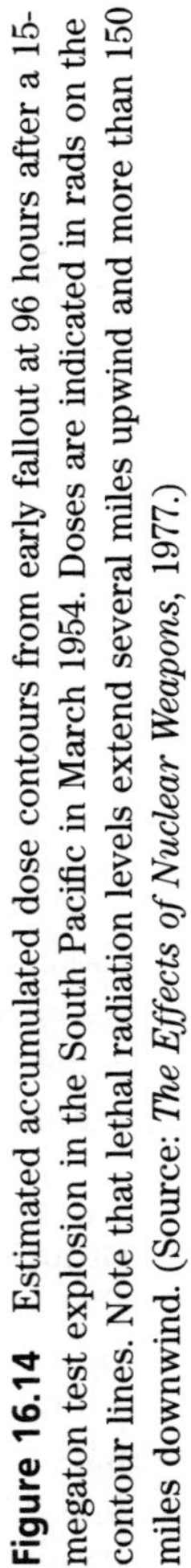

Figure 16.14 Estimated accumulated dose contours from early fallout at 96 hours after a 15-megaton test explosion in the South Pacific in March 1954. Doses are indicated in rads on the contour lines. Note that lethal radiation levels extend several miles upwind and more than 150 miles downwind. (Source: *The Effects of Nuclear Weapons,* 1977.)

Table 16.5 FALLOUT GAMMA-RAY DOSE SHIELDING FACTORS FOR VARIOUS STRUCTURES

Structure Type	Shielding Factor
3 ft underground	5000
Frame house	1.6–3.3
Basement	10–20
Multistory building	
Upper stories	100
Lower stories	10
Concrete blockhouse shelter	
9-in. walls	10–150
12-in. walls	30–1000
24-in. walls	500–10,000
Shelter partly above grade	
With 2-ft earth cover	50–200
With 3-ft earth cover	200–1000

Source: *The Effects of Nuclear Weapons*, DOD/DOE, 1977.

term internal source of radiation comes about through the longer lived isotopes finding their way into the food chain. An example would be ^{90}Sr, with a half-life of 28.5 years, which behaves chemically very much like calcium. If cows graze on land contaminated with this isotope, their milk will have the strontium present in it, and those who drink the milk will have some of this radioactive ^{90}Sr incorporated into their bone structure. The effect is more pronounced for children whose bones are still forming. We all now have some of this strontium, as well as measurable amounts of ^{137}Cs, in our bodies as a result of atmospheric weapons tests years ago. The current underground testing programs do not present this problem.

ELECTROMAGNETIC PULSE

The electromagnetic pulse (EMP) is one of the few immediate effects of nuclear weapons that does not directly harm human beings. The electromagnetic pulse is a short burst of electromagnetic energy just as might come from a super-powerful radio transmitter suddenly turned on, then off. Or it might be compared to the electromagnetic radiation from a nearby lightning strike. It arises following a nuclear explosion from the rapid acceleration outward of large numbers of free electrons released under the influence of gamma rays and x-rays. If the bomb burst occurred in a perfectly symmetric situation, there would be no electromagnetic pulse, because as many electrons would be accelerated upward as downward, as many to the left as to the right, and so forth. There would be no net electron current in any direction, and there would be no EMP radiation. Under real conditions, however, this symmetry is not present. A ground burst has greater opportunity to accelerate electrons upward into the atmosphere than down into the ground, and an air burst senses the lower atmospheric density in the upward direction. Because of these asymmetries, a nuclear explosion does produce a net electron flow in a given direction, thereby radiating electromagnetic energy.

The effects of this electromagnetic radiation are felt by electrical and electronic equipment to distances hundreds of miles from ground zero. In 1962, 30 loops of street lights failed simultaneously on the island of Oahu, Hawaii, 800 miles from ground zero of a high-altitude test. These failures were traced to fuses installed in the system for protection against sudden current surges. At this time the concern is over the potential for EMP to knock out communications systems, satellites, and electronic guidance and surveillance systems throughout the nation. Some studies have shown that it may be possible for an extremely large weapon exploded hundreds of miles above the center of the United States to disrupt communications over the entire continent. The threat of equipment failure has actually become more severe with the development of modern, compact, computer-controlled systems. The older electronic systems that operated with vacuum tubes were substantially more immune than are the transistor-based systems. As our electronics have become more sensitive and have been designed to operate at lower power levels, they have also become vulnerable to damage by the high-power levels expected from electromagnetic radiation.

16.5 WEAPONS INVENTORY AND DELIVERY SYSTEMS

One can imagine nuclear weapons being delivered to their intended targets by every means ranging from suitcases to satellites. There is nothing about the size or weight of the weapon that would preclude the former, and there is no technical barrier to the latter. The smaller nuclear warheads in our arsenal are in the kiloton-yield range and can be fired out of a 155 mm (6.1-in. diameter) artillery piece; the largest are nearly 10 megatons and deliverable by Titan intercontinental ballistic missiles. Between these extremes we have the possibilities of delivering our more than 20,000 deployed warheads by free fall from aircraft, rockets launched from aircraft, rockets launched from ground, rockets launched from submarines, and even the unmanned jet aircraft known as cruise missiles. Each of these possibilities has numerous variations, and we have not even mentioned nuclear depth charges, ship-launched nuclear missiles, or satellite vehicles.

Delivery systems are often classified according to whether they are strategic or tactical in purpose. The strategic label is commonly reserved for the larger weapons with intercontinental or at least international range; weapons described as tactical are those of the smaller size, applicable on the battlefield to more local conditions of warfare; however, the distinctions between the two are blurred to a large extent when considering weapons of intermediate range. In either category the missiles are often ballistic, that is, they travel on a trajectory determined only by the force of gravity, once their initial speed and direction have been achieved through some method of propulsion, usually a rocket motor.

Example 16.6

Estimate the approximate flight time for a ballistic missile to travel halfway around the globe when it is in a fairly low-altitude orbit. The apparent force (centrifugal force) acting outward on such a missile is given by mv^2/r, where m is the missile's mass, v its speed, and r its radius measured from the center of its circular arc, which is the same as its distance from the center of the earth.

Solution

While the missile is traveling in its near-circular trajectory, the apparent outward force, mv^2/r, must be approximately equal to the missile's weight, mg, which is the inward force. If either force were appreciably greater than the other, the missile would either steadily gain or lose altitude. Under conditions such that these two forces are equal

$$mv^2/r = mg$$

or

$$v = (rg)^{1/2}$$

Using values of $r = 6.4 \times 10^6$ m for the radius of the earth and $g = 9.8$ m/sec^2, we get

$$v = (6.4 \times 10^6 \times 9.8)^{1/2} = (6.27 \times 10^7)^{1/2}$$
$$= 7900 \text{ m/sec}$$

This is the same as 17,000 mph. By using the distance halfway around the globe to be $^1/_2 \times \pi \times$ diameter, the flight time is

$$t = \frac{d}{v} = \frac{^1/_2\, \pi\, (12.8 \times 10^6)\text{m}}{7.9 \times 10^3 \text{m/sec} \times 60 \text{ sec/min}} = 42 \text{ min}$$

Note that this result is consistent with the approximate 90-minute period we observe for complete revolutions of low-altitude satellites in orbit around the earth.

In an ordinary artillery piece, the projectile attains a high velocity as it is forced out the barrel by the pressure of gases liberated by a rapidly burning explosive in the closed end of the barrel. The velocities achieved with such devices are limited, as the entire acceleration must take place before the projectile emerges from the open end of the barrel. Artillery of this traditional type has a maximum range of only a few tens of miles, restricting its usefulness in delivering nuclear warheads. Rocket-propelled missiles, however, carry their fuel aboard the missile itself, allowing the fuel burning, and consequent acceleration, to take place over a considerable fraction of the trajectory, thus attaining much higher velocities and correspondingly greater ranges. In a sense, a rocket is the reverse of a gun; a gun recoils backward as the bullet is accelerated forward and out of the barrel. A rocket recoils in the forward direction as the burnt fuel is expelled out of nozzles at the rear. This recoil is the thrust that continues until the fuel is completely burned. The range of a rocket is limited only by the amount of fuel that can be carried aboard and can be enhanced by using multiple-stage rockets, which discard the first-stage motor after its fuel

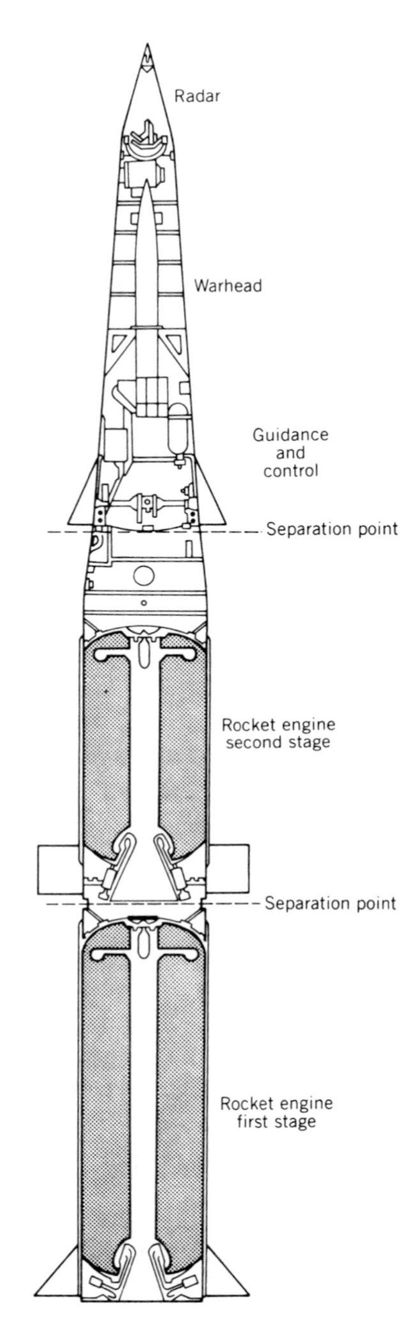

Figure 16.15 The Pershing-II, a U.S. intermediate-range ballistic missile for use in Europe. It is fired from a mobile field launcher, has a range of 1500 km, and is propelled by solid rocket fuel.

is exhausted, so that the fuel in the second and later stages will have less mass to accelerate. Rockets have sufficient range to reach any point on earth and beyond, as demonstrated by several recent interplanetary flights.

An example of a modern intermediate-range ballistic missile is shown in Figure 16.15. The accuracy of this missile, the Pershing-II, is sufficiently improved over earlier versions to permit use of a smaller warhead with the same assured destruction against selected targets.

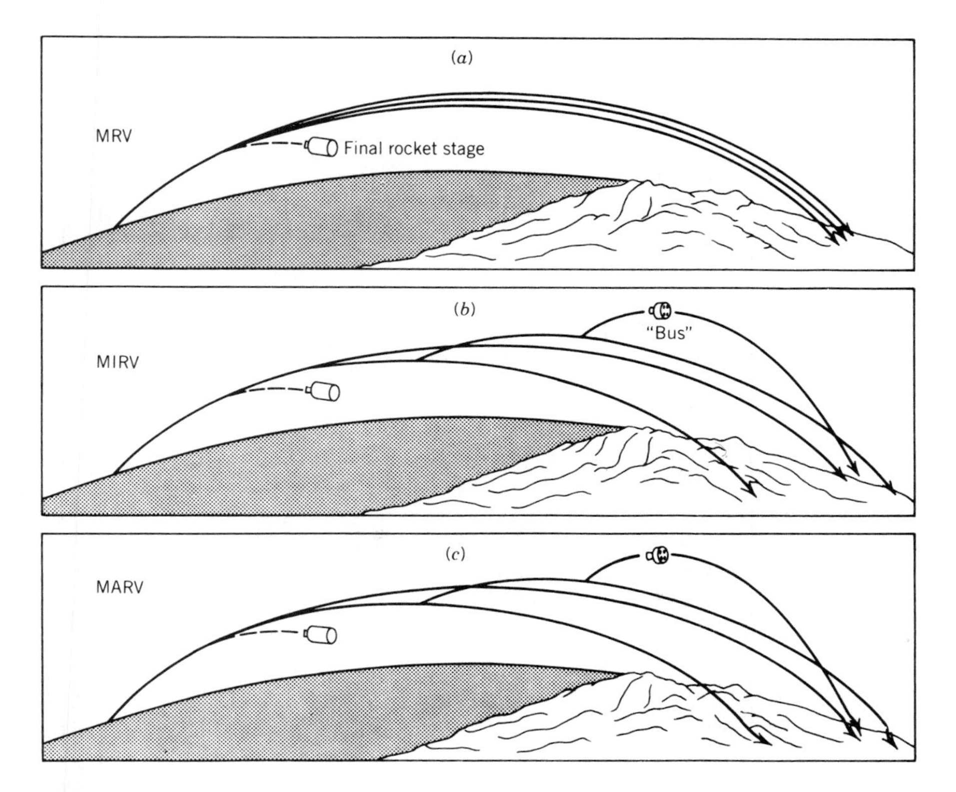

Figure 16.16 Three types of multiple warheads are shown in this figure. In (*a*) a multiple reentry vehicle (MRV) releases its warheads simultaneously at a point rather early in the missile's flight. The warheads then continue on unpowered and unguided ballistic trajectories to fall in a tight pattern. A more advanced version is the multiple independently targeted reentry vehicle (MIRV), shown in (*b*.) With the MIRV, the warheads are carried on a "bus," which after being released from the primary missile can change its direction in flight and release the individual warheads at designated points in flight. The warheads then follow ballistic trajectories to widely separated targets. The United States has developed maneuverable reentry vehicles (MARV), shown in (*c*), in which the individual warheads employ terminal guidance, changing their direction as they recognize local terrain features, thereby achieving greatly improved accuracy. (Source: Barry Carter, *Nuclear Strategy and Nuclear Weapons*. Copyright © May 1974 by *Scientific American*, Inc. All rights reserved.)

Intercontinental missiles are designed to travel over most of their trajectory at very high altitudes, outside most of the earth's atmosphere. As they reenter the atmosphere over their selected targets, the warheads continue on their ballistic trajectory, or, in the case of the more modern devices, they maneuver on to the target under the guidance of a terminal guidance system that recognizes features of the terrain surrounding the target. Multiple-warheaded devices, shown in Figure 16.16, are able to attack several targets simultaneously; the Poseidon missile, for example, which is launched from submarines, carries 10 warheads.

Table 16.6 gives some idea of the number of nuclear weapons at the end of 1991 in the strategic nuclear forces of the United States and the former

Table 16.6 ESTIMATED STRATEGIC NUCLEAR FORCES AT THE END OF 1991

Type	Name	Launchers/ SSBNs	Year Deployed	Warheads × Yield (Megaton)	Total Warheads	Total Megatons[a]
			United States			
ICBMs						
LGM-30G	Minuteman III:	500			1,500	404
	Mk-12	(200)	1970	3 × 0.170 (MIRV)	(600)	(102)
	Mk-12A	(300)	1979	3 × 0.335 (MIRV)	(900)	(302)
LGM-118A	MX/Peacekeeper	50	1986	10 × 0.300 (MIRV)	500	150
Total		550			2,000	554
SLBMs						
UGM-96A	Trident I C-4	384/20	1979	8 × 0.100 (MIRV)	3,072	307
UGM-133A	Trident II D-5	96/4	1990	4–8 × 0.475 (MIRV)	400	190
Total		480/24			3,472	497
Bomber/weapons[b]						
B-1B		84	1986	ALCM 0.05–0.150	1,600	240
B-52G		45	1958	Bombs 0.75 average	1,600	1,200
B-52H		80	1961	ACM 0.05–0.150	100	15
Total		209			3,330	1,455
Grand total		**1,239**			**8,772**	**~2,500**
			Commonwealth of Independent States			
ICBMs						
SS-18 M4/M5/M6	Satan	308	1979	10 × 0.550/0.750 (MIRV)	3,080[c]	1,688
SS-19 M3	Stiletto	210	1979	6 × 0.550 (MIRV)	1,260	693
SS-24 M1/M2	Scalpel	36/56	1987	10 × 0.550 (MIRV)	920	506
SS-25	Sickle	315	1985	1 × 0.550	315	173
Total		925			5,575	3,060

SLBMs						
SS-N-6 M3	Serb	96 (6)[d]	1973	2 × 0.500 (MRV)	96[e]	48
SS-N-8 M2	Sawfly	280 (22)	1973	1 × 1.5	280	420
SS-N-18 M1	Stingray	224 (14)	1978	3 × 0.500 (MIRV)	672[f]	336
SS-N-20 M1/M2	Sturgeon	120 (6)	1983	10 × 0.200 (MIRV)	1,200	240
SS-N-23	Skiff	112 (7)	1986	4 × 0.100 (MIRV)	448	45
Total		832			2,696	1,089
Bomber/weapons						
Tu-95MS6	Bear H6	27	1984	6 AS-15A ALCMs or bombs	162	40
Tu-95MS16	Bear H16	57		16 AS-15A ALCMs or bombs	912	228
Tu-160	Blackjack	16	1987	12 AS-15B ALCMs or 12 As-16 SRAMs, or 12 bombs	192	48
Total		100			1,266	316
Grand total		**1,857**			**9,537**	**~4,500**

[a] Numbers may not add due to rounding.

[b] Bomber numbers reflect Primary Authorized Aircraft and exclude an additional thirteen B–1B and 10 B–52 test aircraft and spares. Bombers are loaded in a variety of ways depending on mission, but most aircraft carry SRAMs and bombs. B–1Bs do not carry ALCMs or ACMs.

[c] Some SS-18s carry a single warhead, although under START all will be counted as carrying ten.

[d] Numbers in parentheses refer to submarines.

[e] SS-Mod 3 MRV warheads counted as one.

[f] Under START, the number of warheads on the SS-N-18 will be counted as three.

ACM—advanced cruise missile; **ALCM**—air-launched cruise missile; **ICBM**—intercontinental ballistic missile, range greater than 5,500 km; **MIRV**—multiple independently targetable reentry vehicles; **SLBM**—submarine-launched ballistic missile; **SRAM**—short-range attack missile; **SSBN**—nuclear-powered ballistic missile submarine; **AS**—air-to-surface missile.

Source: Bulletin of Atomic Scientists January/February and March 1992. Data compiled by Robert S. Norris and William M. Arkin.

Soviet Union, now the Commonwealth of Independent States (CIS). In addition to the thousands of warheads listed here, the United States has another 10,000 or more deployed for tactical use and more in reserve. The CIS has a similar number of tactical weapons.

As the accuracy of delivery systems is improved, there is less need for warheads of high yield. As recently as the 1980s, the United States deployed strategic warheads as large as 9 megatons, the Soviets 20 megatons. The warheads now are mostly under 1 megaton, and are getting smaller. The total megatonnage in the world's nuclear arsenals is substantially lower than in the 1950s and 1960s. The United States has only one quarter the strategic megatons it had in 1958. This trend reduces the human casualties and radioactive fallout expected from an all-out nuclear exchange.

Major changes have taken place since 1985, however, with the dissolution of the Soviet Union into independent republics, the Strategic Arms Reduction Treaty, and reductions by both sides in both strategic and tactical nuclear weapons. The number of strategic warheads actually peaked in 1987; since then, the United States has reduced its arsenal by about 4300 warheads, and CIS has eliminated a similar number. President Bush in September 1991 ordered 450 Minuteman II ICBMs taken off alert, 19 Poseidon submarines taken out of service, and the strategic bombers taken off ground alert. The Strategic Arms Reduction Treaty (START) (which had not yet been ratified when this book went to press), calls for the dismantling of the warheads from the above mentioned weapons in addition to a number of others from the United States and the CIS.

With these significant reductions in strategic weapons and more to come in the future, if present trends continue with the end of the Cold War, it is too easy to become complacent and ignore the reality of thousands still available by both sides. We are now in an era where a number of other countries have the technical capabilities to enter the arms race and to make reductions in strategic weapons a complex international problem. The enormity of the current arsenal can be gauged by considering that only 200 warheads on target would destroy all American cities with more than 100,000 persons. The justification given by each side for the thousands of warheads deployed has been that it is necessary for their safety against attack to ensure absolutely an effective retaliatory force in case of a first strike by the other side. The destructive power of just a fraction of the present inventories between the two countries is considered in the next section.

16.6 THE NATIONAL EFFECTS OF A NUCLEAR WAR

The consequences of an exchange of nuclear missiles between the United States and another country are difficult to calculate because many details are unknown. The nature of the attack is the most variable factor, as possible scenarios include a demonstration in which one city was obliterated, an all-out attack on the ICBM silos and other weapon systems, or perhaps an all-out first strike against military targets, as well as the large cities and targets of economic importance. Another variable factor is the way in which the nuclear weapons are detonated. As we have seen, air bursts and ground bursts have quite different consequences. The number of people killed would also depend on the amount of warning time given.

No nation has an effective defense against ICBM attacks on nonhardened targets, and even hardened missile sites can do nothing more than absorb the impact. In the 1960s the United States developed a prototype anti-ballistic missile (ABM) system near Grand Forks, North Dakota, but it has since been decommissioned. That system used a powerful radar to detect incoming ICBMs and to direct interception missiles (Sprint and Spartan) having nuclear warheads to destroy them. The reasons for decommissioning this system are related both to international arms agreements and to technical shortcomings. Later the Strategic Defense Initiative (SDI), often known as Star Wars, was proposed by the Reagan Administration, but this program has not achieved a credible defense against ICBM attack. Research continues on ABM systems with no clear promise of an effective defense. Our nation, along with all others, is now completely open to attack by long-range missiles.

In light of this vulnerability, it is revealing to make estimates of the magnitude of the possible catastrophe. The Office of Technology Assessment, the Department of Defense, and other groups have made such estimates. Only the case of an all-out attack on urban, industrial, strategic, and military targets in the United States is discussed here.

In summary, the deaths in the United States within 30 days would range from 155 million to 165 million, if no civil defense measures were taken and if all the weapons were used in ground bursts. If defense and evacuation procedures were effective and if 60% of the weapons were air bursts, these numbers could possibly be reduced to 20 million to 55 million casualties.

The question of the treatment of the injured, who would number in the range of 12 million to 30 million, with medical facilities that have been largely obliterated, has not been extensively analyzed; clearly a major problem would exist, particularly with the treatment of extensive burns and radiation sickness from fallout. It is equally disturbing to ponder how our complex society would function with its transportation, communications, fuel and food supply, water and sanitary systems, and industrial capacity seriously disrupted. For those not immediately killed or injured by the blasts, disease and starvation would soon become major problems.

The long-term effects of radiation exposure from the nuclear war, namely genetic defects and the increase in the incidence of cancer, would eventually be felt. The increase in cancer deaths from such an attack have been estimated to be in the range of 1 million to 5.5 million. Abortions due to chromosomal damage would range from 0.15 million to 6 million, and other serious genetic defects would number from 0.4 million to 9 million.

The environmental effects are less well known. As has been discussed, the possible depletion of the ozone layer with the subsequent elevation of solar ultraviolet intensity is a threat because of the increased likelihood of skin cancer. There has also been speculation that the extensive fire storms that would occur in the cities and possibly the forests could cause serious oxygen depletion and add significantly to atmospheric particulates. The psychological effects on the survivors and the disruption of our political and social systems, such as the agencies for law and order, are all great concerns.

The dimensions of such a nuclear holocaust are clearly far beyond anything ever experienced by the United States or any other country. The technological and environmental problems discussed in earlier chapters do, indeed, seem dwarfed by this nuclear threat to civilization as we know it.

16.7 NUCLEAR WEAPONS IN THE WORLD TODAY

The developments in the last several years with respect to the end of the cold war and the reduction in nuclear weapons by the two major adversaries present the most hopeful turn toward peace since the end of the Second World War. How much the nuclear weapon inventories of the United States, France, the United Kingdom, and the Commonwealth countries of the former Soviet Union will ultimately be reduced is, of course, not known, but the reduction is already very significant. There remains, however, a series of international problems that must be addressed before the threat of nuclear weapons is brought down to a level that could be called at all satisfactory.

- The nuclear weapons of the Soviet Union were distributed throughout what are now the 15 independent republics. Some of these republics are small and unstable, and given to frequent conflicts with their neighbors. It is hoped that Russia, as the major component of the Commonwealth, can bring about sufficient control of these nuclear weapons to oversee their safe storage and dismantling.
- The amount of highly enriched ^{235}U and ^{239}Pu on hand and to be recovered from the dismantling of nuclear weapons in the United States and Russia is disturbingly large. In the United States alone there are about 100 tons of weapons-grade ^{239}Pu and about 600 tons of highly enriched ^{235}U that must be carefully disposed of, perhaps by use as reactor fuel. Control of this weapons-grade fissionable material is essential for long-term stability.
- The acquisition of nuclear weapons by additional countries appears to be a major threat to world peace. The Gulf War brought out the extent to which Iraq, as an example, had gone to build surreptitiously gas centrifuges and calutrons for the enrichment of uranium for nuclear weapons. There are suspicions that several countries are attempting to do the same in spite of the Nuclear Nonproliferation Treaty.
- It is highly unlikely that the number of nuclear weapons in the world will be reduced to zero, even though the number may eventually be far lower than it is at the present time. However many there are, some system of control or stable deterrence must be initiated so that the world is assured that they will not be used. Removing the causes of war and strengthening the organizations, such as the United Nations, that can bring about the necessary control and oversight are both necessary for long-term peace. It now seems essential to improve reconnaissance by satellites to assure all sides that there will be no surprises, and it is also essential to continue improvement in the safety and security for all nuclear weapons so that they cannot be stolen or launched unintentionally.

PROBLEMS

1. Compute the ratio of the power outputs (bomb to reactor) for a 1-megaton nuclear explosion and a 1000-MW_e (3000-MW_{th}) reactor. Assume the explosion lasts for 1 microsecond. (Given: 1 ton TNT $= 4.3 \times 10^9$ J.)

2. Verify, by starting with the basic reactions, the statement "fusion produces many times more neutrons per ton of yield as does fission."
3. Estimate the number of tons TNT equivalent in the American nuclear arsenal on a per capita basis for the world's population. Do the same for the population of the United States.
4. A person standing facing a nuclear blast presents a frontal area of 6 ft^2 to the blast wave. How many pounds of force will this blast wave exert on this person at an overpressure of 5 pounds per square inch (psi)?
5. A group of people is confined to a well-shielded fallout shelter that completely protects them from fallout radiation. They know that 48 hours after the explosion of a nuclear weapon nearby, the radiation level outside was 100 rem per hour. What would the expected radiation level be outside at 14 days after the explosion? At 98 days after the explosion? Estimate in each of the two cases whether the people would be in immediate danger of radiation injury if they left the shelter and spent 1 hour exposed to the radiation outside while being evacuated from the area.
6. Estimate the force exerted by the blast wave on the wall of a house 40 ft long by 16 ft high in a distance of 5 miles from a 1-megaton explosion. (See Figure 16.8.)
7. Estimate how many nuclear warheads have been manufactured per day over the past few decades.
8. In the United States the prevailing winds blow from west to east. Discuss the reasons why many of our ICBMs are based in the Midwest, upwind from major population centers, rather than on the eastern coast. Consider that attacks against ICBM bases are likely to be ground bursts, with intense fallout patterns.
9. Estimate the length (in miles) of a train carrying an amount of TNT equivalent in explosive energy to the combined nuclear arsenals of the United States and the CIS. Would this train reach around the earth at the equator? How many times? For comparison, a train carrying all of the explosives detonated in World War II would be about 300 miles long.
10. How many kilowatt-hours of electricity could be generated by using 100 tonnes of ^{239}Pu as fuel for a nuclear reactor power plant with 30% efficiency? (Hint: calculate the number of ^{239}Pu nuclei and assume each yields 200 MeV of energy. How many years would this supply the fuel for a 1000 MW_e reactor?
11. Describe briefly the effects to be expected at radii of 1.0 and 5.0 miles from the epicenter of a 1-megaton nuclear weapon exploded at about 8000 ft. above the center of a city such as Denver.
12. Describe what you believe will be the world situation with regard to nuclear weapons in the year 2000.

SUGGESTED READING AND REFERENCES

1. Broyles, Arthur A.; Drell, Sidney D.; and Wigner, Eugene P. "Civil Defense in Limited War—A Debate." *Physics Today* **29** 4 (April 1976), pp. 44–57.
2. Broyles, A. A. "Nuclear Explosions." *American Journal of Physics* **50** 7 (July 1982), pp. 586–594.
3. Choppin, G. R., and Rydberg, J. *Nuclear Chemistry, Theory and Applications.* New York: Pergamon Press, 1980.
4. Cowan, George A. "A Natural Fission Reactor." *Scientific American* **235** 1 (July 1976), pp. 36–47.
5. Fetter, Steven A., and Tsipis, Kosta. "Catastrophic Releases of Radioactivity." *Scientific American* **244** 4 (April 1981), pp. 41–47.

6. Forsberg, Randall. "A Bilateral Nuclear-Weapon Freeze." *Scientific American* **247** 5 (November 1982), pp. 52–61.

7. Glasstone, Samuel, and Dolan, Philip J. *The Effects of Nuclear Weapons.* United States Department of Defense and Department of Energy, 1977.

8. Inglis, David R. *Nuclear Energy: Its Physics and Its Social Challenge.* Reading, Mass.: Addison-Wesley, 1973.

9. Kaplan, Fred M. "Enhanced-Radiation Weapons." *Scientific American* **238** 5 (May 1978), pp. 44–51.

10. Lewis, Kevin N. "The Prompt and Delayed Effects of Nuclear War." *Scientific American* **241** 1 (July 1979), pp. 35–47.

11. Schell, Jonathan. *The Fate of the Earth.* New York: Alfred A. Knopf, 1982.

12. "The Effects of Nuclear War." Office of Technology Assessment, Congress of the United States, May 1979.

13. *Physics Today* **36** 3 (March 1983), pp. 24–59 (a special issue on Nuclear Arms Education). Weisskopf, Victor F., "There is Still Hope for Hope"; York, Herbert F., "Arms-Limitation Strategies"; Sartori, Leo, "Effects of Nuclear Weapons"; Levi, Barbara G., "The Nuclear Arsenals of the U.S. and the U.S.S.R."; Schroeer, Dietrich, "Teaching About the Arms Race"; and Bethe, Hans A., "Nuclear Arms Education: The Urgent Need."

14. Rhodes, Richard. *The Making of the Atomic Bomb.* New York: Simon and Schuster, 1986.

15. Pilat, Joseph F. "Iraq and the Future of Nuclear Nonproliferation: The Roles of Inspections and Treaties." *Science* **255** 5049 (March 1992), pp. 1224–1229.

16. Albright, David, and Hibbs, Mark. "Iraq's Bomb: Blueprints and Artifacts." *Bulletin of Atomic Scientists* **48** 1 (January/February 1992).

17. Paine, Christopher, and Cochran, Thomas B. "So Little Time, So Many Weapons, So Much to Do." *Bulletin of Atomic Scientists* **48** 1 (January/February 1992).

18. Richelson, Jeffrey T. "The Future of Space Reconnaissance." *Scientific American* **264** 1 (January 1991), pp. 38–44.

19. Blair, Bruce G., and Kendall, Henry W. "Accidental Nuclear War." *Scientific American* **263** 6 (December 1990), pp. 53–58.

20. Schroeer, Dietrich. *Science and Technology of the Nuclear Arms Race.* New York: John Wiley and Sons, 1984.

21. Craig, Paul P., and Jungerman, John A. *Nuclear Arms Race—Technology and Society.* New York: McGraw-Hill Publishing Co., 1990.

22. Cochran, Thomas B.; Arkin, William M.; and Hoenig, Milton M. *Nuclear Weapons Databook.* Natural Resources Defense Council, 1984.

23. von Hippel, Frank; Albright, David H.; and Levi, Barbara G. "Stopping the Production of Fissile Material for Weapons. *Scientific American* **253** 3 (September 1985), pp. 40–47.

APPENDIX A

POWERS OF TEN NOTATION AND CONVERSION OF UNITS

Because the numerical quantities of interest in this text vary from the extremely small to the extremely large, it is hopeless to try to express them by the use of many zeros either before or after a decimal point. For example, the number of Btus consumed in the United States in 1975 was about 72×10^{15}. To try to write this as 72 with 15 zeros after it would be very time-consuming and probably would lead to many mistakes. Similarly, the size of a proton is about 1×10^{-15} m. Now we could write this as a decimal point followed by 14 zeros and a 1, but there is no reason to do so when we can use the efficient notation 1×10^{-15}.

When using powers of 10 in a calculation, there are some simple rules to follow for multiplication and division. For multiplication, say of 4.3×10^8 by 6.2×10^3, we multiply in the usual way the 4.3 by the 6.2 and obtain 26.66. The powers of 10 in any multiplication are added. For example, in our case, $10^8 \times 10^3 = 10^{11}$, with the final result

$$(4.3 \times 10^8) \times (6.2 \times 10^3) = 26.66 \times 10^{11}$$

In the division of 3.5×10^5 by 2.1×10^3 the 3.5 and 2.1 are divided, in the usual way, to obtain 1.67. The power of 10 of the result is the power of 10 in the numerator minus the power of 10 of the denominator. In our example, $10^5/10^3 = 10^2$, with the final result

$$\frac{3.5 \times 10^5}{2.1 \times 10^3} = 1.67 \times 10^2$$

Example A.1

Find the result for the following operations:

(a) $(6 \times 10^{-3}) \times (8.3 \times 10^8)$

(b) $\dfrac{14.1 \times 10^{13}}{5.1 \times 10^8}$

(c) $\dfrac{(4.8 \times 10^9) \times (3.6 \times 10^5)}{2.8 \times 10^{10}}$

(d) $\dfrac{(6.2 \times 10^{-2})(2.7 \times 10^5)}{(3.8 \times 10^{-4})(9.1 \times 10^7)}$

Solutions

(a) 49.8×10^5

(b) 2.76×10^5

(c) 6.17×10^4

(d) $0.48 \times 10^0 = 0.48$

In the last example (part d), the power of 10 is zero, but since any number raised to the zero power is equal to 1, $10^0 = 1$.

When adding or subtracting numbers with various powers of 10, the numbers must first be converted to the same power of 10 and then added in the normal fashion. For example, $4.8 \times 10^6 + 1.2 \times 10^7$ should be written $4.8 \times 10^6 + 12.0 \times 10^6$, with the result 16.8×10^6. In a similar way, $4.8 \times 10^{-4} - 3.6 \times 10^{-5}$ should be written $48 \times 10^{-5} - 3.6 \times 10^{-5}$, with the result 44.4×10^{-5}.

Example A.2

Find the result of the following operations:

(a) $4.6 \times 10^4 + 5.1 \times 10^3 + 6.2 \times 10^5$

(b) $4.6 \times 10^4 - 1.8 \times 10^3 - 0.82 \times 10^5$

Solutions

(a) $0.46 \times 10^5 + 0.051 \times 10^5 + 6.2 \times 10^5 = 6.71 \times 10^5$

(b) $0.46 \times 10^5 - 0.018 \times 10^5 - 0.82 \times 10^5 = -0.378 \times 10^5$

The subject matter covered in this book involves a great assortment of units. Some of the units are British and some are metric (SI). There has been no effort made to present all of the material with a consistent set of metric units since this is not the present practice in the United States for various applied areas. Unfortunately, this plethora of units means that conversions frequently have to be made from one system to another. The conversion factors and energy equivalents are listed on the inside of the front and back covers. There is a very simple scheme for converting units that assures that the process will be carried out correctly. It involves writing down a ratio that has a value of 1 and has the units of the numerator being those we want to go to and the units of the denominator being those we start with. This ratio is then multiplied times the original quantity. For example, if we want to express 4.2 hr in seconds, we would multiply 4.2 hr by (3600 sec/1 hr). The hours cancel, leaving seconds as the final unit:

$$4.2 \text{ hr} \times \left(\frac{3600 \text{ sec}}{1 \text{ hr}}\right) = 1.51 \times 10^4 \text{ sec}$$

If we had used the inverse of the conversion factor, the hours would not cancel and we would know that we had the factor the wrong way. In some conversions, more than one conversion factor is needed. For example, if we want to convert 512 ft • lb into kW • hr, we may know that 1 Btu = 778 ft • lb and 1 kW • hr = 3413 Btu. The conversion would then be

$$512 \text{ ft} \cdot \text{lb} \times \left(\frac{1 \text{ Btu}}{778 \text{ ft} \cdot \text{lb}}\right) \times \left(\frac{1 \text{ kW} \cdot \text{hr}}{3413 \text{ Btu}}\right) = 1.92 \times 10^{-4} \text{ kW} \cdot \text{hr}$$

Example A.3

Make the following conversions:

(a) 49.6×10^6 Btu into joules

(b) 10.1 yd into meters

Solutions

(a) $$49.6 \times 10^6 \text{ Btu} \times \left(\frac{252 \text{ cal}}{1 \text{ Btu}}\right) \times \left(\frac{4.184 \text{ J}}{1 \text{ cal}}\right) = 5.23 \times 10^{10} \text{ J}$$

(b) $$10.1 \text{ yd} \times \left(\frac{36 \text{ in.}}{1 \text{ yd}}\right) \times \left(\frac{2.54 \text{ cm}}{1 \text{ in.}}\right) \times \left(\frac{1 \text{ m}}{100 \text{ cm}}\right) = 9.24 \text{ m}$$

APPENDIX B

LOGARITHMIC AND EXPONENTIAL FUNCTIONS

In studying the growth patterns of certain quantities such as energy consumption, it has been noted that the exponential function is sometimes useful. It is also of use in understanding the decay of radioactive nuclei. Logarithms were used in Chapter 15 in discussing noise and the decibel intensity scale. There is a direct relationship between logarithmic and exponential functions, and certain properties of these functions are rather easily understood and are very useful in thinking about problems concerning energy and the environment.

By definition, the logarithm of a number N to the base a is the exponent or the power to which a must be raised to obtain N. In the form of equations, the definition becomes more clear:

$$N = a^x \quad \text{and} \quad x = \log_a N$$

Two systems of logarithms are in use: the natural or Naperian system, which uses the base $e = 2.71828$, and the common system, which uses the base 10. It is customary to use the symbols ln for natural logarithms and log for the common (base 10) system.

Example B.1

What is the value of log 1000?

Solution

Since it is known that $10^3 = 1000$, it is clear that $3 = \log 1000$.

There are three properties of logarithms to any base that are useful:

(a) $\log M \times N = \log M + \log N$

(b) $\log M/N = \log M - \log N$

(c) $\log M^p = p \log M$, where p can be any number, negative, positive or fractional.

It is also useful to note $\log_b b = 1$ and $\log_b 1 = 0$.
The values for log N for integers from 0 to 100 are presented in Table B.1.

Example B.2

Determine the values of the logarithms of the following numbers:

(a) 49
(b) 4900
(c) $3.6 \times 10^{+6}$
(d) 2.4×10^{-3}

Solutions

(a) $\log 49 = 1.6902$ (directly from table)

(b) $\log 4900 = \log(49 \times 100) = \log 49 + \log 100 = 1.6902 + 2.0000 = 3.6902$

(c) $\log 3.6 \times 10^6 = \log 36 + \log 10^5 = 1.5563 + 5.0000 = 6.5563$

(d) $\log 2.4 \times 10^{-3} = \log 24 + \log 10^{-4} = 1.3802 + (-4.0000) = -2.6198$

In performing calculations of interest to problems of energy and the environment, there is little occasion to use natural logarithms directly. Tables of values of ln N are available in numerous handbooks, and many hand calculators include such functions. What is widely used in various sections of the present text is the exponential function

$$N = e^x$$

with

$$x = \ln N$$

Table B.1 VALUES FOR COMMON (BASE 10) LOGARITHMS

N	log N	N	log N	N	log N	N	log N
1	0	26	1.4150	51	1.7076	76	1.8808
2	0.3010	27	1.4314	52	1.7160	77	1.8865
3	0.4771	28	1.4472	53	1.7243	78	1.8921
4	0.6021	29	1.4624	54	1.7324	79	1.8976
5	0.6990	30	1.4771	55	1.7404	80	1.9031
6	0.7782	31	1.4914	56	1.7482	81	1.9085
7	0.8451	32	1.5051	57	1.7559	82	1.9138
8	0.9031	33	1.5185	58	1.7634	83	1.9191
9	0.9542	34	1.5315	59	1.7709	84	1.9243
10	1.0000	35	1.5441	60	1.7782	85	1.9294
11	1.0414	36	1.5563	61	1.7853	86	1.9345
12	1.0792	37	1.5682	62	1.7924	87	1.9395
13	1.1139	38	1.5798	63	1.7993	88	1.9345
14	1.1461	39	1.5911	64	1.8062	89	1.9494
15	1.1761	40	1.6021	65	1.8129	90	1.9542
16	1.2041	41	1.6128	66	1.8195	91	1.9590
17	1.2304	42	1.6232	67	1.8261	92	1.9638
18	1.2553	43	1.6335	68	1.8325	93	1.9685
19	1.2788	44	1.6435	69	1.8388	94	1.9731
20	1.3010	45	1.6532	70	1.8451	95	1.9777
21	1.3222	46	1.6628	71	1.8513	96	1.9823
22	1.3424	47	1.6721	72	1.8573	97	1.9868
23	1.3617	48	1.6812	73	1.8633	98	1.9912
24	1.3802	49	1.6902	74	1.8692	99	1.9956
25	1.3979	50	1.6990	75	1.8751	100	2.0000

Table B.2 VALUES OF e^x AND e^{-x} FOR x FROM 0 TO 100

x	e^x	e^{-x}	x	e^x	e^{-x}
0	1.0000	1.0000	2.0	7.3891	0.1353
0.10	1.1052	0.9048	2.5	12.182	0.0821
0.20	1.2214	0.8187	3.0	20.086	0.0498
0.30	1.3499	0.7408	3.5	33.115	0.0302
0.40	1.4918	0.6703	4.0	54.598	0.01831
0.50	1.6487	0.6065	4.5	90.017	0.0111
0.60	1.8221	0.5488	5.0	148.41	6.73×10^{-3}
0.6931	2.0000	0.5000	5.5	244.69	4.09×10^{-3}
0.70	2.0138	0.4966	6.0	403.43	2.48×10^{-3}
0.80	2.2255	0.4493	7.0	1096.6	9.12×10^{-4}
0.90	2.4596	0.4066	8.0	2981.0	3.36×10^{-4}
1.00	2.7183	0.3679	9.0	8103.1	1.23×10^{-4}
1.20	3.3201	0.3012	10.0	2.20×10^{4}	4.54×10^{-5}
1.40	4.0552	0.2466	50.0	5.18×10^{21}	1.93×10^{-22}
1.60	4.9530	0.2019	100.0	2.69×10^{43}	3.72×10^{-44}
1.80	6.0496	0.1653			

In Table B.2 values of e^x are listed for values of x and $-x$ ranging from 0 to 100. The rapidity with which e^x becomes very large and e^{-x} very small as x increases toward 100 is evident from the listed values. As discussed in the text, a particularly useful value of e^x and e^{-x} occurs when $x = 0.6931$ at which time these two functions become equal exactly to 2.0 and 0.50, respectively. It is at this value of x that the doubling time, t_D, is taken for $e^{\lambda t}$ and the radioactive half-life, $t_{1/2}$, for $e^{-\lambda t}$. Thus

$$\lambda t_D = 0.693 \quad \text{and} \quad t_D = 0.693/\lambda$$

$$\lambda t_{1/2} = 0.693 \quad \text{and} \quad t_{1/2} = 0.693/\lambda$$

One of the useful aspects of exponential functions is that there is a simple graphical procedure for projecting trends for either $e^{\lambda t}$ or $e^{-\lambda t}$. If we have, for example, a constant growth behavior in which a quantity equal to N_0 at $t = 0$ grows exponentially to the value N at time t:

$$N = N_0 \, e^{\lambda t}$$

From what has been said so far

$$\ln N = \ln N_0 + \ln e^{\lambda t}$$

or

$$\ln N = \ln N_0 + \lambda t$$

Since $\ln N_0$ is simply a constant, we have a function, $\ln N$, that is a linear function of time. That is, it is directly proportional to time. It is possible to take values of $\ln N$ from a table or a calculator and plot them versus time, t, and a straight-line or linear relationship will result. Happily, as noted in the text, there exists semilogarithmic graph paper that is ruled on one axis to coincide with the logarithmic functions. By directly plotting data on such semilog paper, one can avoid the labor of converting N to $\ln N$, and still produce a plot where exponential growth or exponential decay will result in a straight line. A number of such plots are included in the text.

Some specific characteristics of an exponential growth pattern are:

1. The trend is represented by a straight line on a semilogarithmic plot. Such trends are easy to extrapolate on such a plot.
2. The quantity doubles at regular time intervals.
3. The percent growth rate is constant.
4. For a resource being consumed at an exponentially growing rate, during each doubling time period more of the resource is consumed than in all previous time. This is exactly analogous to each entry in a geometric progression being larger than the sum of all previous entries.
5. The doubling time and growth rate are approximately related by

$$\text{doubling time} = \frac{70}{\%\ \text{growth per unit time}}$$

where the time units must be the same on both sides of the equation.

These characteristics can be used for tests of exponential growth; if a growth pattern corresponds to any one of the first three listed characteristics, the growth is exponential and the other four characteristics then also apply.

If a resource such as coal is being consumed with a constant growth rate, λ, we have already noted in Chapter 1 that the amount consumed per year at any time, t, will be

$$N = N_0 e^{\lambda t}$$

where N_0 was the amount consumed per year at time $t = 0$. It is sometimes of interest to examine the total amount consumed in a given period of time. One could calculate how much is used the first year, the second year, and so forth, and then add the sequence of contributions for the time desired. This process is not only laborious, but it is not quite correct, because the time and the coal consumed are changing continuously, not just with each yearly increment. With integral calculus, the correct answer can be obtained in a far simpler and precise fashion because the increments in time used in the calculation become vanishingly small. The result for the total quantity used or produced, A, for a period of T years, beginning with a rate of consumption of N_0 per year and growth constant λ, is

$$A = \frac{N_0}{\lambda}(e^{\lambda T} - 1)$$

Example B.3

Calculate the total amount of coal used during the 97-year period starting now if 600×10^6 tons per year are being used now and the growth rate is 5%.

Solution

$$A = \frac{N_0}{\lambda}(e^{\lambda T} - 1) = \frac{600 \times 10^6}{0.05}(e^{0.05(97)} - 1)$$

$$= 12.0 \times 10^9 (127.74 - 1)$$

$$= 1.52 \times 10^{12} \text{ tons of coal}$$

As noted in Chapter 2, 1.52×10^{12} tons is close to the total coal resource available in the United States.

ANSWERS TO SELECTED END-OF-CHAPTER PROBLEMS

1.1 (a) 490 J
(b) 490 J

1.5 20 yr

1.6 235.2 J

1.8 0.0756 kW

1.9 4.5×10^5 J

1.12 (a) 10 ft • lb/sec
(b) 0.0136 kW
(c) 0.0182 hp
(d) 650 cal

1.14 125 W

1.17 1.22 kW

1.19 (a) 58 bbl
(b) 14.5×10^6 Btu

1.21 2.50×10^8 kW • hr

1.24 1.35×10^3 kcal

2.1 Wood: $6.67/day
Gas: $8.05/day

2.2 (a) 5.9 gal/day • person
(b) about 12 yr

2.10 (a) Coal: 0.102 lb
Oil: 0.0099 gal
(b) $0.42, using oil at $20/bbl
(c) Oil about 80 gal
Coal about 1000 lb

2.11 4.8 yr

2.13 $3.45 \times 10^6 \frac{\text{Btu}}{\text{ton}}$, or 13%

2.15 $t = 0.48$ ft

2.17 7.91×10^6 ft • lb or 0.2%

2.19 2.75 tons of CO_2

3.1 20°C; 293°K

3.8 (a) 35%

3.11 26.8%

3.13 (a) 5.27×10^5 kW_e
(b) 2.95×10^7 kW • hr

3.16 The gas-burning water heater uses only 0.6 as much fuel as the electric water heater.

3.17 (a) 9324 tons of coal/day
(b) 34,000 tons of CO_2/day
(c) 1.6×10^{11} Btu/day

3.20 76%

3.22 4

3.24 4.79×10^7 Btu

3.26 (b) About 10 yr
(c) 7%/yr
(d) About 6.2×10^{12} kW • hr

4.1 (a) 8.1×10^{13} J per kg of ^{235}U
(b) 5.8×10^{11} J per kg of natural uranium

4.5 20 yr

4.6 2.22 MeV

4.10 (a) 0.85 MeV per nucleon
(b) 0.0009 or 0.09%

4.13 8.31 MeV

4.15 17.6 MeV

5.2 About 50 MW_e • yr per bomb

5.5 0.5%

5.6 0.313 curies

5.9 8×10^{-4} per reactor year

5.10 1415 bombs

6.1 1.59 m^2 or 17.2 ft^2

6.3 (a) 35.6 ft^3
(b) 3.29 ft
(c) 266 gal

6.7 47 mi^2

6.9 28 ft^3, 4032 lb

6.12 56,000 QBtu, 1.46×10^{-3}

7.2 (a) 7.06×10^{15} J
(d) $98 million

7.3 (a) 265 J
(b) 251 J

7.5 (a) 60.5 W, 484 W, 3.87 kW
(b) 1, 8, and 64 60-W bulbs
(c) 530 kW • hr, $26.50

7.8 1.70%

7.10 (a) 7777 m^3/sec
(b) 1111 m^2

8.3 $\frac{\text{Wood (Btu)}}{\text{Gasoline (Btu)}} = 160$

$\frac{\text{Wood (\$)}}{\text{Gasoline (\$)}} = 100$

8.6 Dynamite 4.7×10^5 J
Jelly doughnut 15.7×10^5 J

8.8 1722 m^3

8.9 (a) 3.024×10^6 J
(b) 3600 W

8.12 4.7×10^4 J for 4 ft^3 of air at 120 psi
4.3×10^4 J for 10 g TNT

9.1 4930 Btu

9.2 2500 degree days

9.5 About 180×10^6 Btu

9.7 The savings is $102 per heating season; the simple payback time is 4 years; and the annual return on investment is 25%.

9.10 2.99×10^6 Btu

9.14 $5.18—cost of keeping electric blanket on; $32—approximate saving on annual fuel bill. Blanket would help.

10.1 1548 kcal

10.4 Total of 16 lb

10.7 The total cropland equals 1.5 hectares, about equal to three or four football fields. Of this, 1.3 hectares is devoted to producing beef and 0.2 hectares to producing vegetables.

10.9 About 71 persons at 168 W each

10.16 About 4000 km^2 or 1560 m^2. This is an area of 63 by 63 km.

11.1 2.5×10^{-6} A

11.2 195 rem; 1.95 Sievert

11.5 About 125 serious genetic defects per generation

11.6 200 hr

11.7 4200 excess cancer deaths

11.11 395 excess cancer deaths

12.1 $a = 2.2$ ft/sec^2; distance = 1320 ft

12.3 256 lb

12.5 F_{total} = 802 lb; power = 128 hp

12.7 6812 Btu per mile; 18.5 mpg

12.11 14.9 mph, about the same as for a jump without a parachute from a height of 7.5 ft.

12.18 104 mph

13.1 909 m

13.3 2.7×10^{25} molecules/m^3

13.5 0.13 ppm

13.7 107 $\mu g/m^3$

13.9 1114 kg

13.10 2033

14.1 (a) 2.24×10^{23} cal

14.3 3.26×10^5 gal

14.5 $3.62

14.8 10^6 gal/min, 6×10^7 gal/hr, 60 gal/person

14.10 1.65×10^8 gal/day

15.1 16.6 cm in air; 72.5 cm in water

15.3 3.16×10^{-6} W/m^2

15.5 88.98 dB at 50 m

15.8 111.4 dB

15.12 2069 Hz

15.15 30°C 349.2 m/sec
−30°C 312.7 m/sec

16.1 Bomb to reactor power ratio is 1.4×10^{12}.

16.4 4320 lb

16.5 10 rem/hr at 14 days; 1 rem/hr at 98 days; no acute effects in either case.

16.9 More than 40 times

16.10 76.8 year

PHOTO CREDITS

Cover Otto Rogge/The Stock Market

Chapter 1 Opener: Mark Antman/The Image Works. Figure 1.11: Courtesy NASA.

Chapter 2 Opener: Alexander Lowry/Photo Researchers. Figure 2.8: Courtesy Gulf Oil. Figure 2.10: Courtesy Tenneco Oil Co. Figure 2.11: Courtesy U.S. Department of Energy. Figure 2.17: Steve Northrup/Black Star.

Chapter 3 Opener: Jack Spratt/The Image Works. Figure 3.7: Michael Collier/Stock, Boston.

Chapter 4 Opener: Courtesy U.S. Atomic Energy Commission. Figure 4.7: Courtesy U.S. Department of Energy. Figure 4.12: Courtesy Princeton University Physics Laboratory.

Chapter 5 Opener: Dan Miller/Woodfin Camp & Associates. Figure 5.3: Courtesy U.S. Department of Energy.

Chapter 6 Opener: Charles Schneider/FPG International. Figure 6.1: Courtesy Mesa Verde National Park, Colorado. Figure 6.4: Courtesy NASA. Figure 6.13: Robert Perron/ Photo Researchers. Figure 6.15: Peter Menzel/Stock, Boston. Figure 6.17: Courtesy Georgia Power Co. Figure 6.18: Courtesy Solar Energy Research Institute. Figure 6.22: Courtesy U.S. Department of Energy.

Chapter 7 Opener: Courtesy U.S. Department of Energy. Figure 7.2: Courtesy Solar Energy Research Institute. Figure 7.6: (top) Courtesy Metropolitan Museum of Art, Dick Fund, 1925; (bottom) courtesy U.S. Department of Energy. Figure 7.11: Courtesy Solar Energy Research Institute. Figure 7.13: (top) William Keith; (center) courtesy Geological Survey, U.S. Department of Interior; (bottom left) Peter Menzel/Stock, Boston; (bottom right) courtesy William S. Keller, National Park Service. Figure 7.14: Courtesy Pacific Gas and Electric Co. Figure 7.15: Courtesy U.S. Department of Energy.

Chapter 8 Opener: Joe Sohm/The Image Works. Figure 8.3: Courtesy Detroit Edison Co. Figure 8.4: Courtesy Pacific Gas and Electric Co.

Chapter 9 Opener: George Gardner/Stock, Boston. Figure 9.4: Owen Franken/Stock, Boston.

Chapter 10 Opener: Courtesy U.S. Department of Agriculture. Figure 10.1: John Urban/Stock, Boston. Figure 10.3: Peter Menzel/Stock, Boston.

Chapter 11 Opener: Courtesy National Cancer Institute. Figure 11.4: Courtesy M.A. Bender, Brookhaven National Laboratory.

Chapter 12 Opener: FPG International. Figure 12.2: Joel Gordon. Figure 12.8: Spencer Grant/Stock, Boston. Figure 12.10: Daniel S. Brody/Stock, Boston.

Chapter 13 Opener: U.S. Environmental Protection Agency. Figure 13.6: Richard Laird/Leo de Wys. Figure 13.8: Everett Johnson/Leo de Wys.

Chapter 14 Opener and Figure 14.2: Courtesy U.S.D.A. Figure 14.4: Joe Munroe/ Photo Researchers.

Chapter 15 Opener: Joel Gordon. Figure 15.2: Michael Philip Manheim/Photo Researchers. Figure 15.6: Jeff Albertson/Stock, Boston.

Chapter 16 Opener, Figures 16.2 and 16.4: Courtesy U.S. Air Force/Defense Nuclear Agency. Figure 16.5: Courtesy U.S. Department of Energy, Figure 16.10: Courtesy Defense Nuclear Agency.

INDEX